MW00366980

Chemical Oceanography and the Marine Carbon Cycle

The principles of chemical oceanography provide insight into the processes regulating the marine carbon cycle. These topics are essential to understanding the role of the ocean in regulating the carbon dioxide content of the atmosphere and climate on both human and geological time scales.

Chemical Oceanography and the Marine Carbon Cycle provides both a background in chemical oceanography and a description of how chemical elements in seawater and ocean sediments can be used as tracers of physical, biological, chemical and geological processes in the ocean. The book begins with a description of ocean circulation and biological processes, and then moves on to discuss the chemicals that are dissolved in seawater. Subsequent chapters focus on why the ocean has the chemistry that it does, rather than on details of what is there. The first seven chapters present basic topics of thermodynamics, isotope systematics and carbonate chemistry, and explain the influence of life on ocean chemistry and how it has evolved in the recent (glacial–interglacial) past. This is followed by topics essential to understanding the carbon cycle, including organic geochemistry, air–sea gas exchange, diffusion and reaction kinetics, the marine and atmosphere carbon cycle and diagenesis in marine sediments. The many figures in the book (including full-color versions) are available for download at www.cambridge.org/9780521833134.

Developed by two well-known professors of oceanography, *Chemical Oceanography and the Marine Carbon Cycle* is an ideal textbook for upper-level undergraduates and graduates in oceanography, environmental chemistry, geochemistry and earth science. It is also a valuable reference for researchers in oceanography.

Steven Emerson is Professor of Oceanography at the University of Washington, specializing in inorganic geochemistry. He is a Fellow of the American Geophysical Union and has worked on both air–sea interaction and sediment geochemistry.

The late John Hedges was Professor of Oceanography at the University of Washington, specializing in organic geochemistry. He was the recipient in 2000 of the Geochemical Society's Alfred R. Treibs Award for lifetime achievement.

Chemical Oceanography and the Marine Carbon Cycle

Steven Emerson
John Hedges
School of Oceanography, University of Washington, USA

CAMBRIDGE UNIVERSITY PRESS
Cambridge, New York, Melbourne, Madrid, Cape Town, Singapore, São Paulo, Delhi

Cambridge University Press
The Edinburgh Building, Cambridge CB2 8RU, UK

Published in the United States of America by Cambridge University Press, New York

www.cambridge.org
Information on this title: www.cambridge.org/9780521833134

First published 2008
Reprinted with corrections 2009

Printed in the United Kingdom at the University Press, Cambridge

A catalog record for this publication is available from the British Library

ISBN 978-0-521-83313-4 hardback

Contents

Preface

The field of chemical oceanography has evolved over the past several decades from one of discovery to an interdisciplinary science that uses chemical distributions to understand physical, biological, geological and chemical processes in the sea. The study of chemical oceanography includes much of the background required to understand the global carbon cycle on all time scales because of the primary role of the marine carbonate system. Thus, we present this book about *Chemical Oceanography and the Marine Carbon Cycle* as a natural outgrowth of the evolution of our scientific field and a necessary background for building intuition to manage the anthropogenic intrusion into the global carbon cycle.

After a long deliberation about whether we had the time, stamina and personalities to write a book about our subject, John Hedges and I decided to do it, using as a guide, the notes we had compiled from teaching Chemical Oceanography together in the School of Oceanography at the University of Washington. During the first three years of the new century we used sabbatical leaves and time borrowed from teaching and research to compile about half of the book. Then, in 2003 John died suddenly and unexpectedly. Everyone John touched was thrown into a state of shock at the loss of a good friend, reliable colleague and brilliant organic geochemist. At this point we had put so much of ourselves into this undertaking that I felt there was no turning back, and I continued to complete what you see here.

The first part of the book (Chapters 1–7) covers a one-quarter-long course for beginning graduate students. Because of the backgrounds of students in this class, we taught the course so that little previous knowledge of oceanography or chemistry was required. All one needed is some experience in thinking scientifically and the desire to learn. We feel this part of the book should also be appropriate for senior-level undergraduate courses on this subject. The final five chapters of the book are compiled from parts of other, more advanced seminars and should serve well as a guide for research and more advanced courses in *Chemical Oceanography and the Carbon Cycle*.

Steven Emerson

Acknowledgements

Many people played a role in seeing this book to the finish. I'd first like to thank Michael Peterson, who drafted all the figures, edited the text and equations and helped to mold the very different styles of the two authors into a common voice. Bruce Frost, while he was director of the School of Oceanography, helped support this effort with the Karl Banse Fund of the School of Oceanography. Parts of this book were written on sabbaticals, where freedom from normal activities and our generous hosts made life conducive to writing. These include time spent at the Water Research Laboratory of the Swiss Federal Institute of Technology (EAWAG-ETH); Cambridge University; the Ocean Research Laboratory (LOCEAN) of the University of Pierre and Marie Curie (SE); and the Hanse Institute for Advanced Studies in Delmenhorst (JH). We appreciate the hospitality of our colleagues, Werner Stumm, Harry Elderfield, Lillian Merlivat and Juergen Rullkoetter, respectively. At home we both took advantage of the writing-friendly environment of the Whiteley Center at the Friday Harbor Laboratories of the University of Washington.

Many of the chapters were reviewed by colleagues, post-docs and former students. We owe special acknowledgement to Kenia Whitehead, who, as co-author of Chapter 8, was responsible for taking a partly finished manuscript left by John Hedges and molding it into a comprehensive chapter on organic geochemistry. Dieter Imboden helped with the discussion about diffusion in Chapter 9. Others who played valuable roles in reviewing individual chapters are Curtis Deutsch, Burke Hales, Roberta Hamme, David Hastings, Taka Ito, Jennifer Morford, Jim Murray, Paul Quay, Amelia Schevenell and Stuart Wakeham. We would like to thank the editors at Cambridge University Press for their patience and skill in presenting the book. Mistakes that persist after these conscientious efforts should be attributed solely to the authors.

John and I believed that the unlikely birth of a textbook about oceanography from two people raised on farms in Ohio resulted from the influence of our early mentors. Our curiosity and approach to science was instilled at the beginning of our careers by Wallace Broecker of Columbia University and Werner Stumm and Dieter Imboden at EAWAG-ETH (SE), and by Pat Parker of the University of Texas and Tom Hoering of the Carnegie Geophysical Laboratory, Washington D.C. (JH). We feel that collaborations with our PhD graduate students have stimulated our discoveries in the field of chemical oceanography and taught us valuable lessons about science and life along the way. These people are: Rick Jahnke, Lucinda Jacobs, John Ertel, Dan McClorkle, David Archer, Greg Cowie, Miguel Goñi, David Hastings, Ann Russell, Brian Bergamaschi, Burke Hales, Matt

McCarthy, Jennifer Morford, Peter Hernes, Anthony Aufdenkamp, Kenia Whitehead, Roberta Hamme, Angie Dickens, Susan Lang and David Nicholson.

Finally, we dedicate the book to our wives, Julie Emerson and Joyce Hedges, whose support, encouragement and love made it possible to complete this endeavor.

1

Introduction to chemical oceanography

1

Oceanography background: dissolved chemicals, circulation and biology in the sea

1.1 A chemical perspective

This book describes a chemical perspective on the science of oceanography. The goal is to understand the mechanisms that control the distributions of chemical compounds in the sea. The "chemical perspective" uses measured chemical distributions to infer the biological, physical, chemical and geological processes in the sea. This method has enormous information potential because of the variety of chemical compounds and the diversity of their chemical behaviors and distributions. It is complicated by the requirement that one must understand something about the reactions and time scales that control the chemical distributions. Chemical concentrations in the sea "remember" the mechanisms that shape them over their oceanic lifetime. The time scales of important mechanisms range from seconds or less for very rapid photochemical reactions to more than 100 million years for the mineral-forming reactions that control relatively unreactive elements in seawater. The great range in time scales is associated with an equally large range in space scales, from chemical fluxes associated with individual organisms to global processes like river inflow and hydrothermal circulation.

Studies of chemical oceanography have evolved from those focused on discovering what is in seawater and the physical–chemical interactions among constituents to those that seek to identify the rates and mechanisms responsible for distributions. Although there is still important research that might be labeled "pure marine chemistry," much of the field has turned to the chemical perspective described here, resulting in a fascinating array of new research frontiers. We mention just a few of the exciting areas that are presently mature. Chemical alterations associated with hydrothermal processes at mid-ocean ridges have ramifications for whole-ocean mass balance of some elements and are regions of redox reactions catalyzed by microbial processes that may have been the origin of life on Earth. The ocean's role in the global carbon cycle is an important process controlling the fate of anthropogenic CO_2 added to the atmosphere, which has ramifications for global climate, today and in the future. The study of mechanisms by which dissolved metals limit marine biological production has been demonstrated by large-scale iron addition experiments in regions where the surface ocean is rich in phosphorus and nitrate. Investigations of isotope and chemical tracers in calcite shells and the structure of individual organic compounds buried in marine sediments provide analytical constraints for understanding how the ocean influenced atmospheric CO_2 and climate during past glacial ages. These are just a few examples that are relevant to oceanography and the global environment in a field that is continuously developing new research avenues.

Because the science of chemical oceanography is focused on distributions of chemical constituents in the sea, its evolution has been controlled to some extent by analytical developments. It is not our goal to dwell on analytical methods; however, discoveries of new mechanisms and processes often follow the development of better techniques to make accurate measurements. Probably the most recent example has been the evolution of a variety of mass spectrometers capable of precisely determining extremely low concentrations of metals, isotopes on individual organic compounds, and atmospheric gas ratios on small samples. There have been a host of other breakthroughs that are too numerous to mention that have had a great influence on our ability to interpret the ocean's secrets.

The evolution of analytical methods has been accompanied by increased sophistication and organization in sampling the ocean.

Table 1.1. *Areas, volumes and heights of the ocean and atmosphere*

Atmosphere inventory	1.77×10^{20} mol (all gases)
Earth surface area	5.10×10^{14} m^2
Ocean surface area	3.62×10^{14} m^2 (71% of Earth's area)
Ocean mean depth	3740 m
Ocean volume	1.35×10^{18} m^3
Ocean mass	1.38×10^{21} kg
River flow rate	3.5×10^{13} m^3 y^{-1}

Pilson (1998); the river flow rate is from Broecker and Peng (1982).

This trend has so far involved primarily the effective use of research vessels to mount global sampling programs such as the geochemical sections (GEOSECS) program in the 1970s, and the joint geochemical ocean flux study (JGOFS) and world ocean circulation experiment (WOCE), both in the 1990s. All of these programs were international in scope, employing scientists and research vessels from the world community in a coordinated effort to determine global chemical distributions and processes. International collaboration has been particularly important in the last two of these programs and will continue to grow in order to solve problems that are increasingly complex and expensive to tackle. The promise of remotely determining chemical concentrations by using instruments that operate *in situ* on moorings or unmanned vehicles is real, but at the time of writing this book it is only beginning to have a major impact, primarily because of the limited capability of chemical sensors to maintain long-term stability and accuracy.

We begin the book with a brief discussion of background information about the chemical constituents of seawater, the basics of ocean circulation and marine biological processes. Some important information about the volumes and areas of the ocean and atmosphere are presented in Table 1.1. The goal of this chapter is to create a foundation for the discussion of mechanisms later in the book.

1.2 Constituents of seawater

Chemical concentrations in the ocean and atmosphere have been presented over the years in a variety of units, some of which originated in the field of chemistry and others that gained prominence in the geologic literature (Table 1.2). The modern practice in chemical oceanography is to present concentrations in units of moles or equivalents per kilogram of seawater. Moles and equivalents are more meaningful than mass units because reaction stoichiometry is presented on an atomic or molecular basis. Mass is used in the denominator because it is conservative at all depths of the ocean, whereas volume changes because of the compressibility of water.

Table 1.2. *(a) Concentration units encountered in oceanography*

Equivalents, eq, is equal to moles × absolute value of the charge of the species. Units indicated as "seawater units" are those preferred in oceanography. Molality, molarity, normality and volume ratio all have a long history of use in classical chemistry because of their convenience for laboratory preparations.

Name	Basis	Dimensions	Symbol	Definition
Concentrations in aqueous solution				
Molal	mass	mol kg^{-1}	*m*	Moles per kilogram of solvent
Molar	volume	mol l^{-1}	M	Moles per liter of solution
Normal	volume	eq l^{-1}	N	Equivalents per liter of solution
Weight ratio	mass	g kg^{-1}		Mass of solute per mass of solution
Volume ratio	volume	ml l^{-1}		Volume of solute per volume of solution
Seawater units	mass	mol kg^{-1}		Moles per kilogram of solution
Seawater units	mass	eq kg^{-1}		Equivalents per kilogram of solution
Concentrations in the atmosphere				
Mole fraction	moles	mol mol^{-1}	X	Moles of gas per moles of dry air (= volume fraction, e.g. ppmv, for ideal gas)
Fugacity	pressure	bar bar^{-1}	*f*	Gas pressure per atmospheric pressure (= partial pressure, p, for ideal gas)

(b) Exponential terminology used in oceanography

Prefix (symbol)	peta- (P)	tera- (T)	giga- (G)	mega- (M)	milli- (m)	micro- (μ)	nano- (n)	pico- (p)	femto- (f)	atto- (a)
Unit multiplier	10^{15}	10^{12}	10^{9}	10^{6}	10^{-3}	10^{-6}	10^{-9}	10^{-12}	10^{-15}	10^{-18}

Before launching into a detailed discussion of individual constituents, we would like to introduce the total quantity of dissolved material in seawater, salinity, and the processes that determine it.

1.2.1 The salinity of seawater

Salinity is a measure of the total mass in grams of solids dissolved in a kilogram of seawater, a mass ratio. It is composed almost entirely of elements that do not measurably change concentration geographically owing to chemical reactivity. It is thus used as a property against which individual chemical species can be compared to determine their stability in the sea; conservative (unreactive) elements have constant or nearly constant ratios to salinity everywhere in the ocean. Relatively small changes in salinity are important in determining the density of seawater and thermohaline circulation. It can also be useful as a tracer for the mixing of different water masses since salinity values that are determined at the ocean's surface can be traced for great distances within the ocean interior.

For all these reasons it is essential to have a relatively rapid and accurate measurement of seawater salinity. The obvious method would be to dry seawater and weigh the leftover residue. This approach does not work very well because high temperatures (*c.* 500 °C) are required to drive off the tightly bound water in salts such as magnesium chloride and sodium sulfate. At these temperatures some of the salts of the halides, bromides and iodides are volatile and are lost, while magnesium and calcium carbonates react to form oxides, releasing CO_2. Some of the hydrated calcium and magnesium chlorides decompose, giving off HCl gas. The end result of weighing the dried salts is that you come up "light" because some of the volatile elements are gone. Although there were schemes created to obviate these problems, for many years the preferred method for determining salinity was titration of the chloride ion by using silver nitrate

$$Ag^+ + Cl^- \rightleftarrows AgCl_{(s)}, \tag{1.1}$$

which is quantitative. The chloride concentration, $[Cl^-]$, could then be related to salinity, S, via a constant number,

$$S(\text{ppt}) = 1.80655 \times [Cl^-](\text{ppt}), \tag{1.2}$$

where ppt indicates parts per thousand ($g_{\text{solute}}\ kg^{-1}_{\text{seawater}}$). The exact relationship between chlorinity and salinity, however, has evolved over the years, and it is not as accurate and universal as the present method.

Salinity is presently determined by measuring the conductance of seawater by using a salinometer. The modern definition of salinity uses the *practical salinity scale*, which replaces the chlorinity–salinity relationship with a definition based on a conductivity ratio (Millero, 1996). A seawater sample of salinity $S = 35$ has a conductivity equal to that of a KCl solution containing a mass of 32.435 6 g KCl in 1 kg of solution at 15 °C and 1 atm pressure. No units are necessary on the practical salinity scale; however, in practice, one often sees parts per thousand, ppt, or the abbreviation "psu." New salinometers using this method are capable of extremely high precision so that the salinity ratio can be determined to 1 part in 40 000. At a typical salinity near 35 this procedure enables salinities to be determined to an accuracy of 35.000 ± 0.001. This is much better than most chemical titrations, which, at best, achieve routine accuracy of ± 0.5 parts per thousand.

The distribution of salinity in surface waters of the ocean is presented in Fig. 1.1. Because the concentrations for many major seawater constituents are unaffected by chemical reaction on the time scale of ocean circulation, local salinity distributions are controlled by a balance between two physical processes, evaporation and precipitation. This balance is reflected by low salinities in equatorial regions that result from extensive rain due to rising atmospheric circulation (atmospheric lows) and high salinities in hot dry subtropical gyres that flank the equator to the north and south (20–35 degrees of latitude) where the atmospheric circulation cells descend (atmospheric highs).

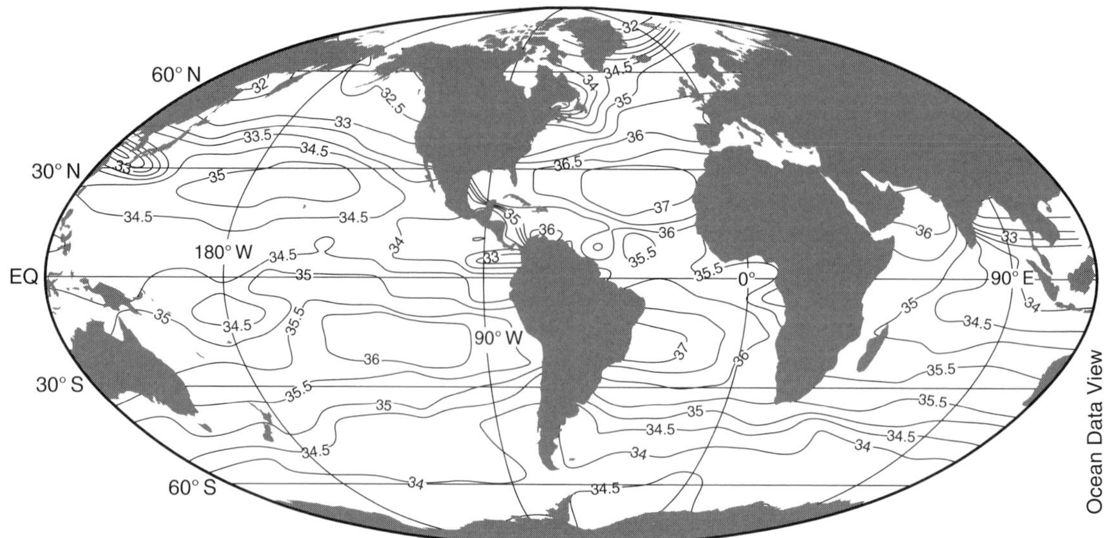

Figure 1.1. Annual mean surface salinity of the world's ocean. (Plotted by using Ocean Data View (Schlitzer, 2001), and surface salinity in Levitus *et al.* (1994).)

Salinity and temperature are the primary factors that determine the density of seawater. The densities of most surface seawaters range from 1024 to 1028 kg m^{-3}, and it is possible to evaluate density to about ± 0.01 of these units. In order to avoid writing numbers with many significant figures, density is usually presented as the Greek letter sigma, σ, which has the following definition

$$\sigma = (\rho/\rho_0 - 1) \times 1000, \tag{1.3}$$

where ρ is the density of the sample (kg m^{-3}) and ρ_0 is the maximum density of water at 3.98 °C (999.974 kg m^{-3}). (Note that the numerical value of this expression is only slightly different from, $\sigma = \rho$ - 1000, which appears in many texts.) Density is calculated from temperature, salinity and pressure (because of the compressibility of water) by using the international equation of state of seawater (Millero, 1996). The expression above represents the density *in situ* of a seawater sample determined from the measured temperature, salinity and depth. Because all water acquired its temperature and salinity while it was at the ocean surface, it is convenient to know the density corrected to one atmosphere pressure, which is indicated by sigma with a subscript t (sigma-tee), σ_t. By the same reasoning, it is often advantageous when tracing the source of a water parcel to calculate density by using temperature corrected for increases caused by water compression under the influence of pressure. The potential temperature, θ, is the temperature the water sample would have if it were raised to the surface with no exchange of heat with the surroundings, i.e., if it changed pressure adiabatically. At the depths of the ocean this is a large effect. A water parcel gains *c.* 0.5 °C when it sinks from the ocean surface to 4000 m depth (*c.* 400 atm). Potential temperature is the temperature it had before sinking. Density calculated at one atmosphere and the potential temperature is called sigma–theta, σ_θ.

Table 1.3. *Temperature, salinity, and flow rate of major deep-ocean water masses*

Water mass	Temperature[e] (°C)	Salinity[e]	Flow estimate[f] (Sverdrups)
AABW[a]	−2.0–0.0	34.6–34.7	5–10
NADW[b]	2.0–3.0	34.9–35.0	15–20
MW[c]	12.0	36.6	—
AAIW[d]	2.0–3.0	34.2	5–10

[a] AABW, Antarctic Bottom Water
[b] NADW, North Atlantic Deep Water
[c] MW, Mediterranean Water
[d] AAIW, Antarctic Intermediate Water
[e] *T* and *S* characteristics from Picard and Emery (1982)
[f] Flow rates are in Sverdrups (10^6 m^3 s^{-1}).

Note that the North Atlantic surface water is nearly 2 salinity units saltier than North Pacific surface water. At first this seems counterintuitive because more large rivers drain into the Atlantic. The reason for this difference has to do with the relative rates of evaporation in the high latitudes of the two oceans. North Atlantic surface water is on average warmer (10.0 °C) than North Pacific surface water (6.7 °C). Warmer water leads to warmer air, which has a higher specific humidity (the mass of water per mass of dry air) and increases evaporation and consequently salinity as well. The temperature difference is due to the warm Gulf Stream waters that flow north along the east coast of North America having a greater impact at high latitudes than their Pacific counterpart, the Kuroshio current. The resulting salinity difference has very important consequences for the nature of global thermohaline circulation. Because salt content (along with temperature) influences the density of seawater, the higher salt content of North Atlantic surface waters gives them greater densities at any given temperature than North Pacific waters. This is the main reason for massive downwelling, all the way to the ocean bottom, in the North Atlantic where the water is cold and salty, but no deep water formation in the North Pacific. There is no North Pacific Deep Water in Table 1.3.

This explanation for the surface salinity differences between the Atlantic and Pacific does not provide the whole story because it overlooks the need to budget atmospheric water transport on a global basis. In fact, the only way to cause a net salinity change in an ocean due to evaporation is via net transport of water vapor to another region on a time scale that is short with respect to the residence time (decades to centuries) of the surface water in question. Simply removing water from an ocean to the atmosphere or to an adjacent landmass is insufficient if that same water rapidly returns to the source ocean. To create a salinity difference between oceans, water must be removed across a continental divide so that it precipitates either directly on another ocean or into the drainage basin of a river discharging into another ocean.

This budgetary constraint makes global salinity patterns the net result of local evaporation, wind patterns, and continental placement and topography. An ideal "vapor export window" from an ocean would be through a region where initially dry prevailing winds blow continuously over warm ocean surface waters and then across a low continental divide. Inspection of the North Atlantic Ocean shows such a window at about 20° N, where the North East Trade Winds blow westward across the Sahara desert, subtropical Atlantic, and then over the relatively low continental divide of Central America. The surface Atlantic Ocean expresses its highest salinity (*c.* 37.5) at this latitude, and high rainfall over western Panama and Costa Rica indicates substantial vapor export to the subtropical Pacific. In contrast, the expansive subtropical Pacific Ocean has few upwind deserts, and a Trade Wind window that is effectively blocked by Southeast Asia. Thus the percentage of net water loss is much less in the bigger ocean.

In the large perspective, the North Atlantic Ocean is now saltier than the North Pacific as a result of the present distribution of ocean and atmosphere currents and continents over the surface of the Earth. Other distributions, as occurred in the past owing to different distributions of ice, deserts or continental topography, would produce very different water balances and global current systems.

Temperature differences in the sea are large, ranging between 30 °C in equatorial surface waters and −2 °C in waters that are in contact with ice. By comparison, salinity is remarkably constant, $S = 33.0\text{–}37.0$, necessitating very accurate determinations in order to distinguish differences. The average temperature and salinity of the sea are 3.50 °C and 34.72, and 75% of all seawater is within ±4 °C and ±0.3 salinity units of these values. Cross sections of the potential temperature and salinity of the Atlantic and Pacific Oceans (Fig. 1.2) demonstrate how water masses can be identified with distinct origins at different densities and hence different depths. The water masses are characterized by the temperature and salinity that is determined at the surface ocean in the area of their formation (Table 1.3). The deepest waters, Antarctic Bottom Water (AABW) and North Atlantic Deep Water (NADW), are formed at the surface in polar regions. AABW is dense because it is formed under the ice in the Weddell and Ross Seas and is thus extremely cold. NADW is not particularly cold, but is highly saline because of the source waters from the Gulf Stream and high evaporation rates in the North Atlantic. Antarctic Intermediate Water (AAIW) is both warmer and less saline than either of the deeper-water masses and thus spreads out in the ocean at a depth of about 1000 m.

1.2.2 Element classification

Chemicals in seawater can be classified into four groups based mainly on the shapes of their dissolved concentration distributions

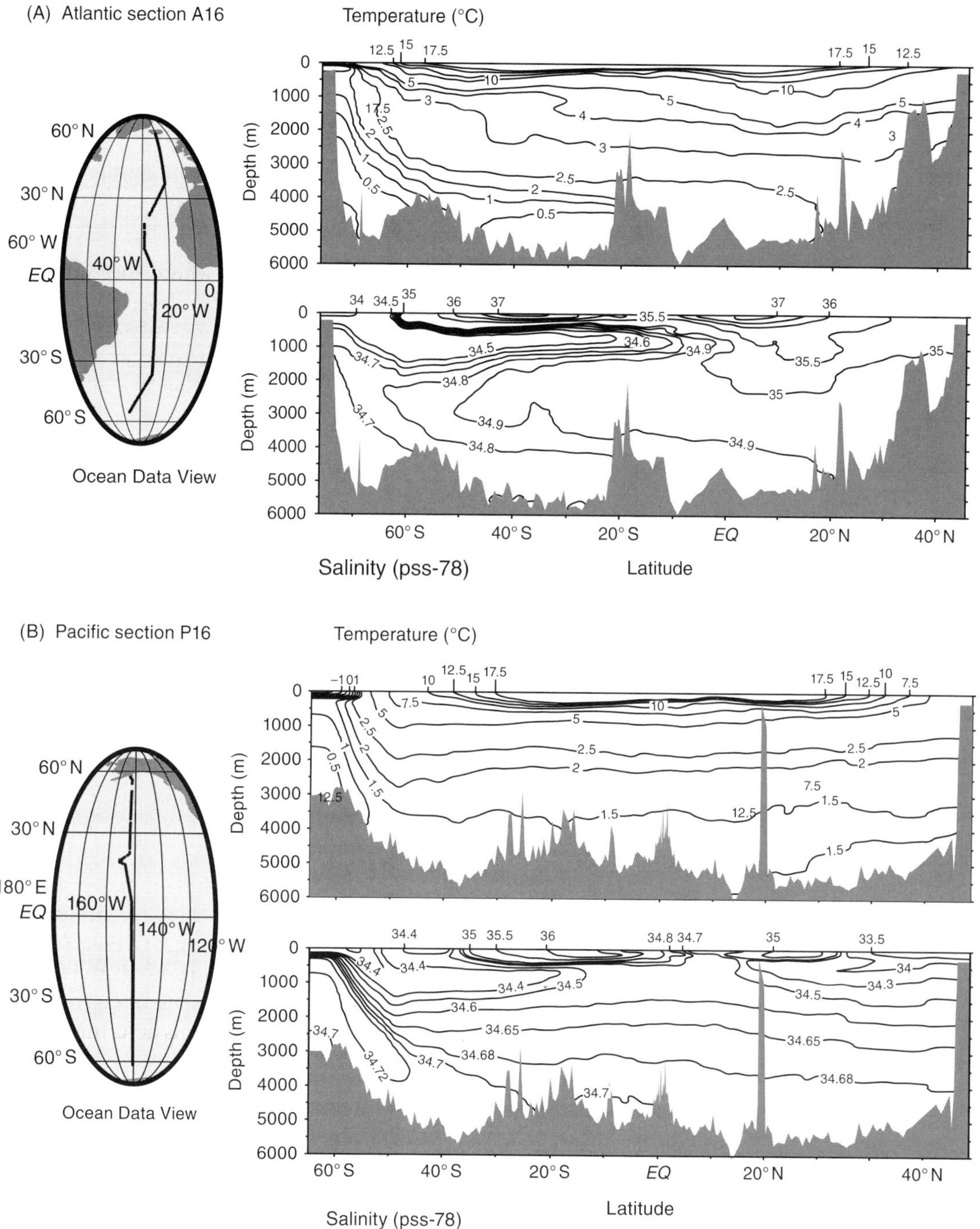

Figure 1.2. Latitudinal cross section of the potential temperature and salinity of the Atlantic (A) and Pacific (B) oceans. Different water masses are definable by their characteristic temperature and salinity (Table 1.3). (Plotted by using Ocean Data View and WOCE hydrographic data (Schlitzer, 2001).)

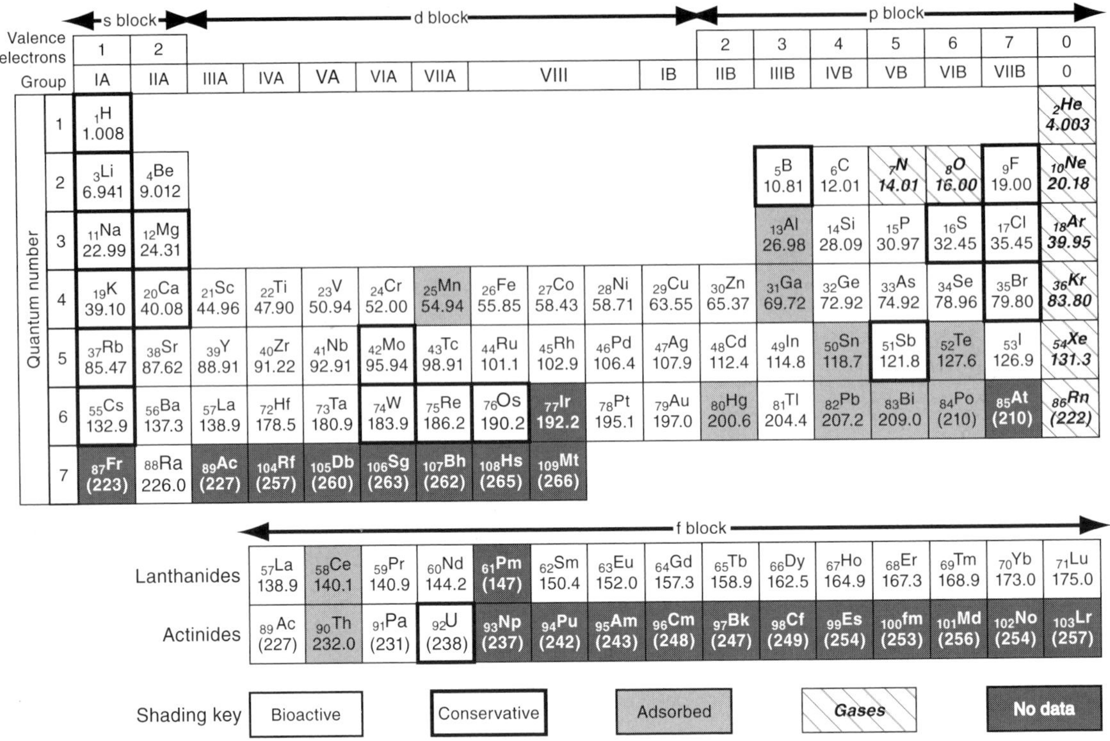

Figure 1.3. Periodic table of the elements with the categories of conservative, bioactive, adsorbed, and gaseous elements indicated.

with depth. Most measurements are made on unfiltered samples, but experiments designed to operationally define "dissolved" and "particulate" phases by using 0.45 μm pore-size filters indicate that, especially below the surface 100 m, nearly all the total mass of any element is in the dissolved phase. Chemical species in seawater are classified into four categories in the Periodic Table of Fig. 1.3: *conservative* elements, *bioactive* elements, *adsorbed* or *scavenged* elements, and *gases*. Because these categories are based primarily on the relative importance of reactivity and mixing to elemental ocean distributions, the boundaries between them are sometimes vague. For example many trace metals, like Fe and Co, are both *bioactive* and *adsorbed* and fall into both categories.

Conservative elements

To within a few percent, conservative elements in seawater have constant concentration : salinity ratios. That is, their concentrations are not greatly affected by processes other than precipitation and evaporation: the same processes that control salinity in the ocean. This definition is of course operational since the ability to determine the effect of biological and chemical processes on concentration depends on the accuracy and precision of the measurement method. Elements of high concentration tend to be conservative because they are relatively unreactive; however, conservative elements are present in all concentration ranges because some of them are both low in crustal abundance and relatively unreactive. There are of course

Table 1.4. *Major ions[a] in surface seawater at salinity S = 35, and their role in the calculation of alkalinity[b]*

Major ions are defined here as those charged constituents with concentrations greater than 10 μmol kg^{-1}, excluding the nutrient nitrate, which varies in concentration.

Cations			Anions					
			Insignificant proton exchange			Significant proton exchange (A_T)[a]		
species	mmol kg^{-1}	meq kg^{-1}	species	mmol kg^{-1}	meq kg^{-1}	species	mmol kg^{-1}	meq kg^{-1}
Na^+	469.06	469.06	Cl^-	545.86	545.86	HCO_3^-	1.80	1.80
Mg^{2+}	52.82	105.64	SO_4^{2-}	28.24	56.48	CO_3^{2-}	0.25	0.51
Ca^{2+}	10.28	20.56	Br^-	0.84	0.84	$B(OH)_4^-$	0.11	0.11
K^+	10.21	10.21	F^-	0.07	0.07			
Sr^{2+}	0.09	0.18						
Li^+	0.02	0.02						
Σcations[b]		605.67	Σanions		603.25			2.42

Concentrations are from DoE (1994).

[a] The concentration cut-off for the definition of major ions traditionally consists of elements with concentrations greater than 1 mg kg^{-1}. The concentration of Li^+ is below this threshold, but it is added here to achieve the charge balance definition of alkalinity (see Chapter 4).

[b] The difference between the total concentrations of cations and anions, (Σcations − Σanions), in the bottom row, and the sum of the constituents of the last column equals the value of the alkalinity, 2.42 meq kg^{-1}. See the discussion in Chapter 4.

some caveats to our classification of conservative ions. Ca^{2+}, Mg^{2+} and Sr^{2+} are not strictly conservative, as changes on the order of one percent in their concentration : salinity ratios have been identified. This property was discovered by making very accurate titration and mass spectrometric measurements. If equally precise methods were applied to the other elements, changes with respect to salinity would probably be found for some.

We define the major ions in seawater (Table 1.4) as those with concentrations greater than 10 μmol kg^{-1}. Most of the major ions are conservative (exceptions are Sr^{2+}, HCO_3^- and CO_3^{2-}) and these ions make up more than 99.4% of the mass of dissolved solids in seawater. Na^+ and Cl^- account for 86% and Na^+, Cl^-, SO_4^{2-}, and Mg^{2+} make up 97%. Conservative elements with concentrations less than 10 μmol kg^{-1} are found in rows 5 and 6 of the periodic table where elements with lower crustal abundances occur.

Bioactive elements

These dissolved constituents of seawater have concentration versus depth profiles characterized by surface water depletion and deep water enrichment caused by plant consumption in the euphotic zone and release at depth when the biological material dies, sinks and degrades. Examples of these elements are nutrients required for phytoplankton growth (P, NO_3^- and HCO_3^-), oxygen consumed during

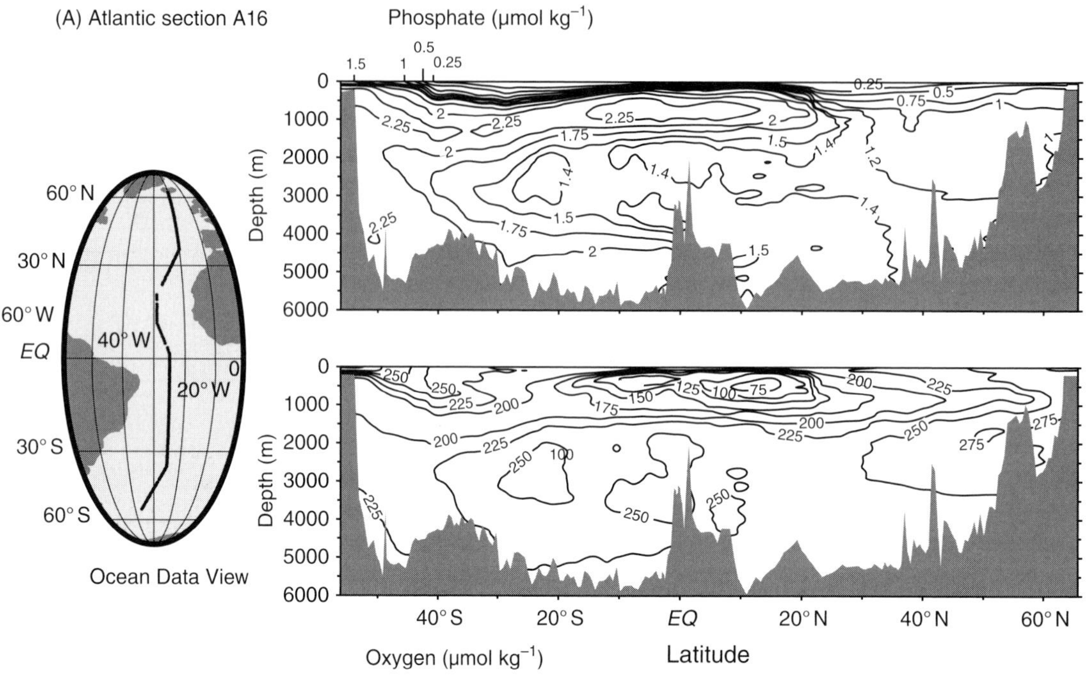

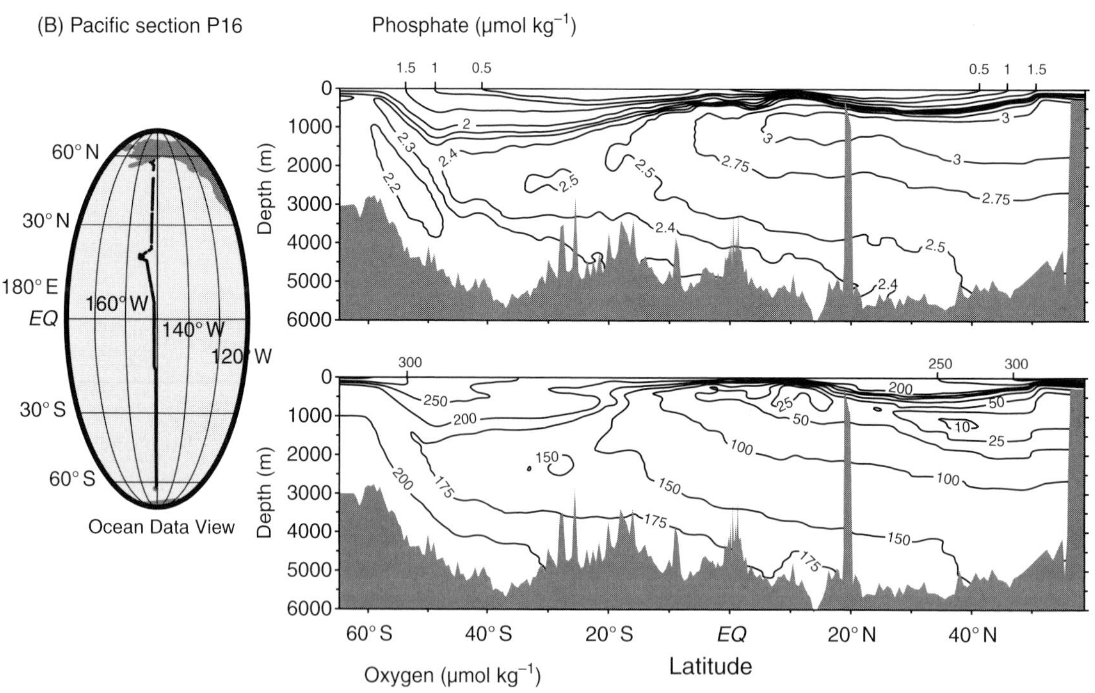

Figure 1.4. Latitudinal cross sections for phosphate and the concentration of oxygen in the Atlantic (A) and Pacific (B) Oceans. (Plotted by using Ocean Data View and WOCE hydrographic data (Schlitzer, 2001).)

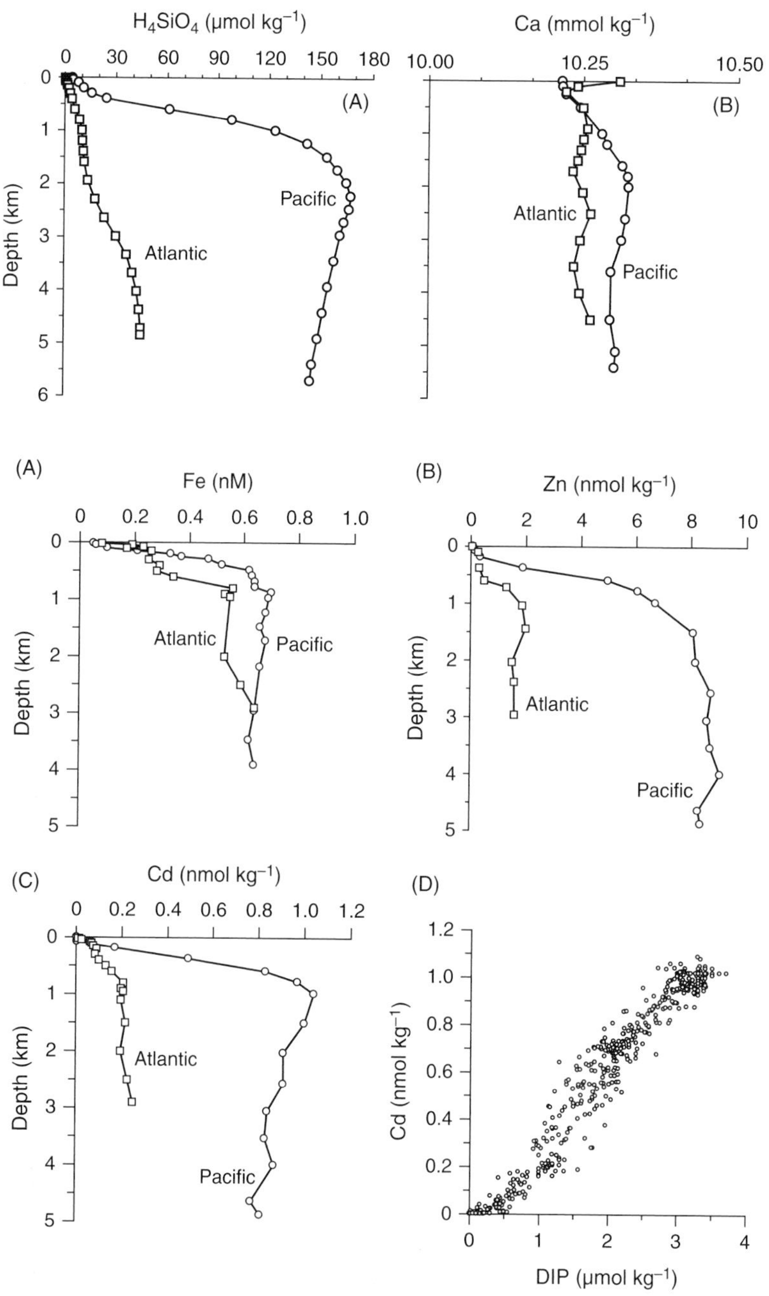

Figure 1.5. Vertical profiles of silicic acid (A) and calcium (B). These elements are released to the dissolved phase of seawater upon the dissolution of opal and carbonate tests. (The silicic acid profile was plotted by using Ocean Data View from WOCE data; Ca data are from de Villiers (1994).)

Figure 1.6. Vertical profiles of the trace metals (A) Fe, (B) Zn and (C) Cd in the Atlantic and Pacific Oceans; and (D) the global dissolved Cd versus P relationship. The shape of the depth distributions indicates that these metals behave like nutrients in the ocean. (Data from Johnson *et al.* (1997), Bruland (1983) and Boyle (1988).)

respiration (Fig. 1.4), constituents of shells made by some plankton species (Si and Ca) (Fig. 1.5), and trace metals (Fig. 1.6) necessary for plankton growth (e.g. Fe) or incorporated into the plankton for uncertain reasons (e.g. Cd and Zn). Both HCO_3^- and CO_3^{2-} are major ions in seawater (Table 1.4), while the micronutrients P, N, and Si have concentrations that range from nano- to micromoles per kilogram. The bioactive metals are true trace elements, having concentrations in the range of nano- or picomoles per kilogram. The fact that biological

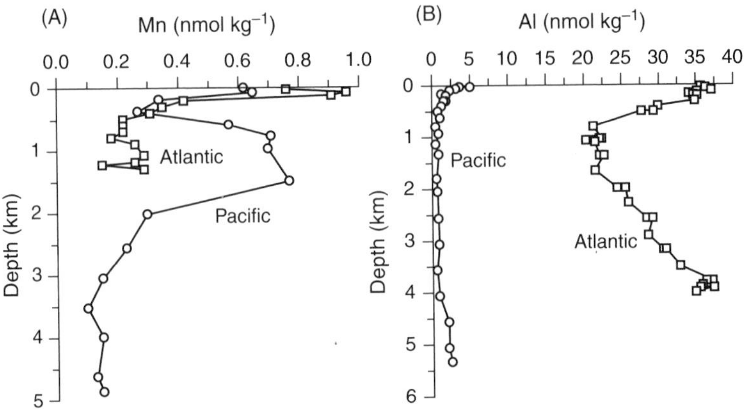

Figure 1.7. Profiles of the total concentrations of (A) Mn and (B) Al. These metals show some of the characteristics of metals that are adsorbed from solution in seawater in that their concentrations decrease with depth and from the Atlantic to Pacific Oceans. (Mn data from Bruland (1983), and Al from Pilson (1998).)

uptake and recycling at depth are the dominant processes altering vertical concentration profiles is indicated not only by the large number of elements that are affected (Fig. 1.3), but also by their ability to alter the concentrations of some major ions such as HCO_3^- and CO_3^{2-}.

Adsorbed elements

These elements have depth profiles that are reversed from those in the bioactive category. Concentrations are higher in surface waters and decrease with depth as the elements are adsorbed to particles that fall through the ocean. In some cases the surface ocean enrichment is also a result of metal input by dust particles transported to the sea by wind. Adsorbed elements are exclusively low in concentration, meaning that this mechanism is not pervasive enough to alter the concentrations of elements with higher concentrations. Examples of metals that have concentration profiles influenced by this process are Mn and Al (Fig. 1.7).

Gases

Gases dissolved in seawater are either chemically inert (the noble gases in the last column of Fig. 1.3) or bioactive (e.g. O_2 and CO_2). We give gases a category of their own because conservative behavior in this case means that the concentrations in seawater are at or near equilibrium with their respective atmospheric gas concentrations rather than having a constant ratio with salinity. Equilibrium, or gas saturation, is primarily a function of temperature (cold water can hold more gas at equilibrium), but salinity also plays a role (saltier water can hold less gas at equilibrium). Dissolved gases at the ocean's surface are at or near saturation with respect to their atmospheric partial pressures and, if they are unreactive, maintain this concentration as they are subducted into the ocean's interior. Thus, conservative gases have concentrations that vary with the temperature of the ocean's surface water. Because temperature decreases with depth in the ocean, inert gas concentrations increase with depth. Gases also have the special property of being responsible for the transfer of certain elements, primarily oxygen

Table 1.5. *The major gases of the atmosphere excluding water vapor, which has a concentration of a few percent at saturation in the atmosphere*

Seawater equilibrium concentrations were calculated from the Henry's Law coefficients at 20 °C and $S = 35$.

Gas	Atmospheric mole fraction (atm)	Seawater equilibrium concentration (μmol kg^{-1})
N_2	7.808×10^{-1}	4.18×10^{2}
O_2	2.095×10^{-1}	2.25×10^{2}
Ar	9.34×10^{-3}	1.10×10^{1}
CO_2	3.65×10^{-4}	1.16×10^{1}
Ne	18.2×10^{-6}	7.0×10^{-3}
He	5.24×10^{-6}	2.0×10^{-3}
Kr	1.14×10^{-6}	2.0×10^{-3}
Xe	0.87×10^{-7}	3.0×10^{-4}

and carbon, between the atmosphere and ocean. Atmospheric pressures and concentrations at saturation equilibrium are presented in Table 1.5. More about the utility of gases as tracers and the processes of atmosphere–ocean exchange is presented in Chapters 3 and 10.

Of the 89 elements that have known seawater concentrations, 54 are bioactive (including the three gases that are involved in biological cycles), 23 are conservative (including the rare gases in the last column, except for radon, which is radioactive), and nine are in the adsorbed category. While the conservative elements make up more than 99% of the dissolved solids in seawater, the characterization of all measurable elements by using the vertical-profile classification scheme indicates that about two thirds are noticeably affected by biological processes and about one tenth are primarily controlled by adsorption.

1.3 Ocean circulation

Chemical dynamics and mechanisms of reactions in the ocean–atmosphere system on time scales equal to and less than that of ocean circulation are evaluated by studying the distribution of chemical compounds within the sea. In order to understand the processes controlling the chemical distributions and their rates, one must know something about how the ocean circulates. The following brief descriptive overview describes the main wind-driven and thermohaline current distributions.

1.3.1 Wind-driven circulation

Circulation in the near-surface ocean is driven by friction of wind on the atmosphere–ocean interface, whereas in deeper waters it is mostly density-driven. Unequal heating of the Earth's surface creates

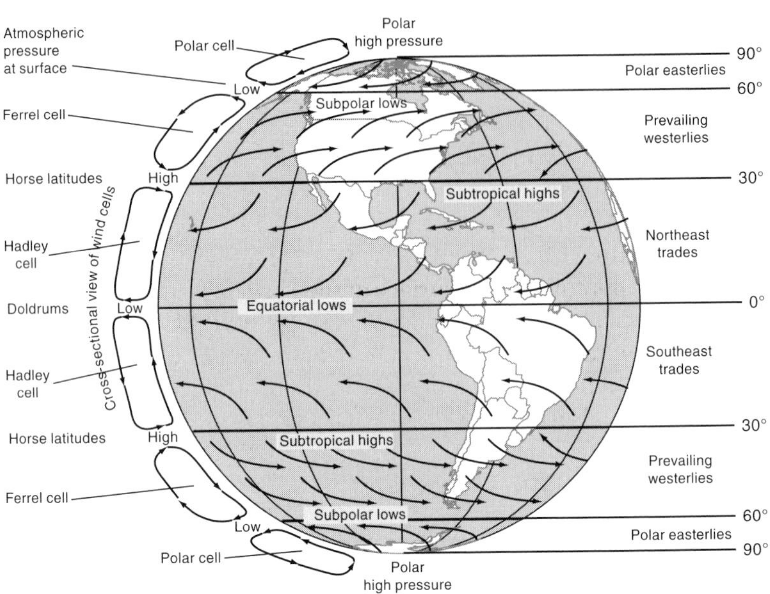

Figure 1.8. The latitudinal distribution of the wind directions on the Earth's surface. (Redrafted from Pinet (1994).)

winds in the atmosphere that impart frictional energy to ocean surface waters. The mean global atmospheric wind pattern consists of east to westward flowing Trade Winds immediately north and south of the Equator, the Westerlies in mid-latitudes and the Easterlies at high latitudes (Fig. 1.8). Atmospheric pressure lows occur near the Equator and 60° latitude due to rising flow of air. Atmospheric highs are created by downward flow at about 30° in mid-latitudes. The lows are accompanied by higher than average precipitation as rising air is cooled and decreases in its moisture-carrying capacity. Highs are characteristically dry because water-poor cool air sinks, is warmed and increases its capacity to carry water vapor. These atmospheric circulation patterns contribute to the overall surface salinity distribution presented at the outset of the chapter in Fig. 1.1.

Friction on the ocean surface drags the surface water in the direction of the wind. The resulting mean flow in the upper 10–100 m, however, is not in the direction of the wind, but 90° to the right of the wind in the Northern Hemisphere and 90° to the left of the wind in the Southern Hemisphere. This flow is called *Ekman transport*. The deviations from the direction of the wind are due to the *Coriolis force*, which is not a true force but a device used to compensate for the fact that all measurements and forces are made relative to a rotating Earth. To understand the reason for this deviation from the direction of the wind forcing, assume an hypothetical particle is accelerated northward in the ocean surface somewhere near the Equator. The particle has the special properties that keep it at the same depth in the surface waters and lacks friction in the direction it is moving, allowing it to maintain the speed it was given at the beginning of the journey. When the particle leaves the equator it has both the northward component of velocity and an eastward component imparted by the rotation of the Earth. Because the Earth is a sphere, the eastward velocity on its surface is faster at the equator (455 km s^{-1}) than that at 30 °N or S (402 km s^{-1}), which is in turn faster than that at 45 °N or S (326 km s^{-1}). As the particle travels northward the Earth beneath it rotates more slowly. From the perspective of the Earth's surface the particle appears to move to the right. What is really happening is that the particle travels in a straight line northward with respect to a rotating cylinder with a diameter of the Earth's equator, but the surface of the Earth lags behind at higher latitudes because its diameter decreases poleward. Similarly, a particle in motion to the south from the Equator would veer to the east. An eastward tendency for a particle in the Northern Hemisphere is movement to the right of the direction in which the particle is moving. In the Southern Hemisphere it is to the left.

Since surface winds change directions, so do surface ocean currents. For example, the northwest-flowing Trade Winds in the Southern Hemisphere and southwest-flowing Trade Winds in the Northern Hemisphere converge at the equator. Since the resulting mean Ekman transport in surface currents is 90° to the right of the wind in the north and 90° to the left in the south, the surface waters in this region flow away from the Equator. This creates a divergence (Fig. 1.9A) in flow that is "filled in" by upwelling of water from below (*Ekman suction*). Conversely, in the subtropics (*c.* 30°) surface water flow converges from the north and south, causing waters to "pile up" and creating a location of general downwelling (*Ekman pumping*). A similar effect is caused by flow of winds along the coasts (Fig. 1.9B, C). For example, the northward flow of winds along the Pacific coast of South America creates a surface water flow to the left (west), which draws water away from the continent (Fig. 1.9B). Surface water transport away from the land is compensated for by upwelling of water along the coast. Vertical movement of water caused by Ekman transport resulting from prevailing winds has important consequences for chemical and biological oceanography because in regions of upwelling (near the Equator and on some continental margins) nutrient-rich waters are brought from below into the sunlight resulting in high productivity and important fisheries. In locations of downwelling (i.e. the subtropical gyres) nutrient concentrations in surface waters are nearly below detection limits.

Although Ekman transport is concentrated in the upper 10–100 m of the ocean, the consequences of upwelling and downwelling set up local circulation patterns that are felt much deeper. Accumulation of water in areas of Ekman convergence and depletion of water in Ekman divergences cause horizontal gradients in water height of a meter or so over basin-scales. As the water flows from the "hills" to the "valleys" it is also influenced by the Coriolis force, creating large-scale gyre transport that is felt to much greater depths. This is called *Sverdrup transport*. An example is the Northern Hemisphere subtropical convergence zone. As water flows downhill from its high level in

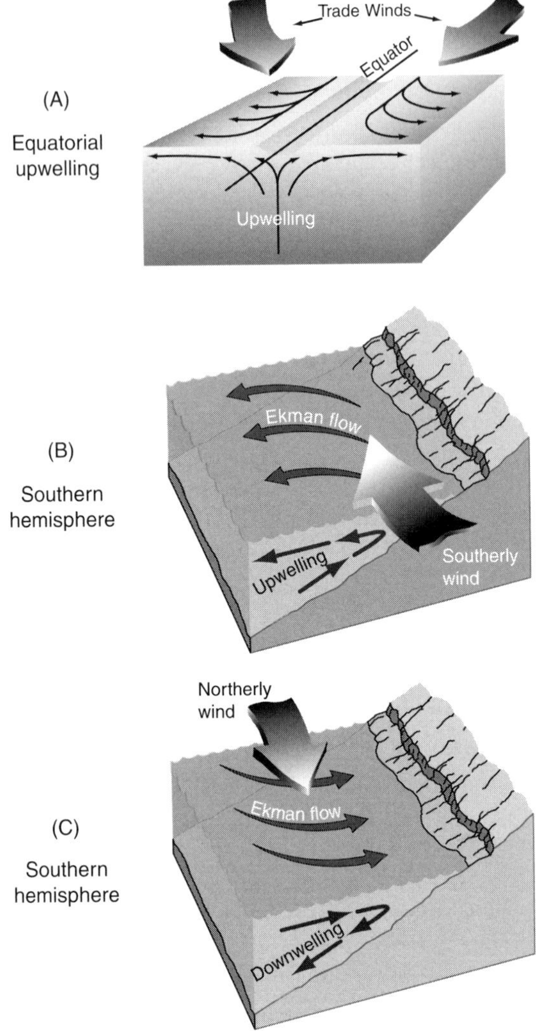

Figure 1.9. Schematic diagrams illustrating Ekman transport in response to wind forcing at the air–water interface (A) at the Equator and (B, C) near the coastline. (Redrafted from Thurman (1994).)

the center of the gyre it is forced to the right by the Coriolis force, creating a large-scale anticyclonic gyre (clockwise flow in the Northern Hemisphere) (Fig. 1.10). In the area of Ekman divergence between 50 and 70 °N in the subarctic oceans, flow of water into the trough is forced to the right, causing a cyclonic gyre (counterclockwise flow in the Northern Hemisphere). In theory the Sverdrup transport extends to the entire depth of the ocean; this maximal theoretical transport is called the *barotropic* component of the Sverdrup transport. In reality, however, ocean stratification at depth weakens the barotropic Sverdrup transport; this weakening is called the *baroclinic* component of the Sverdrup transport.

Warming of the surface ocean by solar heating and turbulence induced by wind stress compete to create a surface mixed layer over most of the ocean that is 10–100 m deep except in some high-latitude areas, where the ocean is mixed more deeply. In the subtropical and

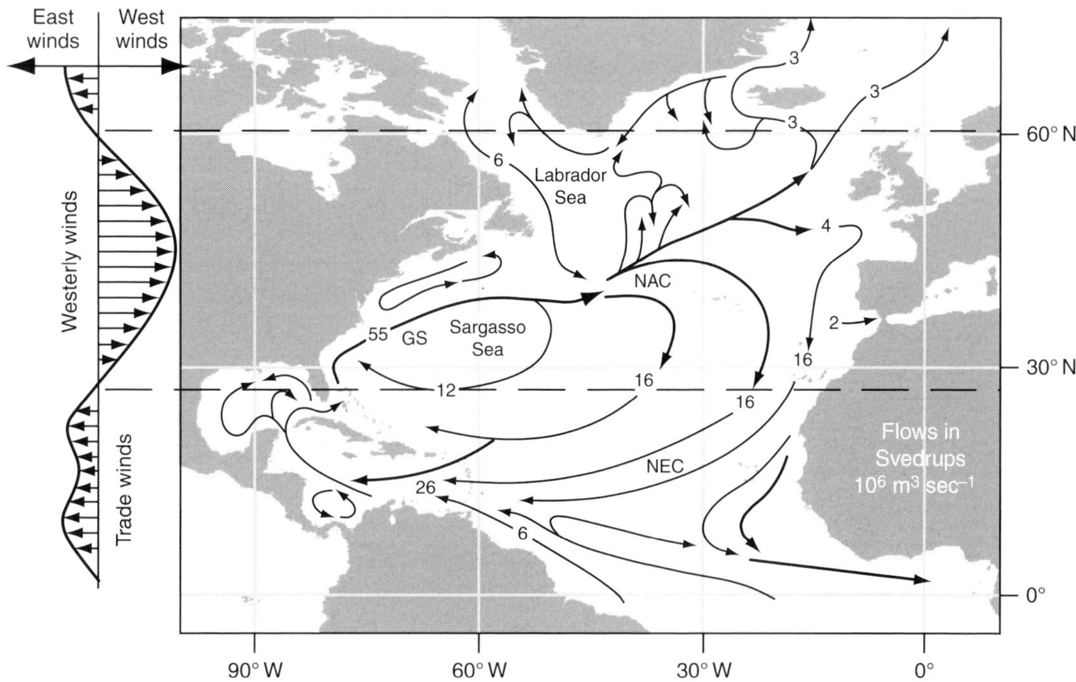

Figure 1.10. Schematic diagrams of the Sverdrup transport in the North Atlantic, driven by sea surface topography and the Coriolis force. Arrows indicate flow direction and the thickness indicates magnitude. Numbers are the flow rate in Sverdrups (10^6 m^3 s^{-1}). GS, Gulf Stream; NAC, North Atlantic Current; NEC, North Equatorial Current. (Redrafted from Pinet (1994).)

subarctic oceans there is a "seasonal" thermocline in the upper 100 m of the water column, which shoals and strengthens in summer owing to solar heating (Fig. 1.11). The surface mixed layer gives way to density stratification in the "permanent" *pycnocline* (density gradient) that separates the upper and deep oceans and occupies the depth range from the winter mixed-layer depth (100–200 m) to 1000–1500 m in most of the ocean.

1.3.2 Thermohaline circulation

Below 1000–1500 m, temperature gradients are small and Sverdrup transport is weak. In this region large-scale transport is caused by *thermohaline* circulation. The overall water balance of the ocean below 1500 m consists of sinking of water in the polar regions (with salty Mediterranean water (Table 1.2) also mixed in from the side), which is balanced by upwelling and return flow from ocean depths to the surface. What sounds like a vertical balance is in reality a complex layered structure of water masses that can be traced in three dimensions throughout the ocean by salinity, oxygen and nutrient differences. The cross sections of salinity in Fig. 1.2 indicate deep water masses with the T and S properties in Table 1.2. Southward-flowing North Atlantic Deep Water (NADW) is bounded above and below by northward-flowing southern-source waters. Antarctic Bottom Water (AABW) flows beneath the NADW and Antarctic Intermediate Water (AAIW) flows on top. Water masses that reach the Southern Ocean are mixed in the Antarctic Circumpolar Water (ACW) that flows around Antarctica and is more vertically homogeneous than in other parts

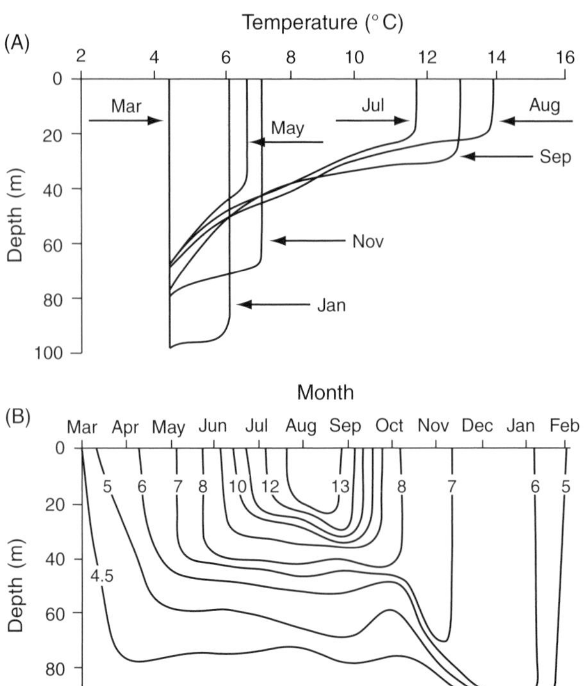

Figure 1.11. Typical growth and decay of the seasonal thermocline in the subarctic Pacific Ocean (50° N, 145° W). (A) Temperature versus depth for different months. (B) Temperature contours on a depth versus time plot. (Replotted from Knauss, 1978).

of the ocean. A deep water mass sometimes called Common Water, because it is a mixture of many water masses, enters the Indian and Pacific Oceans (Fig. 1.12). In the Indian Ocean the water flows through the Crozet basin south of Madagascar into the Arabian Sea. Common Water enters the Pacific south of New Zealand and flows northward along the western boundary of the basin at about 4 km into the Northeast Pacific. Carbon-14 dates of the dissolved inorganic carbon of deep waters reveal that the "oldest" water in the deep sea resides in the Northeast Pacific. This means that this water has been isolated from the surface ocean (where the clock is reset or partly reset for ^{14}C) longer than any other water in the deep sea.

The original theories of deep-ocean circulation assumed that the return flow from deep water formation was uniform global ocean upwelling through the thermocline. This rising flow created the concave upward temperature and salinity profiles of the deep ocean below 1000 m. More modern concepts and tracer interpretation suggest that bottom water flows north in the Pacific Ocean, upwells, and returns south as the North Pacific Deep Water (NPDW) between 2000 and 3000 m depth. This water eventually makes its way into the thermocline and mixed layers and back to high-latitude regions of deep-water formation in the North Atlantic.

The deep water flow in the ocean is often depicted as a "conveyor belt" in which water that originates at the surface in the North Atlantic Ocean (NADW) flows through the Atlantic, Indian and Pacific Oceans before it upwells and returns (Fig. 1.12). The analogy

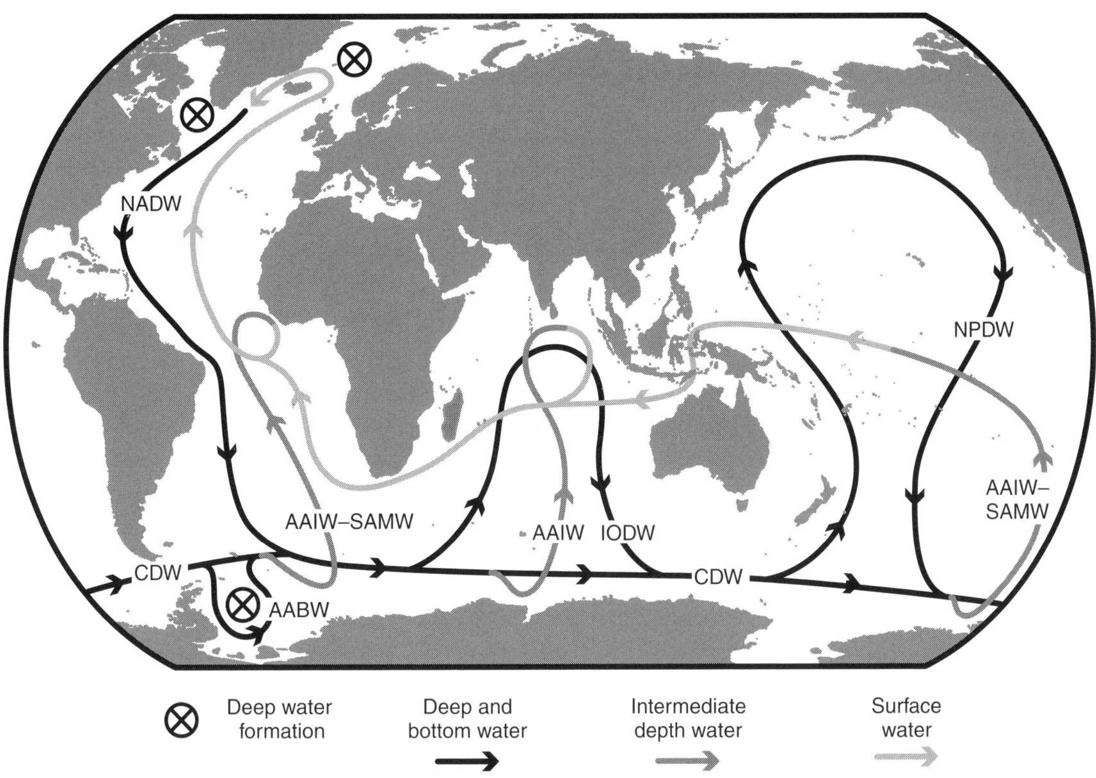

Figure 1.12. The flow directions of the major deep and intermediate waters of the oceans, indicated by the black and dark gray lines, respectively. Shallower return flow is indicated by the light gray line. Locations of deep water formation are indicated by circled Xs. The North Atlantic Deep Water (NADW) occupies 1000–4000 m in the Atlantic and flows from the Norwegian and Greenland Seas south to the Antarctic, where it joins the Circumpolar Deep Water (CDW). Antarctic Bottom Water (AABW) is formed under Antarctic ice shelves, flows north in the Atlantic Ocean and becomes entrained in the CDW, where it joins NADW in the path around Antarctica and into the Indian Ocean, where it becomes Indian Ocean Deep Water (IODW), and ultimately to the Pacific Ocean, where it eventually becomes North Pacific Deep Water (NPDW). The ocean above about 1000 m is ventilated by Intermediate and Mode Waters. The paths of the Antarctic Intermediate (AAIW) and Subantarctic Mode Waters (SAMW) are shown in the figure. (Modified from Gnanadesikan and Halberg (2002).)

is that deep water accumulates metabolic residue from degradation of organic matter and dissolution of opal and calcite that originates from near-surface biological processes just as a conveyor belt accumulates the rain of dust as it proceeds through a factory. Thus, the chemical properties of the NADW become progressively more removed from those of surface waters as it moves at depth from its origin. This is nicely illustrated by the progressive increase in concentration of nutrient phosphate from the North Atlantic to South Pacific to North Pacific in deep waters (Fig. 1.4). It is important to recognize that the conveyor belt analogy ignores the fact that bottom waters have another surface origin in Antarctica (AABW) as depicted in Fig. 1.12. The reason that AABW is not part of the conveyor belt circulation is that these waters do not originate with the chemical properties of low-latitude surface waters as they do in the NADW.

Antarctic Bottom Water, AABW, is formed from water that upwells to the surface in the Antarctic, but it doesn't reside there long enough for biological processes and gas exchange to "reset" nutrient and oxygen concentrations. The circumpolar Southern Ocean acts more as a mixing region for all the water masses that flow there than as a location of renewal of surface water properties to nutrient-poor and oxygen-rich values. We must keep in mind when using the simple conveyor belt analogy that it refers primarily to NADW and does not represent all deep water flow. In reality water masses mix in the deep ocean and upwelling occurs along many regions of the flow path that ultimately leads to the North Pacific.

1.4 Ocean biology

Nearly all chemical reactions in the sea take place via biological metabolism or are catalyzed by biologically produced enzymes. Overwhelmingly, the most important of these processes are photosynthesis and respiration. Photosynthesis uses the sun's energy to create ordered organic compounds that consist of roughly 65% proteins, 20% lipids and 15% carbohydrates. Most of the organic matter produced during marine photosynthesis is in the form of very small (diameter $<100\,\mu m$) particles, but some is in the form of dissolved organic matter (DOM). Some of the microscopic plants and animals create shells in the form of inorganic minerals that consist primarily of calcite and aragonite ($CaCO_3$) and opal (SiO_2). Between 10% and 30% of the organic matter and most of the mineral armor produced in the euphotic zone escapes by gravitational settling or water transport and is degraded or dissolved at depth, where it joins the conveyor belt circulation in dissolved inorganic form. A small fraction (0.1%–1.0%) of the particulate organic material produced at the surface escapes all the way to the bottom and accumulates in marine sediments; a greater fraction of the minerals escapes dissolution and is deposited in the sediments.

In order to understand the role of plants and animals in distributing chemical compounds in the sea one must understand something about the biological agents that perform these tasks. In this text we will focus on biota with sizes smaller than a few millimeters in diameter, which includes by far the most important community affecting the distribution of chemical compounds in the sea. As mentioned earlier, the distinction between dissolved and particulate material is usually operationally defined as the material that is separated by passing water through a filter with pore size 0.45 μm. Particulate material captured this way in the open ocean generally contains a small fraction of the total carbon and other chemical elements found in seawater. Even for carbon, which is the most abundant element in particulate material, the >0.45 μm fraction contains $\leq 1\ \mu mol\ kg^{-1}$ C below the euphotic zone relative to the *c.* 40 $\mu mol\ kg^{-1}$ C found in the dissolved organic form (DOC).

Suspended particles with a size <20 μm sink very slowly (<1 m d^{-1}) whereas particles as large as 100 μm in diameter, usually fecal pellets, can sink as rapidly as 100 m d^{-1}.

Plankton are categorized by their size, metabolic function, and internal structure. The size classification consists of *picoplankton* (<2 μm), *nanoplankton* (2–20 μm), *microplankton* (20–200 μm), and *mesoplankton* (>200 μm). The classification by metabolic function distinguishes those species that create organic matter from dissolved inorganic matter, the *autotrophs* (phytoplankton and some bacteria), and those that gain energy and nutrition by consuming previously existing organic matter, the *heterotrophs* (some bacteria and zooplankton). There are a few species that can exist by both net autotrophy and net heterotrophy; these are called *mixotrophs*. The final distinction is based on whether single-celled organisms contain a nucleus and internal organelles; those that have none are *prokaryotes* and those that have them are *eukaryotes*.

1.4.1 Plankton

Bacteria

Bacteria are prokaryotes in the range of 1 μm in size and consist of species that are both autotrophic and heterotrophic. Bacterial concentrations range from 10^5 to 10^7 cells per cubic centimeter of seawater, with higher concentrations in the surface ocean. The importance of photoautotrophic bacteria in the role of photosynthesis in the upper ocean has been discovered relatively recently. The cyanobacteria *Synechococcus* and *Prochlorococcus* are now believed to dominate the picophytoplankton in most oceanic regimes. (We use the term phytoplankton because of their autotrophic behavior; see later.) These species are believed to contain most of the green, light-absorbing pigment, chlorophyll, in these regions. Their importance was discovered in the 1970s and 1980s with the development of epifluorescence microscopy as a routine method for studying microscopic living organisms. Species of autotrophic bacteria that generate energy from chemical reactions rather than from sunlight are called "chemoautotrophic" and play an important role as primary producers of organic matter at oxic–anoxic interfaces found in anoxic basins like the Black Sea, and at hydrothermal vents. Organic matter production by this energy pathway may have been important for the origin of life, but in today's ocean it represents only a tiny fraction of that produced by photoautotrophy. Larger prokaryotic bacteria such as *Trichodesmium* form colonies in surface waters and are known to fix N_2 gas into NH_3 in regions of the ocean where other nitrogen compounds are in very low concentration. This opportunistic behavior gives the nitrogen fixers an advantage in regions of the oceans that are poor in dissolved nitrogen relative to other nutrients.

Heterotrophic bacteria exist throughout the ocean both as individual entities in seawater (*c.* 80%) and attached to particle surfaces and sediments. These bacteria consume dissolved organic matter because they have no means other than dissolved transport across membranes to internalize nutrients. Some may secrete enzymes, which break down large molecules so they can be transported across the cell walls. In the euphotic zone of the ocean heterotrophic bacteria play an important role in organic matter recycling. Below the euphotic zone and in sediments heterotrophic bacteria, in collaboration with larger animals, are responsible for the vast majority of organic matter respiration.

Phytoplankton

Phytoplankton are unicellular autotrophs which contain the chlorophyll measured in the particulate matter filtered from seawater. The prokaryotic, autotrophic bacteria discussed above are included in this classification, but it is not limited to them. There are many phytoplankton that are eukaryotes. In addition to the autotrophic bacteria there are three other principal types of phytoplankton: diatoms, coccolithophorids, and photoautotrophic flagellates. Diatoms are nano–micro in size, but often form colonies that can be greater than 1 mm long. Their primary distinction is that they produce silica frustules that can be identified under high magnification (Fig. 1.13). Because of the opaline frustules, diatoms control the dissolved silica cycle in the sea, which is characterized by low concentrations in surface waters that increase with depth (Fig. 1.5). These phytoplankton dominate in regions of the ocean where upwelling is important, such as the Southern Ocean and regions of coastal upwelling. A portion of the silica frustules that leave the surface ocean are preserved in sediments. In areas such as the Southern Ocean around Antarctica the sediments consist predominately of diatom frustules because of their dominance in the surface waters. Observations of diatom frustules in sediments and "diatomaceous earth" deposits for the past 100 million years indicate that they have been important phytoplankton since the Cretaceous Period.

Some coccolithophorids are classified as nanoplankton and produce calcite ($CaCO_3$) plates called coccoliths that protect their protoplasm (Fig. 1.13). Generally coccolithophorids are rarer than diatoms and tend to be most abundant in subtropical gyres and in warm waters. *Emiliana huxleyi* is the most abundant coccolithophorid. Because their calcareous plates reflect light, blooms of coccolithophorids can be seen from space by using satellite images (Brown and Yoder, 1994). A spectacular bloom was observed in September of 1997 in the eastern Bering Sea; this had never been seen before in this region and was probably associated with warmer than usual sea surface temperatures. The calcitic coccoliths make up the bulk of the $CaCO_3$-rich sediments presently preserved in the ocean; nearly pure coccolith deposits such as in the White Cliffs of Dover, England, provide geologic evidence that the coccolithophorids have existed for the past 170 million years.

Photoautotrophic flagellates are pico to nanometer-sized plankton, usually with two flagella, whip-like appendages that beat within grooves in the cell wall. They are found in most regions of the ocean,

(A)

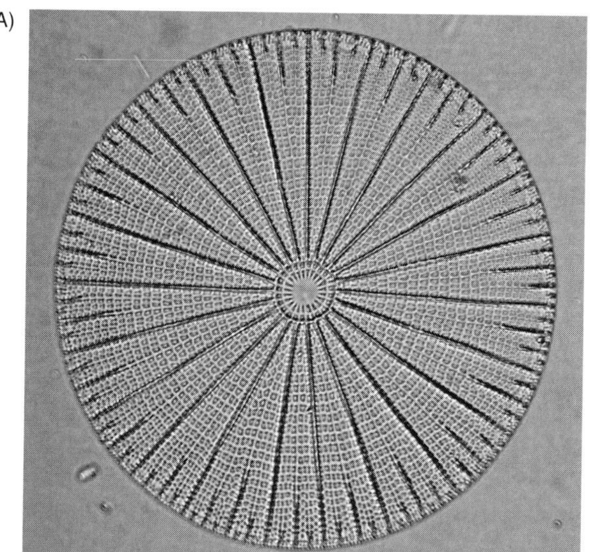

(B)

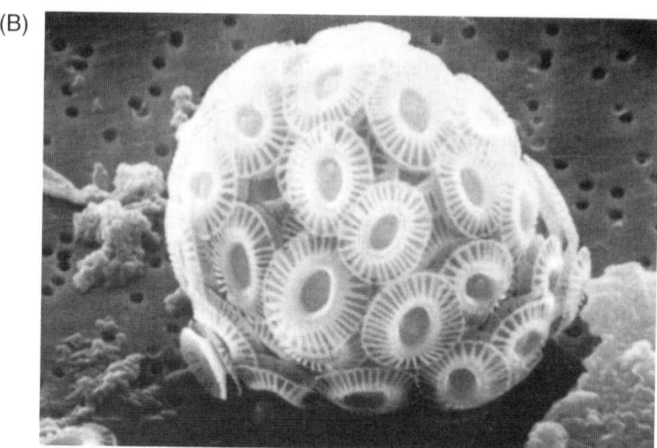

(C)

Figure 1.13. Scanning electron micrographs of (A) a diatom frustule (opal), diameter *c.*20 μm; (B) a coccolithosphere with armor-like coccoliths on a filter paper ($CaCO_3$), diameter *c.*10 μm; and (C) foraminiferan shells ($CaCO_3$), from marine sediments, diameter *c.*100–200 μm. ((C) courtesy of Kathy Newell, University of Washington.)

but they contain no mineral "shell" and are thus poorly preserved in marine sediments. Dinoflagellates are a kind of photoautotrophic flagellate with red to green pigments; they can form large blooms in coastal waters that are sometimes called "red tides."

Zooplankton

Zooplankton obtain their energy heterotrophically by grazing on both smaller phytoplankton and other zooplankton. They are eukaryotic and exist in sizes from nanometers to centimeters. Food webs that contain pico- and nanometer-sized phytoplankton are generally dominated by nanometer-sized flagellate zooplankton with very fast growth rates. They feed by direct interception, sieving, and filtering. Ciliate zooplankton are a little larger than flagellates and graze on flagellates and small diatoms. Heterotrophic dinoflagellates are larger, nano- to micrometer-sized heterotrophs, and graze on nanophytoplankton and nanozooplankton.

The only zooplankton of any abundance that form mineral shells (or tests) are the Foraminifera, Pteropoda and Radiolaria, which form tests of $CaCO_3$ (usually calcite), $CaCO_3$ (aragonite) and SiO_2 (opal), respectively. Of the larger zooplankton (copepods, euphausiids, and crustaceans) copepods play an important role in the flux of chemical species to the ocean interior because they create large (*c.*100 μm) fecal pellets that sink very rapidly (*c.* 100 m d^{-1}) into the deep sea. Sediment trap experiments have shown that fecal pellets often dominate the particle flux in the deep ocean. This can have important consequences for the distribution of metabolic products because their rate of release to the water in dissolved form depends on both the degradation and dissolution rates as well as their sinking velocity. The sinking rate greatly depends on the density difference between the particles and seawater, so fecal pellets that contain the minerals of $CaCO_3$ and SiO_2 will sink much faster than those that only contain organic matter.

1.4.2 Marine metabolism: estimates of abundance and fluxes

The presence and distribution of phytoplankton in the sea is determined primarily by the abundance of chlorophyll. Although this is not a direct measure of the number of cells, because they contain different amounts of chlorophyll under different conditions, it is the most rapid and widely used method of identifying the presence of photosynthetic organisms. Because the color of the ocean can be determined by satellites, it is possible to determine the global content of chlorophyll in the sea over one optical depth, about the first 30 m of surface waters.

The rate of primary production in the sea is determined by incubating samples with ^{14}C-labeled DIC and then measuring the amount of the ^{14}C that is taken up into the particulate material. These experiments are done *in situ* in glass bottles attached at different depths to a floating mooring over a period of 12–24 h or on board ship under controlled temperature and light conditions. It is assumed that the ^{14}C uptake measures only phytoplankton metabolism because marine zooplankton are not abundant enough or rapid enough grazers to affect the results. A complication arises in interpreting the incubation data because photosynthetic organisms are both autotrophic (during the day) and heterotrophic (during both day and night).

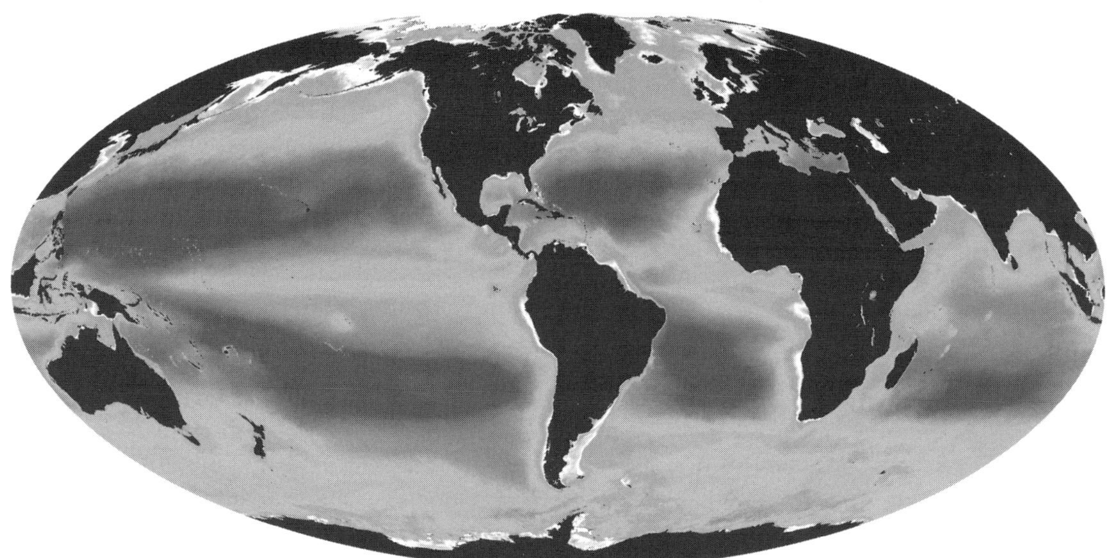

Figure 1.14. A global map of surface-ocean chlorophyll derived from satellite ocean color imagery. Images like this are used to determine relative distributions of ocean primary production; however, these are approximate because the relation between surface ocean color and primary productivity is variable and in some cases uncertain. The figure is the average of *c.* 10 years of SeaWIFS ocean color data, 1997–2007. (The image is from the NASA/MODIS ocean color web site, http://oceancolor.gsfc.nasa.gov.) (See Plate 1.)

Although the net result is autotrophic, they both create organic matter and respire organic matter at different times in their metabolic cycle. The creation of organic compounds from DIC by phytoplankton is called *gross photosynthesis*. The net result of phytoplankton autotrophy and heterotrophy is called *net photosynthesis*. ^{14}C uptake experiments are believed to measure a flux closer to net than to gross photosynthesis (Marra and Barber, 2004). There are other experimental complications in interpreting the results from these experiments, such as differences between *in situ* and on board incubations, DOM production (which is usually not measured), and possible trace metal contamination from chemical spikes and glassware. None the less, this approach is the most used direct estimate of the flux of carbon through marine autotrophic organisms.

Although one does not expect a direct relationship between the standing stock of marine phytoplankton, as inferred from chlorophyll concentrations, and the rate of DIC transformation to organic carbon, as determined by ^{14}C incubation experiments, there are regional correlations. These relations have been used with satellite color measurements to estimate global marine primary production (Fig. 1.14). One caveat to keep in mind, however, is that a satellite measures color over only one optical depth. Although biological oceanographers attempt to compensate for this and other factors, there is presently no accurate way to relate surface chlorophyll concentrations with euphotic zone photosynthetic productivity in vast regions of the ocean.

The biological process that most influences the chemistry of the atmosphere and the sea is the flux of organic matter from the surface to the deep ocean. This flux, called the *biological pump*, is not directly proportional to net photosynthesis, as the latter refers only to phytoplankton production. In a steady-state situation or on an annual average basis, the mean organic carbon flux out of the euphotic zone should equal the *net community production* of organic carbon. Net community production is the difference between net photosynthesis and heterotrophic respiration occurring in the water.

There is no direct measure of net community carbon production, but methods have been devised, by using incubation techniques with nitrogen isotopes, to determine the uptake of nitrogen compounds (NO_3^-, NH_4^+ and urea). During photosynthesis NH_4^+ and urea are preferred by phytoplankton because they do not need to be reduced before incorporation into organic compounds. Because of this they are taken up immediately when released during respiration and they do not accumulate above very low concentrations in the water. NO_3^-, however, accumulates in relatively high concentrations below the euphotic zone and in surface waters at high latitudes. Dugdale and Goering (1967) called the $^{15}NO_3^-$ uptake *new production* to distinguish it from recycled production, which is represented by $^{15}NH_4^+$ (and urea) uptake. At steady state, over an annual cycle, the rate of NO_3^- uptake by photosynthesis should equal the flux of NO_3^- into the euphotic zone, which must also equal the particulate and dissolved organic N flux out of the euphotic zone. The fraction of dissolved nitrogen uptake in the euphotic zone that is delivered by NO_3^- varies from about 0.1 in regions where there is a well-developed food web to recycle phytoplankton production, to 0.8 in bloom regions where grazers have not become well established.

In an ideal steady-state world the flux of organic carbon from the upper ocean would be equal to net community production, which would be related via a stoichiometric ratio of C:N to new production. This relationship was first suggested by Eppley and Peterson (1979) who termed the ratio of new to total production the *f-ratio*. A schematic picture of the processes that lead to the creation and transport of organic matter from the surface to the deep ocean is presented in Fig. 1.15. Nearly all the reaction rates and reservoir fluxes for N, C and O_2 in the figure have been measured. Relating the fluxes of the

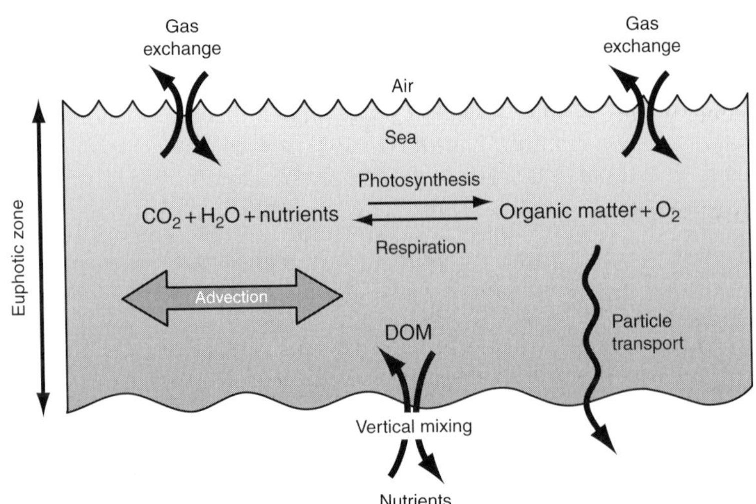

Figure 1.15. A schematic diagram of processes influencing the mass balance of organic matter and oxygen in the euphotic zone of the ocean. Photosynthesis produces organic matter and oxygen; respiration consumes them. Organic matter escapes via settling of particulate organic matter and mixing or advection of dissolved organic matter (DOM) out of the surface ocean. Oxygen leaves by gas exchange between the atmosphere and ocean and mixing or transport to deeper waters.

Table 1.6. *Stoichiometric "Redfield" ratios for consumption of P, N, C and production of O_2 during photosynthesis and the opposite reaction during respiration in the ocean*
All values are relative to a phosphorus value of 1.0.

Source	Organic matter			O_2
	P	N	C	
Redfield *et al.*, 1963[a]	1.0	16	106	138
Anderson and Sarmiento, 1994[b]	1.0	16 ± 1	117 ± 14	170 ± 10
Anderson, 1995[c]	1.0	16	106	141–161
Kortzinger *et al.*, 2001[d]	1.0	17.5 ± 2.0	123 ± 10	165 ± 15
Hedges *et al.*, 2002[e]	1.0	17	106	154

[a] The first and original stoichiometry was determined from observations of the $NO_3^-:PO_4^{3-}$ ratios in ocean deep waters and then assuming a stoichiometry for organic matter.
[b] This value used the same approach as [a] and included DIC and O_2 on dineutral surfaces below 400 m.
[c] These values were determined by using C, H and O content of organic compounds that make up plankton, with the assumption that there are 106 moles of C per mole of P.
[d] These values are based on measurements of DIP, DIN, DIC (corrected for anthropogenic CO_2) and O_2 on constant density surfaces.
[e] These values were determined by chemical and NMR analysis of marine planktonic organic matter. A C:P ratio of 106 is assumed.

nutrients N and P to those of C and O_2 requires some assumptions about the stoichiometry of their production and degradation during photosynthesis and respiration. The ratios at which these elements or compounds are used during marine metabolism are today called *Redfield ratios* after Alfred Redfield, who first suggested that they are constant. His classic values of P:N:C:O_2 have been modified recently (Table 1.6) by modern measurements of the dissolved inorganic ratios on constant density surfaces in the deep ocean and from more accurate knowledge of the elemental makeup of organic matter. The details behind the calculation of Redfield ratios and more about the rates of organic matter transfer between the euphotic zone and deeper waters are discussed in Chapter 6.

References

Anderson, L. A. (1995) On the hydrogen and oxygen content of marine phytoplankton. *Deep-Sea Res.* **42**, 1675–80.

Anderson, L. A. and J. L. Sarmiento (1994) Redfield ratios of remineralization determined by nutrient data analysis. *Global Biogeochem. Cycles* **8**, 65–80.

Boyle, E.A. (1988) Cadmium: chemical tracer in deepwater paleoceanography. *Paleoceanography* **3**, 471–89.

Broecker, W. S. and T.-H. Peng (1982) *Tracers in the Sea.* Lamont-Doherty Geological Observatory.

Brown, C.W. and J.A. Yoder (1994) Coccolithophorid blooms in the global ocean. *J. Geophys. Res.* **99**, 7467–82.

Bruland, K. (1983) Trace elements in seawater. In *Chemical Oceanography*, Vol. 8 (ed. J.P. Riley and R. Chester), pp. 157–220. New York, NY: Academic Press.

de Villiers, S. (1994) The geochemistry of strontium and calcium in coralline aragonite and seawater. Ph.D. thesis, University of Washington, Seattle, WA.

DoE (1994) *Handbook of Methods for the Analysis of Various Parameters of the Carbon Dioxide System in Seawater*, version 2, ed. A.G. Dixon and C. Goyet. ORNL/CDIAC-74.

Dugdale, R. C. and J. J. Goering (1967) Uptake of new and regenerated forms of nitrogen in primary productivity. *Limnol. Oceanogr.* **12**, 196–206.

Eppley, R.W. and B.J. Peterson (1979) Particulate organic matter flux and planktonic new production in the deep ocean. *Nature* **282**, 677–80.

Gnanadesikan, A. and R. Halberg (2002) Physical oceanography, thermal structure and general circulation. In *Encyclopedia of Physical Science and Technology* (ed. R. A. Meyers). San Diego, CA: Academic Press.

Hedges, J. I., J. A. Baldock, Y. Gelinas *et al.* (2002) The biochemical and elemental compositions of marine plankton: a NMR perspective. *Mar. Chem.* **78**, 47–63.

Johnson, K. S., R. M. Gordon and K. H. Coale (1997) What controls dissolved iron concentrations in the world ocean?, *Mar. Chem.* **57**, 137–61.

Knauss, J. A. (1978) *Introduction to Physical Oceanography*. Englewood Cliffs, NJ: Prentice-Hall.

Kortzinger, A., J. I. Hedges and P. D. Quay (2001) Redfield ratios revisited: removing the biasing effect of anthropogenic CO_2. *Limnol. Oceanogr.* **46**, 964–70.

Levitus, S., R. Burgett and T. Boyer (1994) *MOAA Atlas NESDIS 3, World Ocean A5lqw, 1994 v. 3: Salinity*. Washington, DC: US Department of Commerce.

Marra, J. and R. T. Barber (2004) Phytoplankton and heterotrophic respiration in the surface layer of the ocean. *Geophys. Res. Lett.* **31**, LO9314, doi: 10.1029/2005GLO19664.

Millero, F. J. (1996) *Chemical Oceanography*. Boca Raton, FL: CRC Press.

Picard, G. L. and W. J. Emery (1982) *Descriptive Physical Oceanography*. Oxford: Pergamon Press.

Pilson, M. E. Q. (1998) *An Introduction to the Chemistry of the Sea*. Upper Saddle River, NJ: Prentice-Hall.

Pinet, P. R. (1994) *Invitation to Oceanography*. Boston, MA: Jones and Bartlett.

Redfield, A. C., B. H. Ketchum and R. A. Richards (1963) The influence of organisms on the composition of seawater. In *The Sea*, vol. 2 (ed. M. N. Hill). New York, NY: John Wiley and Sons.

Schlitzer, R. (2001) Ocean Data View. www.awi.bremerhaven.de/GEO/ODV.2001.

Thurman, H. (1994) *Essentials of Oceanography*, 7th edn. New York, NY: Macmillan.

Geochemical mass balance: dissolved chemical inflow and outflow from the ocean

Classification of the chemical constituents of seawater into "conservative," "bioactive" and "adsorbed" (Chapter 1) revealed much about the processes that control concentration distributions in seawater of the latter two categories, but little about the conservative elements. Concentrations of the elements that make up most of the salinity of the oceans provide clues to the mechanisms that control their sources and sinks. Thus, the chemical perspective of oceanography revealed by conservative element concentrations is about processes that occur at the ocean boundaries: weathering reactions on land, authigenic mineral formation in marine sediments and reactions with the crust at hydrothermal areas. The amount of time some of the dissolved constituents remain in solution before they are removed chemically is very long, suggesting the possibility for chemical equilibrium between seawater and the minerals in the ocean

Figure 2.1. A schematic representation of the sources and sinks of the dissolved chemical constituents of seawater. Sources are from rivers and wind, and sinks are authigenic mineral formation.

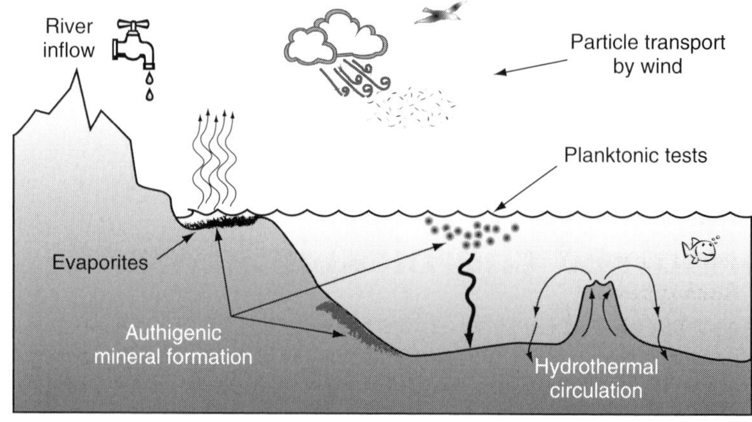

basins as a controlling factor. On the other hand, energy from both the sun and the Earth's interior is constantly driving seawater constituents away from chemical equilibrium.

2.1 Mass balance between input from land and authigenic mineral formation

The main mechanisms for delivery of dissolved constituents to the ocean are river inflow and atmospheric input. Formation of authigenic minerals (those minerals that form *in situ*) is the ultimate sink (Fig. 2.1) for these constituents. Authigenesis primarily involves precipitation of plant and animal shells, chemical reactions in sediments, and high-temperature reactions at hydrothermal regions. We begin with a brief review of the chemical reactions influencing the dissolved ion concentrations of rivers, and end with an attempt to balance the river sources with plausible sinks for the major seawater ions.

2.1.1 Weathering and river fluxes to the ocean

Based on chemical measurements for river water and atmospheric particles, it is clear that river inflow is by far the most important mechanism for the delivery of dissolved major ions and elements to the ocean. This is not the case for all elements; some of the trace metals such as iron and lead have important sources from atmospheric dust, but our discussion will focus on the flux of major elements to the ocean. The concentration and origin of the major ions to river water is presented in Table 2.1. Weathering of rocks on land is the origin of the cations, Na^+, Mg^{2+}, Ca^{2+} and K^+, whereas the source of the anions Cl^-, SO_4^{2-} and HCO_3^- is partly from rock weathering and partly from the gases CO_2, SO_2 and HCl that are delivered to the atmosphere via volcanic emissions over geologic time.

Table 2.1. *The sources of major ions to rivers*

The average concentration of dissolved species in rivers is naturally somewhat uncertain.

	Anions (meq kg^{-1})			Cations (meq kg^{-1})				Neutral (mmol kg^{-1})
Source	HCO_3^-	SO_4^{2-}	Cl^-	Ca^{2+}	Mg^{2+}	Na^+	K^+	H_4SiO_4
Atmosphere	0.58	0.09	0.06	0.01	<0.01	0.05	<0.01	<0.01
Rock types								
Silicates	—	—	—	0.14	0.20	0.10	0.05	0.21
Carbonates	0.31	—	—	0.50	0.13	—	—	—
Sulfates	—	0.07	—	0.07	—	—	—	—
Sulfides	—	0.07	—	—	—	—	—	—
Chlorides	—	—	0.16	0.03	0.01	0.11	0.01	—
Organic C	0.07	—	—	—	—	—	—	—
SUM	0.96	0.23	0.22	0.75	0.35	0.26	0.07	0.22

These values and their sources are from Holland (1978). Some of the total river water concentrations differ from those in Table 2.3, where the numbers are from Li (2000).

Weathering involves the breakdown of rocks on land to create dissolved constituents in solution and altered mineralogy on land. To aid those unfamiliar with geology, a very brief review of important rocks and minerals involved in the weathering process is presented in Appendix 2.1. Weathering reactions involve the reaction of CO_2, which ultimately comes from the atmosphere but also is generated in soils by plant respiration, with aluminosilicate and carbonate minerals. Some of these reactions are illustrated in Table 2.2 by using formulae for pure minerals. Carbonate minerals are represented by calcium carbonate, $CaCO_3$, which is the dominant form of limestone rocks on land and carbonate shells produced in the ocean today. Silicate minerals are represented by the igneous minerals potassium feldspar and mica and the clay mineral kaolinite. Clay minerals are the reaction products of igneous rock weathering. They consist of the fine-grained particles that make up most muds. There are many different clay minerals and lots of mixed versions between the mineralogically pure end members.

The common feature of weathering reactions (Table 2.2) is consumption of CO_2 and the release to solution of HCO_3^- along with the cations of the weathered mineral. During limestone weathering Ca^{2+} and HCO_3^- enter solution. In the case of silicate weathering involving potassium feldspar and biotite mica (Table 2.2), H_4SiO_4, HCO_3^-, Mg^{2+} and K^+ are dissolved. Bicarbonate is by far the most abundant anion in most rivers because it is formed during weathering reactions. According to Table 2.1 about two thirds of the HCO_3^- in rivers has an atmospheric origin and the other one third comes from carbonate rocks. In the simple $CaCO_3$ dissolution reaction in

Table 2.2. Examples of reactions that occur between rocks and in CO_2-rich soil water to form the dissolved composition of river water: (a) the reaction with carbonate rocks; (b, c) reactions with two examples of silicate rocks

Reaction
(a) CO_2 in soils reacts with water to form H^+ that dissolves $CaCO_3$ according to the net reaction (iii):
(i) $CO_2 + H_2O \rightleftarrows HCO_3^- + H^+$
(ii) $CaCO_3 + H^+ + \rightleftarrows HCO_3^- + Ca^{2+}$
(iii) $\underset{\text{Carbonate rocks}}{CaCO_3(s)} + CO_2 + H_2O \underset{\text{dissolution}}{\rightleftarrows} 2\,HCO_3^- + Ca^{2+}$
(b) CO_2 in soils reacts with water to form H^+ that reacts with aluminosilicate rocks to form the clay mineral kaolinite according to the net reaction (vi)
(iv) $2CO_2 + 2H_2O \rightleftarrows 2\,HCO_3^- + 2H^+$
(v) $2KAlSi_3O_8(s) + 2H^+ + 9H_2O \rightarrow Al_2Si_2O_5(OH)_4(s) + 2K^+ + 4H_4SiO_4$
(vi) $\underset{K-\text{feldspar}}{2KAlSi_3O_8(s)} + 2CO_2 + 11H_2O \rightarrow \underset{\text{kaolinite}}{Al_2Si_2O_5(OH)_4(s)} + 2K^+ + 2HCO_3^- + 4H_4SiO_4$
(c) CO_2 in soils reacts with water to form H^+ that reacts with aluminosilicate rocks to form the clay mineral kaolinite according to the net reaction (ix)
(vii) $14CO_2 + 14H_2O \rightleftarrows 14HCO_3^- + 14H^+$
(viii) $2KMg_3AlSi_3O_{10}(OH)_2(s) + 14H^+ + H_2O \rightarrow Al_2Si_2O_5(OH)_4(s) + 2K^+ + 6Mg^{2+} + 4H_4SiO_4$
(ix) $\underset{\text{biotite}}{2KMg_3AlSi_3O_{10}(OH)_2(s)} + 14CO_2 + 15H_2O \rightarrow \underset{\text{kaolinite}}{Al_2Si_2O_5(OH)_4(s)} + 2K^+ + 6Mg^{2+} + 14HCO_3^- + 4H_4SiO_4$

Table 2.2, half of the carbon source for HCO_3^- comes from CO_2 and half from $CaCO_3$. Combination of these two mass balance concepts requires that about half of the atmospheric CO_2 source be involved in $CaCO_3$ dissolution reactions and the other half be consumed by dissolution of aluminosilicates; or, about half of the HCO_3^- in river water that has an atmospheric origin has reacted with aluminosilicates (Fig. 2.2).

2.1.2 Residence times of seawater constituents

The simplest conceptual model of seawater composition would be to assume that seawater contains ions in the same ratios as delivered by rivers, but at higher concentrations due to evaporation. Comparison of the relative ion concentrations (Fig. 2.3) indicates, however, that

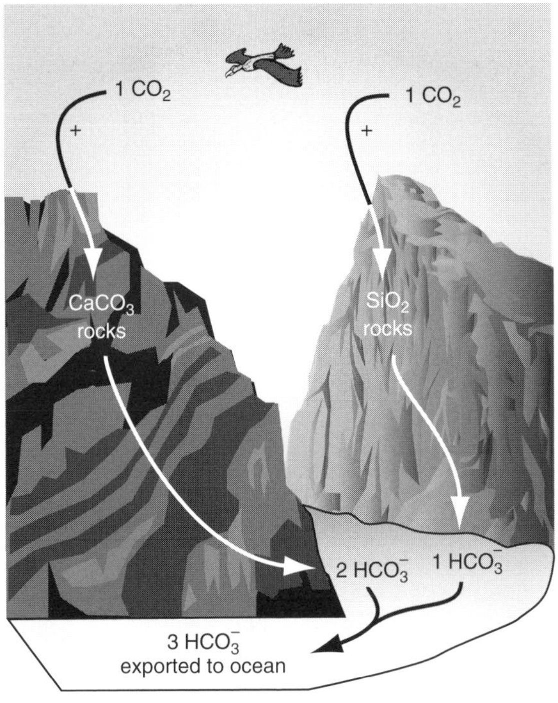

Figure 2.2. A schematic representation of the approximate flow of carbon during weathering. The figure is a representation of the fluxes in Table 2.1, which suggest that about 2/3 of the HCO_3^- produced by weathering via the reactions in Table 2.2 comes from the atmosphere and 1/3 is from dissolution of $CaCO_3$ rocks.

this is clearly not the case. River water contains mostly dissolved Ca^{2+}, Na^+, and HCO_3^-, whereas the ocean contains mainly Na^+ and Cl^-. The reasons for the differences are the varying reactivities of the elements in the ocean. Because the predominant removal mechanism of dissolved ions from the sea is in the form of solid authigenic minerals, those elements that readily combine to form an insoluble mineral, or rapidly adsorb to the surface of solids, are lower in concentration than those that persist in solution until high concentrations are reached. We know from observations of saline lakes and evaporite basins such as the Dead Sea and the Great Salt Lake that NaCl (salt) does not precipitate from water except in regions where there are very high evaporation rates and ions become highly concentrated in a brine. Thus, Na^+ and Cl^- are relatively unreactive. On the other hand, we shall see later that the sea floor in some regions of the ocean is almost purely $CaCO_3$, indicating that solid calcite and aragonite readily form at present-day seawater concentrations of Ca^{2+} and CO_3^{2-}, making these major ions relatively reactive.

The ratio of the inventory (total amount) of an element in the sea to its inflow (or outflow) rate at steady state (unchanging total mass) is a measure of its reactivity. These two terms combine to form the *residence time* of the dissolved constituent with respect to river inflow, τ_R:

$$\tau_{R_i}(\text{years}) = \frac{\text{ocean inventory(mol)}}{\text{river inflow rate(mol y}^{-1})} = \frac{[C_i]\ V_0}{F_{R_i}}. \tag{2.1}$$

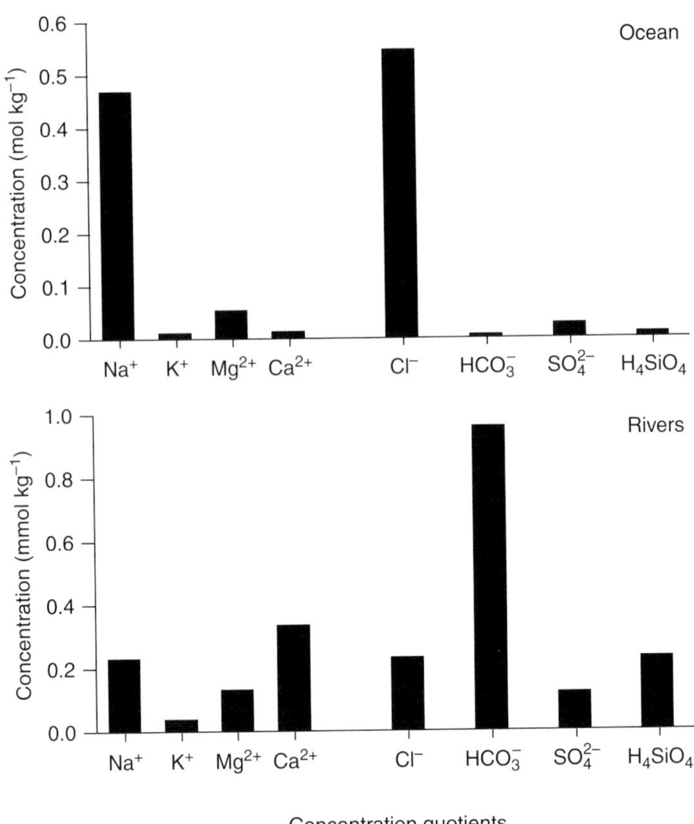

Figure 2.3. Bar graphs showing the relative concentrations of the major ions in seawater and river water along with a comparison of some of the major ion ratios in the ocean and river water. The concentrations in the two solutions are listed in Table 2.3.

Concentration quotients

	Na^+/K^+	Mg^{2+}/Ca^{2+}	Na^+/Ca^{2+}	$(Ca^{2+}+Mg^{2+})/HCO_3^-$
Ocean	46	5.1	46	27
Rivers	6.0	0.39	0.70	0.48

As usual, concentration is indicated by brackets [ ]. V_O is the ocean volume and F_{Ri} is the river inflow rate of the ith constituent. Residence times of the major constituents in the sea, calculated from their seawater concentrations and global average concentrations in rivers (Table 2.3), vary from more than 10^8 y for Cl^- to *c.*10^5 y for dissolved inorganic carbon (DIC $= [HCO_3^-] + [CO_3^{2-}] + [CO_2]$). The renewal rate of water via the river inflow and evaporation cycle is on the order of 40 000 y, and the circulation time of the ocean is on the order of 1000 y. Thus, the most reactive of the major ions, HCO_3^- and CO_3^{2-}, reside in the sea only about three times as long as the water, but long enough to circulate through the ocean roughly 100 times before reacting. The least reactive of the major ions, Cl^-, resides in the sea about 3000 times as long as the water. This is why the seawater concentration of DIC is only about three times that of rivers, but the Cl^- concentration factor is about 3000 times larger. Residence time is a measure of the time required to remove a dissolved species from the ocean via chemical reactions. The actual meaning in mathematical terms is described in more detail in Appendix 2.2.

Table 2.3. *Residence times of some of the major dissolved elements in the ocean*

Constituent	Seawater concentration (mmol kg^{-1})	Inventory[a] (10^{18} mol)	River water concentration (μmol kg^{-1})	River inflow[b] (10^{12} mol y^{-1})	τ (10^6 y)
H_2O	—	—	—	—	0.04
Na^+	469.1	647	231	8.1	80
Mg^{2+}	52.8	72.8	128	4.5	16
Ca^{2+}	10.3	14.2	332	11.6	1.2
K^+	10.2	14.1	38.4	1.3	10
Cl^-	545.9	753	220	7.7	98
SO_4^{2-}	28.2	39	115	4.0	10
DIC[c]	2.3	3.2	958	33.5	0.1

Source: Global average concentrations in rivers are from Li (2000).
[a] Ocean mass $= 1.38 \times 10^{21}$ kg.
[b] River flow rate $= 3.5 \times 10^{16}$ l y^{-1} (Broecker and Peng, 1982).
[c] DIC = dissolved inorganic carbon $= HCO_3^- + CO_3^{2-} + CO_2$.

2.1.3 Mackenzie and Garrels mass balance

A classic approach to determining the importance of the formation of authigenic minerals to the removal of the major elements in seawater was the chemical mass balance of Mackenzie and Garrels (1966). The goal was to calculate a rough mass flux balance between dissolved river inflow and authigenic mineral formation for each major element. They assumed that the composition of rivers has not changed over the longest residence time of the major elements (i.e. that for Cl^-, about 10^8 y) and that the river–ocean system is a steady state over this period of time. This is only a rough calculation because the time interval is long enough to include very different climates and configurations of the continents, and thus probably different weathering and river compositions. Also, many known authigenic minerals are clearly formed intermittently in very restricted locations, like the salt deposits of closed basins and marginal seas. Nevertheless, residence times for all the major ions except Ca^{2+} and DIC are greater than 10^6 y (Table 2.3) so time rates of change will be strongly damped in response to intermittent removal reactions. Mackenzie and Garrels proceeded with the expectation of deriving first-order information about the long-term marine chemical mass balance.

The first step was to calculate the total mass of each of the major ions discharged by rivers over the past 10^8 y and then remove them from the ocean in a normative way by combining cations and anions to form solids with the stoichiometry of sedimentary rocks (Table 2.4). Salts found in sedimentary rocks that are made of the major ions in seawater are halite (NaCl), calcite ($CaCO_3$), anhydrite ($CaSO_4$), pyrite (FeS_2), and opal (SiO_2) (Table 2.4).

Although $CaSO_4$ and FeS_2 are known to be the major sinks of sulfur in sedimentary rocks, their concentrations and accumulation rates are not known well enough to determine burial fluxes. Thus, it

Table 2.4. *(a) A summary of the inventory of major ions in seawater after the normative removal from the ocean by formation of the main sedimentary minerals and (b) the mineral formation reactions*

(a) Inventory

		SO_4^{2-}	Ca^{2+}	Cl^-	Na^+	Mg^{2+}	K^+	H_4SiO_4	HCO_3^-
Major ion :		SO_4^{2-}	Ca^{2+}	Cl^-	Na^+	Mg^{2+}	K^+	H_4SiO_4	HCO_3^-
Mass removed in 10^8 y (10^{18}mol) :		500	1680	1040	1360	740	240	710	4160
Mineral formed	Moles removed	Amount of ion remaining after reaction							
Pyrite, FeS_2	250[a]	250	1680	1040	1360	740	240	710	4160
Anhydrite, $CaSO_4$	250[a]	0	1430	1040	1360	740	240	710	4160
Calcium carbonate, $CaCO_3$	1430		0	1040	1360	740	240	710	1300
Sodium chloride, NaCl	1040			0	320	740	240	710	1309
Opal, SiO_2	71[b]				320	740	240	639[b]	1300

(b) Formation reactions

Pyrite:	$SO_4^{2-} + 2CH_2O(s) \rightleftarrows S^{2-} + 2CO_2 + H_2O$ followed by $Fe^{2+} + S^{2-} + S^0 \rightleftarrows FeS_2$
Anhydrite:	$Ca^{2+} + SO_4^{2-} \rightleftarrows CaSO_4(s)$
Calcium carbonate:	$Ca^{2+} + 2\,HCO_3^- \rightleftarrows CaCO_3(s) + CO_2 + H_2O$
Sodium chloride:	$Na^+ + Cl^- \rightleftarrows NaCl(s)$
Opal:	$H_4SiO_4 \rightleftarrows SiO_2(s) + 2H_2O$

Source: Following Mackenzie and Garrels (1966).
[a] Assume half of the SO_4 is removed by pyrite formation and half by $CaSO_4$ formation.
[b] The amount of opal (SiO_2) formation is based on sediment accumulation rates.

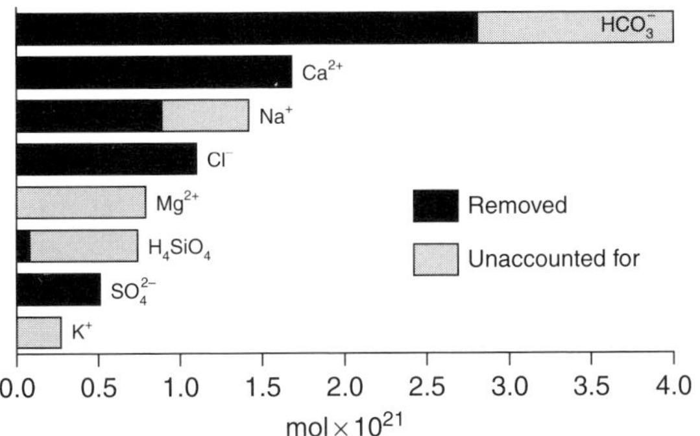

Figure 2.4. A bar graph representing the results of the Mackenzie and Garrels (1966) mass-balance calculation. The bar length is the total number of moles of the ion delivered to the ocean in 100 million years. The dark region of the bar is the portion of that flux accounted for by the normative calculation in Table 2.4.

was simply assumed that at steady state half of the inventory of SO_4^{2-} is removed to the mineral FeS_2 and half to $CaSO_4$. Although iron is only a trace element in seawater and not listed in Table 2.4, it is the third most abundant element in marine sediments (*c*.6% by mass) and it is assumed that there is plenty available as labile iron oxides to supply the Fe necessary to form authigenic pyrite. Because it is assumed the concentrations in the ocean approximate a steady state with respect to inflow and outflow, precipitation of these two minerals is the fate of SO_4^{2-} that flows into the sea via rivers. Sedimentation of the entire inventory of SO_4^{2-} also removes about 15% of the calcium ion inventory. Next, the remaining Ca^{2+} is removed as $CaCO_3$, which also removes about 70% of the HCO_3^-. Then, Na^+ and Cl^- are precipitated as halite, removing all the Cl^- and about 75% of the Na^+. Finally H_4SiO_4 is removed as opal. Since no additional cations or anions are incorporated into opal, the amount of silicate removed is based on estimates of opal accumulation rates in marine sediments. This estimate is extremely rough, but the result is very interesting because it can only account for a small fraction of the H_4SiO_4 input via rivers.

With the exception of some relatively minor adsorption reactions, the calculation is essentially complete, and a summary of the results (Fig. 2.4) indicates some rather embarrassing conclusions. Nearly all the river inflow of Mg^{2+}, K^+, and H_4SiO_4 and about a quarter of that for HCO_3^- and Na^+ remain unexplained by the formation of the major authigenic minerals. The Na^+ imbalance may not be as severe as the others since this ion is known to be removed through ion exchange with Ca^{2+} in clay minerals entering the ocean as riverine suspended material. Also, one might argue that the accumulation rates of opal and chert in marine sediments are not known well enough to become upset about the silicate imbalance. The situations for Mg^{2+}, K^+ and HCO_3^-, however, are not easily accounted for by any of the processes mentioned so far. Even though the HCO_3^- imbalance is only 30%, its excess inflow presents one of the most severe problems because HCO_3^- has the shortest residence time of the major ions, and its concentration is very influential in controlling the content of CO_2 in the atmosphere.

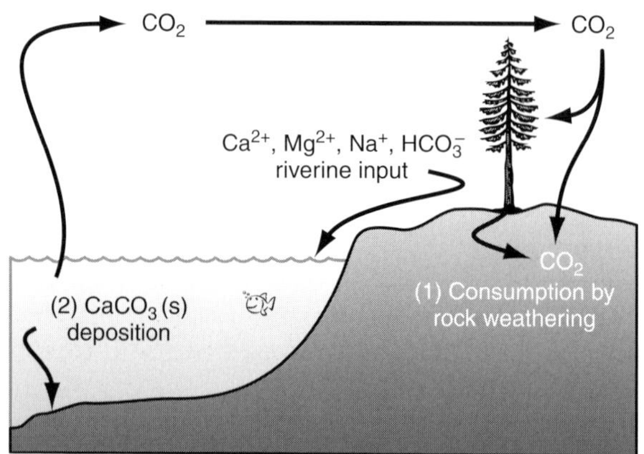

Figure 2.5. A schematic representation of the long-term, global carbon balance between the atmosphere, land and ocean. Flow (1) represents CO_2 consumption from the atmosphere by weathering and the transport of HCO_3^- to the sea. Flow (2) is the return of CO_2 to the atmosphere by precipitation of $CaCO_3$ in the oceans.

The global carbon balance between the atmosphere, ocean and land (Fig. 2.5) is dominated by weathering reactions that consume CO_2 from the atmosphere and $CaCO_3$ precipitation in the ocean that returns CO_2 to the atmosphere. The first is indicated by the reactions in Table 2.2 and the latter by $CaCO_3$ precipitation:

$$2\,HCO_3^- + Ca^{2+} \rightleftarrows CaCO_3(s) \downarrow + CO_2(g) \uparrow + H_2O.$$

The rate of global production of bicarbonate by weathering can be determined because we know approximately the flow of HCO_3^- in the world's major rivers. This represents a drain of the CO_2 of the atmosphere, which must be balanced by resupply to maintain a steady f_{CO_2}. One is tempted to estimate the vulnerability of atmospheric f_{CO_2} to the imbalance between the atmospheric CO_2 drain by weathering and resupply by $CaCO_3$ precipitation by focusing on the CO_2 fluxes from and to the atmosphere. This, however, would be incorrect because the ocean and atmosphere carbon reservoirs are approximately in chemical equilibrium on time scales greater than the circulation of the ocean (see Chapter 11). In order to emphasize the severity of the HCO_3^- imbalance estimated in Fig. 2.4 one should focus on the fluxes of DIC and alkalinity between the land and ocean. Because we have not yet discussed alkalinity and DIC relationships (Chapter 4) a simple approximation can be made by considering the fluxes of bicarbonate and calcium.

The flux of HCO_3^- to the ocean via rivers is:

$$\begin{aligned} F_{HCO_3^-} &= f_{\text{river}} \times \left[HCO_3^-\right]_{\text{river}} = \left(3.5 \times 10^{16}\,\text{l y}^{-1}\right)\left(9.6 \times 10^{-4}\,\text{mol l}^{-1}\right) \\ &= 3.4 \times 10^{13}\ \text{mol y}^{-1}. \end{aligned}$$

At steady state the rate of bicarbonate removal by $CaCO_3$ precipitation in the ocean is given by two times the flux of calcium from rivers, based on the river flow rate and Ca^{2+} concentration in Table 2.3. (This is an upper limit for the removal of HCO_3^- by precipitation of $CaCO_3$ because a small amount of the Ca^{2+} from rivers is precipitated as anhydrite in the Mackenzie and Garrels mass balance, Table 2.4.)

$$F_{\mathrm{CaCO_3,ppt}} = 2f_{\mathrm{river}} \times [\mathrm{Ca}^{2+}]_{\mathrm{river}} = 2(3.5 \times 10^{16}\,\mathrm{l\,y^{-1}})(3.3 \times 10^{-4}\,\mathrm{mol\,l^{-1}})$$
$$= 2.4 \times 10^{13}\ \mathrm{mol\,y^{-1}}.$$

There is a bicarbonate imbalance between weathering and marine $CaCO_3$ precipitation of 3.4 - $2.4 \times 10^{13} = 1.0 \times 10^{13}$ eq y^{-1}. (Note that eq and mol are interchangeable for HCO_3^- because it is singly charged.)

This imbalance stems from the fact that we considered both $CaCO_3$ and silicate rocks during the weathering reactions to create HCO_3^- and remove CO_2 from the atmosphere, but only $CaCO_3$ precipitation during the reaction that consumes HCO_3^- from rivers and returns CO_2 to the atmosphere (Figs. 2.2 and 2.5). If there were no other important reactions removing HCO_3^- from seawater, this imbalance would represent a significant increase in bicarbonate concentration in an astonishingly short time geologically. This can be estimated by calculating the residence time of ocean HCO_3^- with respect to the bicarbonate imbalance we have estimated. The ocean HCO_3^- reservoir is 1.8×10^{-3} eq $kg^{-1} \times 1.38 \times 10^{21}$ kg $= 2.5 \times 10^{18}$ eq, thus:

$$\tau_{\mathrm{atmos.\ change}} = \frac{2.5 \times 10^{18}\ \mathrm{eq}}{1.0 \times 10^{13}\ \mathrm{eq\ y^{-1}}} = 2.5 \times 10^{5}\ \mathrm{y} = 250\,\mathrm{ky}.$$

Changes in the bicarbonate reservoir of the ocean would also be reflected in the stability of atmospheric f_{CO_2} and calcium carbonate preservation. We know from ice core data that the atmosphere f_{CO_2} has gone through very regular 100 000 year cycles over the past 800 kyr in which its concentration has changed by about 40%, but there is not a longer-term change in the mean f_{CO_2} (see Chapter 7). It is thus highly unlikely that the imbalance calculated here can be maintained for periods of hundreds of thousands of years. There must be marine reactions that help maintain the HCO_3^- of the ocean and the f_{CO_2} in the atmosphere relatively constant.

2.2 Reverse weathering

Mackenzie and Garrels (1966) explained the imbalance of Mg^{2+}, K^+ and HCO_3^- and the CO_2 carbon problem by suggesting that a major sink for the remaining ions was to be found in the form of *reverse weathering* reactions in marine sediments. Reactions of the type in Table 2.2 consume cations, silicate, and bicarbonate if they proceed from right to left, which is in the opposite direction to weathering reactions on land, hence the name reverse weathering. This solution to the mass balance conundrum has little experimental evidence to back it up, but it has a basis in thermodynamics. It was proposed by Sillen (1967) that the composition of seawater, and in particular seawater pH, was controlled by heterogeneous thermodynamic equilibrium between the major rock-forming minerals and seawater solution. The original calculation was handicapped by the lack of necessary thermodynamic data; in particular, the equilibrium constants for important clay mineral reactions were unknown. As the

Figure 2.6. A pH – $p\{H_4SiO_4\}$ thermodynamic stability diagram among some of the major silicate mineral phases on the Earth's surface. The gray area is the location of seawater pH and silicate concentration. Notice that the minerals predicted to be stable are those enriched in cations, such as the mixed-layer illite–montmorillonite clay minerals and microcline, rather than the less cation-rich kaolinite. The percent scale is the relative mixture of illite in the illite–montmorillonite solid solution. From Helgeson and Mackenzie (1970).

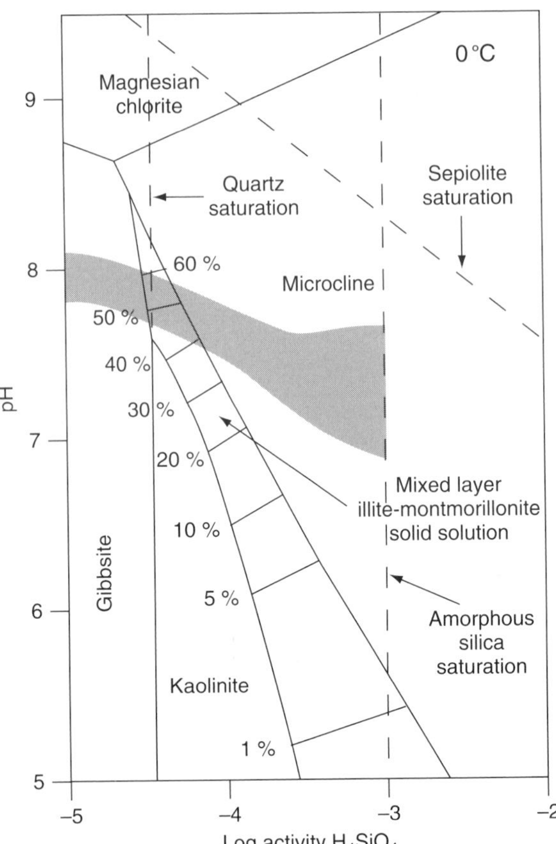

thermodynamic data accumulated, subsequent calculations (see for example, Helgeson and Mackenzie, 1970; Berner, 1971) showed that, among the silicate minerals, Mg- and K-rich clay minerals were thermodynamically stable at the pH of seawater (Fig. 2.6). The implication was that, upon entering seawater, the ion-poor clay mineral kaolinite should undergo alteration by consuming cations, particularly K^+ and Mg^{2+} to form more illite- and montmorillonite-rich minerals.

The problem with this explanation is that X-ray diffraction analysis of clay minerals found in marine sediments shows that they reflect the type of clay minerals that lie within local drainage basins on land and not those that are thermodynamically stable in seawater. On the other hand, only a small fraction (*c.* 10%–20%) (Holland, 1978) of the clay minerals delivered to the ocean by rivers would have to undergo authigenesis to provide the sink for the excess K^+ and Mg^{2+} inventories. There is some experimental evidence that reverse-weathering-type reactions occur over long incubations of clay minerals in contact with seawater. Whitehouse and McCarter (1958) observed that montmorillonite clay minerals become more Mg- and K-rich and more Na-poor after incubation in artificial seawater for five years. Nearly 35 y after the Whitehouse

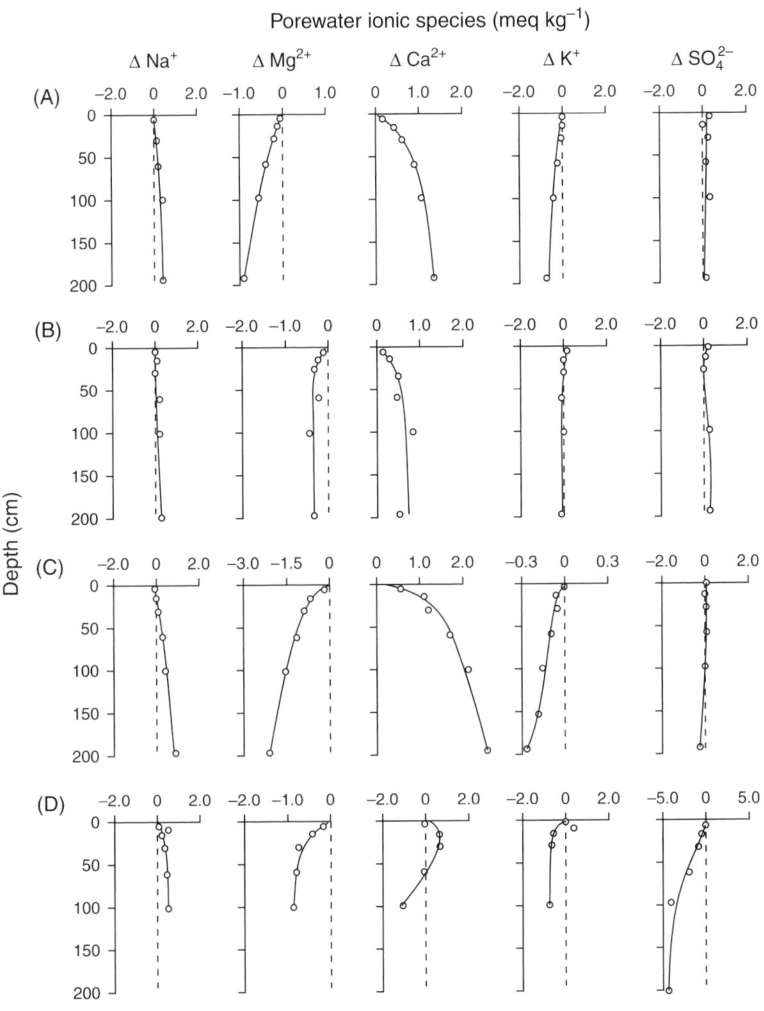

Figure 2.7. The results of measurements of major ions in sediment porewaters collected by *in situ* filtration. Values are presented as the difference in the concentration between bottom water and porewater samples. Stations are from the continental margin and deep sea in the South Atlantic: (A) on the mid-Atlantic Ridge between Rio de Janeiro and Capetown at c.2000 m; (B) from the continental rise off Rio de Janeiro at 2000 m; (C) from the continental slope off Rio de Janeiro at c.1000 m; and (D) from the Venezuela Basin east of Barbados at c.3000 m. Note that the concentrations of both K^+ and Mg^{2+} decrease with depth in almost all cases. From Sayles, (1979).

and McCarter experiment, Michalopoulos and Aller (1995) placed seed materials in anoxic Amazon delta sediments, and found that they accumulated K–Fe–Mg-rich clay minerals after 12–36 month incubations (see Chapter 12). These results together with the thermodynamic calculations leave little doubt that such reactions occur. The question remains as to their importance in the marine mass balance.

The quantitative importance of fluxes of major ions into marine sediments has been addressed by studying gradients of the major ions across the seawater–porewater interface by using an *in situ* sampler (Sayles, 1979). The results of these studies (e.g. Fig. 2.7) revealed that there were nearly always depletions of Mg^{2+} and K^+ in marine porewater relative to seawater values, indicating fluxes to the sediments and implying authigenic reactions. Sayles found that these gradients were greater in near-shore regions where sediments are more clay-mineral-rich. Fluxes calculated from the chemical gradients and the molecular diffusion equation indicate removal rates of Mg^{2+} and K^+ from the ocean that are of the same magnitude as the river inflow!

A check on the plausibility of these calculated fluxes can be made by determining the expected Mg^{2+} and K^+ enrichment in the sediments implied by the porewater gradients. The authigenic enrichment is related to the sedimentation rate; the faster the sedimentation rate, the less authigenic enrichment is necessary to bury a given amount of the component derived from seawater. The implied sedimentary reactions from these porewater studies are feasible in near-shore areas where sedimentation rates are high (tens of cm ky^{-1}) and the predicted steady-state Mg^{2+} enrichment in the sediments is within the lower range of measured values. There is a problem, however, when attempting to reconcile the pore water measurements and authigenic enrichments in deep-sea sediments where sedimentation rates are 0.1–1.0 cm ky^{-1}. In this case the implied authigenic accumulations are greater than the total Mg^{2+} and K^+ in the sediments. This can clearly not be maintained at steady state, and the only realistic explanation seems to be that the porewater gradients were overestimated for some experimental reason.

Although the porewater measurements have become somewhat discredited because of this failure to achieve a balance between predicted reaction rates and sedimentary concentrations, it may be that Sayles's conclusion about the importance of these reactions in near-shore and river-mouth sediments (where the bulk of particulate material from rivers is deposited) is correct. One could assess this possibility by comparing Mg/Al and K/Al quotients in these sediments with those measured in the particulate material of the rivers. It would require a lot of highly accurate determinations (to within a few percent) and an excellent year-round measure of the Mg/Al quotient in the river particulate material to compare with the sediment measurements. Probably for this reason, this research has not been attempted.

2.3 Hydrothermal circulation

The other most likely explanation for the mass imbalance of major ions in Fig. 2.4 is that there are substantial fluxes of seawater into hydrothermal areas where the chemistries of dissolved constituents are amended by contact with basalt at high temperatures and pressures. This phenomenon is described here by first reviewing the most important chemical changes in hydrothermal waters, and then discussing what these changes mean in terms of fluxes to and from the ocean. We shall see that the chemical aspects of this question are pretty well understood. However, as is usual in marine chemistry, estimation of fluxes has proven to be more difficult.

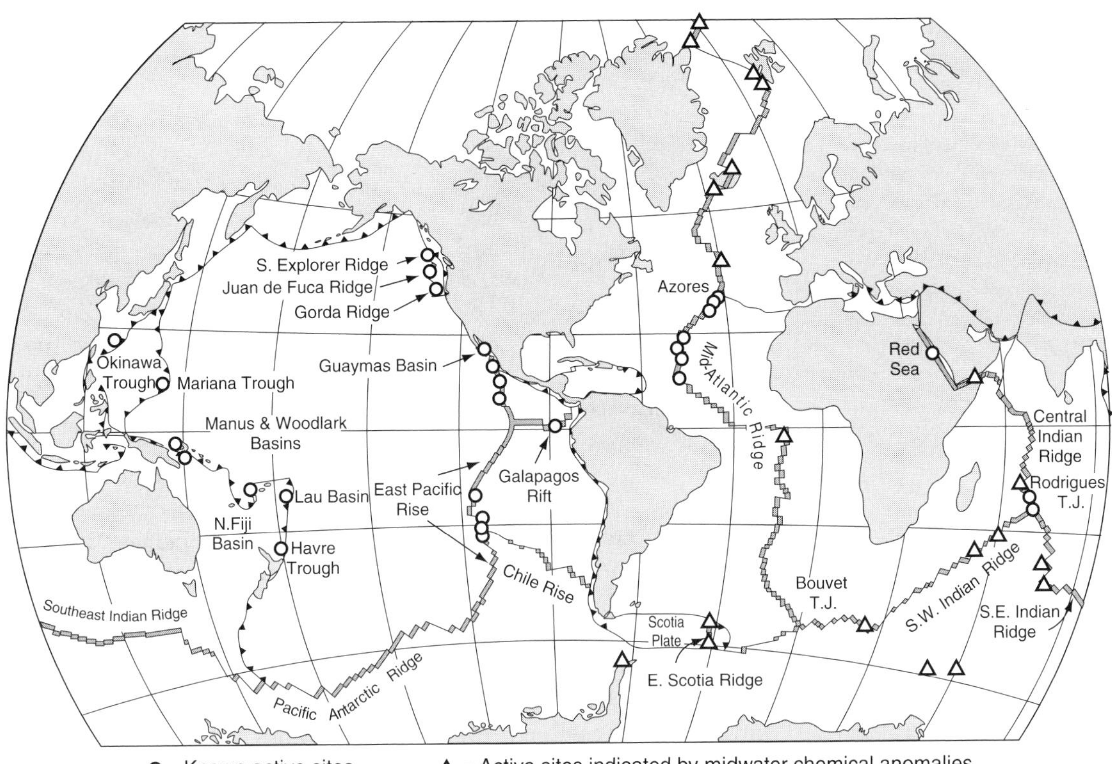

Figure 2.8. Schematic map of the global ridge crest showing active hydrothermal vents that have been discovered (dark circles) and those that are known to exist because of characteristic signals in the overlying water column (dark triangles). From German and Von Damm (2003).

2.3.1 The chemistry of hydrothermal waters

Most hydrothermal regions are found in the mid-ocean ridge system (Fig. 2.8) where new crust is formed. In some of these locations seawater comes in contact with hot, newly-formed basalt (Fig. 2.9). Laboratory experiments reveal that when seawater is heated to high temperatures and pressures (for example 350 °C and 400 atm) in the presence of basalt, Mg^{2+}, SO_4^{2-}, and HCO_3^- are removed, and Ca^{2+} and H_4SiO_4 are released into solution (see, for example, Bischoff and Dickson, 1975; Mottl and Holland, 1978). Some of the important reactions are the precipitation of anhydrite ($CaSO_4$) and magnesium hydroxysulfate hydrate ($MgSO_4 \cdot (1/4)Mg(OH)_2 \cdot (1/2)H_2O$) (Li, 2000):

$$Ca^{2+} + SO_4^{2-} \rightarrow CaSO_4(s). \tag{2.2}$$

$$\frac{5}{4}Mg^{2+} + SO_4^{2-} + H_2O \rightarrow MgSO_4 \cdot \frac{1}{4}Mg(OH)_2 \cdot \frac{1}{2}H_2O(s) + \frac{1}{2}H^+. \tag{2.3}$$

$$\frac{5}{4}Mg^{2+} + CaSO_4 + H_2O \rightarrow MgSO_4 \cdot \frac{1}{4}Mg(OH)_2 \cdot \frac{1}{2}H_2O(s) + \frac{1}{2}H^+ + Ca^{2+}. \tag{2.4}$$

While Ca^{2+} is removed from solution by precipitation of anhydrite (Eq. (2.2)), the subsequent reactions above (Eqs. (2.3) and (2.4)) and others alter the Ca/Mg quotient of these silicates to become more Mg-rich at the expense of Ca, resulting in a net increase in Ca^{2+} in solution. The release of hydrogen ion in the above reactions titrates

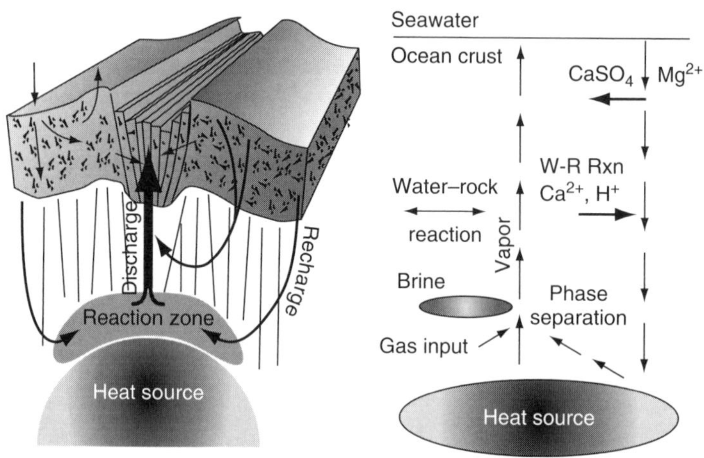

Figure 2.9. Schematic diagram of hydrothermal convection. The left half is a cross section of a spreading center with a shallow heat source indicating the water flow. The right half is a schematic description of the relative locations of water–rock reactions (W-R Rxn) and phase separation along the flow of hydrothermal circulation. Adapted from German and Von Damm (2003).

HCO_3^- and CO_3^{2-} to form CO_2 in hydrothermal solutions, resulting in a much lower pH than in normal seawater. H_4SiO_4 is released to solution by high-temperature silicate reactions, and SO_4^{2-} is reduced to S^{2-} by oxidation of Fe^{2+} to Fe^{3+} within iron-containing silicates.

These general results were originally confirmed by studying the chemistry of wells drilled in the extension of the mid-Atlantic Ridge on Iceland, but entered the chemical perspective of oceanography only when the first hydrothermal areas on the sea floor were discovered, near the Galapagos Islands. John Edmond did the first detailed sampling of the warm (*c*. 20 °C) water exiting the hydrothermal vents and found trends that were similar to those found in Icelandic hydrothermal wells and predicted from laboratory experiments (Edmond *et al*., 1979). Edmond predicted the hydrothermal water end-member concentrations (seawater concentrations in solutions contacting basalt at temperatures exceeding 300 °C), and later he and his students measured these concentrations in the high-temperature "black smokers" (see, for example, Von Damm *et al*., 1985).

The high-temperature measurements were made by using special sampling equipment used on submersible vehicles. Titanium syringes were inserted into the hydrothermal water plume, retrieving samples that were a combination of the hydrothermal solution and the surrounding seawater because of turbulent mixing during venting on the sea floor. An example of the correlation between dissolved Mg^{2+} and H_4SiO_4 in samples from a black smoker at 21° N on the mid-Pacific Rise (Fig. 2.10) reveals measurements that reflect a continuous range of mixing between the seawater and hydrothermal end members. Since nearly all samples taken so far indicate undetectable Mg^{2+} concentrations in the hydrothermal solutions at 350 °C, other constituents are often plotted against the Mg^{2+} concentration so that their hydrothermal end member concentrations can be identified as the point where $[Mg^{2+}]$ extrapolates to zero. Some of the most notable chemical changes in the hydrothermal end member solutions from a single vent on the East Pacific Rise are presented in Table 2.5. It should be noted that

Table 2.5. *The chemical composition of hydrothermal end-member water from the Hanging Garden hydrothermal vent at 21° N in the North Pacific Ocean and the (hydrothermal end member):(seawater) concentration quotients*

Data are from Von Damm *et al.* (1985) and are an example of the values determined for the hydrothermal end member solution. Note there are many more data and concentrations vary among data sets.

Chemical species	Concentration (mmol kg^{-1})	Hydrothermal/seawater
pH (pH units)	3.3	
Alkalinity (meq kg^{-1})	−0.5	
Fe^{2+}	2.4	very large
Mn^{2+}	0.878	very large
Si	15.6	97.5
Li^{+}	1.32	50.8
Rb^{+}	0.033	25.4
K^{+}	23.9	2.44
Ca^{2+}	11.7	1.14
Na^{+}	433	0.95
Cl^{-}	496	0.92
Sr^{2+}	0.065	0.74
SO_4^{2-}	0	0
Mg^{2+}	0	0
H_2S	8.7	—

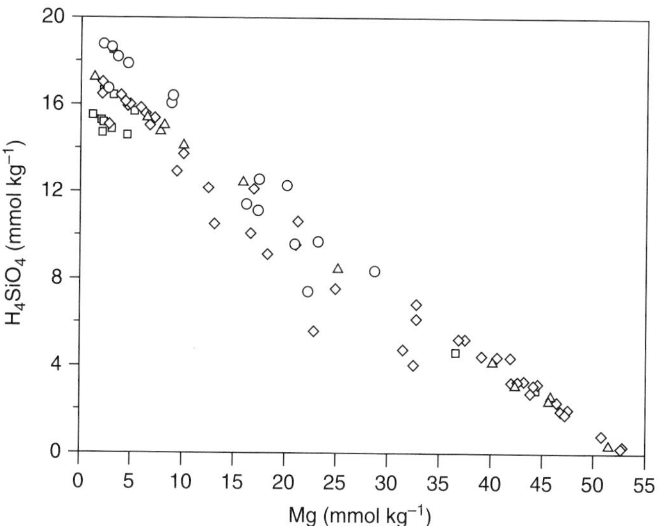

Figure 2.10. H_4SiO_4 and Mg^{2+} concentrations measured in the neck of a "hot smoker" at the East Pacific Rise at the 21° N hydrothermal area in the North Pacific Ocean. Symbols indicate data from different vents. The range of concentrations varies from the seawater end member on the right-hand side with 53 mmol kg^{-1} Mg^{2+} and near zero H_4SiO_4 to the "hydrothermal end member" with near zero Mg^{2+} concentration and between 16 and 20 mmol kg^{-1} H_4SiO_4. From Von Damm *et al.* (1985).

concentrations of the hydrothermal end member are variable from vent to vent partly because of phase separation in those vents that have low enough pressures and high enough temperatures for the solutions to separate into "brine" and "vapor" phases (German and

Von Damm, 2003) (Table 2.5). This complicates interpretation of the data, which are sometimes presented on a constant Cl^- basis to distinguish phase separation effects from changes due to chemical reactions.

In general, high-temperature vent fluids are almost totally depleted of Mg^{2+}, SO_4^{2-}, and HCO_3^-. The pH is low enough that the alkalinity is sometimes negative (excess of H^+ over HCO_3^-). The vent fluids are enriched in Ca^{2+} and K^+ and extremely enriched in Fe^{2+}, Mn^{2+} and other trace metals (German and Von Damm, 2003). Thus, hydrothermal solutions are a potential sink for Mg^{2+}, HCO_3^- and SO_4^{2-} and they are a source for Ca^{2+} and K^+. The hydrothermal Ca^{2+} source represents an additional sink for HCO_3^- that enters the ocean via rivers. This is true because hydrothermal Ca^{2+} is introduced to seawater without accompanying HCO_3^-, so it removes HCO_3^- from another source (rivers) when it ultimately precipitates as $CaCO_3$ via the reverse of reaction (iii) in Table 2.2.

Many of these reactions are in the direction needed to close the marine mass balances for major ions (Fig. 2.4). The exceptions are that they supply an unnecessary additional sink for SO_4^{2-}($CaSO_4^{2-}$ precipitation) and a vast additional source of K^+. The additional sink for SO_4^{2-} does little damage to the marine SO_4^{2-} mass balance in Fig. 2.4 because its removal affects only Ca^{2+} and only at the level of about 15% of the Ca^{2+} riverine inflow. The hydrothermal source for K^+ cannot be rationalized as easily, because there is no adequate sink in the marine environment. Research into the sources and sinks of alkali metals reveals that K^+ (and other alkali metals) that are released from basalts at high temperature are reincorporated back into basaltic rock on the sea floor at low temperature. Thus, K^+ is "recycled" in the vicinity of hydrothermal vents. The rates of release and incorporation are uncertain enough to obscure whether the net K^+ flux is into or from the ocean in these regions. It is possible that the low-temperature removal of K^+ to basalt represents a net sink large enough to accommodate the river inflow.

Determining the influence of hydrothermal circulation on the mass balance of chemical constituents in seawater boils down to knowing how much water flows through the hydrothermal areas and how important the reactions at high temperature are to the total hydrothermal flux. There is presently a lot of debate about these issues. The rest of this chapter is devoted to an explanation of how one goes about interpreting hydrothermal fluxes from observed changes in hydrothermal chemistry.

2.3.2 Estimating the hydrothermal heat and water fluxes

Geochemical mass balance

Chemical oceanographers have developed a host of mass balance methods for determining the importance of hydrothermal solutions to the marine mass balance of dissolved elements in the ocean. The idea is that if one knows the flow of water through hydrothermal areas and the change in concentration of the element in question then one can calculate the hydrothermal sink. The classic example of this is the mass balance for magnesium. Since we know that the hydrothermal solution end member contains essentially undetectable Mg^{2+} (Table 2.5), and to a first approximation, there is no other significant sink for Mg^{2+} in sedimentary rocks, the marine mass balance for Mg^{2+} is simply that the river inflow equals the hydrothermal removal as seawater circulates through the system:

$$f_{\text{river}}[\text{Mg}^{2+}]_{\text{river}} = f_{\text{hydro}}\left([\text{Mg}^{2+}]_{\text{SW}} - [\text{Mg}^{2+}]_{\text{hydro}}\right), \tag{2.5}$$

where f indicates river and hydrothermal flow rates (kg y^{-1}) and square brackets as usual indicate concentrations (mol kg^{-1}).

Hydrothermal circulation is viewed as flow of seawater with its ambient concentration into the hydrothermal zone to replace water that is heated and rises to exit on the sea floor (Fig. 2.9). Consequently, the right-hand side of Eq. (2.5) expresses the difference in Mg^{2+} concentration between seawater, $[Mg^{2+}]_{SW}$, and the hydrothermal solution, $[Mg^{2+}]_{hydro}$. In the case of Mg^{2+} the cool limb takes seawater to the hot zone, and the altered seawater rises with no Mg^{2+} left in solution at all. Since the river inflow of Mg^{2+} is about 4.5×10^{12} mol y^{-1} (Table 2.3), the hydrothermal flow is simply:

$$\begin{aligned} f_{\text{hydro}} &= \frac{4.5 \times 10^{12}\ \text{mol y}^{-1}}{53\ \text{mol m}^{-3}} = 8.5 \times 10^{10}\text{m}^3\ \text{y}^{-1} \\ &\approx 8.5 \times 10^{13}\ \text{kg y}^{-1}. \end{aligned} \tag{2.6}$$

We shall see that present estimates of the hydrothermal water fluxes at high-temperature circulation regions indicate that this value is too large! Similar kinds of clever and elegant calculations have been made for helium, strontium, and lithium isotopes as well as germanium : silica ratios (see Elderfield and Schultz, 1995, for a review). In every case the mass balances are flawed because it is assumed that the entire hydrothermal sink or source has the chemistry of the 350 °C end member. The reason for this assumption is simply that most of our knowledge of hydrothermal fluid chemistry is from the concentrations of the 350 °C end member. We know now that there is extensive hydrothermal water flow through low-temperature regions and on the flanks of ridge crests where the chemical alteration is less pronounced. Thus the simple expression above is an overestimate of the high-temperature hydrothermal flow because it is incorrect to assume that Mg^{2+} is removed from seawater in hydrothermal waters of different regions and temperatures to the same extent as it is at 350 °C and 400 bar. In order to constrain the importance of the hydrothermal water fluxes one must begin by discussing the magnitude and location of hydrothermal heat flow.

Heat flow and seawater convection at hydrothermal regions

The phenomenon of hydrothermal water flow (or convection) through the crust and sediments in regions where new sea floor is

Figure 2.11. A schematic cross section of a sea floor crustal plate showing the relative sediment thickness away from the Mid-Ocean Ridge (MOR) and the location of heat flow measurements.

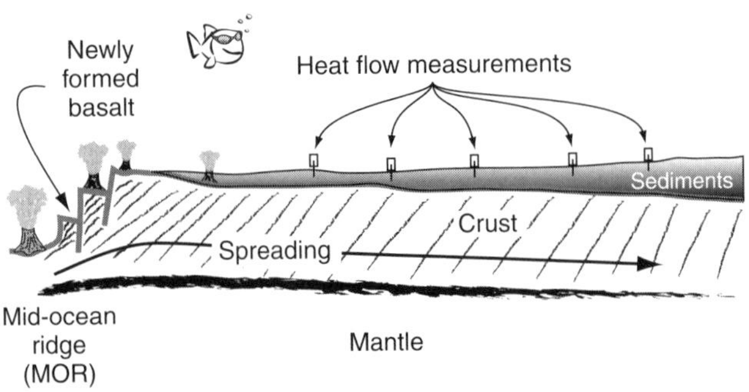

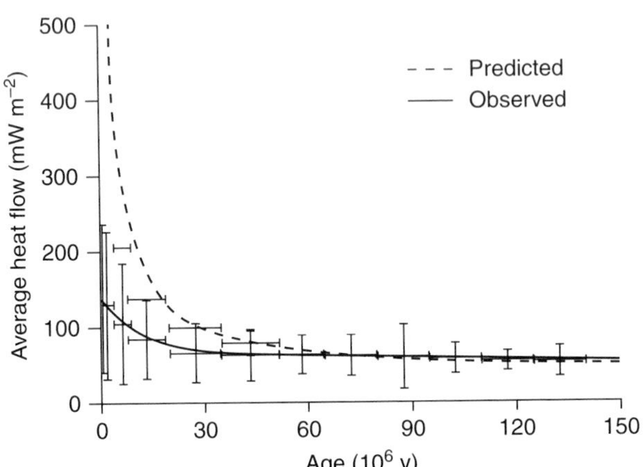

Figure 2.12. Heat flow (in mW m^{-2}) from the ocean floor as a function of sediment age (in millions of years, which is proportional to distance from the ridge crest, where age = 0). The dashed line is the predicted heat flow from a crustal cooling model and the solid line is the observed conductive heat flow based on sediment thermal gradients and the heat conduction equation. From Elderfield and Schultz (1995).

being emplaced by plate tectonics was discovered by making heat flow measurements in regions of sea floor spreading (Fig. 2.11). As the basaltic crust spreads away from the ridge crust, it cools and becomes covered with sediment. Thermistors strapped to piston corers and driven into the sediments were used to determine the temperature gradient as a function of depth (dT/dz) near the sediment–water interface. Diffusive (or conductive) heat flow, q_{cond}, from the sediments is determined from the measured temperature gradients as a function of depth:

$$q_{\text{cond}} = D_{\text{H}}\left(\frac{\text{d}T}{\text{d}z}\right)C_{\text{p}}\rho = \left(\frac{\text{cm}^2}{\text{s}}\right)\left(\frac{\text{K}}{\text{cm}}\right)\left(\frac{\text{J}}{\text{g}\cdot\text{K}}\right)\left(\frac{\text{g}}{\text{cm}^3}\right) \\ = \text{J cm}^{-2}\,\text{s}^{-1} = \text{W cm}^{-2}, \qquad (2.7)$$

where D_{H} is the thermal molecular diffusivity (*c.* 10^{-3} cm^2 s^{-1}), C_{p} is the heat capacity of water (*c.* 4 J g^{-1} K^{-1} and temperature-dependent) and ρ is the density of seawater (*c.* 1.023 g cm^{-3}). Values determined from these heat flow measurements and Eq. (2.7) are less than those predicted for a cooling slab moving away from its origin of formation (Fig. 2.12). The theoretical estimates of heat flow from a cooling slab were considered to be too accurate to be off by the amount suggested by the difference between the curves in Fig. 2.12. Rather, the reason for the difference is that the heat flux determined from the conductivity equation above considers only the diffusive (conductive) component. The difference between the predicted and measured values was thus attributed to the heat lost to the ocean by convection (advection) of water through the crust and sediments near the ridge crests:

$$q_{\text{total}} = q_{\text{cond}} + q_{\text{conv}} = D_{\text{H}}\left(\frac{\text{d}T}{\text{d}z}\right)C_{\text{P}}\rho + uT\rho C_{\text{P}}, \tag{2.8}$$

where u is the convection velocity (cm s^{-1}), which is positive.

Convection should create curvature in the measured temperature–depth profiles in the sediments, but because of the relatively large thermal diffusion coefficient the curvature is slight, and at the time of the first heat flow measurements it could not be detected. As thermistors improved, detailed temperature vs. depth measurements in sediments near active ridge crest areas indicated non-linearity caused by convection with both concave upward profiles indicating upwelling and concave downward profiles caused by downwelling of water in the sediments.

By integrating the difference between the predicted (total) and observed (conduction) curves in Fig. 2.12 for many areas on the sea floor, geophysicists have determined a hydrothermal convective heat flow of 11×10^{12} W, which is about 34% of the total convective and conductive hydrothermal heat flux. Notice that the heat flow anomaly in Fig. 2.12 extends from the ridge axis to a crustal age of about 65×10^6 y. If the spreading rate is 1 cm y^{-1} (a slow rate) this translates to 600 km from the ridge crest, indicating that much of the conductive heat loss at spreading areas is far from the ridge axis.

It is difficult to partition the heat flux into axial and off-axis components by using measurements of the conductive heat loss because temperature gradients are measured in sediments and the ridge axes are usually not sediment-covered. Mottl and Wheat (1994) estimated a maximum axial heat loss by assuming that all of the heat from new crust formation is carried away by water flow (convection) through the crust. The calculation requires knowing the amount of new crust formed each year (which can be determined from sea floor spreading rates), the latent heat of crystallization released when liquid basalt changes phase to a solid, and the difference in temperature between hot basalt and seawater at the hydrothermal end member. If one assumes that the area of newly formed basalt is 3.6 km^2 y^{-1}, it is 6 km thick and has a density of 2.8 g cm^{-3}, then the number of grams formed per unit time is:

$$\begin{aligned}\text{formation rate} &= 3.6\ \text{km}^2\ \text{y}^{-1} \times 6\ \text{km} \times 10^{15}\text{cm}^3\ \text{km}^{-3} \times 2.8\ \text{g cm}^{-3} \\ &= 60 \times 10^{15}\ \text{g y}^{-1} = 1.9 \times 10^9\ \text{g s}^{-1}.\end{aligned} \tag{2.9}$$

Liquid basalt has a temperature of about 1200 °C and it cools to 350 °C at the hydrothermal end member, resulting in a heat release of 1130 J g^{-1} (850 °C times the heat capacity of basalt of 1.3 J g^{-1} K^{1}). The other source of heat is the latent heat of crystallization, which is estimated to be 400–700 J g^{-1} for basalt. Multiplying the above heat sources (cooling plus the latent heat of crystallization) by the amount of crust created results in a total heat release of $2.9–3.4 \times 10^{12}$ W ($1530–1830\,J\,g^{-1} \times 1.9 \times 10^9\,g\,s^{-1} = 2.9–3.4 \times 10^{12}\,J\,s^{-1}$). Thus, a maximum of about one third of the 11×10^{12} W in the convective heat flow anomaly can be attributed to convection at the ridge axis through crustal rock that is less than about 0.1×10^6 y in age. The rest of the heat must be released to the ocean at locations where the crust is cooler.

Hydrothermal water fluxes

The relationship between heat flow and water flow is given through the heat capacity of water and the change in water temperature:

$$F_{H_2O\,conv} = \frac{\Sigma q_{conv}}{\Delta T\, C_p} = \frac{J\,s^{-1}}{K\,J\,g^{-1}\,K^{-1}} = g\,s^{-1}, \tag{2.10}$$

where Σq_{conv} is now the heat flow over the entire heat flow anomaly and has units of J s^{-1} (or W). The heat capacity of seawater is $4.0\,J\,g^{-1}\,K^{-1}$ at 5.0 °C and 350 bar and $5.8\,J\,g^{-1}\,K^{-1}$ at 350 °C and 350 bar. There is some danger of error here because the heat capacity becomes very large as one approaches the phase boundary for seawater, increasing the uncertainty of this calculation (German and Von Damm, 2003). If all the axial heat flow calculated above were released from black smokers, which are the vents debouching *c.* 350 °C water, then the water flow at the axis would be:

$$F_{H_2O\,conv} = \frac{\Sigma q_{conv}}{\Delta T\, C_p} = \frac{2.9-3.4 \times 10^{12}\,J\,s^{-1}}{350\,K \times 5.8\,J\,g^{-1}\,K^{-1}} \approx 1.5 \times 10^9 g\,s^{-1}$$
$$= 4.8 \times 10^{13} kg\,y^{-1}. \tag{2.11}$$

Since all the Mg^{2+} is consumed at high temperature, this flow would account for an uptake of 2.5×10^{12} mol Mg y^{-1}:

$$f_{Mg\ hydro} = (4.8 \times 10^{13}\ kg\,y^{-1})(53 \times 10^{-3} mol\ kg^{-1}) = 2.5 \times 10^{12} mol\ y^{-1}, \tag{2.12}$$

or about half of the Mg^{2+} inflow to the ocean from rivers (Table 2.3). This is probably a very high upper limit, however, because it has been suggested that only a small fraction of the heat released at the axis is at these high temperatures (Elderfield and Schultz, 1995). Even near the ridge axis there are many more low-temperature hydrothermal vents than black smokers and the reaction temperature at these locations is uncertain. These authors suggest that approximately 10% of the axial heat flux is released at 350 °C and 90% at a temperature of 10 °C. Repeating the above water flow calculation for these conditions results in a high-temperature axial flow of one tenth the above value ($F_{H_2O,axial,350} = 0.5 \times 10^{13}\,kg\,y^{-1}$) and a low-temperature axial flow of $F_{H_2O,axial,10} = 440 \times 10^{13}\,kg\,y^{-1}$. It requires much more water flow to carry the heat away if the difference in water temperature is 10 °C rather than 350 °C. This observation creates a large uncertainty in the calculation of the amount of Mg^{2+} removed by axial convection.

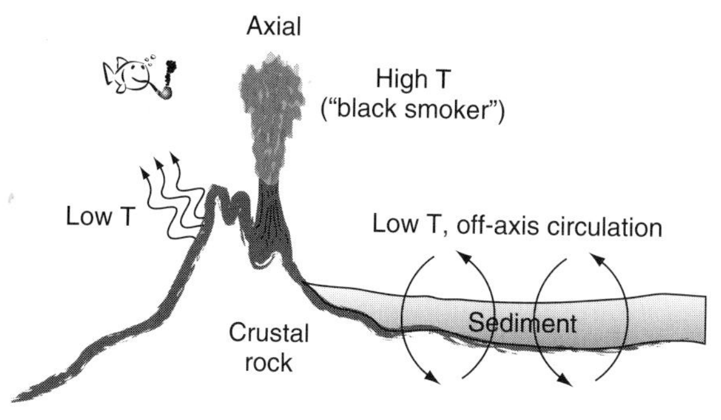

Figure 2.13. A schematic representation of the categories of heat and water convection through the sediments and crust. Axial (ridge crest) heat convection is divided into that from high-temperature (*c.* 350 °C) and lower-temperature vents, and the rest is off-axis flow through the thin veneer of sediments on the ridge flank.

It was suggested above that the off-axis heat flux must be about two thirds of that on axis because cooling of new crust can account for a maximum of one third of the hydrothermal heat anomaly. By this reasoning the off-axis heat flux becomes about 8×10^{12} W. If we assume an exit temperature of 10 °C and use these values in Eq. (2.10), it results in a water flux of $F_{H_2O,\text{ off-axis, }10} = 630 \times 10^{13}\ \text{kg y}^{-1}$.

Now, we can present a summary of the three categories of hydrothermal water flow (Fig. 2.13) with that for river flow thrown in for perspective:

$$F_{H_2O,\text{ axis, }350} = 0.5 \times 10^{13}\ \text{kg y}^{-1}; \quad (2.13)$$

$$F_{H_2O,\text{ axial, }10} = 440 \times 10^{13}\ \text{kg y}^{-1}; \quad (2.14)$$

$$F_{H_2O,\text{ off-axis, }10} = 630 \times 10^{13}\ \text{kg y}^{-1}; \quad (2.15)$$

$$F_{H_2O,\text{ river}} = 3500 \times 10^{13}\ \text{kg y}^{-1}. \quad (2.16)$$

These flux estimates demonstrate the problem involved in estimating hydrothermal chemical fluxes. Chemical concentrations for hydrothermal waters are known only for the high-temperature axial category, which by this estimate is only *c.* 0.05% of the hydrothermal flow.

2.3.3 Estimates of water chemistry changes on the ridge flanks

If the off-axis hydrothermal flow is as large as suggested from heat flow estimates, then the most important uncertainty concerning the role of hydrothermal processes in the geochemical mass balance is accurately determining the concentration changes associated with this type of flux. At the time of writing this chapter there is not enough quantitative information about chemical changes on ridge flanks to determine chemical fluxes, but there are some studies that reveal trends. These studies involve determinations of porewater chemistry in sediments that cover the flanks of ridge crests.

In the first example, porewater measurements at different locations on the eastern flank of the Juan de Fuca Ridge near 48° N in the North East Pacific Ocean indicate a decrease in Mg^{2+} with depth (Mottl and Wheat, 1994) (Fig. 2.14), indicating chemical removal. This result is a typical example of porewater profiles that have been

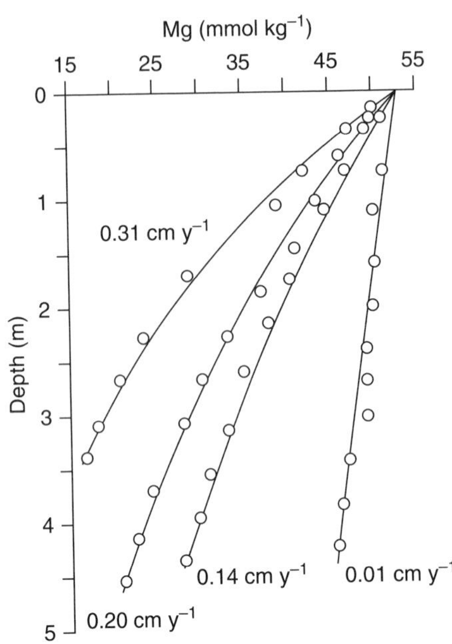

Figure 2.14. Magnesium concentrations as a function of depth (meters below the sea floor) in sediment porewaters from the western flank of the Juan de Fuca Ridge near 48° N in the North Pacific Ocean. The Mg^{2+} concentration decreases with depth because it is removed from solution by reaction with crustal rocks at the sediment–crustal boundary. The curves are convex upward because of porewater upwelling along the upward-flowing limb of a convection cell. Velocities of the upwelling are determined by using a one-dimensional advection–diffusion model and are indicated by the numbers on the curves. Redrafted from Wheat and Mottl (2000).

determined in many different areas of the oceans where hydrothermal convection is expected. It is assumed that Mg^{2+} is unreactive in the sediments, but is removed from solution at the sediment–crust boundary as basalt and seawater interact at elevated temperatures. Upward convection of water through the sediments brings Mg-poor water to the sediment surface, causing the curvature observed in these porewater profiles. A one-dimensional, advection–diffusion model matched to the Mg^{2+} results in the flow rates presented in Fig. 2.14. If future studies of this type reveal some consistency in Mg^{2+} flow in different off-axis regions of the ocean's hydrothermal areas, it may be possible to estimate global fluxes.

The second example of ridge flank sediment porewater investigations is a study in which the porewater was actually sampled at the crustal–sediment boundary by drilling to the basement (crust) through 50–500 m of sediments in a transect away from the Juan de Fuca Ridge at about 48° N in the Eastern Basin of the North Pacific Ocean (Elderfield *et al.*, 1999). Data presented in Fig. 2.15 show the temperatures and concentrations of some major ions at the sediment–crustal contact as a function of distance from the ridge crest. The first thing to note is that the temperature at the sediment–crustal boundary increases away from the ridge crest. At first, this trend seems opposite from what one might expect because the hottest rocks are at the ridge crest. Note, however, that the transect begins 20 km away from the crest where the sediment thickness is about 50 m. At this location circulation of seawater through the crustal–sediment boundary is rapid enough to cool the rocks and maintain chemistry near seawater values. As the sediment thickness increases away from the ridge crest, the sediment–crustal contact is less well ventilated with seawater and conditions begin to become more altered by hydrothermal reactions.

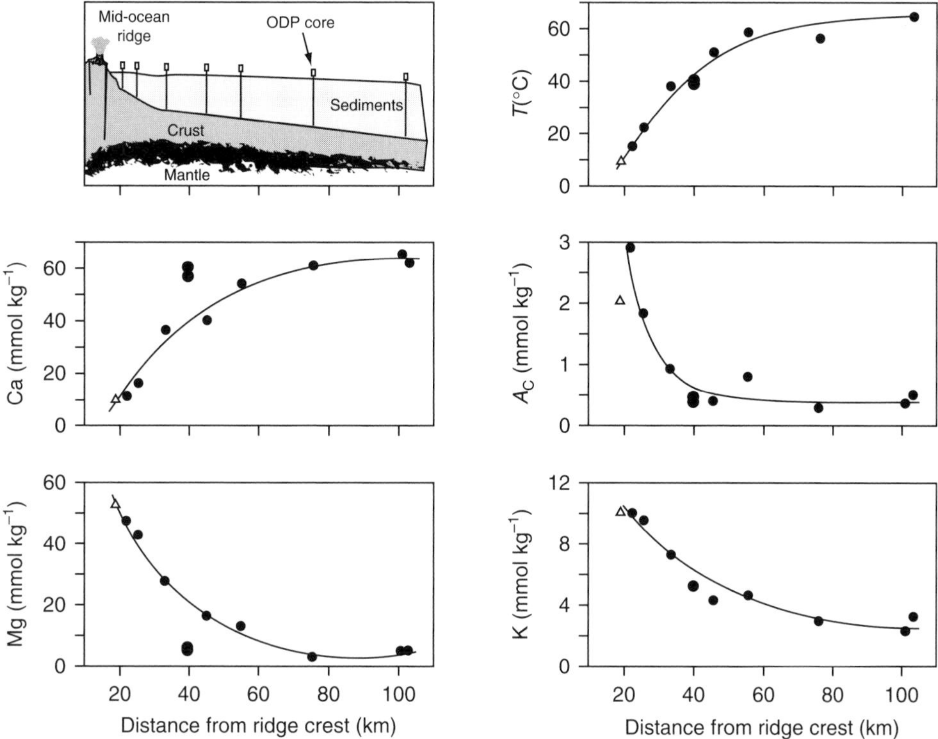

Figure 2.15. The temperature and concentration of seawater constituents in sediment porewaters at the sediment–crustal boundary as a function of distance from the Juan de Fuca Ridge crest in the North Pacific Ocean. The cartoon inset indicates how samples were taken from the sediment–crustal boundary in cores progressively distant from the ridge crest. Dots in the figures are values measured on porewater from holes drilled to the crust by the Ocean Drilling Program (ODP). The triangles are seawater values. From Elderfield *et al.* (1999).

Chemical trends with temperature are, for the most part, the same as seen in the hydrothermal end member waters (Table 2.5): Mg^{2+} and alkalinity decrease and Ca^{2+} increases as temperature increases. The exception is that the concentration of K^+ decreases with increasing temperature. This is consistent with the assumption that K^+ is released from hydrothermal systems and taken up as crustal rocks cool.

Both examples of the hydrothermal influence on seawater chemistry indicate substantial chemical alterations on the ridge flanks away from the axis. The current challenge for improving the marine mass balance is to devise a way to determine fluxes in enough areas so that some generalizations can be made about the role of the off-axis hydrothermal circulation in the geochemical mass balance.

2.4 Summary and conclusions

The geochemical mass balance for the major ions of seawater (Mackenzie and Garrels, 1966) demonstrated the necessity for reactions that could remove dissolved constituents Mg^{2+}, K^+, and HCO_3^- that have insufficient sinks based on sedimentary rock-forming minerals. Until the discovery of hydrothermal vents it was assumed that reverse weathering reactions must be the answer, even though there were few convincing experimental observations that these reactions take place. Following measurements of the chemistry of hydrothermal solutions and evaluation of the high-temperature, end member concentrations, chemical fluxes were calculated based on geochemical mass balances that suggested that these high-temperature regions are sinks (or in some cases, sources) of major ions, and that they are of the same magnitude as river inflow to the ocean.

With growing appreciation of the heat flow and chemical complexity of hydrothermal regions, it is now believed that hydrothermal convection takes place mainly through areas that are less spectacular both visually and chemically than the high-temperature, axial vents. The chemical changes and magnitudes of water flow through these widely dispersed, off-axis hydrothermal regions is poorly known, so their contribution to the marine chemical mass balance is presently uncertain. Recent evidence for reverse weathering reactions, particularly in near-shore areas where most sediments are buried (see Chapter 12), has reopened the possibility that some of the sinks for the cations Mg^{2+} and K^+ may indeed be widely dispersed, low-temperature "reverse weathering" sedimentary reactions.

We will leave this subject without resolving the issues of chemical mass balance for the major ions, even though it has been with us since nearly the beginning of modern chemical oceanography. The significance of reverse weathering reactions for the mass balance of major ions lies somewhere between an interesting curiosity and a very important sink for some ions. Hydrothermal processes certainly play a role in the mass balance of Mg^{2+}, K^+, HCO_3^-, and SO_4^{2-}; however, their quantitative importance as a sink for these elements is still uncertain.

Appendix 2.1 An extremely brief review of rocks and minerals

To help understand the discussion on weathering, a very brief review of the names and chemical formulae for some of the important rock-forming minerals is presented here. The silicate discussion is mainly from Drever (1982).

Igneous rocks

The ultimate source of most cations as well as the silicate dissolved in rivers and the ocean is igneous rock. Granites are light-colored acidic rocks; basalts are dark-colored with high concentrations of metal ions. These rocks originate from deep within the Earth, where at one time they were in a molten state. They are made of minerals like *feldspar*, *mica*, and *quartz*. *Feldspars* are the pink, green, and white minerals visible in granite:

Potassium feldspar = $KAlSi_3O_8$ (potassium aluminosilicate)
Albite feldspar = $NaAlSi_3O_8$ (sodium aluminosilicate).

Micas are the shiny flakes in igneous rocks that catch your eye by reflecting sunlight while you pick your path on a mountain hike:

Biotite mica = $KMg_3AlSi_3O_{10}(OH)_2$.

Quartz is the glassy looking parts of granitic rocks:

Quartz = SiO_2.

Clay minerals

Clay minerals are formed when igneous rocks weather. These minerals are the main constituent of fine-grained (< 63 μm) particles in mud. In general these minerals are less cation-rich than their igneous precursors. *Kaolinite* has the simplest clay mineral formula because it is pure aluminosilicate. It is the mineral that held the secret to making porcelain, which was greatly valued by the emperors of China before AD 1000, after they discovered how hard and clear kaolin becomes when heated to 1300–1400 °C. Other, more complicated clay minerals, e.g. *illite* and *montmorillonite*, have various amounts of cations added to their structures.

Kaolinite = $Al_2Si_2O_5(OH)_4$ (aluminosilicate)
Illite = similar to *biotite mica*, but less cation-rich
Montmorillonite = $(Na,Ca)(Al,Mg)_6(Si_4O_{10})_3(OH)_6\text{-}nH_2O)$.

Authigenic minerals

Minerals that precipitate from solution at Earth's surface temperatures and pressures are termed *authigenic* minerals. These are the minerals that form the shells of plants and animals that live primarily in ocean surface waters, minerals that form in sediments (some clay minerals are authigenic), and evaporite minerals that form in places like the Dead Sea and the Great Salt Lake. Some examples of authigenic minerals are:

Calcite and aragonite = $CaCO_3$ (shells of coccoliths and foraminifera, corals)
Opal = $SiO_2\ (H_2O)_n$ (shells of diatoms and radiolaria)
Pyrite = FeS_2 (forms in SO_4^{2-} reducing sediments)
Halite = NaCl (salt, formed in evaporite basins)
Gypsum and anhydrite = $CaSO_4(H_2O)_n$, $CaSO_4$ (formed in evaporite basins).

Appendix 2.2 The meaning of residence time

Two analogies are presented to describe the meaning of residence time. The simplest is that of filling a reservoir. Consider a child's swimming pool with a volume, V_{pool} (m^3) being filled with a water hose of inflow rate, f ($m^3\,h^{-1}$). How long does it take to fill the pool? From time $t = 0$ to $t = \tau$ the rate of change in volume is:

$$dV = f\ dt. \qquad (2A2.1)$$

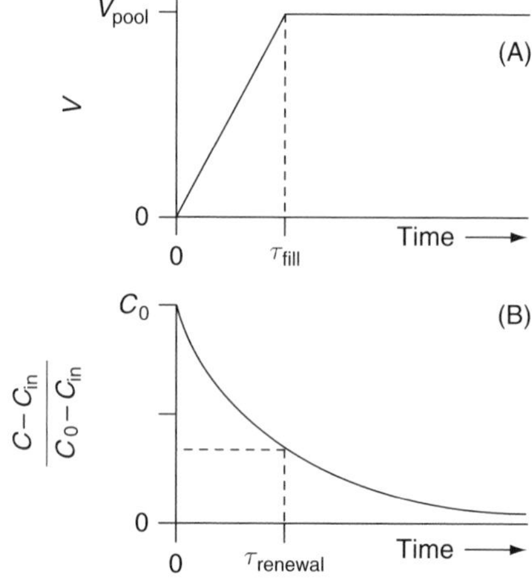

Figure 2A2.1 The solutions to Eqs. (2A2.1) and (2A2.7). (A) The volume of water, V, in the pool as a function of time; V_{pool} is the final value when the pool is full. (B) The relative concentration, C, of *E. coli* in the pool as a function of time. C_{in} is the concentration in the inflow and C_0 is the concentration at time $t = 0$. Residence times, τ, are indicated for both models.

We define the residence time, τ_{fill}, as the time when the pool starts overflowing, as illustrated in Fig. 2A2.1

$$\tau_{fill} = \frac{V_{pool}}{f}. \tag{2A2.2}$$

In this analogy the residence time is the time at which the volume of the system reaches steady state (i.e. the volume at overflow).

Now let's make the analogy a little more realistic. The pool is full, but the water has gotten dirty. It has *E. coli* bacteria of concentration C (mol m^{-3}), and you want to renew the water in the pool with clean water that has an *E. coli* concentration of C_{in}. At $t = 0$ you begin the inflow again. Your dog is swimming in the pool during renewal to keep the water well mixed, and we assume that the concentration of *E. coli* does not change because of biological growth or death. How long will it take to renew the old water with fresh water?

Since the concentration of *E. coli* in the pool asymptotically approaches that of the inflowing water, the pool water will theoretically never be *totally* renewed, so one defines a time, τ_{renew}, when a certain fraction of the old water has been replaced. What is the time τ_{renew}, and how much water has been replaced at this time?

The concentration change from the beginning is now;

$$V_{pool}\frac{dC}{dt} = C_{in}f - Cf \quad \text{with initial conditions at } t = 0,\ C = C_0. \tag{2A2.3}$$

These equations have the solution below that is illustrated in Fig. 2A2.1(B):

$$C - C_{in} = (C_0 - C_{in})\exp -\left(\frac{f}{V}\cdot t\right); \tag{2A2.4}$$

at $t = \tau_{renew}$

$$\frac{C - C_{in}}{C_0 - C_{in}} = \exp -\left(\frac{f}{V_{pool}}\cdot \tau_{renew}\right). \tag{2A2.5}$$

Now, if we assume the renewal residence time is equal to the final, steady-state inventory in the pool divided by the rate of inflow, as in the definition of the residence time in Eq. (2A2.2), then τ_{renew} is equal to τ_{fill}:

$$\frac{V_{\text{pool}} C_{\text{in}}}{f\, C_{\text{in}}} = \frac{V_{\text{pool}}}{f} = \tau_{\text{renew}}. \qquad (2A2.6)$$

Substituting Eq. (2A2.6) into (2A2.5) gives the concentration at the residence time, τ_{renew}:

$$\frac{C - C_{\text{in}}}{C_0 - C_{\text{in}}} = \exp - \left(\frac{f}{V_{\text{pool}}} \times \frac{V_{\text{pool}}}{f} \right) = \exp(-1) = \frac{1}{\text{e}} = \frac{1}{2.7} = 0.37 \qquad (2A2.7)$$

or

$$\frac{C_0 - C}{C_0 - C_{\text{in}}} = 1 - 0.37 = 0.63. \qquad (2A2.8)$$

The residence time (or renewal time) is defined by the e-folding time, 1/e. In this example it is the time required for the difference between the concentration and its initial value to reach 63% of the difference between the initial and final values. The process of renewing the water in the pool is 63% of the way to completion.

References

Berner, R. (1971) *Principles of Chemical Sedimentology*. New York, NY: McGraw-Hill.

Bischoff, J. L. and F. W. Dickson (1975) Seawater-basalt interaction at 200 °C and 500 bars: implications for the origins of sea-floor heavy metal deposits and regulation of seawater chemistry. *Earth Planet. Sci. Lett.* **25**, 385–97.

Broecker, W. S. and T.-H. Peng (1982) *Tracers in the Sea*. Palisades, NY: Eldigio Press.

Drever, J. I. (1982) *The Geochemistry of Natural Waters*. Englewood Cliffs, NJ: Prentice-Hall.

Edmond, J. M. *et al.* (1979) Ridge crest hydrothermal activity and the balances of the major and minor elements in the ocean: the Galapagos data. *Earth Planet. Sci. Lett.* **46**, 1–18.

Elderfield, H. and A. Schultz (1995) Mid-ocean ridge hydrothermal fluxes and the chemical composition of the ocean. *A. Rev. Earth Sci.* **24**, 191–224.

Elderfield, H. *et al.* (1999) Fluid and geochemical transport through oceanic crust: a transect across the eastern flank of the Juan de Fuca Ridge. *Earth Planet. Sci. Lett.* **172**, 151–65.

German, C. and K. Von Damm (2003) Hydrothermal processes. In *The Oceans and Marine Chemistry* (ed. H. Elderfield), vol. 6, *Treatise on Geochemistry* (ed. H. D. Holland and K. K. Turekian), pp. 181–222. Amsterdam: Elsevier.

Helgeson, H. and F. Mackenzie (1970) Silicate-seawater equilibria in the ocean system. *Deep-Sea Res.* **17**, 877–92.

Holland, H. D. (1978) *The Chemistry of the Atmosphere and Oceans*. New York, NY: John Wiley.

Li, Y.-H. (2000) *A Compendium of Geochemistry*. Princeton, NJ: Princeton University Press.

Mackenzie, F. T. and R. M. Garrels (1966) Chemical mass balance between rivers and the ocean. *Am. J. Sci.* **264**, 507–25.

Michalopoulos, P. and R. Aller (1995) Rapid clay mineral formation in the Amazon delta sediments: reverse weathering and ocean elemental cycles. *Science* **270**, 614–17.

Mottl, M. J. and H. D. Holland (1978) Chemical exchange during hydrothermal alteration of basalt by seawater, I. Experimental results for major and minor components of seawater. *Geochim. Cosmochim. Acta* **42**, 1103–15.

Mottl, M. and J. Wheat (1994) Hydrothermal circulation through mid-ocean ridge flanks: fluxes of heat and magnesium. *Geochim. Cosmochim. Acta* **58**, 2225–37.

Sayles, F. (1979) The composition and diagenesis of interstitial solutions, I. Fluxes across the seawater-sediment interface in the Atlantic Ocean. *Geochim. Cosmochim. Acta* **43**, 527–46.

Sillen, L. G. (1967) Gibbs phase rule in marine sediments. In *Equilibrium Concepts in Natural Water Systems* (ed. R. F. Gould), pp. 57–69. Washington, DC: American Chemical Society.

Von Damm, K. L., J. M. Edmond, B. Grant and C. I. Measures (1985) Chemistry of submarine hydrothermal solutions at 21 N, East Pacific Rise. *Geochim. Cosmochim. Acta* **49**, 2197–220.

Wheat, J. and M. Mottl (2000) Composition of pore and spring waters from Baby Bare: global implications of geochemical fluxes from ridge flank hydrothermal system. *Geochem. Cosmochim. Acta* **64**, 629–42.

Whitehouse, U. G. and R. S. McCarter (1958) Diagenetic modification of clay mineral types in artificial seawater. In *Clays and Clay Minerals* (ed. A. Swineford), Natl. Res. Council Pub. 566, pp. 81–119.

3

Thermodynamics background

One of the great advantages of the chemical perspective on ocean processes is the ability to predict whether specific reactions between molecules and ions might occur in aqueous media. Given the extreme complexity of natural systems, this type of fundamental constraint is invaluable. Such predictions are based on concepts and energetic information that have been painstakingly generated over the past several centuries and assembled into the discipline known as thermodynamics. The purpose of this chapter is to present an introduction to the properties of water and ions, and the basic concepts of the thermodynamics of chemical reactions. Rather than cover the breadth of thermodynamics, we seek to demonstrate the applications of free energy to the prediction of equilibrium distributions of chemical species among gaseous, liquid, and solid phases. The goal is to establish the conceptual foundation and tools that can be applied to the following chapters on ocean processes.

3.1 The properties of water and ions

3.1.1 The structure of water

Water accounts for approximately 96.5 mass percent of seawater. The innate characteristics of water affect almost all the properties of the ocean (e.g. density, salinity, and gas solubility) and the processes (e.g. circulation, heat exchange, chemical reactions, and biochemical transformations) that occur within it. Water is so much a part of our world and daily lives that it is easy to overlook how unusual this substance is in its physical and chemical properties. A major source of this uniqueness originates at the molecular level from the much greater affinity (electronegativity) of oxygen versus hydrogen for shared electrons. In the two covalent bonds of an H_2O molecule, oxygen attracts the bulk of the electron density it shares with hydrogen and thereby acquires a partial negative charge that is redistributed into its two opposite non-bonding electron pairs (Fig. 3.1). As a result, each of the hydrogen atoms linked in the two bonding orbitals exhibits a partial positive charge. Because of the tetrahedral orientation of the four orbitals about the central oxygen, symmetric cancellation of this unequal electron distribution is not possible within individual H_2O molecules. Each H_2O molecule therefore exhibits a net separation of charge in space (Fig. 3.1), which is referred to as a *dipole*. Because the two electron pairs in the non-bonding orbitals about the oxygen atom are attracted primarily by the oxygen nucleus, their electron densities are more expansive, resulting in a repulsive force on the bonding orbitals that causes a bond angle compression. The H–O–H bond angle is compressed from the value for a perfect tetrahedron (109.5°) to 104.5°.

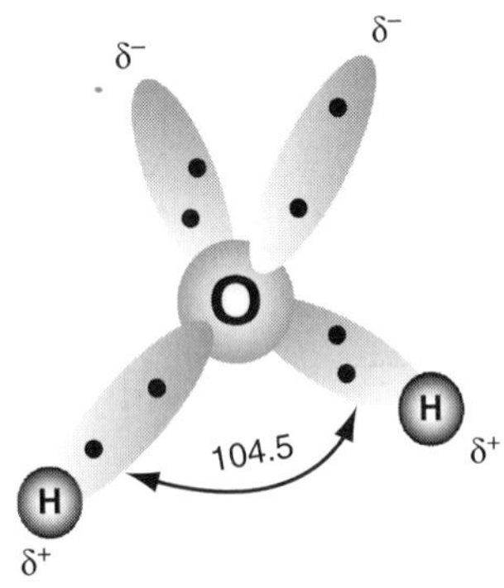

Figure 3.1. The three-dimensional structure of a water molecule. The sphere in the middle represents the oxygen atom and the dark circles are hydrogen atoms. Ovals represent electron orbitals of the outer shell electrons; δs indicate net charges.

This unequal distribution of electronic charge within the H_2O molecule has huge ramifications for how these molecules interact and react. As a result of their polarity, H_2O molecules orient in solution in such a way as to minimize the proximity of like partial charges by (on average) turning hydrogens toward oxygens of neighboring molecules. Adjacent H_2O molecules undergo an additional form of intermolecular interaction known as *hydrogen bonding*, which is unique to molecules having hydrogen bonded directly to highly electronegative elements such as N, O, F and occasionally Cl. Hydrogen bonding can be thought of as an intermolecular attraction of the hydrogen atoms of water molecules. The hydrogen atoms of each individual water molecule interact along an O–H bond that is aligned with one of the two non-bonding orbitals of the oxygen in an adjacent H_2O molecule. Hydrogen bonds are weaker than ordinary intramolecular bonds but are generally the strongest type of intermolecular attraction. The resulting intermolecular interaction has a characteristic distance of 1.8 Angstroms ($1\,\text{Å} = 10^{-10}$ cm) between the oxygen of one water molecule and the hydrogen of an adjacent water molecule and exhibits a strength of *c.* 19 kJ mol^{-1}. By contrast, in an H–O bond the covalent bond distance is approximately 1.0 Å and has a strength of 463 kJ mol^{-1} (1 kcal = 4.184 kJ). Because the joined H_2O molecules must be aligned along the axes of their bonding orbital, hydrogen bonding causes H_2O molecules to exist in strongly ordered assemblies.

Pervasive hydrogen bonding is evident in all physical properties that reflect the strengths of association among H_2O molecules in both liquid water and ice. For example, liquid water has one of the highest heat capacities, heats of vaporization, surface tensions, and dielectric constants (ion insulating capability) of all substances. In addition, the amount of heat necessary to go from the solid to the liquid phase is greater for water than for all substances except ammonia. The unusually high temperatures needed to melt ice and boil water relative to dihydrides of other elements in the oxygen family are not readily predictable by the usual method of extrapolating physical property trends as a function of molecular mass (Fig. 3.2). Clearly, hydrogen bonding makes H_2O a peculiar molecule. One oceanographically important outcome of hydrogen bonding is that water is excellent thermostating material. For example, 1 g of water absorbs over 3000 J of energy as it is heated from ice at −50 °C to steam at +150 °C (Fig. 3.3). The same amount of heat would raise the temperature of 10 g of granite (heat capacity ≈0.8 J g^{-1}) to over 380 °C. This difference

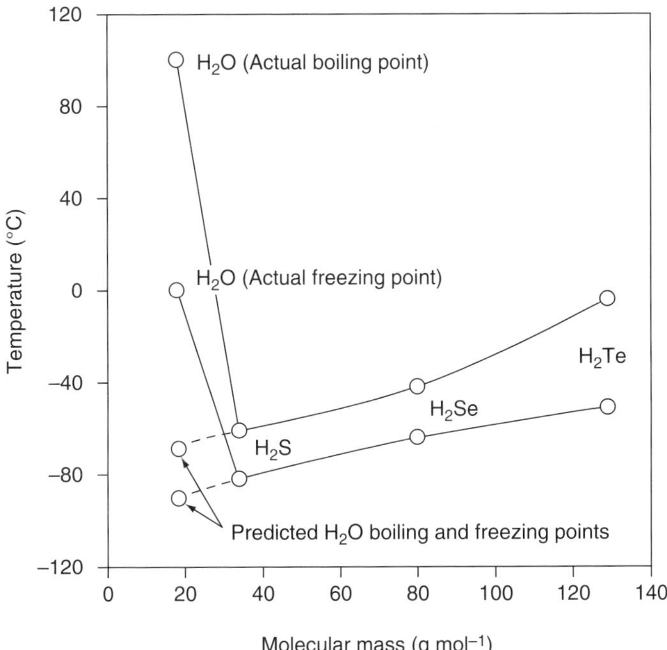

Figure 3.2. Freezing and boiling temperature of dihydride molecules of elements in column VI of the periodic table as a function of their molecular mass. The actual values for H_2O are connected by a solid line. Extrapolated melting and boiling points for water are indicated by the dashed line. Modified from Libes (1992).

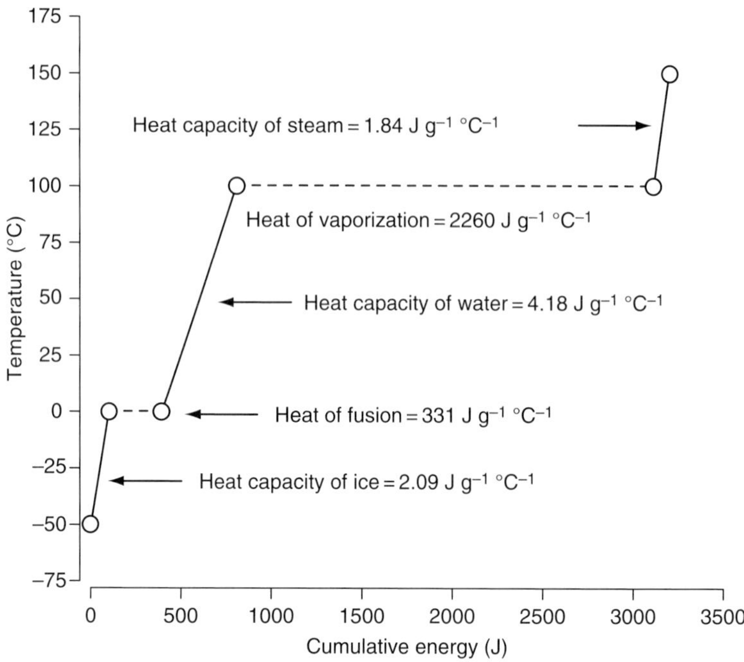

Figure 3.3. The cumulative heat energy needed to heat water from −50 to +150 °C. Modified from Libes (1992).

occurs largely because water has five times the heat capacity of granite and undergoes two (versus none for granite) phase transitions between −50 °C and +150 °C. It is easy to see why the sea is able to moderate temperature changes on adjacent continental margins and transport huge amounts of heat energy from low to high latitudes in the form of ocean currents and water vapor (Chapter 1).

The best defined structural properties of water are at the thermal extremes for vapor and ice. Water vapor consists primarily of individual H_2O molecules with negligible intermolecular attractions holding individual H_2O molecules together. Hence water vapor has essentially no structure. At the other temperature extreme, ice has a remarkably ordered structure. The structure of Ice-I, the stable form at atmospheric pressure, represents complete accommodation of individual molecules to maximal collective hydrogen bonding. In Ice-I (Fig. 3.4), all hydrogen atoms are located along the axes between the oxygen atoms of H_2O molecules that have been stretched into perfect tetrahedra. Full hydrogen bonding can be accomplished only by increasing the spacing of the water structural units in Ice-I, resulting in a solid of roughly half the density (0.92 g cm^{-3}) that would be exhibited for the closest packed structure (1.7 g cm^{-3}).

The specific structure of liquid water is poorly defined, but can be thought of as a slush of ice-like clumps floating in a pool of relatively unassociated H_2O molecules. This type of mixture helps explain many of the maxima and minima in such physical properties as density and viscosity that are often observed when liquid water is cooled or pressurized. The best known of these trends is the maximum in liquid water density near 4 °C (Fig. 3.5). This phenomenon

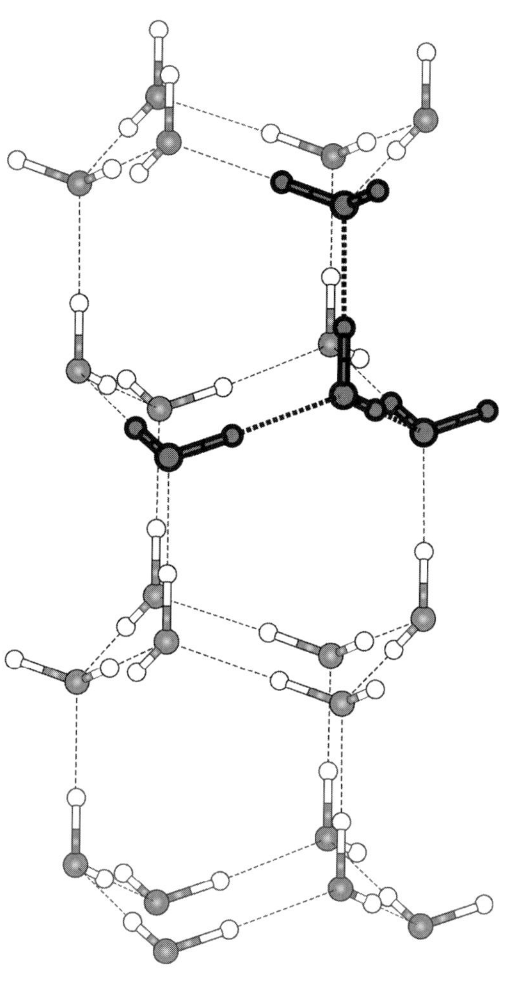

Figure 3.4. The structure of Ice-I, the predominant form of ice at one atmosphere pressure. Shaded spheres indicate oxygen atoms and open spheres hydrogen atoms. The darker shading represents four closest neighbor water molecules.

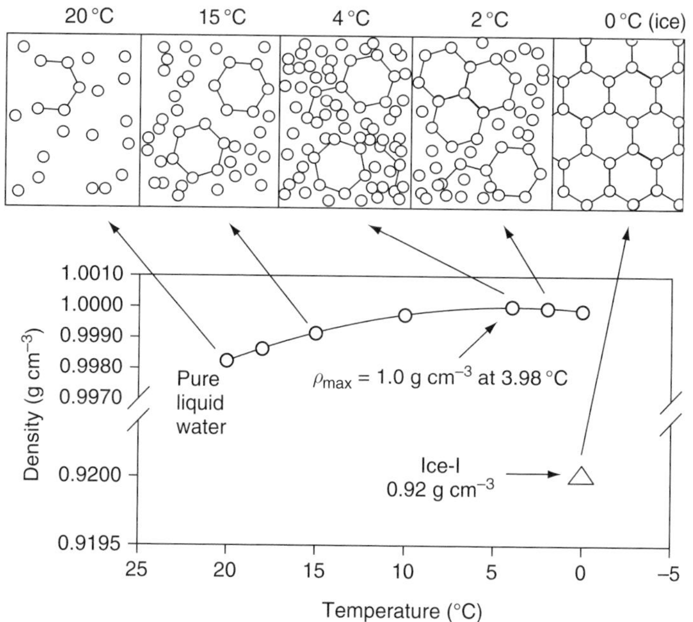

Figure 3.5 The density of liquid water as a function of temperature. The relative proportions of free molecules (unconnected circles) and ordered molecules (circles connected by lines) are illustrated for selected temperatures (°C) in the schematic illustration. Modified from Libes (1992).

apparently represents the net result of opposing density responses by associated and relatively free H_2O molecules to changing temperature. Lowering temperature toward the freezing point increases density by allowing slower-moving, free H_2O molecules to pack closer together, but will simultaneously decrease density through formation of more, less dense, ice-like clusters. These two opposing effects balance each other near 4 °C, below which the formation of ice-like structures predominates. Independent energy-related evidence for a high degree of association among H_2O molecules in the liquid state is that the amount of heat energy given off when liquid water freezes into ice (the heat of fusion) is only *c.*15% of the total that would theoretically be released by forming all the component hydrogen bonds from initially unassociated H_2O molecules. It is as if most liquid H_2O molecules were already in the frozen configuration. This high degree of structure in liquid water is not observed by most spectrographic-based measurements, which are too slow to resolve H_2O molecules as they rapidly interconvert between different association patterns.

3.1.2 Water–ion interactions

Given that H_2O molecules have strong dipoles, it is not surprising that they interact even more strongly with charged, dissolved ions than with each other. One of the most evident indications of this strong association is *electrostriction*, the reduction of water volume upon the addition of charged ions. This effect is illustrated in Table 3.1 for the addition of NaCl to water at concentrations similar to those in seawater. In this case, the total volume of the water–ion system shrinks by about 0.5% owing to the addition of *c.*3 wt% NaCl. This contraction results because H_2O molecules immediately surrounding the added ions are more closely packed than in bulk liquid owing to strong ion–dipole interactions (Fig. 3.6). The H_2O molecules of hydration will be packed with the negative (O) portions of the molecules oriented toward dissolved cations and the positive (H) regions directed toward dissolved anions. This reorientation capability partly nullifies the electrostatic charge of the hydrated ions and contributes toward the great ability of water to insulate, and hence dissolve, ions. Another common effect of dissolving salt in water is an increase in the viscosity of the solution, because the highly structured ion–dipole assemblies act as “lumps” that inhibit flow. The magnitudes of such electrostriction and viscosity effects increase with the charge density (charge/volume) and number of each type of dissolved ion. These effects are also a function of the temperature of the system. Because liquid water at higher temperatures has less structure, there is more potential for net structural increases as temperature increases. At low temperatures, large monovalent ions with low charge densities (e.g. K^+ and Cl^-) can act as “hydrogen bond structure breakers” because the electrostatic interaction with immediately adjacent H_2O molecules is greater than the strength of the hydrogen-bonded structure of pure water (Fig. 3.6).

Table 3.1. *Effects of ions on the density and structure of liquid water*

Ingredient	g	ρ (g cm^{-3})	cm^3
H_2O (25 °C)	970.78	0.993	973.5
NaCl	29.22	2.165	13.5
Simple sum	1000.00	—	987.0
Actual value	1000.00	—	982.0
Difference	0	—	5.0

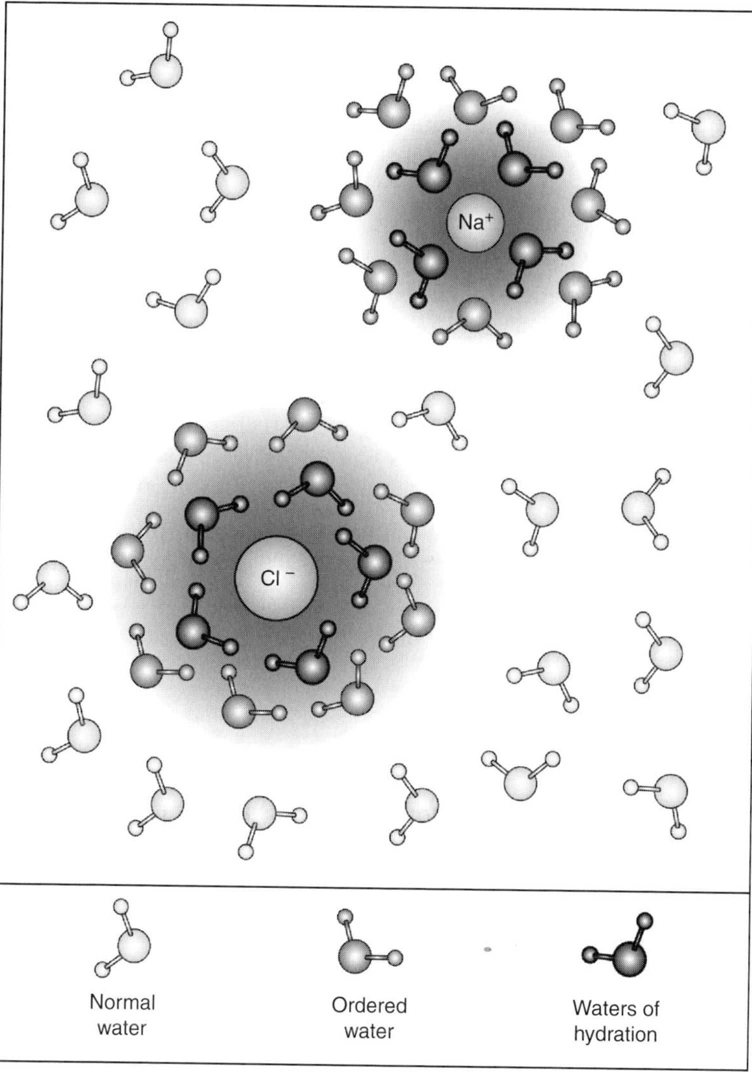

Figure 3.6. A cartoon of the structure of water in the vicinity of dissolved sodium and chloride ions. Waters of hydration are closest to the ions and indicated by the darkest water molecules. Layers of ordered water molecules outside the waters of hydration are indicated by shaded molecules.

3.2 Ion–ion interactions and activity coefficients

In spite of the great insulating properties of water, ions still interact electrostatically with each other in aqueous solutions. The electrostatic force, F_z, between any two ions is expressed by the equation

$$F_z = \frac{Z_1 \times Z_2}{r^2}, \tag{3.1}$$

where Z_1 and Z_2 are the charges on the individual ions and r is the distance between them. In electrolytic solutions containing many ions, the average distance between ions decreases exponentially with their increasing concentration. Because ions interfere with each other in an electrolytic solution, their ability to react chemically is diminished except at infinite dilution, where large r values in the above equation drive F_z toward zero. In natural waters, containing finite concentrations of dissolved salts, an ion's *concentration* (amount per unit volume or mass) almost always overestimates its *activity* (reaction potential) versus conditions at infinite dilution. Thus, it is practical to have a correction factor that can be used to adjust the concentration of a dissolved substance, which is readily measured, to the corresponding chemical activity that actually expresses its reactivity. This correction factor is known as an *activity coefficient* (γ), and for any chemical species (i) is defined as

$$\gamma_i = \frac{a_i}{M_i} \tag{3.2}$$

where a_i and M_i indicate the activity and concentration of the ith species, respectively. In this book the chemical activities will usually be indicated within parentheses and concentrations within brackets, so that for sodium ion the relation between activity and concentration is written as

$$(\mathrm{Na^+}) = \gamma_{\mathrm{Na^+}}[\mathrm{Na^+}]. \tag{3.3}$$

As discussed in Chapter 1, concentrations are normally expressed in the oceanographically convenient form of moles of solute per kilogram of solution ($\mathrm{mol\,kg^{-1}}$) or in molar (M) units. Because activity coefficients are dimensionless correction factors, they are equally applicable to all concentration units.

3.2.1 Activity–ionic strength relations

The challenge of correcting concentrations to activities for natural waters is that the activity coefficients vary non-linearly, often in complex relations to bulk ion concentrations. For dilute electrolyte solutions, such as some lake and river waters, it is practicable to estimate the activity coefficient of an individual ion theoretically based on that ion's charge and a general measure of the effective total ion concentration of the bulk solution. The latter measure is called the *ionic*

Table 3.2. *Methods of calculating activity coefficients from ionic strength*

I is ionic strength in units of molarity, M. In the expressions, $A \approx 0.5$ in water at 25 °C, $B = 0.33$, and a is an adjustable parameter ranging from 3 to 9 (see Stumm and Morgan, 1996).

Approximation	Equation	I (M)
Debye–Hückel	$\log \gamma_i = -Az_i^2 \sqrt{I}$	$<10^{-2}$
Extended Debye–Hückel	$\log \gamma_i = -Az_i^2 \frac{\sqrt{I}}{1+Ba\sqrt{I}}$	$<10^{-1}$
Davies	$\log \gamma_i = -Az_i^2 \left(\frac{\sqrt{I}}{1+\sqrt{I}}\right) - 0.2I$	$<5 \times 10^{-1}$

strength, I, of the solution, which is simply half the sum of the molar, M, concentrations of each ion multiplied by its charge squared

$$I = \tfrac{1}{2}\sum M_i z_i^2. \tag{3.4}$$

(In older texts molality is used instead of molarity, but we adopt the modern convention of using the more environmentally familiar units.) Thus, a 1.0 M solution of NaCl has an I of 1.0 M, whereas a 1.0 M solution of $MgCl_2$ has an I of 3.0 M. Charges are squared in determinations of I because this parameter seeks to compensate for the electrostatic forces between ions that vary with charge squared (Eq. (3.1)). (This calculation for ionic strength is only exactly correct for the most dilute solutions, less than approximately 0.1 M. The calculation for more concentrated solutions is further complicated by incomplete dissociation and ion pairing; see later.) The ionic strength of seawater is $I \approx 0.7$.

The simplest expression relating the activity coefficient of a particular ion to the ionic strength of the solution is the Debye–Hückel equation (Table 3.2), which provides a useful approximation of γ_i at ionic strengths of $<10^{-2}$ M. The extended Debye–Hückel and Davies equations relate $\log \gamma_i$ to more complex expression of I and are useful for slightly more concentrated solutions (up to 0.1 M ionic strengths). All these equations are theoretically based and involve simplifying assumptions, such as treatments of ions as point charges (ignoring ionic dimensions), that fail at higher ionic strengths. Unfortunately, none of these equations can be applied at the ionic strength of seawater ($I \approx 0.7$ M), making other methods for calculating activity coefficients necessary.

3.2.2 The mean salt method

Single-ion activity coefficients at higher ionic strength are estimated by using an empirical approach in which complex ion and dipole interactions are accounted for by direct laboratory measurements. One such experimentally based approach that can be applied to seawater is the *mean salt method* (MSM). This procedure is based upon measurements of the mean activity coefficients $\gamma_{\pm M_aX_b}$ of simple salt solutions that usually contain only two types of ion. For example, determination of the mean activity coefficient for the dissolution of a salt, M_aX_b,

$$M_aX_b \rightleftarrows aM^{b+} + bX^{a-} \tag{3.5}$$

is done by measuring a property of the salt solution (e.g. electrical conductivity) that is known as a function of the salt activity and then comparing this with the value measured as a function of concentration. Another approach is to determine the activity of the ion in solution directly by using ion-sensitive electrodes.

The first step in the MSM is to measure the $\gamma_{\pm M_aX_b}$ of a KCl solution whose concentration is adjusted to be identical to the I of the electrolyte solution for which the activity coefficients are desired. For seawater a KCl solution of $I \approx 0.7$ would be used. One begins with KCl because the K^+ and Cl^- ions interact almost exclusively in an electrostatic manner, without formation of interfering ion pairs and other complexes. Next one assumes that $\gamma_{\pm M_aX_b}$ measured for any solution is the geometric mean of the individual activities of the component ions. The general formula for the geometric mean is

$$\gamma_{\pm M_aX_b} = \left[\gamma^a_{M^{b+}} \times \gamma^b_{X^{a-}}\right]^{1/(a+b)}, \tag{3.6}$$

which in the specific case of KCl becomes

$$\gamma_{\pm KCl} = \left[\gamma^1_{K^+} \times \gamma^1_{Cl^-}\right]^{1/2}. \tag{3.7}$$

It is further assumed that $\gamma_{K^+} = \gamma_{Cl^-}$, which simplifies Eq. (3.7) to

$$\gamma_{\pm KCl} = \gamma_{K^+} = \gamma_{Cl^-}. \tag{3.8}$$

This assumption is based on a variety of observations indicating that K^+ and Cl^- have similar effects on such properties of electrolyte solutions as viscosity and density. Once these two single-ion activity coefficients are established it is possible to make "bootstrap" estimates of the activity coefficients for all other ions by preparing simple solutions of individual salts containing either K^+ or Cl^- at the desired I. For example, to determine $\gamma_{Ca^{2+}}$, one would make up a $I = 0.7$ solution of $CaCl_2$, for which

$$\gamma_{\pm CaCl_2} = \left[\gamma_{Ca^{2+}} \times \gamma^2_{Cl^-}\right]^{1/3}. \tag{3.9}$$

Once $\gamma_{\pm CaCl_2}$ is determined experimentally, the only unknown would be $\gamma_{Ca^{2+}}$. For an anion, the same procedure would be followed by using a K salt. In this manner, single-ion activity coefficients can be determined for all the major ions in an electrolytic solution, such as seawater. In general, activity coefficients in seawater fall in the range of 0.6–0.8 for monovalent ions and 0.1–0.2 for divalent ions, and are near 0.01 for trivalent ions. Clearly, multiple charges lead to disproportionate attenuation of activity at a given electrolytic concentration, as would be expected given the strong dependency of electrostatic forces on charge (Eq. (3.1)).

Although useful, activity coefficients calculated by the MSM can only be applied under restricted conditions. One limitation is that empirically determined γ_i values hold only for the temperature, pressure and ionic strength of electrolyte solutions used in the laboratory measurements. A second complication for a mixed electrolyte like seawater is that γ_i values do not account for *ion pairing*. This term refers to a specific direct association of two ions that can occur in addition to general electrostatic interactions. An example of an ion pairing reaction that occurs in seawater is

$$Mg^{2+} + F^- \rightleftharpoons MgF^+. \tag{3.10}$$

The effects of this specific reaction would not be represented in MSM calculations. In fact, it is assumed that no ion pairs are formed at $I = 0.7$ in the experiments with individual electrolytes used to calculate activity coefficients in the MSM. This is true mainly for salts of potassium and chlorine, which are the ones primarily used in the MSM calculations. Because any ions that join to form ion pairs in a mixed electrolyte solution are effectively removed from solution with respect to reaction potential, an accurate determination of the extent of ion pairing in addition to the "free" ion activity coefficient must be made to determine the "total" activity coefficient in seawater.

The percentage of the total concentration of a specific ion in seawater that is involved in ion pairs can be determined from thermodynamic equilibrium information and will be demonstrated later in the chapter. For now, suffice it to say that the individual ion concentration is determined by multiplying the total ion concentration by a fraction between 0 and 1.0 representing the total ion that is free of ion pairs. The product, $(\%\text{free}_i/100)\, M_i$, can then be multiplied by the *free ion activity coefficient* to determine the activity. In general form, the total correction equation is

$$a_i = (\%\text{free}_i/100) \times \gamma_i \times [\ ]_i = \gamma_{i,T}[\ ]_i \tag{3.11}$$

where the overall correction, $((\%\text{free}_i/100) \times \gamma_i)$, is frequently referred to as the *total activity coefficient*, $\gamma_{i,T}$. The two correction factors are kept separate here so that it is clear that both ion pairing and general electrostatic inhibition factors are being addressed.

3.3 Thermodynamic basics

After chemical concentrations are converted to activities, the latter can be used to predict the probabilities and extents of specific chemical reactions based on the concept of *free energy*, the energy available in a chemical system to do work. This thermodynamically based method is founded on energetic relations that can be established for a chemical species or reaction system. The fundamental energetic property of a given chemical species is its *standard free energy of*

Table 3.3. *ΔG_f^0 values of different forms of hydrogen*

Compound	ΔG_f^0(kJ mol^{-1})
H_2 (gas)	0
H_2 (aqueous)	17.57
H^+ (aqueous)	0
H_2O (liquid)	−237.18
H_2O (gas)	−228.57

From Stumm and Morgan (1996).

formation (the symbol used in modern texts is ΔG_f^0), where the superscripted zero indicates the standard state and the subscript *f* indicates free energy of formation. By definition, ΔG_f^0 is the free energy change when one mole of a substance in its standard state is formed from stable elements under standard conditions.

All free energy changes are relative to each other and rely on definitions of values at *standard conditions* and *standard states*. These states are largely a matter of convenience, with *standard conditions* corresponding to the temperature and pressure occurring in a laboratory. *Standard states* correspond either to pure substances or to gas and liquid activities that are readily recreated and mathematically convenient. There are three important conventions to be observed: (1) *standard conditions* are set to 1 atm of pressure and 25 °C; (2) the *standard states* of dissolved ions, as well as pure solids, liquids, gases, and solutions have activities equal to 1; and (3) by definition, the ΔG_f^0 values at standard conditions for pure elements in their most stable forms, as well as the proton (H^+) and electron (e^-), are all equal to zero. The standard free energies of formation for various physical states of hydrogen and its oxide, water, are presented in Table 3.3. Note that ΔG_f^0 values vary with physical as well as chemical form, and normally only the pure element has a value of zero.

The usual unit of ΔG_f^0 is kilojoules per mole (kJ mol^{-1}). Tables of standard free energies of formation are collected in books with compilations of thermodynamic data (see, for example, Stumm and Morgan, 1996; Morel and Herring, 1993; *Handbook of Chemistry and Physics*, 1970).

3.3.1 Free energy change during reaction

The purpose of defining standard free energies of formation is to use these individual reference values to determine standard free energies of reaction, ΔG_r^0, where the subscript *r* indicates reaction. By definition, ΔG_r^0 is the free energy *change* attending a balanced chemical reaction that involves substances at their standard states

$$\Delta G_r^0 = \sum n\left(\Delta G_f^0\right)_{\text{products}} - \sum n\left(\Delta G_f^0\right)_{\text{reactants}}, \tag{3.12}$$

where n represents stoichiometric coefficients. Thus, for a generalized reaction (at standard conditions)

$$bB + cC \rightleftarrows dD + eE \tag{3.13}$$

$$\Delta G_r^0 = d\Delta G_{f,D}^0 + e\Delta G_{f,E}^0 - b\Delta G_{f,B}^0 - c\Delta G_{f,C}^0. \tag{3.14}$$

Establishing ΔG_r^0 for a specific reaction of constituents in their standard states is a useful step toward developing a general expression for the free energy of reaction, ΔG_r, at *any* combination of reactant and product activities. The general expression relating the free energy of reaction to the activities of the reactants and products is

$$\Delta G_r = \Delta G_r^0 + RT \ln \frac{(D)^d (E)^e}{(B)^b (C)^c}, \tag{3.15}$$

where $R = 0.008\,314\ \mathrm{kJ\,mol^{-1}\,deg^{-1}}$ and T is temperature in Kelvin [$K = °C + 273.15$]. Converting the natural logarithm to a base ten logarithm ($\ln(z) = 2.30 \log_{10}(z)$) gives

$$\Delta G_r = \Delta G_r^0 + 2.30RT \log_{10} \frac{(D)^d (E)^e}{(B)^b (C)^c}. \tag{3.16}$$

The quotient of activities in Eq. (3.15) is referred to as the *ion activity product*, Q, and is the only variable (at a fixed temperature and pressure) that determines ΔG_r. Note that if the activities of all the reactants and products in Eq. (3.16) are equal to unity, then $Q = 1$ and the last term becomes zero. Under these conditions, $\Delta G_r = \Delta G_r^0$, which makes sense because the standard state is defined as involving all chemical species at unit activity. Although activities of pure solids are always equal to unity regardless of the amount present in the chemical system, this is not true for gases (even if pure) and aqueous chemical species. Gas activities, called *fugacity* (see later), are presented in units of pressure, whereas activities of dissolved chemical species are given in units such as moles of solute per liter (or kilogram) of solution. In both cases, measured concentrations have to be corrected to corresponding activities by using the appropriate activity coefficients.

3.3.2 Thermodynamic equilibrium

Thermodynamic equilibrium is by definition the state of minimal free energy for a reaction system at a given temperature and pressure. Under these conditions $\Delta G_r = 0$ and no free energy is available in the system to do work. It follows from Eq. (3.16) that at equilibrium

$$\Delta G_r^0 = -2.30RT \log_{10} \frac{(D)^d (E)^e}{(B)^b (C)^c}. \tag{3.17}$$

Since ΔG_r^0, R and T are constants at any given temperature, so is Q. Thus, in the particular case of equilibrium, Q is equal to a fixed value that is referred to as the *thermodynamic equilibrium constant* (K). Therefore it follows by identity that

Table 3.4. *ΔG_r, Q, K, and Ω values before, at, and after thermodynamic equilibrium*

State	ΔG_r	Q vs. K	Ω	Prediction
Before equilibrium	<0	$Q<K$	<1	Reaction *might* occur as written
At equilibrium	0	$Q \equiv K$	1	No net reaction can occur
After equilibrium	>0	$Q>K$	>1	Reaction cannot spontaneously occur as written

$$\Delta G_r^0 = -2.30RT \log_{10} K. \tag{3.18}$$

At 25 °C (298.15 K), Eq. (3.18) becomes

$$\Delta G_r^0 = -5.70 \times \log_{10} K. \tag{3.19}$$

Note that even though K is a constant in Eqs. (3.18) and (3.19), the individual activities of D, E, B, and C can vary. The major advantage of these equations is that the K for any balanced chemical reaction can be directly calculated from ΔG_r^0, which in turn can be determined by plugging ΔG_f^0 values from thermodynamic tables into Eq. (3.14). In a chemical system at equilibrium the individual activities of reactants and products are closely constrained by this equation.

The sign alone of ΔG_r is a useful indicator of whether a specific chemical reaction can spontaneously occur (i.e. without addition of free energy) in the direction *as written*. The reaction direction that is being tested must be clearly defined because it determines which expressions occur in the numerator and denominator of Q, and hence the sign of ΔG_r. Given this constraint, a negative ΔG_r value indicates that the reaction can occur spontaneously as written, whereas a positive sign indicates that the spontaneous reaction of reactants to products is energetically impossible. A ΔG_r value of zero indicates that the chemical system is in a state of *equilibrium*, where no free energy is available and no *net* reaction possible. The *reaction quotient*, Ω, is defined as Q/K and is helpful in indicating whether there is a deficit ($\Omega < 1$) or surplus of reactants versus products in a chemical system. These three conditions of ΔG_r, and the corresponding relationships of Q, K, and Ω, are summarized in Table 3.4. The main caveat to keep in mind when making predictions for the feasibility of a chemical reaction is that a thermodynamically favored reaction may not in fact occur because of unfavorable reaction kinetics that are not predictable. Thus the most definitive application of ΔG_r is to identify chemical reactions that cannot occur spontaneously as written at the given reactant and product activities.

Free energy relations are illustrated in terms of reaction extent in Fig. 3.7 using the same generalized reaction of B and C to form D and E as in Eq. (3.13). Equation (3.16) can be rewritten in parallel construction as a function of K and Q:

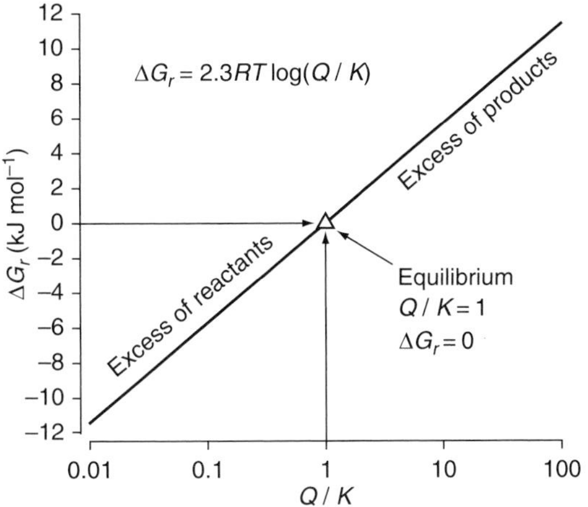

Figure 3.7. A schematic illustration of the change in free energy of a reaction, ΔG_r, as a function of the reaction progress. The *x*-axis is the quotient of the ion product, Q, and the equilibrium constant, K.

$$\Delta G_r = -2.30\ RT \log_{10} K + 2.30\ RT \log_{10} \frac{(\mathrm{D})^d (\mathrm{E})^e}{(\mathrm{B})^b (\mathrm{C})^c}. \tag{3.20}$$

The two terms to the right can be combined to generate the alternate form

$$\Delta G_r = 2.30\ RT \log_{10}(Q/K). \tag{3.21}$$

This relation is plotted in Fig. 3.7. Equilibrium ($\Delta G_r = 0$) corresponds to the point at which $Q = K$. Prior to equilibrium, there is a deficit of products versus reactants ($Q < K$) that corresponds to a negative ΔG_r and the potential to react until Q increases to equal K. If the system is in a state where $Q > K$ and ($\Delta G_r > 1$), reaction to the right will only increase Q and move the reaction system toward an even greater free energy excess versus the reference point of ΔG_r^0, so this cannot occur. Rather, the reaction is driven to the left until $Q = K$.

3.4 Equilibrium constraints on chemical activities

3.4.1 Speciation of ions in seawater

Because reactions among ionic species in solution are rapid, thermodynamic calculations are used to constrain the activities of dissolved chemical species at equilibrium. Garrels and Thompson (1962) were the first to calculate the speciation of the major ions in seawater by determining the extent to which each species is involved in ion pairing with each counter-ion. This information is necessary to establish the percentages of free major ions available in chemical equilibrium calculations. This section presents an example of how such multiple equilibrium systems can be constrained.

The example system consists of ions released by the complete dissolution of 1.00 mol each of $MgSO_{4(s)}$ and $CaF_{2(s)}$ in 1.00 l of solution at 25 °C and 1 atm of pressure, resulting in concentrations of total Mg, SO_4 and Ca of 1 molar, M, and total F of 2 M. The goal is to use thermodynamic information to determine the concentrations of the four dissolved free ions, Mg^{2+}, Ca^{2+}, SO_4^{2-}, and F^-, plus all four dissolved ion pairs that they form, $MgSO_4^0$, MgF^+, $CaSO_4^0$, and CaF^+. We will assume that these ions form no precipitates or complexes in addition to simple ion pairs (meaning there are no entities like MgF_2^0 or CaF_2^0). Given that there are eight unknown concentrations, it will be necessary to generate eight independent equations describing the amounts and interactions of these ions. Half of the needed constraints can be derived from mass balance for concentrations of the four ions (excluding oxygen).

$$[Mg]_t = [Mg^{2+}] + [MgSO_4^0] + [MgF^+] \tag{3.22}$$

$$[Ca]_t = [Ca^{2+}] + [CaSO_4^0] + [CaF^+] \tag{3.23}$$

$$[SO_4]_t = [SO_4^{2-}] + [MgSO_4^0] + [CaSO_4^0] \tag{3.24}$$

$$[F]_t = [F^-] + [MgF^+] + [CaF^+]. \tag{3.25}$$

The other needed constraints can be obtained from equilibrium equations for the formation of each of the four ion pairs:

$$Mg^{2+} + SO_4^{2-} \rightleftarrows MgSO_4^0 \tag{3.26}$$

$$Mg^{2+} + F^- \rightleftarrows MgF^+ \tag{3.27}$$

$$Ca^{2+} + SO_4^{2-} \rightleftarrows CaSO_4^0 \tag{3.28}$$

$$Ca^{2+} + F^- \rightleftarrows CaF^0. \tag{3.29}$$

For example, the equilibrium constant for the formation of $[MgSO_4^0]$ can be determined by entering standard free energies of formation for Mg^{2+}, $MgSO_4^0$, and SO_4^{2-}, into Eq. (3.12) to obtain the standard free energy of the ion pair formation reaction, and then by using that ΔG_r^0 value in Eq. (3.18) to determine a $K_{MgSO_4^0} = 10^{2.36}$.

$$K_{MgSO_4^0} = \frac{(MgSO_4^0)}{(Mg^{2+})(SO_4^{2-})} = \frac{[MgSO_4^0]\gamma_{MgSO_4^0}}{[Mg^{2+}]\gamma_{Mg^{2+}}[SO_4^{2-}]\gamma_{SO_4^{2-}}}. \tag{3.30}$$

Note that single-ion activity coefficients are used to relate concentrations to the equilibrium constant. The only interactions inhibiting the activities of an ion and ion pair from achieving values equal to their concentration are the long-range charge interactions. In this calculation it is assumed for simplicity that the ions have the same single-ion activity coefficients as in seawater (see Table 3.5). Activity coefficients for neutral ion pairs are $\gamma = 1.13$, and values for singly charged ion pairs are similar to individually charged cations, $\gamma = 0.68$. Fluoride ion is assumed to have an activity coefficient equal to that of Cl^-, $\gamma_{F^-} = 0.63$. With the activity coefficients evaluated, one can determine the concentration ratio on the right side of Eq. (3.30):

Table 3.5. *Concentrations, single-ion activity coefficients, γ_i, percent of the ion that is free of ion pairing, and the total ion activity coefficient, $\gamma_{i,T}$, of the major seawater ions using the ion pairing model*

Ion	mol kg^{-1} [a]	γ_i	% free	$\gamma_{i,T}$
Na^+	0.4691	0.71	98	0.69
K^+	0.01021	0.63	98	0.62
Mg^{2+}	0.05282	0.29	89	0.26
Ca^{2+}	0.01028	0.26	88	0.23
Cl^-	0.54586	0.63	100	0.63
HCO_3^-	0.00186	0.68	79	0.54
SO_4^{2-}	0.02824	0.22	38	0.085
CO_3^{2-}	0.00023	0.21	14	0.029

From Millero (1996).
[a] Values are per kg of seawater.

$$\frac{[MgSO_4^0]}{[Mg^{2+}][SO_4^{2-}]} = \frac{(0.29)(0.22)}{1.13} \times 10^{2.36} = 12.9 \tag{3.31}$$

and therefore

$$[MgSO_4^0] = 12.9 \times [Mg^{2+}][SO_4^{2-}]. \tag{3.32}$$

The same procedure can be used to determine the concentrations of the other three ion pairs in terms of the known K for their formation and the concentrations of the two component ions

$$[MgF^+] = 10.2 \times [Mg^{2+}][F^-] \tag{3.33}$$

$$[CaSO_4^0] = 10.3 \times [Ca^{2+}][SO_4^{2-}] \tag{3.34}$$

$$[CaF^+] = 1.8 \times [Ca^{2+}][F^-]. \tag{3.35}$$

At this point, the above four equilibrium-based relations, Eqs. (3.32) to (3.35), and the four earlier mass balance relations, Eqs. (3.22) to (3.25), can be combined to give the following four equations, within which the unknown concentrations of the ion pairs have been eliminated:

$$[Mg]_t = [Mg^{2+}] + 12.9 \times [Mg^{2+}][SO_4^{2-}] + 10.2 \times [Mg^{2+}][F^-] = 1.0\,\text{M} \tag{3.36}$$

$$[Ca]_t = [Ca^{2+}] + 10.3 \times [Ca^{2+}][SO_4^{2-}] + 1.8 \times [Ca^{2+}][F^-] = 1.0\,\text{M} \tag{3.37}$$

$$[SO_4]_t = [SO_4^{2-}] + 12.9 \times [Mg^{2+}][SO_4^{2-}] + 10.3 \times [Ca^{2+}][SO_4^{2-}] = 1.0\,\text{M} \tag{3.38}$$

$$[F]_t = [F^-] + 10.2 \times [Mg^{2+}][F^-] + 1.8[Ca^{2+}][F^-] = 2.0\,\text{M}. \tag{3.39}$$

Because the numbers of unknowns and equations now are equal, the above equations can be solved for the four unknown ion concentrations. This can be done by iteration, where one makes an initial guess at the concentration of the anions F^- and SO_4^{2-} and then resets the concentration of free ion with successive loops. For more complicated systems a computer program to find the solution to simultaneous linear equations is more appropriate. The results for the free ions, $[Mg^{2+}] = 0.07$ M, $[Ca^{2+}] = 0.17$ M, $[SO_4^{2-}] = 0.30$ M and $[F^-] = 1.0$ M, and for the ion pairs, $[MgSO_4^0] = 0.26$ M, $[CaSO_4^0] = 0.52$ M, $[MgF^+] = 0.67$ M and $[CaF^+] = 0.31$ M, indicate that ion pairs dominate in this very concentrated solution and that the cations form ion pairs much more readily than anions.

Garrels and Thompson (1962) performed essentially the same calculation for the major ions in seawater. Their calculations were based on the assumption that the major ions in seawater associate only into ion pairs consisting of 1:1 cation : anion complexes. The activity coefficients used in this calculation and the resulting percentages of ion pairing (Table 3.5) have largely survived the test of time and can be broadly applied in thermodynamic descriptions of the major seawater ions and their reactions.

In complex electrolytes with relatively uniform major ion concentration, such as seawater, there are two commonly used types of equilibrium constant. In addition to the thermodynamic constant described above, apparent equilibrium constants, K', are often used. Apparent equilibrium constants are defined in terms of the *total concentrations* of chemical species. The apparent equilibrium expression for the reaction of A and B to give AB is

$$K' = \frac{[AB]_T}{[A]_T[B]_T}. \tag{3.40}$$

It follows that

$$K = \frac{\gamma_{T,AB}[AB]_T}{\gamma_{T,A}[A]_T\gamma_{T,B}[B]_T} = \frac{\gamma_{T,AB}}{\gamma_{T,A}\gamma_{T,B}} \times \frac{[AB]_T}{[A]_T[B]_T} = \frac{\gamma_{T,AB}}{\gamma_{T,A}\gamma_{T,B}} \times K', \tag{3.41}$$

where gammas, γ_T, are total activity coefficients that consider both electrostatic interactions and ion pairing. Although K and K' values for a given chemical reaction can be related by knowledge of the total activity coefficients of all chemical species involved, the advantage of apparent constants is that they can be determined experimentally in terms of concentrations and used without having to evaluate total activity coefficients. The disadvantage is that K' determined by experiment is valid only for the temperature, pressure and ionic composition of the experimental solution. Because the major ions in seawater have relatively constant concentrations, and highly complex interactions among themselves and with minor ions, equilibrium equations based on measured concentrations and apparent equilibrium constants are most often used in oceanography. This convention is used in the following chapter on the carbonate system, but the subscript $_T$ on the concentration brackets, [ ], is deleted for convenience.

3.4.2 Equilibrium among coexisting phases: the phase rule

So far we have considered chemical equilibrium in solution only. The thermodynamic constraints can be applied to reactions among all phases, including gas partitioning between vapor and liquid phases, and reactions between solid and dissolved phases. A thermodynamically related tool for determining the quantitative constraints among phases in an equilibrium system is the *Gibbs phase rule*,

$$F = C + 2 - \bar{P}. \tag{3.42}$$

In this equation F represents the *degrees of freedom* within the system, that is the number of constraining independent variables such as concentrations, temperature, and pressure. $\bar{P}$ represents the *total number of phases* such as a gas, a homogeneous solution, or a homogeneous solid. Variable C corresponds to the *number of components* in the solution, which can be thought of as the minimal number of minerals, molecules and ions that are needed to fully create the mixture. In the previous ion pairing example, $C = 9$ (the eight different ions plus water), and $\bar{P} = 1$ (the homogeneous liquid solution without a gaseous phase being included), so that $F = 10$. If the temperature and pressure of the solution are fixed, F is reduced by two to a value of eight. This means that eight independent equations are necessary to fix each chemical component at a given activity (or concentration). In the previous ion pairing example, those constraints came from mass balance and equilibrium relations. Had another component, for example CaF_2^0, existed in appreciable amounts in the solution at equilibrium, the previous calculation results would have been in error. This problem could be addressed by using thermodynamic data to calculate the equilibrium constant for formation of CaF_2^0, which (along with the corresponding activity coefficient) would add the needed additional constraint. In effect, fixing the temperature and pressure of the system locks in the values of the equilibrium constants.

Another example that illustrates the constraining effect of adding additional phases is that for pure water. If only pure water is present in a system ($C = 1$) in only one phase (either solid, liquid or gas) ($\bar{P} = 1$), then $F = 1 + 2 - 1 = 2$. In this case, the two degrees of freedom are system temperature and pressure (T and P). When only one pure water phase is present, the system is confined only to T and P values within the liquid water region of the phase diagram (Fig. 3.8). If, however, the number of phases is increased to two, then only one degree of freedom remains. On the phase diagram for water, the simultaneous presence of two phases would constrain all T and P values to fall on one of the three lines separating the gas–liquid, gas–ice or ice–liquid regions. Along any of these lines, setting T fixes P and vice versa. If all three water phases are present at one time, the Gibbs phase rule indicates zero degrees of freedom, and hence fixes both T and P values. On the phase diagram, this unique

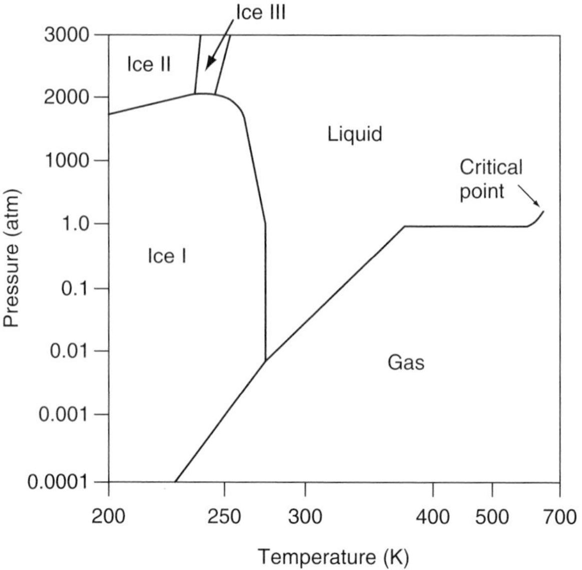

Figure 3.8. The phase diagram for pure water, demonstrating the relationships among the degrees of freedom in T and P space and the number of possible phases. See text.

circumstance corresponds to the triple point of water, which occurs at $T = 273.2$ K and $P = 0.006$ atm vapor pressure (Fig. 3.8).

3.4.3 Solid–solution equilibrium: $CaCO_3$

An example of how a free energy calculation can be used to predict whether a spontaneous reaction between a solution and solid phase is possible is the oceanographically important process of calcite ($CaCO_3(s)$) mineral dissolution. The question is whether this reaction

$$CaCO_3(s) \rightleftarrows Ca^{+2} + CO_3^{2-} \tag{3.43}$$

is possible in warm surface seawater (1 atm and 25 °C). To address this question it is first necessary to look up ΔG_f^0 values for all three chemical species and calculate the corresponding ΔG_r^0. In this case

$$\begin{aligned}\Delta G_r^0 &= 1 \times \Delta G^0_{f,Ca^{2+}} + 1 \times \Delta G^0_{f,CO_3^{2-}} - 1 \times \Delta G^0_{f,CaCO_3(s)} \\ &= -553.54 - 527.9 + 1128.8 = +47.4\,\text{kJ}.\end{aligned} \tag{3.44}$$

Once this value of ΔG_r^0 is substituted into the equation for the free energy of reaction,

$$\Delta G_r = 47.4 + 5.70 \log_{10} \frac{(Ca^{2+})^1 (CO_3^{2-})^1}{(CaCO_3(s))^1}, \tag{3.45}$$

the only remaining unknowns are the activities of the three chemical species. These must be obtained from the corresponding concentrations, activity coefficients, and percentages of free ion in seawater (see Table 3.5). Because the activity of ($CaCO_3(s)$) is by definition equal to unity (assuming this solid mineral is pure), the denominator of Q in Eq. (3.45) drops out. At this point, the activities (calculated from the information in Table 3.5) can be substituted into Eq. (3.45) to give

$$\Delta G_r = 47.4 + 5.70 \log_{10}\left[\left(\frac{\%\text{free}}{100}\right)_{Ca^{2+}} \times \gamma_{Ca^{2+}} \times [Ca^{2+}] \times \left(\frac{\%\text{free}}{100}\right)_{CO_3^{2-}} \times \gamma_{CO_3^{2-}} \times [CO_3^{2-}]\right]$$

$$= 47.4 + 5.70 \log_{10}\left[\left(\frac{88}{100}\right) \times 0.26 \times 1.03 \times 10^{-2} \times \left(\frac{14}{100}\right) \times 0.21 \times 2.3 \times 10^{-4}\right]$$

$$= 47.4 + 5.70 \log_{10}(1.59 \times 10^{-8}) = 3.0 \text{ kJ mol}^{-1}. \quad (3.46)$$

Because ΔG_r is positive ($Q > K,\ \Omega > 1$), it is impossible that the reaction (Eq. (3.43)) will go spontaneously to the right and that calcite will dissolve in surface seawater under the specified conditions (coccolithophorides can relax). In fact, surface seawater has an excess of reactants versus the equilibrium point (Fig. 3.7) and is "supersaturated" with respect to calcite. Although the reverse reaction of calcite precipitation is energetically favorable, it too does not occur readily in surface seawater because of kinetic constraints.

3.4.4 Solid–solution adsorption reactions

Adsorption thermodynamics is distinct from the solid–solution reactions discussed previously or the solution-gas equilibrium reactions to be discussed in the following section because it deals with *exchange* equilibrium between ions in solution and on the solid surface, as opposed to *reaction* equilibrium between bulk phases. Adsorption of ions at solid surfaces plays an important role in controlling the concentration of trace metals in seawater. In fact, it has been suggested that surface adsorption reactions are the predominant mechanism controlling the chemistry of the sea (Whitfield and Turner, 1979; Li, 1981). There are several different models of the metal oxide surface. One of the most successful is the Stumm-Schindler model, in which the metal oxide surface in nature is treated with the same acid–base equilibrium equations used for solutions. This model (Fig. 3.9) envisions a metal oxide surface in which metal ions are partly coordinated, leaving a residual positive charge. H_2O molecules adsorb to the surface, orienting their more negative oxygen end toward the surface. Ultimately each H_2O molecule loses a proton by *dissociative chemisorption*, resulting in a hydroxylated surface (Fig. 3.9c). The result is that the surface, designated as $\equiv S$, is complexed by OH^- ions to form, $\equiv SOH_n$, which can exchange protons just like acids in water. The acid–base (*amphoteric*) character of the surface can then be described by the same equations used for acid–base behavior in solution (see Chapter 4):

$$\equiv SOH_2^+ \overset{[K_{a,1}]^s}{\rightleftarrows} \equiv SOH + H^+ \quad (3.47)$$

$$\equiv SOH \overset{[K_{a,2}]^s}{\rightleftarrows} \equiv SO^- + H^+. \quad (3.48)$$

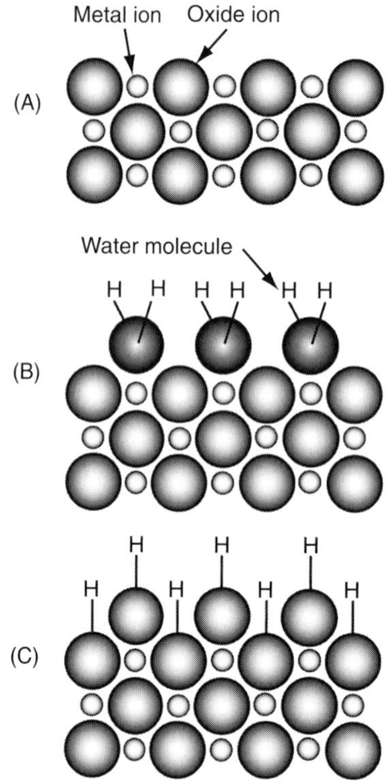

Figure 3.9. Illustration of the surface of a metal oxide where small circles are the metal atoms and large circles are oxygen atoms. (A) Surface metal atoms are not totally coordinated. (B) In water, surface metal ions coordinate water molecules. (C) Dissociative chemisorption leads to a hydroxylated surface. From Schindler and Stumm (1987).

Equilibrium constants for these reactions are indicated by K. The total capacity of the surface for adsorption of H^+ ions and the equilibrium constants are determined by potentiometric titration, which involves addition of a known amount of acid or base and then measuring the change in solution pH. The difference between the amount of titrant added and its concentration in the solution represents the amount of H^+ that has adsorbed onto, or reacted with, the surface.

It has been shown (Schindler and Stumm, 1987) that equilibrium constants characterizing the amphoteric behavior of metal oxides (the surface acidity) are linearly correlated with the equilibrium constants for the amphoteric behavior of the same metal ions in solution (acidity in solution). This finding supported the model in which the metal oxide surface is described with the same acid–base equations as the metal ions in solution, and allows the equilibrium constants for the oxide surfaces to be estimated from those determined in solution.

The next step in describing the importance of surface adsorption in this model involves the exchange affinity of the surface sites for metal ions as well as protons. In other words, metals in solution compete with protons for the oxide surface sites. Equations for this process are exactly analogous to those for similar solution reactions. Again it was demonstrated (see, for example, Schindler and Stumm, 1987) that the equilibrium constants determined for the competition between protons and metals for the metal oxide surface are linearly correlated with the equilibrium constants describing the same reactions in solution. Further discussion of the adsorption model for oxide surfaces is beyond our purposes here and presented in elegant detail in books about aquatic chemistry (see, for example, Stumm and Morgan, 1996). These arguments are used later in the discussion of the dissolution rates of minerals (Chapter 9).

3.4.5 Gas equilibrium between the air and water

Thermodynamic equilibrium relations are used to define the partitioning of gases between the vapor and liquid phases. The amount of gas in the vapor phase is most often expressed as pressure (in units of atmospheres, bars or Pascals). One atmosphere is equal to 1013.25 millibars pressure (mbar) and 101.325 kilopascals (kPa; $1\ \text{bar} = 10^5\ \text{Pa}$). In a mixture of gases the *partial pressure*, p_i, of an individual gas, i, is its fraction of the total gas pressure. The total pressure of gases in the atmosphere, P_{atm}, is equal to the sum of the partial pressures of the individual gases

$$P_{atm} = p^a_{N_2} + p^a_{O_2} + p^a_{H_2O} + p^a_{Ar} + p^a_{CO_2} + \cdots \tag{3.49}$$

where the superscript ^{a} indicates atmosphere.

The *mole fraction*, X, of a gas in the atmosphere is defined as the number of moles of that gas per total moles of the atmospheric gases ($mol_g\ mol_a^{-1}$) in the absence of water vapor so that it does not depend on altitude. For an ideal gas the mole fraction and volume fraction are identical and the units are frequently presented as ($cm_g^3\ m_a^{-3}$) or parts per million by volume (ppmv). The atmospheric pressure and mole fraction of gas, C, are thus related by the partial pressure of water vapor, p_{H_2O},

$$p^a_C = X_C \times (P_{atm} - p_{H_2O}). \tag{3.50}$$

In a dry atmosphere, the partial pressure and mole fraction are equal. Mole fractions for the major atmospheric gases are presented in Table 3.6, and the temperature dependence of p_{H_2O} at equilibrium with seawater, $p^s_{H_2O}$, in Fig. 3.10. At saturation equilibrium, water vapor has the third highest partial pressure in the atmosphere.

The amount of gas that the solvent water will accommodate at thermodynamic equilibrium is represented by *Henry's Law*, in which the concentration in the water, C ($mol\ kg^{-1} atm^{-1}$) and *fugacity*, f, in the gas phase are related via the *Henry's Law coefficient*, K_H ($mol\ kg^{-1}\ atm^{-1}$),

$$C = K_{H,C} \times f_C. \tag{3.51}$$

The fugacity and partial pressure of a gas are related in the same way as activities and concentrations in solution. Interaction of molecules with each other in a real gas diminishes the reactivity of an individual gas slightly, creating an effective partial pressure called the *fugacity*. As the gas pressure approaches zero, the pressure and fugacity are equal. The interference effect on gases in the atmosphere, however,

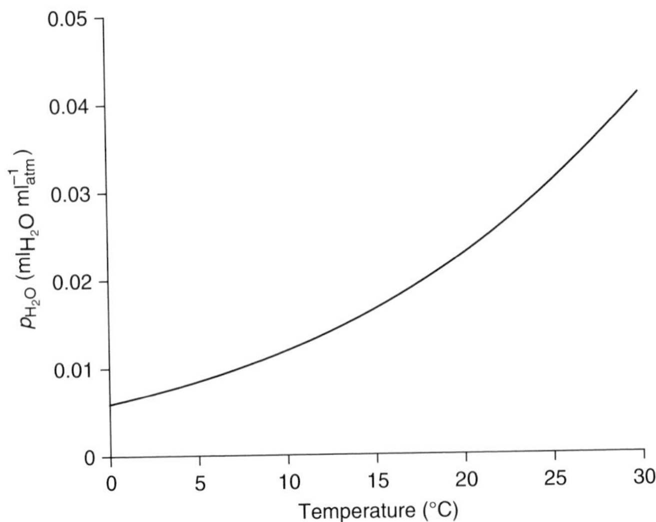

Figure 3.10. The partial pressure of water vapor in equilibrium with pure water as a function of temperature. From DoE (1994).

is much less than with ions in water. Except for CO_2, the fugacities of the major gases in the atmosphere are greater than 99.9% of their respective partial pressures. Klots and Benson (1963) suggest a value for $(f_i - p_i)/p_i$ for N_2 of 0.000 14–0.000 4 between 2 and 27 °C. Weiss (1974) calculates this value for CO_2 to be much larger but still less than 1%: 0.003–0.004 4 between 0 and 30 °C. For this reason fugacity and partial pressure tend to be used interchangeably for most atmospheric gases.

Interpreting the Henry's Law coefficient can be confusing because of the number of different units used in the literature. In this text it is presented in units of moles per kilogram so that the effect of pressure on volume in the ocean is normalized. Other units that are often used are molar (mol l^{-1}) and volume fraction at standard temperature (0 °C) and pressure (1 atm) (ml l^{-1}, STP). Be careful: STP for gases and "standard conditions" for free energies are not the same! The pressure terms are identical, but the temperatures are 0 and 25 °C, respectively. This is one of the casualties of an old science that evolved from many different laboratories. Since the volume of a mole of ideal gas at STP is exactly 22.414 l, there is a direct relation between moles of gas and milliliters (STP) of gas. Another potential confusion is that the Henry's Law relation is sometimes referred to as the reciprocal of the value given here, e.g. $1/K_H$. We can only say that the bulk of marine literature follows the definition used in Eq. (3.51), and one should make careful note of the units when using this constant.

Values of K_H in seawater at 20 °C and 1 atm are presented for several gases in Table 3.6. The values for the six most concentrated gases in the atmosphere were derived by using the regression equations presented in Table 3A1.1; others are taken from the literature. Note that the values of K_H are all within a factor of 10 of each other except for CO_2 and N_2O, which are much more soluble than the other gases.

Table 3.6. *Solubilities of the major atmospheric gases in seawater (s = 35) at one atmosphere pressure and 20 °C*

X_C is the mole fraction in a dry atmosphere (Glueckauf, 1951); $K_{H,C}$, the Henry's Law coefficient; C^s, the concentration in seawater at saturation equilibrium with the atmosphere. Saturation concentrations and Henry's Law coefficients for N_2, O_2, Ar, CO_2, Ne and He are calculated by using the equations in Table 3A1.1. Values for K_r, CH_4 and N_2O are from the compilation in Wanninkhof (1992). No correction was made here for the difference between the volume of the solvent and the solution in β and α (see text).

Gas	X_C	$K_{H,C}$	$C^{s\,a}$	β^b	α^c
	$(mol_g\ mol_{atm}^{-1})$	$(mol\ kg^{-1} atm^{-1})$	$(\mu mol\ kg^{-1})$	$(cm_g^3)/(cm_{sw}^{-3})$	
N_2	7.8084×10^{-1}	5.51×10^{-4}	4.21×10^{2}	1.27×10^{-2}	1.36×10^{-2}
O_2	2.0946×10^{-1}	1.10×10^{-3}	2.25×10^{2}	2.53×10^{-2}	2.71×10^{-2}
Ar	9.34×10^{-3}	1.21×10^{-3}	1.10×10^{1}	2.78×10^{-2}	2.98×10^{-2}
CO_2	3.65×10^{-4}	3.24×10^{-2}	1.16×10^{1}	7.44×10^{-1}	7.98×10^{-1}
Ne	1.818×10^{-5}	3.84×10^{-4}	6.83×10^{-3}	8.82×10^{-3}	9.47×10^{-3}
He	5.24×10^{-6}	3.29×10^{-4}	1.66×10^{-3}	7.47×10^{-3}	8.01×10^{-3}
Kr	1.14×10^{-6}	2.20×10^{-3}	2.44×10^{-3}	5.05×10^{-2}	5.42×10^{-2}
CH_4	1.6×10^{-6}	1.21×10^{-3}	1.89×10^{-3}	2.78×10^{-2}	2.97×10^{-2}
N_2O	5.0×10^{-7}	2.34×10^{-2}	1.14×10^{-2}	5.37×10^{-1}	5.77×10^{-1}

[a] $C^s = K_{H,i} f_C$; the fugacity is assumed equal to the partial pressure, p, except for CO_2.
[b] The Bunson coefficient, $\beta = K_H(RT_{STP})\rho$, where $R = 0.082\,0571$ atm deg^{-1} mol^{-1}; $T_{STP} = 273.15$; ρ is the density of seawater (at 20 °C and 35 ppt, $\rho = 1.0248$ kg l^{-1}).
[c] The Ostwald solubility coefficient, $\alpha = K_H\, RT\rho = \beta(T/T_{STP}) = \beta T/273.15$.

Equation (3.51) refers to the general relation between the fugacity of a gas in solution and its concentration. We give f a superscript, w, to indicate that it refers to the water phase.

$$[C] = K_{H,C} \times f_C^{w}. \tag{3.52}$$

The fugacity of a gas in water is usually calculated from measurements of the gas concentration by using the above relation. The concentration of a gas in surface waters is at solubility equilibrium with the atmosphere (saturation) when the fugacities of the atmosphere and water are equal

$$f^{a} = f^{w} \tag{3.53}$$

so that

$$C^{sat} = K_{H,C} \times f_C^{a}. \tag{3.54}$$

The superscript, sat, indicates that this is the concentration of the gas at saturation equilibrium with the atmosphere and has sometimes been called the air solubility (Weiss, 1971). Concentrations at saturation equilibrium with air are presented for the major atmospheric gases at 20 °C and 1 atm in Table 3.6.

The solubilities of gases are sometimes presented as the *Bunson coefficient*, β, or the *Ostwald solubility coefficient*, α. These values are directly related to the Henry's Law coefficient (see Table 3.6), but they present the solubility in units that have some advantages conceptually. The Bunson coefficient is defined as the volume of gas at STP dissolved in a unit volume of solvent at some temperature, T, when the total pressure of the gas and its fugacity are 1 atm. The Ostwald solubility coefficient is the same as the Bunson coefficient except that the gas volume is not at STP, but rather at the same temperature as the solvent water. These quantities represent equilibrium between the pure gas and water where the solvent and solute are presented in the same units. One can envision these constants as air–water partition coefficients for a pure gas at equilibrium with water, when the gas phase is one atmosphere and the gas and water reservoirs have equivalent volumes. Bunson and Ostwald solubility coefficients for the major atmospheric gases are presented in the last two columns of Table 3.6. At 20 °C, when overlying gas and liquid water phases have equal volumes, only about 1% of pure N_2 gas and 2%–3% of either Ar or O_2 gas exists in the water at equilibrium. The solubility of the rare gases in these units increases with molecular mass from 0.75% for He to 5% for Kr and brackets the values for the most abundant gases N_2 and O_2. The anomalies are CO_2 and N_2O. On an equal volume basis the amount of CO_2 residing in the water at 20 °C is about 80% of that residing in the atmosphere. (This is the value in the absence of chemical reaction with water.)

The only tricky thing about Bunson and Ostwald solubility coefficients is that they represent a volume of gas per volume of solvent (not solution). Because gases increase the volume of the solution when they dissolve into it, a correction has to be made for this difference. The correction is significant and on the order of 0.14% (Weiss, 1971). The values presented in Table 3.6 have not been corrected for this effect, and since this is a potential point of confusion, we will use the Henry's Law coefficient most often in this book.

The temperature dependences of the Henry's Law coefficients of the different gases listed in Table 3.6 are quite variable (Fig. 3.11). Helium, the least soluble noble gas, has very little solubility temperature dependence between 0 and 30 °C. On the other hand, Kr, the second most soluble of the non-radioactive noble gases, is much less soluble at higher temperatures. More details about gas solubilities are presented in the chapter on air–sea gas exchange (Chapter 10). Another notable aspect of the temperature dependence of the gas solubilities is that they are not linear. Thus, mixing between parcels of water of different temperatures at saturation equilibrium with the atmosphere results in a mixture that is supersaturated. This effect has been observed for noble gases in the ocean and may ultimately have a utility as a tracer of mixing across density horizons.

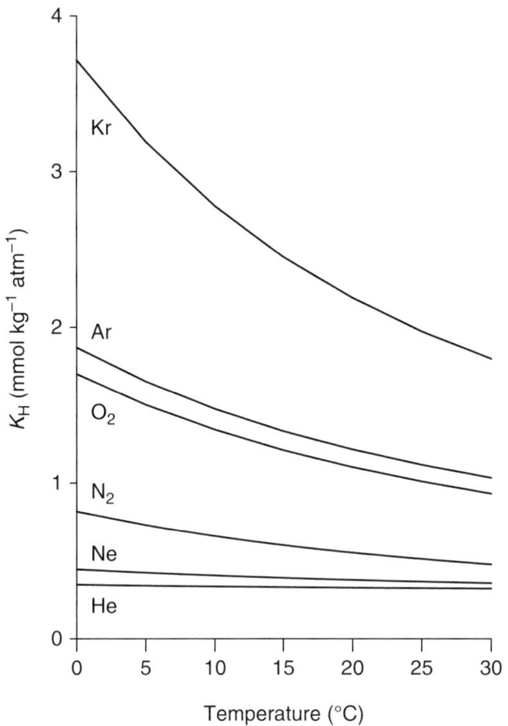

Figure 3.11. The Henry's Law coefficient, K_H (mol kg^{-1} atm^{-1}) for the noble gases, O_2 and N_2 in seawater (1 atm and $S = 35$) as a function of temperature. From the equations in Table 3A1.1.

3.5 Redox reaction basics

Chemical processes involving simultaneous *reduction* (electron gain) and *oxidation* (electron loss) are referred to as reduction–oxidation reactions, or *redox reactions* for short. Redox reactions are of wide interest because they include both photosynthesis and respiration and thus fuel essentially all life processes. Redox reactions are discussed here separately from other chemical reactions, because they involve an additional step in derivation of their free energy expression and are especially prone to slow approaches to equilibrium.

3.5.1 The standard electrode potential, E_h^0, and pe

A simple redox reaction in which dissolved copper-II ion is converted to elemental copper and elemental zinc is converted to the corresponding zinc-II cation is

$$Cu^{2+} + Zn^0 \rightleftharpoons Cu^0 + Zn^{2+}. \tag{3.55}$$

This complete reaction can be thought of as consisting of two simultaneous half-reactions

$$Zn^0 \rightarrow Zn^{2+} + 2e^- \tag{3.56}$$

$$Cu^{2+} + 2e^- \rightarrow Cu^0. \tag{3.57}$$

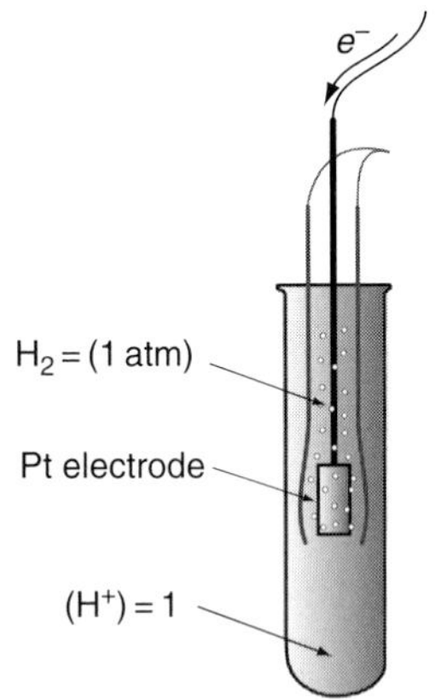

Figure 3.12. A schematic diagram of the Standard Hydrogen Electrode (SHE), which consists of a platinum electrode immersed in an aqueous solution at a pH of 1 in equilibrium with 1 atmosphere of hydrogen gas. The activity of H_2 is 1 atm and by convention the activity of $(H^+) = 1$.

In the first half-reaction Zn^0 loses electrons and by definition is oxidized. Note that the process of oxidation is defined in terms of electron loss, and does not necessarily involve the element oxygen in any form. In the second half-reaction Cu^{2+} accepts electrons and is said to be reduced. The two chemical species that make up each half-reaction are referred to as a *couple* (i.e. Zn^0/Zn^{2+} and Cu^{2+}/Cu^0) that interconvert by gain or loss of electrons. The separation of full redox reactions into half-reactions is largely conceptual because electrons are directly exchanged among reacting species and thus an oxidation cannot occur without a simultaneous reduction.

Nevertheless, it is convenient to be able to compare the relative affinities of redox couples for electrons as a tool for predicting reaction directions. This comparison is typically done by comparing the electron affinity of a couple to that of the Standard Hydrogen Electrode (SHE). As illustrated in Fig. 3.12, the SHE consists of a platinum electrode (redox reaction site) that is immersed in a water solution at 25 °C containing H^+ ions at an activity of 1 (pH = 0). Pure hydrogen gas is bubbled around the platinum electrode at a partial pressure of 1 atm, so that H_2 also has an activity of 1. The reaction occurring in the SHE is the reduction of H^+ to gaseous H_2:

$$2H^+ + 2e^- \rightleftarrows H_2\,(g). \tag{3.58}$$

By convention, the SHE is defined as having a *standard electrode potential*, E_h^0, of zero volts (V) at standard conditions (25 °C and 1 atm). The standard electrode potentials of other couples are similarly determined as reduction half-reactions at unit activity versus the SHE. If the E_h^0 for a given half-reaction is > 0, that couple has the potential to oxidize the SHE. A negative E_h^0 indicates a couple that can reduce the SHE. Tables of redox half-reactions and the corresponding E_h^0 values can be found in Stumm and Morgan (1996). Table 3.7 gives E_h^0 values and related parameters from these sources for a dozen environmentally important redox reactions.

The general free energy expression (Eq. (3.16)) can be extended to the description of redox reactions by establishing the relations between E_h^0 and ΔG_r^0. This can be done by the simple equation

$$\Delta G_r^0 = -nFE_h^0, \tag{3.59}$$

where ΔG_r^0 here is in joules (not kJ) and F is Faraday's constant (96 500 coulombs per mol) and equal to the electrical charge of one mole of electrons. The coefficient n represents the number of moles of electrons transferred in the balanced reaction. In essence, Eq. (3.59) can be thought of as representing the amount of free energy needed (or released) when nF electrons are passed through a standard electrode potential of E_h^0 for the couple in question. Dividing all terms in Eq. (3.15) by $-nF$ gives

$$-\frac{\Delta G_r}{nF} = -\frac{\Delta G_r^0}{nF} - \frac{RT}{nF} \times \ln \frac{(\mathrm{D})^d(\mathrm{E})^e}{(\mathrm{B})^b(\mathrm{C})^c}. \tag{3.60}$$

Table 3.7. *Common redox half-reactions and corresponding E_h^0, $E_{h,water}^0$, pe^0 and pe_{water}^0 [a] values*

Half-reaction	E_h^0	$E_{h,water}^0$ [a]	pe^0	pe_{water}^0
$\frac{1}{4}O_2(g) + H^+ + e^- \longrightarrow \frac{1}{2}H_2O$	+1.23	+0.81	+20.75	+13.75
$\frac{1}{5}NO_3^- + \frac{6}{5}H^+ + e^- \longrightarrow \frac{1}{10}N_2(g) + \frac{3}{5}H_2O$	+1.25	+0.75	+21.05	+12.65
$\frac{1}{2}MnO_2 + 2H^+ + e^- \longrightarrow \frac{1}{2}Mn^{2+} + H_2O$	+1.29	+0.46	+21.80	+7.80
$\frac{1}{2}NO_3^- + H^+ + e^- \longrightarrow \frac{1}{2}NO_2^- + \frac{1}{2}H_2O$	+0.84	+0.42	+14.15	+7.15
$\frac{1}{8}NO_3^- + \frac{5}{4}H^+ + e^- \longrightarrow \frac{1}{8}NH_4^+ + \frac{3}{8}H_2O$	+0.88	+0.36	+14.90	+6.15
$FeOOH(s) + 3H^+ + e^- \longrightarrow Fe^{2+} + 2H_2O$	+0.94	− 0.30	+16.0	−5.0
$\frac{1}{2}CH_2O + H^+ + e^- \longrightarrow \frac{1}{2}CH_3OH$	+0.24	− 0.18	+3.99	− 3.01
$\frac{1}{8}SO_4^{2-} + \frac{9}{8}H^+ + e^- \longrightarrow \frac{1}{8}HS^- + \frac{1}{2}H_2O$	+0.25	− 0.22	+4.25	− 3.75
$\frac{1}{8}CO_2(g) + H^+ + e^- \longrightarrow \frac{1}{8}CH_4(g) + \frac{1}{4}H_2O$	+0.17	− 0.24	+2.87	− 4.13
$\frac{1}{6}N_2(g) + \frac{4}{3}H^+ + e^- \longrightarrow \frac{1}{3}NH_4^+$	+0.28	− 0.28	+4.68	− 4.68
$H^+ + e^- \longrightarrow \frac{1}{2}H_2(g)$	0.00	− 0.41	0.00	− 7.00
$\frac{1}{4}CO_2(g) + H^+ + e^- \longrightarrow \frac{1}{4}CH_2O + \frac{1}{4}H_2O$	− 0.071	− 0.48	− 1.20	− 8.20

[a] Values are calculated for unit activities of oxidant and reactant in neutral water of pH = 7.0 at 25 °C and 1 atm. Half-reactions are listed in order of decreasing pe_{water}^0, and hence of decreasing oxidizing power at pH 7. E_h values are in volts.

Substituting from Eq. (3.59) gives the electrode potential, E_h, counterpart of the free energy expression

$$E_h = E_h^0 - \frac{RT}{nF} \times \ln \frac{(D)^d (E)^e}{(B)^b (C)^c}. \tag{3.61}$$

At 25 °C and on a $\log_{10}$ basis, with $R = 8.314\,J\,mol^{-1}\,deg^{-1}$, the above expression becomes

$$E_h = E_h^0 - \frac{0.0592}{n} \times \log_{10} \frac{(D)^d (E)^e}{(B)^b (C)^c}. \tag{3.62}$$

Expressions (3.61) and (3.62) are different forms of the Nernst equation, which gives the electrode potential in volts of a *reduction* half-reaction as a function of E_h^0 for that reaction and the activities of the oxidized (reactants) and reduced (products) species raised to the power of their respective coefficients in the balanced redox equation. Note that if reactants and products were reversed (as is the case in some texts) the sign on the right side would be different. If the resulting E_h is equal to zero, there is no potential to do work and the system is at equilibrium. It then follows from Eq. (3.62) that

$$E_h^0 = \frac{0.0592}{n} \times \log_{10} K. \tag{3.63}$$

Standard electrode potentials can be calculated from the balanced half-reaction, thermodynamic tables of ΔG_f^0 (to yield ΔG_r^0) and Eq. (3.59). Equation (3.63) can then be used to determine the equilibrium constant for the half-reaction. In addition, the E_h^0 for a specific half-reaction can be used in the Nernst equation, along with the activities of the various reactants and products, to determine the sign of the corresponding E_h and whether the reaction can occur spontaneously as written. A positive E_h corresponds to a negative ΔG_r (Eq. (3.59)), indicating the potential for a spontaneous reduction half-reaction (Table 3.7). The E_h for a particular half-reaction under natural conditions, however, may be quite different from those under standard conditions. This difference results primarily because environmental pHs are much higher than the value of 1 used to define standard conditions.

Since all redox reactions are written with equations that have electrons in them, one can think of the activity of the electron (e^-) as a master variable, just as the activity of the hydrogen ion (pH) is the master variable for acid–base (H^+ exchange) reactions (see Stumm and Morgan, 1996). An example is the reaction in the SHE (Eq. (3.58)):

$$E_h = E_h^0 - \frac{RT}{nF} \times \ln \frac{(H_2)^{1/2}}{(H^+)(e^-)}. \tag{3.64}$$

In the standard state the activities of H_2 and H^+ are one and at equilibrium $E_h = 0$, thus

$$\frac{E_h^0 F}{2.3RT} = -\log_{10}(e^-) = pe^0. \tag{3.65}$$

The above equation defines the *standard* pe (pe^0). The general expression for the generic reaction in Eq. (3.13) by analogy to Eq. (3.16) is

$$pe = pe^0 - \frac{1}{n}\log_{10}\frac{(D)^d(E)^e}{(B)^b(C)^c}. \tag{3.66}$$

This relation is useful because it defines the activity of electrons in the same manner as pH defines the activity of the H^+ ion. The two parameters, pe and pH, can be used as master variables to describe the stability of environmental reactions. Plots of pe versus pH are typically used to describe the environmental conditions within which specific chemical species are thermodynamically stable under a range of natural conditions (see, for example, Stumm and Morgan, 1996 and later discussion).

Redox conditions in aquatic systems are bounded by the reactions of potential electron donors and acceptors with water, much as water buffers acid/base reactions by accepting or donating protons (Chapter 4). For example, in oxic marine systems, where dissolved O_2 concentrations are measurable, the controlling redox couple (written as a reduction) is O_2–H_2O:

$$\tfrac{1}{2}O_2\ (g) + 2H^+ + 2e^- \rightleftarrows H_2O. \tag{3.67}$$

The E_h^0 for this reaction is +1.23 V ($pe^0 = 20.8$). The E_h of this half-reaction in warm (25 °C) surface seawater can be calculated by plugging typical activity values into the Nernst equation (Eq. (3.62)). In this case the activity of H_2O would be 1.0 and the activity of dissolved O_2 would be approximately 0.2 atm. The latter value can be estimated by assuming that the partial pressure of oxygen gas dissolved in surface seawater is equal to that in the atmosphere (by convention, gas concentrations are given in atmospheres and not moles). Surface seawater has a pH of ≈8 or an H^+ activity of $\approx 10^{-8}$ molar. Entering these values into the Nernst equation produces

$$E_h = E_h^0 - \frac{0.0592}{2} \times \log_{10} \frac{(H_2O)^1}{(O_2)^{1/2}(H^+)^2} \tag{3.68}$$

$$E_h = 1.23 - \frac{0.0592}{2} \times \log_{10} \frac{(1)^1}{(0.2)^{1/2}(10^{-8})^2} = 1.23 - 0.48$$
$$= +0.75 \text{ V}. \tag{3.69}$$

Note that the coefficients in reaction (3.67) could just as easily have been all divided by two to scale the stoichiometry to the exchange of a single electron (as is done in Table 3.7), without causing any change in the calculated E_h value. The E_h of water varies non-linearly with the concentration of oxygen and is not very sensitive to the O_2 concentration. For example, a 99% decrease in $[O_2]$ causes only a 4% drop in the E_h of the water. (Saturated water, 0.2 atm, has an E_h of 0.75 V; water with only 1% O_2 saturation, 0.002 atm, has an E_h of 0.72 V.) Thus, as long as measurable amounts of dissolved O_2 remain, the E_h of seawater will continue to be very positive.

The lower redox threshold for aqueous systems is also established by a gas-generating reaction involving a water constituent. In this case, the controlling half-reaction is the H^+–H_2 couple that also occurs in the SHE (Eq. (3.58)) and thus has an E_h^0 of zero. At this extreme, a stronger reducing agent (electron donor) than water will spontaneously convert protons to H_2 gas. Because the partial pressure of H_2 in the atmosphere is near zero, any generated hydrogen gas will tend to escape the liquid phase. As a simplification, and to include the fact that the partial pressure of O_2 in surface waters is roughly 0.2 atm, a concentration of one atmosphere is generally taken as the practical upper limit for both H_2 and O_2. At this partial pressure either gas forms bubbles and escapes, as occurs in electrolytic cells when strongly positive or negative voltages are introduced at platinum electrodes such as the SHE (Fig. 3.12).

Because each of the bounding redox reactions (3.58) and (3.67) involve H^+ as a reactant, this term shows up in the denominator of the Nernst equation (Eq. (3.62)) and therefore affects the E_h of both half-reactions, as is illustrated in Fig. 3.13. The three lines to the right in this figure indicate the atmospheres of O_2 present at equilibrium for different *pe* values at pH 4, 7, and 10. Similarly, the three lines to the left indicate the atmospheres of H_2 present at equilibrium that correspond to varying *pe* at the same three pH values. Beginning with the pair of lines for pH = 7 it can be seen that a *pe* greater than *c*.14 (an E_h greater than + 236 V; Eq. (3.65)) corresponds to more than 1 atm of O_2, which is an upper limit for oxidizing conditions. Likewise, a *pe* less than −7 (an E_h less than −118 V) is

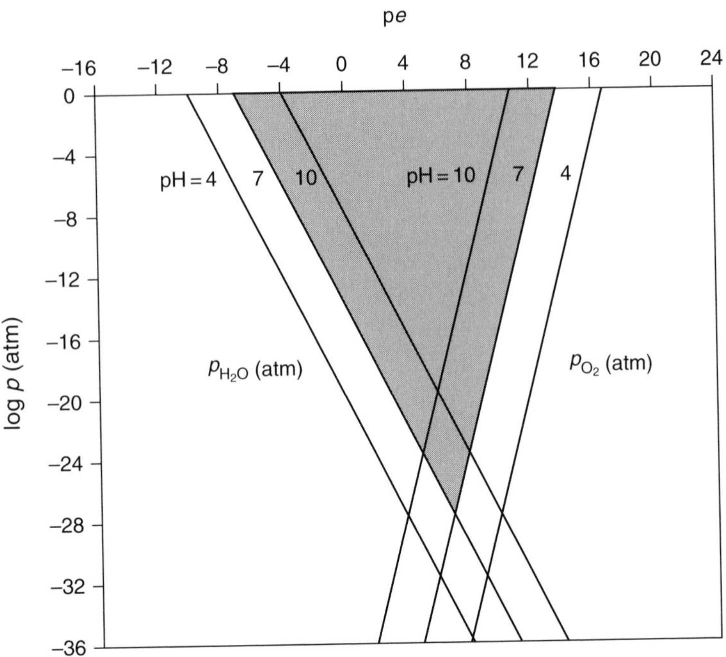

Figure 3.13. The log of the partial pressures of O_2 and H_2 as a function of pH for the half-reactions $\frac{1}{2}O_2 + 2H^+ + 2e^- = H_2O$ and $H^+ + e^- = ½\ H_2(g)$, illustrating the field of pe that can be occupied by redox species in water at a given pH.

unstable at pH = 7 because of spontaneous evolution of H_2. The thermodynamic stability field for liquid water with respect to the two bounding redox reactions must fall in the wedge-shaped region between the two pH = 7 lines. For any chemical species to be thermodynamically stable in an aqueous system, the E_h for its half-reaction must fall within the stability field of water under the prescribed pH, T and P conditions.

3.5.2 Environmental redox reactions

What redox reactions occur in marine systems within the extremes of the oxygen and hydrogen gas evolution reactions? And what are the processes that determine the overall electron richness of a natural environment? To address these questions it is helpful to first consider the major source of redox imbalance at the Earth's surface, oxygenic photosynthesis, the simultaneous production of organic matter and O_2 gas by green plants. Although this process will be covered in much more detail within Chapter 6, the simplest chemical representation for photosynthesis is the unimolar equation

$$CO_2 + H_2O \rightleftarrows CH_2O(s) + O_2. \tag{3.70}$$

In this reaction solar energy captured by plants is used to "split water" and thereby produce organic matter (represented generically as $CH_2O(s)$) and molecular oxygen (O_2). These two products are, respectively, the most abundant reducing and oxidizing agents in the environment and have no other quantitatively important source in addition to photosynthesis.

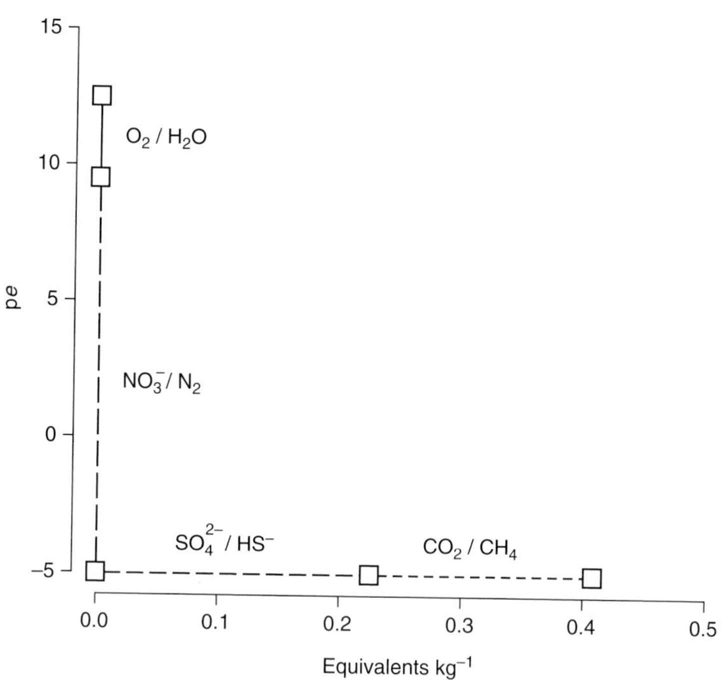

Figure 3.14. The redox buffering capacities of seawater illustrated by the ranges of pe in which electron acceptors are stable (ordinate) plotted against the concentration of the electron acceptor in seawater. Although O_2 and NO_3^- reduction dominate the redox reactions and the range of pe in seawater and sediments of the ocean, the most abundant electron acceptors are SO_4^{2-} and CO_2 and they occupy a relatively small range of pe.

The tendency is for organic matter produced by energy from the sun to break down into thermodynamically stable forms. The general pattern during the organic matter oxidation reaction is for electrons to pass from organic matter to the strongest oxidizing agent (electron acceptor) that is present at an appreciable concentration. Thus in general, electron acceptors will be utilized sequentially in the order of the corresponding half-reactions as listed in Table 3.7 (see also, Chapter 12). This is also the order of decreasing free energy yield, beginning with O_2 and then continuing with NO_3^-, $MnO_{2(s)}$, $FeOOH_{(s)}$, SO_4^{2-} and finally CO_2. Multicellular organisms are only capable of O_2-based respiration. The remaining oxidizing agents are exclusively utilized by microorganisms, many of which specialize in catalyzing only a particular redox reaction. The capacity of seawater constituents to serve as oxidizing agents can be assessed by multiplying their dissolved concentrations by the number of electrons that they take up during the conversion to their stable reduced forms. As is illustrated in Fig. 3.14, dissolved O_2 and NO_3^- have a very limited capacity for electrons before they are completely reduced to water and N_2 gas. In contrast, dissolved SO_4^{2-} ion and CO_2 have huge capacities for electrons.

Also illustrated in Fig. 3.14 is the constraint that as long as measurable amounts of O_2 and NO_3^- are present in seawater, the redox potential of the system will be poised near the very positive *pe* values (*c*.12.5) of the corresponding O_2–H_2O and NO_3^-–N_2 couples (see Table 3.7). When both oxidants are depleted, seawater cascades through trace levels of intermediate oxidizing agents to sulfate. The half-reaction that occurs at this point

$$SO_4^{2-} + 9H^+ + 8e^- \rightarrow HS^- + 4H_2O \tag{3.71}$$

has an E_h^0 of $+0.25$ V. The E_h (and hence pe) of seawater at this redox stage can be estimated by the Nernst equation, which for reaction (3.71) becomes

$$E_h = E_h^0 - \frac{0.0592}{8} \times \log_{10} \frac{(HS^-)}{(SO_4^{2-})(H^+)^9}. \tag{3.72}$$

If it is assumed as a useful approximation that $[HS^-] = [SO_4^{2-}]$ and $\gamma_{SO_4^{2-}} = \gamma_{HS^-}$, then E_h becomes independent of the activities of the two sulfur species and

$$E_h = 0.25 - \frac{0.0592}{8} \times \log_{10} \frac{(1)^1}{(10^{-8})^9} = 0.25 - 0.53 = -0.28 \text{ V}. \tag{3.73}$$

The corresponding pe is $-0.28 \times F/2.3RT$ ($2.3RT = 16.9$ at 25 °C), or -4.7. Thus in going from control by the O_2–H_2O couple to the $SO_4^{2-}/SO_4^{2-}HS^-$ couple, seawater undergoes an abrupt decrease in redox potential of over 17 pe units – nearly the whole environmental range! Whether or not these couples actually control the pe or E_h of the environment depends on the lability of the electron transfer, which varies among chemical half-reactions (Stumm and Morgan, 1996).

This concept has been tested by measuring concentrations of redox couples in waters that transition from oxic (O_2-containing) to reducing (HS^--containing) conditions. An example is the water column of a fjord in Vancouver, BC, where water is trapped behind a sill and oxygen is totally depleted in the deeper waters. Depth profiles of seven redox couples measured simultaneously are plotted in Fig. 3.15, and pe values are calculated (Table 3.8) from thermodynamic data and the concentrations of the redox species in the 125–135 m depth range by using Eq. (3.66). The calculated pe ranges from 12.6 to -3.5 over the very short distance of a few meters. If the environment were in chemical equilibrium one would expect the values to be equal to the predominant (most concentrated) redox couples in the water, which are O_2–H_2O for the oxic layer and HS^-–SO_4^{2-} for the deeper reducing waters. To a first approximation the calculated pe values for SO_4^{2-} and Fe^{2+} are in the same range; however, this is not the case for the rest of the couples, which vary from 6.6 to 12.6. The main reason for the wide range of pe values in the more oxidizing waters is that the rates of oxidation of reduced species (Fe^{2+}, Mn^{2+}, Cr(III) and I^-) are slow compared with transport. These species are produced in the reducing deep waters and they mix to shallower waters that contain oxygen and nitrate, where they persist even though they are thermodynamically unstable. This example illustrates that chemical species in the environment are often not at redox equilibrium.

Perhaps the most obvious example of redox thermodynamic disequilibrium in the environment is nitrogen gas in oxic surface

Table 3.8. *pe values calculated from the concentration of redox pairs in Saanich Inlet*

pe^0 values are from Table 3.7 or Emerson *et al.* (1979). Values of pe were calculated by using Eq. (3.66) and the concentrations of the redox couples in the 125–135 m depth range of Fig. 3.15. pH was 7.4, $[SO_4^{2-}]$ is 28 mmol kg^{-1}.

Reaction	pe^0	pe
$\frac{1}{4}O_2(g) + H^+ + e^- \longrightarrow \frac{1}{2}H_2O$	20.8	12.6
$\frac{1}{6}IO_3^-(g) + H^+ + e^- \rightarrow \frac{1}{6}I^- + \frac{1}{2}H_2O$	18.4	10.5
$\frac{1}{8}NO_3^- + \frac{5}{4}H^+ + e^- \longrightarrow \frac{1}{8}NH_4^+ + \frac{3}{8}H_2O$	14.9	5.7
$\frac{1}{2}MnO_2 + 2H^+ + e^- \longrightarrow \frac{1}{2}Mn^{2+} + \frac{1}{2}H_2O$	21.8	9.8
$\frac{1}{3}CrO_4^{2-} + 2H^+ + e^- \rightarrow \frac{1}{3}Cr(OH)_2^+ + \frac{2}{3}H_2O$	22.0	6.6
$\frac{1}{2}FeOOH(s) + 2H^+ + e^- \rightarrow \frac{1}{2}Fe^{2+} + H_2O$	8.25	−2.7
$\frac{1}{8}SO_4^{2-} + \frac{9}{8}H^+ + e^- \longrightarrow \frac{1}{8}HS^- + \frac{1}{2}H_2O$	4.25	−3.5

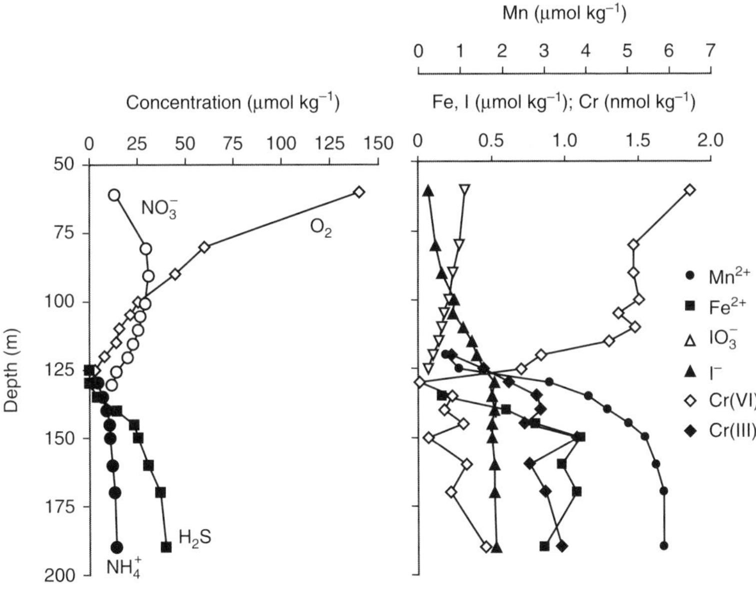

Figure 3.15. Concentrations of redox species (O_2, NO_3^-, NH_4^+, HS^-, Fe(II), Mn(II), IO_3^-, I^-, Cr(III) and Cr(VI)) with depth in the water column of Saanich Inlet, Vancouver, BC. The concentrations above and below the O_2–HS^- boundary are used to calculate the pe for each couple in Table 3.8. Modified from Emerson *et al.* (1979).

seawater. The reaction in question is listed on the second line of Table 3.7, which has an E_h^0 of $+1.25$ V and upon conversion to a non-fractional coefficient for NO_3^- becomes

$$NO_3^- + 6H^+ + 5e^- \longrightarrow \tfrac{1}{2}N_2 + 3H_2O. \qquad (3.74)$$

If this half-reaction were at equilibrium with molecular oxygen in the atmosphere, then its E_h would be essentially the same as that of oxic seawater, which is $+0.75$ V. Plugging the appropriate terms for reaction (3.74) into the Nernst equation gives

Table 3A1.1. *Coefficients used in the fitting equations for air saturation (C^s) and Henry's Law coefficients (K_H) of gases in seawater (Table 3.6)*

The coefficients and fitting equations in the footnotes are for saturation values of O_2, N_2, Ar, Ne, and He in units of μmol kg^{-1} and ml kg^{-1}. Values can be transformed between these units by using the real gas molar volumes calculated from Van der Waals constants (22.385 9, 22.391 9, 22.386 9, 22.422 4 and 22.436 9 mol^{-1} for O_2, N_2, Ar, Ne, and He, respectively). The fitting equation for CO_2 is for the Henry's Law coefficient, K_H (mol kg^{-1} atm^{-1}) instead of the saturation concentration.

Coefficient	O_2[a]	N_2[b]	Ar[b]	Ne[b]	He[c]	K_{H,CO_2}[d]
		(μmol kg^{-1})		(nmol kg^{-1})	(ml kg^{-1})	(mol kg^{-1} atm^{-1})
A_0	5.808 710	6.432 41	2.791 63	2.181 40		
A_1	3.202 910	2.927 58	3.177 14	1.289 31	−67.217 8	−60.240 9
A_2	4.178 870	4.303 51	4.136 58	2.122 35	216.344 2	93.451 7
A_3	5.100 060	4.266 73	4.866 32		139.203 2	23.358 5
A_4	−0.098 664				−22.620 2	
A_5	3.803 690					
B_0	−0.007 016	−0.007 443 16	−0.006 963 17	−0.005 947 22		
B_1	−0.007 700	−0.007 999 36	−0.007 683 87	−0.005 093 70	−0.044 781	0.023 517
B_2	−0.013 86	−0.001 529 48	−0.001 190 78		0.023 541	−0.023 656
B_3	−0.009 515				−0.0034266	0.0047035
C_0	$-2.759\ 150 \times 10^{-7}$					
$[C]^s$ at 20 °C 35 ppt	225.5	420.5	11.08	6.826	3.729×10^{-5}	0.0324

[a] Garcia and Gordon (1992): $\ln C^s = A_0 + A_1T_s + A_2T_s^2 + A_3T_s^3 + A_4T_s^4 + A_5T_s^5 + S(B_0 + B_1T_s + B_2T_s^2 + B_3T_s^3) + C_0S^2$;
where $T_s = \ln\{(298.15 - t)(273.15 + t)^{-1}\}$ and t is temperature (°C).

[b] Hamme and Emerson (2004): same equation as in [a].

[c] Weiss (1971): $\ln C^s = A_1 + A_2(100/T) + A_3\ln(T/100) + A_4(T/100) + S\{B_1 + B_2(T/100) + B_3(T/100)^2\}$, where T is absolute temperature.

[d] Weiss (1974): $\ln K_{H,CO_2} = A_1 + A_2(100/T) + A_3\ln(T/100) + S\{B_1 + B_2(T/100) + B_3(T/100)^2\}$, where T is absolute temperature.

$$+0.75 = +1.25 - \frac{0.0592}{5} \times \log_{10} \frac{(0.2)^{1/2}}{(NO_3^-)(10^{-8})^6}. \tag{3.75}$$

Solving this equation (assuming concentrations equal activities) gives log $(NO_3^-) = +5.3$, which is equivalent to $(NO_3^-) \approx 10^5$ M. Given the stoichiometry for the full oxidation reaction,

$$N_2 + 2.5O_2 + H_2O \longrightarrow 2HNO_3 \tag{3.76}$$

and the 4:1 molar excess of N_2 versus O_2 in the atmosphere (Chapter 1), it is apparent that the atmosphere would be stripped of essentially all oxygen before such a concentrated nitric acid solution could be formed. If this thermodynamically feasible reaction actually occurred, the ocean would become a highly oxidizing nitric acid bath in which carbon-based life forms would immediately perish. There would be no safety on land either in a thermodynamically spontaneous world because all organic matter would immediately combust to carbon dioxide and water. For living creatures and textbooks alike, slow kinetics in a thermodynamically imperfect world pose some distinct advantages.

References

DoE (1994) *Handbook of Methods for Analysis of the Various Parameters of the Carbon Dioxide System in Seawater*; version 2, ed. A. G. Dickson and C. Goyet. ORNL/CDIAC-74.

Emerson, S., R. Cranston and P. Liss (1979) Redox species in a reducing fjord: equilibrium and kinetic considerations. *Deep-Sea Res.* **1**, 26pp.

Garcia, H. E. and L. I. Gordon (1992) Oxygen solubility in seawater: better fitting equations. *Limnol. Oceanogr.* **37**, 1307–12.

Garrels, R. M. and M. E. Thompson (1962) A chemical model for sea water at 25 °C and one atmosphere total pressure. *Am. J. Sci.* **260**, 57–66.

Glueckauf, E. (1951) The composition of atmospheric air. In *Compendium of Meteorology*, Malone, T. F. (ed.), pp. 3–11. Boston, MA: American Meteorological Society.

Hamme, R. and S. Emerson (2004) The solubility of neon, nitrogen and argon in distilled water and seawater. *Deep-Sea Res.* I, **51**, 1517–28.

Handbook of Chemistry and Physics (1970) Cleveland, OH: The Chemical Rubber Publishing Co.

Klots, C. E. and B. B. Benson (1963) Solubilities of nitrogen, oxygen, and argon in distilled water. *J. Mar. Res.* **21**, 48–57.

Li, Y.-H. (1981) Ultimate removal mechanisms of elements from the ocean. *Geochim. Cosmochim. Acta* **45**, 1659–64.

Libes, S. M. (1992) *An Introduction to Marine Biogeochemistry*. New York, NY: John Wiley and Sons.

Millero F. (1996) *Chemical Oceanography*. Boca Raton, FL: CRC Press.

Morel, F. M. and J. G. Herring (1993) *Principles and Applications of Aquatic Chemistry*. New York, NY: Wiley Interscience.

Schindler, P. W. and W. Stumm (1987) The surface chemistry of oxides, hydroxides, and oxide minerals. In *Aquatic Surface Chemistry* (ed. W. Stumm), pp. 83–107. New York, NY: Wiley-Interscience.

Stumm, W. and J.J. Morgan (1996) *Aquatic Chemistry*. New York, NY: Wiley Interscience.

Wanninkhof, R. (1992) Relationship between wind speed and gas exchange over the ocean. *J. Geophys. Res.* **97**, 7373–82.

Weiss, R.F. (1971) The solubility of helium and neon in water and seawater. *J. Chem. Engin. Data* **16**, 235–41.

Weiss, R.F. (1974) Carbon dioxide in water and seawater: the solubility of a non-ideal gas. *Mar. Chem.* **2**, 203–15.

Whitfield, M. and D.R. Turner (1979) Water-rock partition coefficients and the composition of river and seawater. *Nature* **278**, 132–6.

4

Carbonate Chemistry

One of the most important components of the chemical perspective of oceanography is the carbonate system, primarily because it controls the acidity of seawater and acts as a governor for the carbon cycle. Within the mix of acids and bases in the Earth-surface environment, the carbonate system is the primary buffer for the acidity of water, which determines the reactivity of most chemical compounds and solids. The carbonate system of the ocean plays a key role in controlling the pressure of carbon dioxide in the atmosphere, which helps to regulate the temperature of the planet. The formation rate of the most prevalent authigenic mineral in the environment, $CaCO_3$, is also the major sink for dissolved carbon in the long-term global carbon balance.

Dissolved compounds that make up the carbonate system in water (CO_2, HCO_3^- and CO_3^{2-}) are in chemical equilibrium on time scales longer than a few minutes. Although this is less certain in the heterogeneous equilibrium between carbonate solids and dissolved constituents, to a first approximation $CaCO_3$ is found in marine sediments that are bathed by waters that are saturated or supersaturated thermodynamically and absent where waters are undersaturated. It has become feasible to test models of carbonate thermodynamic equilibrium because of the evolution of analytical techniques for the carbonate system constituents and thermodynamic equilibrium constants. During the first major global marine chemical expedition, Geochemical Sections (GEOSECS) in the 1970s, marine chemists argued about concentrations of dissolved inorganic carbon, DIC ($= HCO_3^- + CO_3^{2-} + CO_2$), and alkalinity at levels of 0.5%–1%, and the fugacity of CO_2, f_{CO_2}, at levels of $\pm 20\%$. pH (the negative log of the hydrogen ion concentration) was a qualitative property because its accuracy was uncertain when measured by glass electrodes, which could not be adequately standardized. By the time of the chemical surveys of the 1980s and 1990s, the world ocean circulation experiment (WOCE) and the joint global ocean survey (JGOFS), the accuracy of the carbonate system measurements had improved dramatically. Part of the improvement was due to new methods such as coulometry for DIC and colorimetry for pH. Another important advance was the development of certified, chemically stable DIC standards that resulted from both greater community organization, and the wherewithal to make stable standards. Since it was now possible to determine DIC and alkalinity to within several tenths of 1% and f_{CO_2} to within a couple of microatmospheres, it became necessary to improve the accuracy of equilibrium constants used to describe the chemical equilibria among the dissolved and solid carbonate species.

Homogeneous reactions of carbonate species in water are reversible and fast, so they can be interpreted in terms of chemical equilibrium, which is the primary focus of the first section of this chapter. Applications of these concepts to $CaCO_3$ preservation in sediments and the global carbon cycle are presented in Chapters 11 and 12. The following discussion uses terminology and concepts introduced in Chapter 3 on thermodynamics. We deal almost exclusively with *apparent* equilibrium constants (denoted by the prime on the equilibrium constant symbol, K') instead of *thermodynamic* constants, which refer to solutions with ionic strength approaching zero. Since seawater chemistry is for the most part extremely constant (see Chapter 1) it is feasible for chemical oceanographers to determine equilibrium constants in the laboratory in seawater solutions with chemistries that represent more than 99% of the ocean. The equilibrium constants have been determined as a function of temperature and pressure in the seawater medium. With this approach one forgoes attempts to understand the interactions that are occurring among the ions in solution for a more empirical, but also more accurate, description of chemical equilibria. We begin our discussion of the carbonate system by describing acids and bases in water, and then evolve to chemical equilibria and kinetic rates of CO_2 reactions. The chapter concludes with a discussion of the processes controlling alkalinity and DIC in the ocean.

4.1 Acids and bases in seawater

The importance of the many acid–base pairs in seawater in determining the acidity of the ocean depends on their concentrations and equilibrium constants. Evaluating the concentrations of an acid and its conjugate anion (base, Ba^-) as a function of pH ($pH = -\log [H^+]$) requires knowledge of the equation describing the acid/base equilibrium (hydrogen ion exchange), the apparent equilibrium constant, K', and information about the total concentration, $[Ba]_T$, of the acid in solution:

$$HBa \rightleftarrows H^+ + Ba^- \tag{4.1}$$

$$K' = \frac{[H^+] \times [Ba^-]}{[HBa]} \tag{4.2}$$

$$[Ba]_T = [HBa] + [Ba^-]. \tag{4.3}$$

Combining Eqs. (4.2) and (4.3) gives expressions for the concentration of the acid, HBa, and its conjugate base, Ba^-, as functions of the apparent equilibrium constant, K', and the hydrogen ion concentration, $[H^+]$:

$$[HBa] = \frac{[Ba]_T \times [H^+]}{K' + [H^+]} \quad \text{or} \quad \log[HBa] = \log[Ba]_T + \log[H^+] - \log(K' + [H^+]) \tag{4.4}$$

and

$$[Ba^-] = \frac{[Ba]_T \times K'}{K' + [H^+]} \quad \text{or} \quad \log[Ba^-] = \log[Ba]_T + \log[K'] - \log(K' + [H^+]). \tag{4.5}$$

A plot of these logarithmic equations (Fig. 4.1) illustrates that the concentration of the acid dominates the solution concentration below pH = pK' (on the acid side), and in the region where pH is greater than pK' (the basic side), the conjugate base, Ba^-, dominates. At a pH equal to pK' the concentrations of the acid and basic forms are equal, $[HBa] = [Ba^-]$.

The final constraint is that of charge balance, which in this simple solution involves the only two ions:

$$0 = [H^+] - [Ba^-]. \tag{4.6}$$

This equation constrains the system to a single location on the plot (where the lines for these two concentrations cross in Fig. 4.1), which uniquely fixes the pH and concentrations of acids and bases in the system. In this simple system the solution is acidic (pH = 4) because the concentration of the hydrogen ion and anion must be equal.

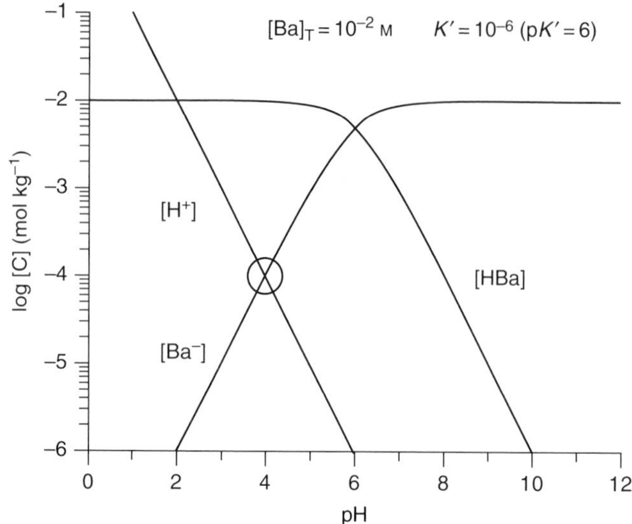

Figure 4.1. Concentrations of the acidic [HBa] and basic [Ba^-] forms of an acid with total concentration $[Ba]_T = 10^{-2}$ mol kg^{-1} and an equilibrium constant $K = 10^{-6}$, as a function of pH. The concentrations are equal at the point where pH = p*K*. When the criterion of charge balance is included in the equations, the system is defined at a single pH where $[H^+] = [Ba^-]$, indicated by the small circle.

These simple equations and ideas provide the basis for describing the carbonate system in terms of the f_{CO_2}, DIC, pH, and alkalinity of seawater. We will build up a plot similar to that in Fig. 4.1 for the important acids and bases in seawater. These are listed along with their concentrations and apparent equilibrium constants in Table 4.1. It will then be demonstrated how the constraint of charge balance (called alkalinity) determines the pH of seawater.

4.1.1 The important acids and bases in seawater

The importance of an acid–base pair in controlling the pH of a solution is determined by its total concentration and pK′. The pH of the solution will be near the pK′ of the acids with the highest concentration. Carbonic and boric acids are the most concentrated hydrogen ion exchangers with pK′ values near seawater pH.

Carbonic acid

In water, inorganic carbon exists in four distinct forms; the gas in solution or aqueous carbon dioxide, $CO_2(aq)$, and the three products of hydration reactions, which are carbonic acid, H_2CO_3, bicarbonate, HCO_3^-, and carbonate, CO_3^{2-}. Chemical equilibria among these species in seawater are described by the apparent constants, which have units necessary to make the dimensions of the equilibrium expressions correct

$$H_2CO_3 \rightleftharpoons CO_2(aq) + H_2O \qquad K'_{CO_2(aq)} = \frac{[CO_2]\,(aq)}{[H_2CO_3]} \qquad (4.7)$$

$$H_2CO_3 \rightleftharpoons HCO_3^- + H^+ \qquad K'_{H_2CO_3} = \frac{[HCO_3^-] \times [H^+]}{[H_2CO_3]} \qquad (4.8)$$

Table 4.1. *Compounds that exchange protons in the pH range of seawater*

Equilibrium constants are for 25 °C and $S = 35$ from the equations in Appendix 4.2 and Millero (1995) for nitrogen and sulfur species. An asterisk (*) indicates the concentration is in the μmol kg^{-1} range and variable. (p$K = -\log K$.)

Species	Reaction	Concentration (mol kg^{-1})	Concentration $-\log C_T$	pK'
H_2O	$H_2O \rightleftharpoons OH^- + H^+$			13.2
DIC	$CO_2 + H_2O \rightleftharpoons HCO_3^- + H^+$	$\approx 2.04 \times 10^{-3}$	2.69	5.85
	$HCO_3^- \rightleftharpoons CO_3^{2-} + H^+$			8.97
B	$B(OH)_3 + H_2O \rightleftharpoons B(OH)_4^- + H^+$	4.16×10^{-4}	3.38	8.60
Si	$H_4SiO_4 \rightleftharpoons H_3SiO_4^- + H^+$	*	*	9.38
P	$H_3PO_4 \rightleftharpoons H_2PO_4^- + H^+$	*	*	1.61
	$H_2PO_4^- \rightleftharpoons HPO_4^{2-} + H^+$	*	*	5.96
	$HPO_4^{2-} \rightleftharpoons PO_4^{3-} + H^+$	*	*	8.79
SO_4^{2-}	$HSO_4^- \rightleftharpoons SO_4^{2-} + H^+$	2.824×10^{-2}	1.55	1.00
F	$HF \rightleftharpoons F^- + H^+$	7.0×10^{-5}	4.15	2.52
Anoxic water				
N	$NH_4^+ \rightleftharpoons NH_3 + H^+$	*	*	9.19
HS^-	$H_2S \rightleftharpoons HS^- + H^+$	*	*	6.98

$$HCO_3^- \rightleftharpoons CO_3^{2-} + H^+ \qquad K_2' = \frac{[CO_3^{2-}] \times [H^+]}{[HCO_3^-]}, \tag{4.9}$$

where the equilibrium constant, K_2', indicates the second dissociation constant of carbonic acid. Because only a few tenths of one percent of the neutral dissolved carbon dioxide species exists as H_2CO_3 at equilibrium, and because it is difficult to analytically distinguish between $CO_2(aq)$ and H_2CO_3, these neutral species are usually combined and represented with either the symbol $[CO_2]$ or $H_2CO_3^*$ (see Chapter 9, Table 9.2). We use the former here:

$$[CO_2] = [CO_2(aq)] + [H_2CO_3]. \tag{4.10}$$

Equations (4.7) and (4.8) can be combined to eliminate $[H_2CO_3]$ and give a new composite first dissociation constant of CO_2 in seawater. If one assumes that $[CO_2(aq)] = [CO_2]$, the first dissociation constant of carbonic acic, K_1', is

$$CO_2 + H_2O \rightleftharpoons HCO_3^- + H^+ \qquad K_1' = \frac{[HCO_3^-][H^+]}{[CO_2]} \cong \frac{K'_{H_2CO_3}}{K'_{CO_2(aq)}}. \tag{4.11}$$

The approximation involved in combining $[CO_2(aq)]$ and $[H_2CO_3]$ as $[CO_2]$ is illustrated by solving Eqs. (4.7), (4.8), (4.10) and (4.11) to derive a relationship among the equilibrium constants, K'_1, $K'_{CO_2(aq)}$, and $K'_{H_2CO_3}$

$$K'_1 = \frac{K'_{H_2CO_3}}{K'_{CO_2(aq)} + 1}. \tag{4.12}$$

Because $K'_{CO_2(aq)} \gg 1$ (the thermodynamic value for $K'_{CO_2(aq)}$ is 350–990) (Stumm and Morgan, 1996):

$$K'_1 \approx \frac{K'_{H_2CO_3}}{K'_{CO_2(aq)}}. \tag{4.13}$$

Since it is the value K'_1 that is measured by laboratory experiments, analytical measurements and theoretical equilibrium descriptions are consistent.

At equilibrium the gaseous CO_2 in the atmosphere, expressed in terms of the fugacity, $f^{a}_{CO_2}$ (in atmospheres, atm), is related to the aqueous CO_2 in seawater, $[CO_2]$ (mol kg^{-1}), via the Henry's Law coefficient, K_H (mol kg^{-1} atm^{-1}) (see Chapter 3):

$$K_{H,CO_2} = \frac{[CO_2]}{f^{a}_{CO_2}}. \tag{4.14}$$

The partial pressure and fugacity are equal only when gases behave ideally; however, Weiss (1974) has shown that the ratio of $f^{a}_{CO_2}$, to its partial pressure, p_{CO_2}, is between 0.995 and 0.997 for the temperature range of 0–30 °C, indicating that the differences are not large. The term p_{CO_2} is often used in the literature because the non-ideal behavior of CO_2 gas in the atmosphere is small.

The content of CO_2 in surface waters is often presented as the fugacity (or partial pressure) in solution, $f^{w}_{CO_2}$. An example of this application is that the difference in the fugacities of CO_2 between the atmosphere and the ocean ($f^{a}_{CO_2} - f^{w}_{CO_2}$) is often used in gas exchange rate calculations (Chapter 10). The fugacity of CO_2 in water is calculated by using Eq. (4.14).

With the above equilibria we are now prepared to define the total concentration of dissolved inorganic carbon and construct a diagram of the variation of the carbonate species concentrations as a function of pH. For simplicity we begin by assuming there is no atmosphere overlying the water, so Eq. (4.14) is not necessary to describe the chemical equilibria in this example. The total concentration, C_T, for inorganic carbon in seawater is called dissolved inorganic carbon (DIC) or total CO_2 (ΣCO_2). As the first term is more descriptive, we adopt it here. The DIC of a seawater sample is the sum of the concentrations of the dissolved inorganic carbon species:

$$\mathrm{DIC} = [HCO_3^-] + [CO_3^{2-}] + [CO_2]. \tag{4.15}$$

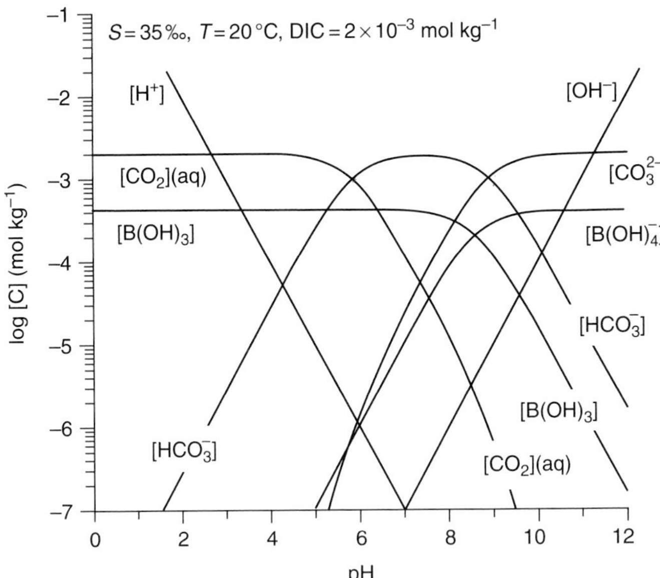

Figure 4.2. Concentrations of the species of the acid–base pairs of carbonate, borate and water in seawater as a function of pH. (Salinity, $S = 35$, temperature, $T = 20\,^{\circ}$C and DIC $= 2.0 \times 10^{-3}$ mol kg^{-1}.)

Since this is a total quantity, it has the advantage that it is independent of temperature and pressure, unlike the concentrations of its constituent species. Experimentally, DIC is determined by acidifying the sample, so that all the HCO_3^- and CO_3^{2-} react with H^+ to become CO_2 and H_2O, and then measuring the amount of CO_2 gas evolved. To create a plot of the concentrations of the three dissolved carbonate species as a function of pH we assign the DIC its average value in seawater (Table 4.1). Combining Eq. (4.15) with Eqs. (4.9) and (4.11) yields separate equations for the carbonate species as a function of equilibrium constants, DIC and pH:

$$[CO_2] = \frac{\mathrm{DIC}}{1 + \dfrac{K_1'}{[H^+]} + \dfrac{K_1'K_2'}{[H^+]^2}} \tag{4.16}$$

$$[HCO_3^-] = \frac{\mathrm{DIC}}{\dfrac{[H^+]}{K_1'} + 1 + \dfrac{K_2'}{[H^+]}} \tag{4.17}$$

$$[CO_3^{2-}] = \frac{\mathrm{DIC}}{1 + \dfrac{[H^+]^2}{K_1'K_2'} + \dfrac{[H^+]}{K_2'}}. \tag{4.18}$$

The plot in Fig. 4.2 demonstrates the relative importance of the three carbonate species in seawater as a function of pH. At pH $= \mathrm{p}K_1'$ the concentrations of CO_2 and HCO_3^- are equal and at pH $= \mathrm{p}K_2'$ the concentrations of HCO_3^- and CO_3^{2-} are equal. Since we know that the pH of surface waters is about 8.2, it is clear that the dominant carbonate species is HCO_3^-. What has been done so far, however, does not yet explain why the pH of seawater is between 7.6 and 8.2, and we will return to this question.

Boric acid

The acid–base pair with the second highest concentration and a pK′ near the pH of seawater is boric acid (Table 4.1). The carbonate system and boric acid turn out to be by far the most important contributors to the acid–base chemistry of seawater, but they contrast greatly in their reactivity in the ocean: carbon is involved in all metabolic processes and varies in concentration from place to place, whereas borate is conservative and maintains a constant ratio to salinity. The equilibrium reaction and total boron, B_T, equations are:

$$B(OH)_3 + H_2O \rightleftarrows B(OH)_4^- + H^+ \qquad K'_B = \frac{[B(OH)_4^-][H^+]}{[B(OH)_3]} \tag{4.19}$$

$$B_T = B(OH)_3 + B(OH)_4^-. \tag{4.20}$$

The equations for the boron species as a function of pH and K'_B are thus

$$[B(OH)_3] = \frac{B_T \times [H^+]}{[H^+] + K'_B} \tag{4.21}$$

$$[B(OH)_4^-] = \frac{B_T \times K'_B}{[H^+] + K'_B}. \tag{4.22}$$

From the graph of these two equations shown in Fig. 4.2, it is clear why boric acid plays a role as a pH buffer in seawater. The two species that exchange hydrogen ions are equal when the $pK' = pH$, which in this case is $pH = 8.6$ (Table 4.1).

It is now clear that the most important criteria for describing the role of an acid–base pair in determining the pH of seawater are the total concentration, C_T, and the apparent equilibrium constants. For example, hydrochloric acid, HCl, and sulfuric acid, H_2SO_4, are well-known acids because we use them frequently in the laboratory. We know, also, that Cl^- and SO_4^{2-} ions are the two most concentrated anions in seawater. Why, then, are these acid–base pairs not considered in our discussion? The answer is because of their extremely low pK' values; for example, $pK_{HSO_4} = 1.0$ (Table 4.1). The pH where the HSO_4^- and SO_4^{2-} ions are at equal concentration is so low that the SO_4^{2-} ion may be considered totally unprotonated at the pH of seawater.

The rest of the acids in seawater with pK' values in the vicinity of 8–9, silicic acid and phosphoric acid, have low and variable concentrations (1–200 μmol kg^{-1}), but they must be considered in order to have a complete representation of the acid–base components of seawater. From the acid–base plot of Fig. 4.2 one can determine which species are most involved in the exchange of protons in seawater. Any constituent for which the lines are curved in the pH range 7–9 contributes to the seawater buffer system. Before we can answer the question of why the sea has a pH of between 7.6 and 8.2, we must deal with an extremely important but somewhat troublesome constituent of the carbonate system: the modern concept of the alkalinity of seawater.

4.1.2 The alkalinity of seawater

Just as the charge balance had to be identified in order to determine the pH at equilibrium on the simple acid–base plot in Fig. 4.1, so must the charge balance be evaluated to determine the pH at equilibrium on the acid–base plot for seawater (Fig. 4.2). Presently, the system of equations includes the equilibria and total concentrations (Eqs. (4.9), (4.11) and (4.15) for the carbonate species; Eqs. (4.19) and (4.20) for borate, and so on for the minor players in oxic seawater S, F, P, and Si) which describe the predominant acid–base species over the entire pH range. There are as yet insufficient constraints to evaluate the equilibrium position on the plot in Fig. 4.2: one is free to move left and right on the pH scale. For example, in the case of the carbonate system there are five unknowns (DIC, $[HCO_3^-]$, $[CO_3^{2-}]$, $[CO_2]$ and pH) and only three equations. If the concentration of DIC is designated, we are still one equation shy of being able to solve the system of equations uniquely and exactly define the pH. The missing equation is the expression for total alkalinity, A_T, which represents the charge balance of the mixed electrolyte system of seawater. The practical advantage for this new constraint is that it is measurable, and it is a total quantity like DIC, which is independent of temperature and pressure.

The alkalinity in a mixed electrolyte solution is the excess in bases (proton acceptors) over acids (proton donors) in the solution. The alkalinity is measured by adding acid to seawater to an end point where most all proton acceptors have reacted. When one adds acid the hydrogen ion concentration does not increase as much as it would in the absence of alkalinity because some of the added hydrogen ions react with the excess bases (CO_3^{2-}, HCO_3^-, $B(OH)_4^-$, ...). Since it is possible to precisely determine the hydrogen ion concentration change in solution, the difference between the amount of H^+ added and the measured change can be accurately determined by titration. The units of alkalinity are equivalents per kilogram (eq kg^{-1}).

One way of defining the alkalinity is by separating the anions that exchange protons during the titration from those that do not. For neutrality, the alkalinity must equal the difference in charge between cations and anions that do *not* exchange protons to any significant extent during the titration. One can calculate the alkalinity of standard seawater by using the concentrations of conservative ions at a salinity of 35‰ presented in Table 1.4 and Table 4.2. SO_4^{2-} and F^- ions are included among the species that do not exchange protons because their reaction with H^+ is so small during the titration that they are conservative to the five decimal places presented in the table. By this definition, the numerical value for total alkalinity, A_T, is equal to

Table 4.2. *Concentrations of cation and anion species that do not significantly exchange protons in the pH range of seawater (35‰)*

Cation	eq kg^{-1}	Anion	eq kg^{-1}
Na^+	0.469 06	Cl^-	0.545 86
Mg^{2+}	0.105 64	SO_4^{2-}	0.056 48
Ca^{2+}	0.020 56	Br^-	0.000 84
K^+	0.010 21	F^-	0.000 07
Sr^{2+}	0.000 18		
Li^+	0.000 02		
Total cations	0.605 67	Total anions	0.603 25

Source: From the compilation in Table 1.4.
Σcations $-$ Σanions $= 0.605\,67 - 0.603\,25 = 0.002\,42$.

$$A_T = \text{cation charge} - \text{anion charge} = 0.605\,67 - 0.603\,25\ (\text{eq kg}^{-1}) = 0.002\,42\ (\text{eq kg}^{-1}).$$

Acids and bases that make up the total alkalinity must protonate in solution in a way that achieves charge balance. For example, the difference in equivalents evaluated in Table 4.2 determines the relative abundances of $[HCO_3^-]$ and $[CO_3^{2-}]$ that are required for charge balance. As the difference between A_T and DIC increases (becomes a larger positive number) there must be a higher carbonate concentration to achieve charge balance because CO_3^{2-} carries two equivalents and HCO_3^- only one.

The concentrations of the species that make up the charge difference evaluated in Table 4.2 are bases that react with H^+ at pH = 8.2 in Fig. 4.2. The concentrations of the species that make up the bulk of the alkalinity in surface seawater are presented in Table 4.3. Values in this table are for surface seawater, which is low in nutrient concentrations. In regions of the ocean where silicate and phosphate concentrations are measurable, they must also be included in the definition of total alkalinity:

$$A_T = [HCO_3^-] + 2\,[CO_3^{2-}] + [B(OH)_4^-] + [H_3SiO_4^-] + [HPO_4^{2-}] + 2\,[PO_4^{3-}] + [OH^-]. \tag{4.23}$$

Notice that the coefficients on the concentrations on the right-hand side of Eq. (4.23) are equal to the charge of the ions except in the cases of HPO_4^{2-} and PO_4^{3-}. The reason for this is that the precise definition of the alkalinity of seawater is based on the method by which it is determined and the species that exchange protons during the titration.

As stated previously, the alkalinity is determined by adding acid to the seawater solution and measuring the pH during the process. The equivalence point, the pH at which the amount of acid equals the

Table 4.3. *The concentrations of the species that make up the total alkalinity ($A_T = 2420$ μeq kg^{-1}) of seawater at pH c.8.2 ($T = 20°$ C, $S = 35$)*

Since this is the pH of surface seawater, it is presented without the contribution of silicate and phosphate.

	Concentration		
Species	μmol kg^{-1}	μeq kg^{-1}	% of A_T
HCO_3^-	1796	1796	75
CO_3^{2-}	255	510	21
$B(OH)_4^-$	108	108	4
OH^-	6	6	0.2

alkalinity of the solution, is accurately defined so it is possible to state precisely which base species will accept protons in this range. Dickson (1981) describes the alkalinity as,

> The number of moles of hydrogen ion equivalent to the excess of proton acceptors (bases formed from weak acids with a dissociation constant $K \leq 10^{-4.5}$ at 25 °C and zero ionic strength) over the proton donors (acids with $K > 10^{-4.5}$) in one kilogram of sample.

Proton acceptors with $K \leq 10^{-4.5}$ (p$K \geq 4.5$) in Table 4.1 include HCO_3^-, CO_3^{2-}, $B(OH)_4^-$, OH^-, $H_3SiO_4^-$, HPO_4^{2-}, and PO_4^{3-}, but not $H_2PO_4^-$, which means that HPO_4^{2-}, and PO_4^{3-} will be titrated to $H_2PO_4^-$, but not to H_3PO_4. This is the reason that the stoichiometric coefficients of the phosphate species in Eq. (4.23) are one less than the charge. To complete the precise definition of alkalinity, we subtract H^+ and the acids in Table 4.1 with $K > 10^{-4.5}$, HSO_4^-, HF and H_3PO_4:

$$A_T = [HCO_3^-] + 2[CO_3^{2-}] + [B(OH)_4^-] + [H_3SiO_4^-] + [HPO_4^{2-}] + 2[PO_4^{3-}] + [OH^-] - [H^+] - [HSO_4^-] - [HF] - [H_3PO_4]. \quad (4.24)$$

This rather long expression includes all known inorganic proton acceptors and donors in oxic seawater that follow Dickson's definition of the titration alkalinity. It includes two uncharged species at the very end, so it is not exactly consistent with the previous charge balance definition; however, in practice, the concentrations of acidic species in seawater (H^+, HSO_4^-, HF, and H_3PO_4) are too low in the pH range of 7.0–8.0 to be significant and are frequently not included in the alkalinity definition. Including them here demonstrates the fate of protons during the course of acid addition to determine total alkalinity. (These species also play a more important role in more dilute environmental solutions such as rainwater, and in many freshwater lakes.) The concentrations in Table 4.3 indicate that the ions of carbonate and borate define about 99% of the total alkalinity. Thus, calculations are sometimes made which include only these two species, and we define this as the carbonate and borate alkalinity, $A_{C\&B}$,

$$A_{C\&B} = [HCO_3^-] + 2 \cdot [CO_3^{2-}] + [B(OH)_4^-]. \tag{4.25}$$

Another shortened form of the alkalinity consists only of the carbonate species, which make up about 96 % of the total alkalinity, and is termed the carbonate alkalinity, A_C,

$$A_C = [HCO_3^-] + 2 \cdot [CO_3^{2-}]. \tag{4.26}$$

This definition is sometimes used for illustration purposes because of the simplicity of the calculations involved.

In anoxic waters a whole new set of acids are created by the lower redox conditions. The most prevalent are the different forms of sulfide and ammonia (Table 4.1). Clearly, these species meet the criteria to be included in the titration alkalinity and their concentrations can become as high as hundreds of $\mu mol\ kg^{-1}$ in some highly reducing environments. For normal situations in which the water contains oxygen these species are too low in concentration to be important.

4.2 Carbonate equilibria: calculating the pH of seawater

We have now described the system of equations necessary for determining the pH of seawater and the distribution of carbonate species. By including the definition and numerical value of the alkalinity to the system of equations used to determine the curves in Fig. 4.2, we have constrained the location on the plot to a single pH. The equations necessary to determine this location are summarized in Appendix 4.1 for the progressively more complicated definitions using the three forms of the alkalinity, A_C, $A_{C\&B}$, and A_T.

In order to solve the equations and determine pH and the concentrations of the species that make up the alkalinity, the apparent equilibrium constants, K', must be accurately known. These constants have been evaluated and re-evaluated in seawater over the past 50 y. The pH scales and methods of measuring pH during these experiments have been different, and this has complicated comparisons of the data until recently, when many have been converted to a common scale. Equations for the best fit to carbonate system equilibrium constants as a function of temperature and salinity are presented by Luecker *et al.* (2000), DoE (1994) and Millero (1995) (see Appendix 4.2).

The pH and carbonate species distribution for waters from different locations in the ocean (Table 4.4) are calculated by using data for A_T and DIC and the equilibrium constants. The equilibrium equations were solved with the computer program of Lewis and Wallace (1998) using the carbonate equilibrium constants K'_1 and K'_2 of

Table 4.4. *Carbonate system parameters calculated for different conditions in the surface and deep oceans at 35‰ salinity using two different methods*

Column (I)[a] is the calculation utilizing all species in the total alkalinity, A_T. A_{Si} and A_P (bottom row) are the alkalinities due to silicate and phosphate species. Column (II)[b] is the calculation assuming the total alkalinity does not include Si and P species, $A_T = A_{C\&B}$. Concentrations and DIC are in units of μmol kg^{-1} and alkalinity values, A_T, are in μeq kg^{-1}.

Parameter	Surface Water		North Atlantic Deep Water		Antarctic Deep Water		North Pacific Deep Water	
Measured concentrations								
Z (km)	0.0		4.0		4.0		4.0	
T (°C)	20.0		2.0		2.0		2.0	
A_T	2300		2350		2390		2460	
DIC	1950		2190		2280		2370	
[Si]	0.0		60		130		160	
[P]	0.0		1.5		2.2		2.5	
Calculated carbonate parameters (Models I and II)								
	I	II	I	II	I	II	I	II
pH	8.19	8.20	7.95	8.11	7.80	7.98	7.74	7.92
f_{CO_2} (μatm)	256	255	316	333	462	478	562	575
$[HCO_3^-]$	1698	1698	2064	2052	2171	2161	2264	2254
$[CO_3^{2-}]$	244	244	108	118	82	91	73	83
$[CO_2]$	8	8	18	19	27	28	33	33
$[B(OH)_4]$	108	108	67	60	50	46	44	40
A_{Si}	0.0	0.0	1.3	0.0	2.0	0.0	2.1	0.0
A_P	0.0	0.0	1.6	0.0	2.3	0.0	2.5	0.0

[a] Calculated by using the program of Lewis and Wallace (1998) with the K_1 and K_2 of Mehrbach *et al.* (1973) as reinterpreted by Dickson and Millero (1987).
[b] Calculated by using the program in Appendix 4.1, with the K_1 and K_2 of Mehrbach *et al.* (1973) as refitted by Luecker *et al.* (2000).

Mehrbach *et al.* (1973) as redetermined by Dickson and Millero (1987). This program allows one to calculate the carbonate species at equilibrium from any two of the species measured by using the complete description of the alkalinity, A_T, including the contributions from silicate and phosphate. The results are presented in columns labeled I in Table 4.4. We have also solved a simplified version of the equilibrium equations, using the approximation that the total alkalinity includes only the carbonate and borate alkalinity, $A_{C\&B}$. Carbonate species determined by this approach are presented in columns labeled II in Table 4.4, and the program is listed in Appendix 4.1.2. Ideally, the solutions using these two methods would be identical in surface waters because concentrations of Si and P are below detection limits. Indeed, they are only slightly different (compare columns I and II). Other differences between columns I and II may be due as much to slightly different values used for K_1' and K_2' in the different programs as it is to the differences in A_T and $A_{C\&B}$. The values presented by Luecker *et al.* (2000) and presented in Appendix 4.2 are recommended for surface water calculations (see later).

Both DIC and A_T increase from surface waters to the deep Atlantic, Antarctic and Pacific Oceans as one follows the route of the ocean "conveyor belt" (Fig. 1.12). Along this transect pH changes from about 8.2 in surface waters to 7.8 in the deep Pacific Ocean, and CO_3^{2-} decreases from nearly 250 μeq kg^{-1} to less than a third of this value, 75 μeq kg^{-1}. The reason for this change has to do with the ratio of the change in A_T and DIC in the waters and is discussed in the final section of this chapter. Notice that the contribution of the nutrients Si and P to the total alkalinity is only between 0 and 5 μeq kg^{-1} or at most 0.2% of the total alkalinity. Although Si concentrations are much greater than those of P, the two nutrients have nearly equal contributions to the alkalinity (Table 4.4) because the pK' values for two phosphate reactions are closer to the pH of seawater than is the pK for silicate (see Table 4.1).

The present high level of analytical accuracy makes the choice of appropriate equilibrium constants to use for the carbonate system an important consideration. The most rigorous test of how well the carbonate equilibrium in seawater is known is to calculate a third parameter from two known values and compare the calculated value with an independent measurement of that parameter. Millero (1995) compared the estimated accuracy of measured and calculated values of carbonate system parameters; his results are summarized in Table 4.5. In addition to the error associated with the accuracy of the analytical measurements, there are two estimates of calculation errors listed in the table. The first row (I) is the error to be expected from compounding the errors of the analytical measurements used to calculate the parameter, assuming the equilibrium constants are perfectly known. The second row (II) is the error determined from compounding the errors of the equilibrium constants, which Millero estimates to be accurate to within ± 0.002 for pK_1' and ± 0.005 for pK_2'. This analysis assumes that there are no systematic offsets in the estimation of K_1' and K_2' other than this scatter about the mean. There are two clear messages from Table 4.5. The first is that the contributions of the analytical uncertainty and the errors in the equilibrium constants to the uncertainty in calculated parameters are nearly equal. The second is that one can measure and calculate the individual parameters about equally well if one can choose the correct measured values.

Although the accuracies of all the parameters are impressive (approaching 0.1% in the cases of DIC and A_T), one's ability to calculate carbonate system concentrations varies depending on which species are measured. For example, $f_{CO_2}^{w}$ and pH are presently the most readily determined, continuous measurements of the carbonate system by unmanned moorings and drifters. This is good for gas exchange purposes because it will become less expensive to derive large data bases of surface ocean $f_{CO_2}^{w}$, but very poor for defining the rest of the carbonate system by using remote measurements because

Table 4.5. *Estimates of the errors in measurement and calculation of the carbonate system parameters*

All values are standard deviations about the mean. Measurement error is based on comparison to standard values. Calculated error is determined either by: (I) compounding errors in the analytical accuracy of the input values assuming equilibrium constants are perfect; or (II) compounding errors in the first and second dissociation constants, assuming the measurements are perfect. The total error of the calculated estimate would involve compounding these two errors.

Parameter	Calculation method	pH	A_T (μeq kg^{-1})	DIC (μmol kg^{-1})	f_{CO_2} (μatm)
Measurement error		0.0020	4.0	2.0	2.0
Calculated error (methods I and II)					
pH – A_T	I			3.8	2.1
	II			2.4	1.7
pH – DIC	I		2.7		1.8
	II		2.6		1.6
pH – $f^{w}_{CO_2}$	I		21	21	
	II				
$f^{w}_{CO_2}$ – DIC	I	0.0025	3.4		
	II	0.0019	2.6		
$f^{w}_{CO_2}$ – A_T	I	0.0026		3.2	
	II	0.0019		2.1	
A_T – DIC	I	0.0062			5.7
	II	0.0036			2.9

From Millero (1995).

of the large errors in calculating A_T and DIC from this analytical pair (Table 4.5). The error analysis in Table 4.5, also, is not the whole story, because it does not address the possibility of systematic errors in the equilibrium constants. This has been assessed recently by comparing the $f^{w}_{CO_2}$ measured in seawater solutions at equilibrium with standard gases with $f^{w}_{CO_2}$ calculated from A_T and DIC (Luecker *et al.*, 2000). They found that the constants of Mehrbach *et al.* (1973), reinterpreted to the "total" pH scale (Appendix 4.2), were most accurate if the $f^{w}_{CO_2}$ was less than 500 μatm kg^{-1}. The $f^{w}_{CO_2}$ calculated from A_T and DIC, with accuracies of 1 μmol kg^{-1} and 2 μeq kg^{-1}, respectively (about 0.05 and 0.1%), agreed with measured $f^{w}_{CO_2}$ values to within ±3 μatm. However, the ability to distinguish the correct equilibrium constants by comparing measured and calculated values deteriorated as the $f^{w}_{CO_2}$ increased above 500 μatm kg^{-1}.

At the time of writing this book we are in the situation where it has been demonstrated that there is one set of preferred constants for calculating surface water $f^{w}_{CO_2}$ from A_T and DIC, but these values are not necessarily preferred for deeper waters where $f^{w}_{CO_2}$ exceeds 500 μatm. The best agreement is in the most important region from the point of view of air–sea interactions, and errors deeper in the ocean are not very large. The reason for the variability may be that there are unknown organic acids and bases in the dissolved organic matter of seawater that alter the acid–base behavior, but this has not been experimentally demonstrated. Although great advances in our understanding of the carbonate system have occurred relatively recently, it is also true that a version of the carbonate equilibrium constants determined more than 30 years ago (Mehrbach *et al.*, 1973) is still preferred.

4.3 Kinetics of CO_2 reactions in seawater

Although most of the reactions between carbonate species in seawater are nearly instantaneous, the hydration of CO_2

$$CO_2 + H_2O \rightleftarrows H_2CO_3 \tag{4.27}$$

is relatively slow, taking tens of seconds to minutes at the pH of most natural waters. This slow reaction rate has consequences for understanding the processes of carbon dioxide exchange with the atmosphere and the uptake of CO_2 by surface water algae. The rate equation for CO_2 reaction has four terms (Eq. (b) of Table 9.2):

$$\frac{d[CO_2]}{dt} = (k_{CO_2} + k_{OH^-}[OH^-]) \cdot [CO_2] + (k_{CO_2,r}[H^+] + k_{HCO_3}) \cdot [HCO_3^-]. \tag{4.28}$$

The mechanisms for this reaction are discussed in the chapter on kinetics (Chapter 9). It is a combination of first- and second-order reactions, which is not solvable analytically because of the nonlinear terms following the rate constants k_{OH^-} and $k_{CO_2,r}$. The rate constants were determined in the laboratory by choosing the experimental conditions in which one of the two mechanisms predominated. pH values of natural waters, however, often fall in the range 8–10, in which the reaction with both water and OH^- can be important. To determine the life time of CO_2 as a function of pH, one must derive the solution to the reaction rate equation. This is facilitated by employing the DIC and carbonate alkalinity, A_C, (Eqs. (4.15) and (4.26)) to eliminate the concentration of bicarbonate $[HCO_3^-]$, in the CO_2 reaction rate equation. This substitution results in an expression

$$\frac{d[CO_2]}{dt} = -A[CO_2] + B, \tag{4.29}$$

where

$$\begin{aligned} A &= -(k_{CO_2} + k_{OH^-}[OH^-]) + 2 \cdot k_{CO_2,r} \cdot \frac{K_W}{[OH^+]} + 2 \cdot k_{HCO_3^-} \\ B &= (2 \times DIC - A_C) \cdot k_{CO_2,r} \cdot \frac{K_W}{[OH^+]} + k_{HCO_3^-}, \end{aligned} \tag{4.30}$$

that has an analytical solution if we assume that not only A_C and DIC, but also pH, is constant:

$$[CO_2](t) = [CO_2]^0 - \frac{B}{A}\exp(-A \times t) + \frac{B}{A}. \quad (4.31)$$

This is an approximation because the OH^- concentration does change during the reaction, but since the change is not very great the equation is adequate to illustrate the importance of the two reaction mechanisms. Equation (4.31) is the solution for a reversible reaction that begins with an initial concentration of $[CO_2]^0$ and progresses toward an equilibrium value of $[CO_2]^0 + B/A$. The value represented in A is the reciprocal of the residence time of CO_2 with respect to chemical reaction and incorporates both mechanisms of reaction.

The reaction rate constants have been determined as a function of temperature and salinity by Johnson (1982). Values in Table 4.6 are calculated from the best-fit equations for his experiments. After a

Table 4.6. *Temperature dependence of rate constants of CO_2 reaction with H_2O in pure water and seawater*

The values are from the equation which best fit the data of Johnson (1982). His values for $k_{OH^-}K_W$ are reinterpreted as indicated in Emerson (1995). The equilibrium constants necessary to calculate the reverse rate constants are also tabulated. Where two values are presented in column 1 the first is for fresh water ($I = 0$) and the second is for seawater. The exponential notation in column 1 indicates the order of magnitude the variable is multiplied by to equal the number in the table. (For example, $K_1 \times 10^7$ in column 1 means 3.44×10^{-7} was multiplied by 1×10^7 before tabulating it as 3.44 in column 2.)

Temperature (°C)	Pure water					Seawater (35‰)				
	10	15	20	25	30	10	15	20	25	30
Equilibrium constants										
[a]K_1, K_1' (mol kg^{-1}) $\times 10^7$	3.44	3.80	4.15	4.45	4.71	10.0	11.2	12.5	13.9	15.4
[b]K_W, K_W' (mol kg^{-1})^2 $\times 10^{14}$	0.29	0.45	0.68	1.01	1.47	1.4	2.4	3.8	6.1	9.4
Reaction rate constants										
[c]k_{CO_2} (s^{-1}) $\times 10^2$	0.8	1.4	2.4	3.7	5.4	0.8	1.4	2.4	3.7	5.4
[c]$k_{OH^-}K_W$, $k_{OH^-}K_W'\gamma_{H^+}$ (mol kg^{-1}s^{-1}) $\times 10^{11}$	1.2	2.1	3.8	7.1	13.4	2.3	4.1	7.4	13.7	25.6
[d]k_{OH^-} (kg mol^{-1}s^{-1}) $\times 10^{-3}$	4.1	4.7	5.6	7.0	9.1	2.7	2.8	3.2	3.7	4.5
[e]$k_{H_2CO_3}$ (kg mol^{-1}s^{-1}) $\times 10^{-4}$	2.3	3.7	5.8	8.3	11.5	0.8	1.2	1.9	2.7	3.5
[e]$k_{HCO_3^-}$ (s^{-1}) $\times 10^5$	3.5	5.5	9.2	16.0	28.4	3.8	6.1	9.9	16.4	27.7

[a] K_1 ($I = 0$) from Harned and Davis (1943); K_1 (seawater) from DoE (1994).
[b] K_W ($I = 0$) from Harned and Owen (1958); K_W (seawater) from DoE (1994).
[c] From Johnson (1982).
[d] Calculated from (*c*) and (*b*), $\gamma_{H^+} = 0.6$, Millero (1995).
[e] $$\frac{k_{CO_2}}{k_{H_2CO_3}} = \frac{k_{OH^-}K_W}{k_{HCO_3^-}} = K_1\ (I = 0) \quad : \quad \frac{k_{CO_2}}{k_{H_2CO_3}} = \frac{k_{OH^-}K_W'}{k_{HCO_3^-}} = K_1'\ \text{(seawater)}.$$

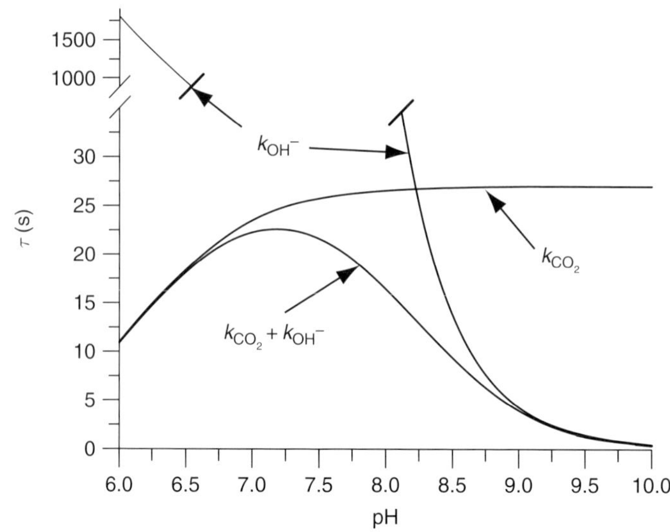

Figure 4.3. The residence time (τ) of CO_2 with respect to hydration and reaction with OH^- as a function of pH. The curves were determined from the coefficient *A* in Eq. (4.30) and the rate constants in Table 4.6. The residence times with respect to the two separate reactions are presented separately and together. CO_2 hydration is indicated by k_{CO_2} and calculated for the case where $k_{OH} = k_{HCO_3} = 0$. Hydroxylation is indicated by k_{OH} and calculated for the case where $k_{CO_2} = k_{CO_{2,r}} = 0$. Together the reactions are indicated by ($k_{CO_2} + k_{OH}$). In the pH range of seawater both reactions are important in determining the reaction residence time.

small correction to the data noted by Emerson (1995) the values in Table 4.6 are consistent with those presented by Zeebe and Wolf-Gladrow (2000). The residence time of CO_2 in seawater, calculated from Eq. (4.31) and the rate constants in Table 4.6, is presented in Fig. 4.3 (at 25 °C). The reaction of CO_2 with water dominates in the lower pH range, <8.0, and the direct combination with hydroxyl ion is most important at $pH > 10$. Between $8 < pH < 10$ both reaction mechanisms are operative.

The most important applications of these rate equations are in calculating the flux of CO_2 across the air–water interface and across the diffusive boundary layer surrounding phytoplankton. In these cases the residence times with respect to CO_2 transport (across a diffusive boundary layer) are similar to the reaction residence times. If there is enough time for reaction, a gradient in HCO_3^- is created across the boundary layer, which enhances the carbon transport over that which would be expected from a linear gradient of CO_2 across the layer. In practice, it is not possible to determine the structure of the concentration gradients across the layer so they must be calculated. We discuss this problem as it applies to CO_2 exchange across the air–water interface in Chapter 10. The excellent book by Zeebe and Wolf-Gladrow (2000) describes the application of the CO_2 reaction and diffusion kinetics to problems of plankton growth.

4.4 Processes that control the alkalinity and DIC of seawater

4.4.1 Global ocean, atmosphere, and terrestrial processes

On the global spatial scales and over time periods comparable to, and longer than, the residence time of bicarbonate in the sea (*c.* 100 ky), the alkalinity and DIC of seawater are controlled by the species composition of rivers, which is determined by weathering. The imbalance of non-protonating cations and anions in seawater is caused by the reactions of rocks with atmospheric CO_2 that are described in Chapter 2. In the generalized weathering reaction, the hydrogen ion reacts with rocks, and when this reaction is combined with the hydration reaction for CO_2 (Eq. (4.4)) bicarbonate is formed

$$\begin{array}{l} \text{rock} + H^+ + H_2O \rightarrow \text{cations} + \text{clay} + SiO(OH)_4(aq) \\ \quad + CO_2 + H_2O \rightarrow HCO_3^- + H^+ \\ \hline \text{rock} + CO_2(aq) + H_2O \rightarrow \text{cations} + \text{clay} + HCO_3^- + SiO(OH)_4(aq) \end{array} \quad (4.32)$$

Bicarbonate is the main anion in river water because of the reaction of CO_2-rich soil water with both calcium carbonate and silicate rocks (see Chapter 2). Thus, neutralization of acid in reactions with more basic rocks during weathering creates cations that are balanced by anions of carbonic acid. In this sense the composition of rocks and the atmosphere determine the overall alkalinity of the ocean.

Seawater has nearly equal amounts of alkalinity and DIC because the main source of these properties is riverine bicarbonate ion, which makes equal contributions to both constituents. The processes of $CaCO_3$ precipitation, hydrothermal circulation, and reverse weathering in sediments remove alkalinity and DIC from seawater and maintain present concentrations at about 2 mmol (meq) kg^{-1}. Reconciling the balance between river inflow and alkalinity removal from the ocean is not well understood, and is discussed in much greater detail in Chapter 2.

4.4.2 Alkalinity changes within the ocean

On time scales of oceanic circulation (1000 y and less) the internal distribution of carbonate system parameters is modified primarily by biological processes. Cross sections of the distribution of A_T and DIC in the world's oceans (Fig. 4.4) and scatter plots of the data for these quantities as a function of depth in the different ocean basins (Fig. 4.5) indicate that the concentrations increase in deep waters (1–4 km) from the North Atlantic to the Antarctic and into the Indian and Pacific Oceans following the "conveyer belt" circulation (Fig. 1.12). Degradation of organic matter (OM) and dissolution of $CaCO_3$ cause these increases in the deep waters. The chemical character of the particulate material that degrades and dissolves determines the ratio of A_T to DIC.

The stoichiometry of the phosphorus, nitrogen, and carbon in OM that degrades in the ocean (see Table 1.5 and Chapter 6) is about

$$P:N:C = 1:16:106. \quad (4.33)$$

Organic carbon degradation and oxidation creates CO_2, which is dissolved in seawater. This increases DIC but does not change the alkalinity of the water. Alkalinity is a measure of charged species;

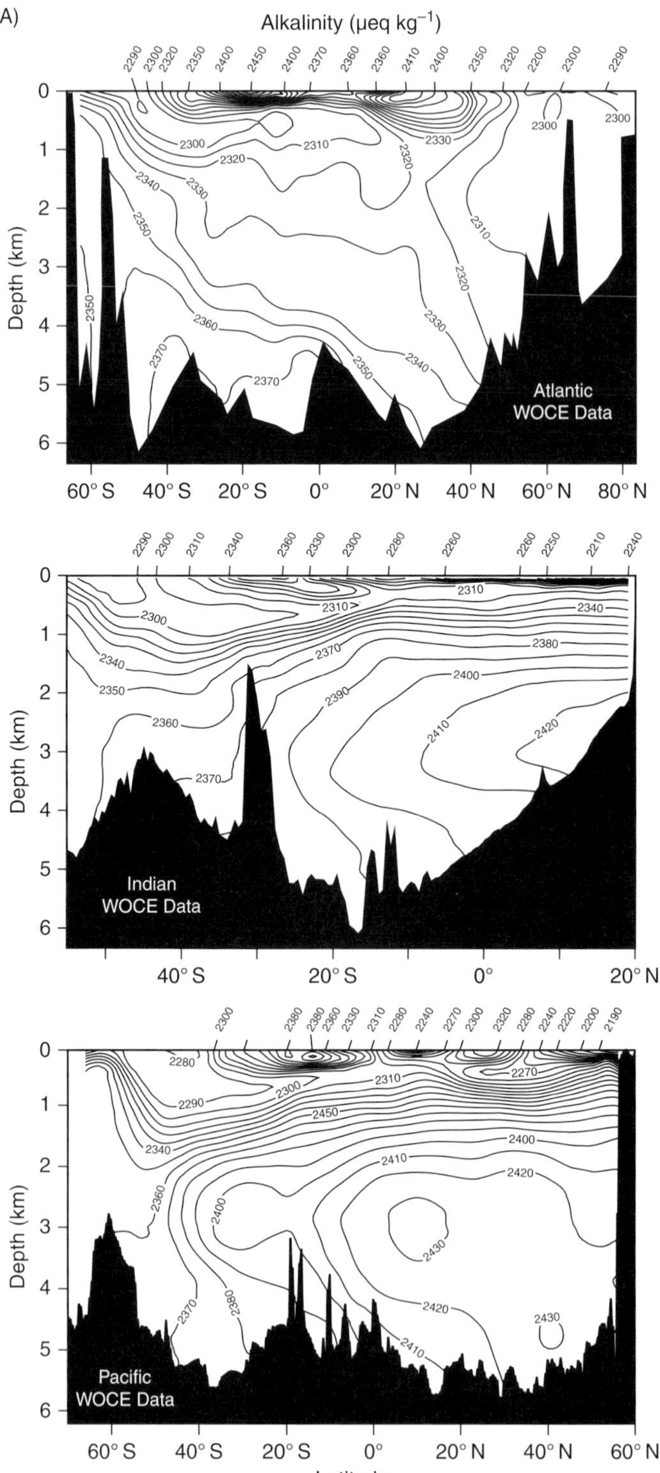

Figure 4.4. Cross sections of total alkalinity (A) and DIC (B) in the Atlantic, Indian and Pacific Oceans. Modified from the figure in Key *et al.* (2004).

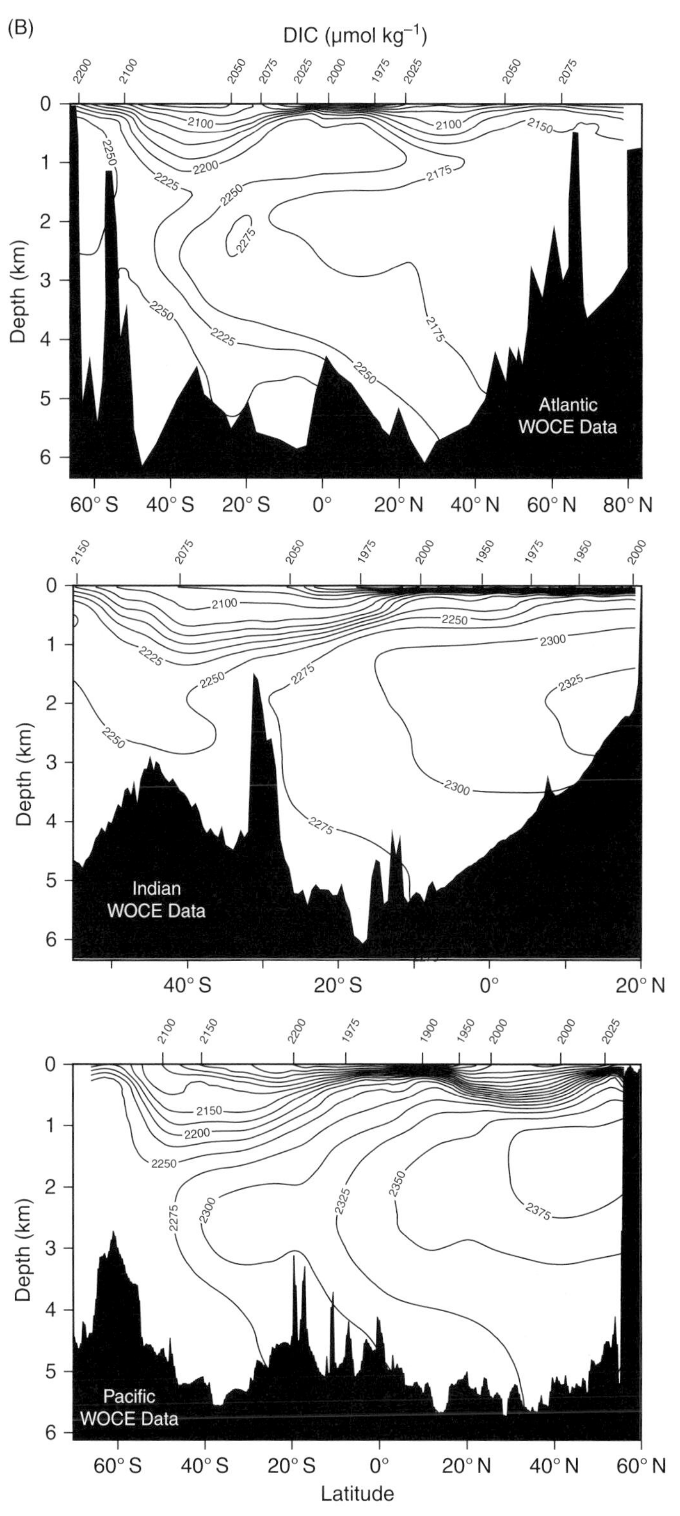

Figure 4.4. Cont.

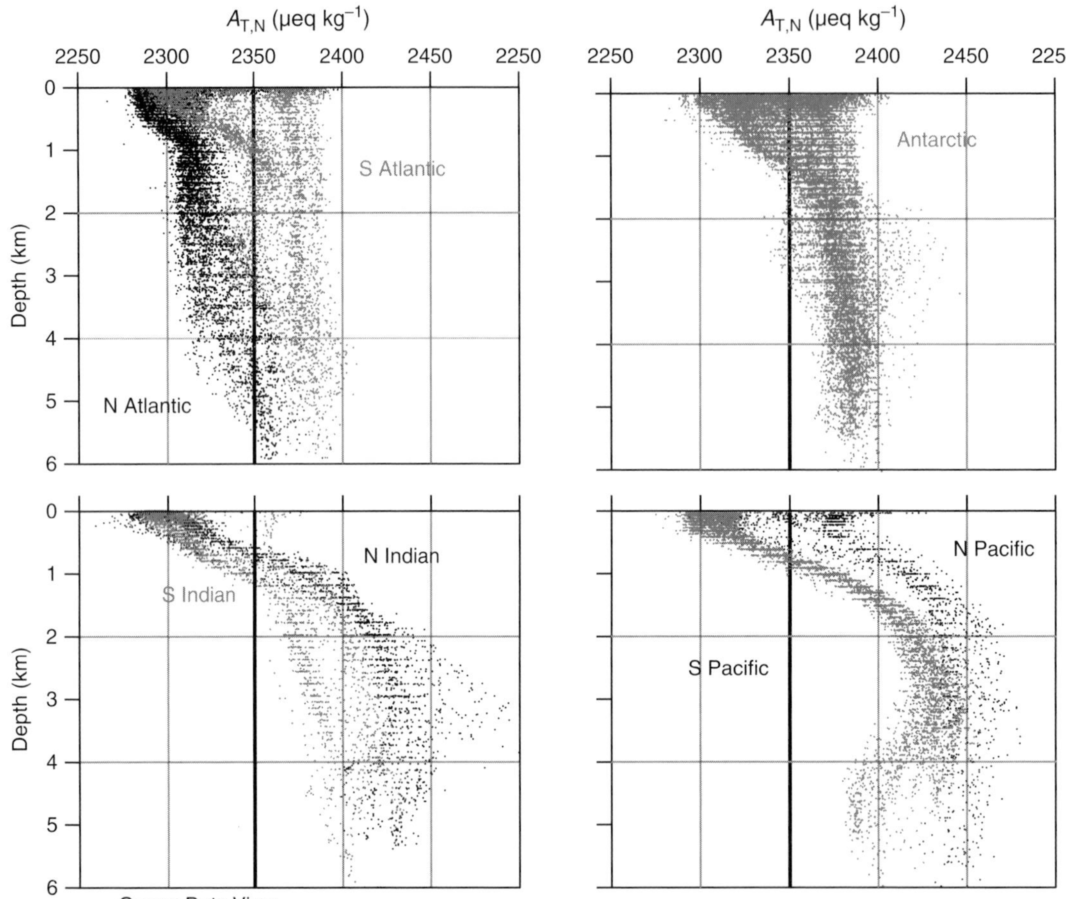

Figure 4.5. Depth profiles of total alkalinity (A_T) in the Atlantic, Antarctic, Indian and Pacific Oceans. Plotted in Ocean Data View, using data from the e-WOCE compilation.

since there is no charge associated with CO_2, its release to solution does not alter the alkalinity. The case for the nitrogen component in organic matter is not so simple because ammonia in OM is oxidized to dissolved NO_3^- during oxic degradation. This is a redox reaction that involves the transfer of hydrogen ions into solution and therefore results in an alkalinity change:

$$NH_3(OM) + 2O_2 \rightleftharpoons NO_3^- + H^+ + H_2O. \tag{4.34}$$

Since a proton is released into solution during this reaction the alkalinity decreases (see Eq. (4.24)). Thus, when a mole of organic carbon as OM is degraded it causes the DIC to increase by one mole and the alkalinity to decrease by 16/106 = 0.15 eq,

$$\Delta DIC_{OM} = 1; \Delta A_T = -0.15. \tag{4.35}$$

The change in DIC and A_T of the solution during $CaCO_3$ dissolution is very different from that resulting from OM degradation and oxidation. One mole of calcium carbonate dissolution

$$CaCO_3(s) + H_2O \rightleftharpoons Ca^{2+} + CO_3^{2-} \tag{4.36}$$

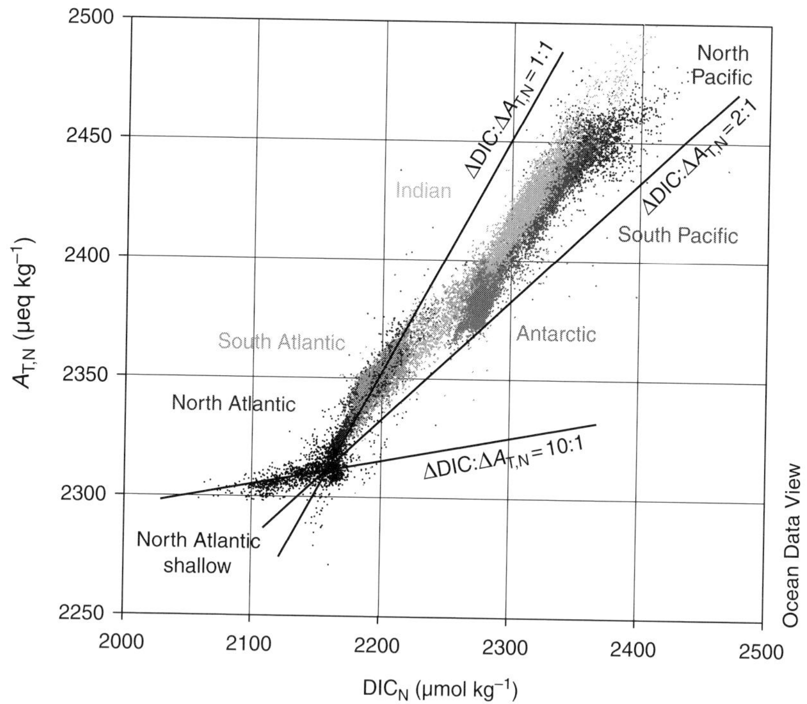

Figure 4.6. Salinity-normalized ($S = 35$) total alkalinity, $A_{T,N}$, versus salinity-normalized dissolved inorganic carbon, DIC_N, for the world's ocean. Data are for the deep ocean at depths >2.5 km except for the section labeled "North Atlantic Shallow," which is 100–1000 m in the North Atlantic Ocean. Lines indicate different $DIC_N : A_{T,N}$ ratios. (See Plate 2.)

causes an increase in alkalinity that is twice that of DIC because CO_3^{2-} introduces two charge equivalents for each mole of carbon change in solution. Thus:

$$\Delta DIC_{CO_3^{2-}} = 1; \Delta A_{T,CO_3^{2-}} = 2.0. \qquad (4.37)$$

It is thus clear that the change in alkalinity and DIC in seawater during degradation and dissolution of algae created in the surface ocean during photosynthesis depends greatly on the chemical character of that particulate material. The ecology in the ocean euphotic zone greatly influences the chemical changes observed in the sea.

Figure 4.6 is a plot of the salinity-normalized alkalinity, $A_{T,N}$, versus salinity-normalized dissolved inorganic carbon, DIC_N, for the ocean between Atlantic surface water and the deep North Atlantic (100–2000 m) and then along the deep water conveyor belt circulation between 2 and 4 km. The lines in the figure illustrate that the $DIC_N : A_{T,N}$ ratios during the "aging" of subsurface seawater are not constant throughout the ocean. Between the surface Atlantic and the base of the thermocline the change in $DIC_N : A_{T,N}$ is about 10:1 whereas in the depth range of 2–4 km, from the deep N Atlantic to deep Indian and Pacific Oceans, the ratio is between 1:1 and 2:1. The difference is due to the high OM : $CaCO_3$ ratio in particles that exit the euphotic zone and more rapid degradation of organic matter than dissolution of $CaCO_3$ as particles fall through the water. More organic matter degrades than $CaCO_3$ dissolves in the upper portion of the ocean. In the deeper waters the $DIC_N : A_{T,N}$ ratio is close to that expected for the addition of HCO_3^- to the water ($DIC : A_T = 1:1$)

except in the Antarctic, where the trend is relatively richer in DIC. Mineral-secreting plankton in the Southern Ocean are dominated by diatoms, which form opal rather than $CaCO_3$ shells. Thus, particle dissolution at depth in this part of the ocean releases DIC and H_4SiO_4 to the water but little alkalinity. The general 1:1 increase in DIC_N and $A_{T,N}$ in ocean deep waters is probably strongly influenced by reactions at the sediment–water interface (see Chapter 6 and Jahnke and Jackson, 1987). In carbonate-rich sediments a large percentage of the CO_2 produced by organic matter degradation reacts with $CaCO_3$ to produce HCO_3^-, which translates to an equal increase in DIC and A_T in solution.

The relation between the relative changes of DIC and A_T in seawater and the OM degradation to $CaCO_3$ dissolution ratio in particulate matter is illustrated in Table 4.7. The DIC : A_T ratio to be expected is calculated assuming one part $CaCO_3$ dissolution and progressively greater parts of OM degradation by using the stoichiometry in Eqs. (4.35) and (4.37). Solid-phase OM : $CaCO_3$ ratios necessary to create the observed ΔDIC : ΔA_T ratios vary from about 8:1 for the transition from the upper ocean through the thermocline in the North Atlantic to about 1.5:1 in the deeper waters of the world's ocean. The higher values are less than the ratio exiting the upper ocean (Sarmiento *et al.* (2002) determine an export flux of DIC: A_T ~15), presumably because much of the organic matter is respired in the top few hundred meters below the euphotic zone; the data along the 10:1 line in Fig. 4.6 are from a much greater depth range (100–2000 m). The deeper values are less than the ratio of 4:1 that derives from box models (Broecker and Peng, 1982) where the entire deep ocean is a weighted average of the data presented in Fig. 4.6.

It was demonstrated in Table 4.4 how the DIC and A_T changes observed in deep waters alter the carbonate system constituents. One can predict the relative change in carbonate ion concentration resulting from solubilization of particulate matter with an OM : $CaCO_3$ molar ratio of between 10 and 1.5 by focusing on the changes in alkalinity and DIC. We use carbonate alkalinity, A_C, in this calculation for simplicity. In all cases of Table 4.7 the composite change

Table 4.7. *Relative changes in DIC and A_T in seawater caused by dissolution of 1 μmol kg^{-1} of $CaCO_3$ along with degradation of 2–8 μmol kg^{-1} of organic carbon*

The ΔDIC : ΔA_T trends in Fig. 4.6 are in accord with OM : $CaCO_3$ ratios of *c.* 8:1 and *c.* 1.5:1.

OM degraded (μmol kg^{-1})	ΔDIC (μmol kg^{-1})			ΔA_T (μeq kg^{-1})			ΔDIC : ΔA_T
	from OM	from $CaCO_3$	composite	from OM	from $CaCO_3$	composite	
2	2	1	3	−0.3	2	1.7	1.8
4	4	1	5	−0.6	2	1.4	3.6
6	6	1	7	−0.9	2	1.1	6.4
8	8	1	9	−1.2	2	0.8	11.2

of DIC is greater than that for A_C. Subtracting the equation for DIC from that for carbonate alkalinity gives:

$$\begin{aligned} A_C - \text{DIC} &= \left(\left[\text{HCO}_3^-\right] + 2 \cdot \left[\text{CO}_3^{2-}\right] \right) - \left(\left[\text{HCO}_3^-\right] + \left[\text{CO}_3^{2-}\right] + \left[\text{CO}_2\right] \right) \\ &= \left[\text{CO}_3^{2-}\right] - \left[\text{CO}_2\right] \\ &\cong \left[\text{CO}_3^{2-}\right]. \end{aligned} \tag{4.38}$$

(Note that the approximation in the last step is only accurate in ocean waters with pH equal to or greater than 8.0. This is seen in Table 4.4, where CO_3^{2-} and CO_2 concentrations are evaluated in different water masses.) The above approximation indicates that addition of more DIC than A_C to the water results in a decrease in carbonate ion concentration ($\Delta A_C - \Delta \text{DIC} = \Delta CO_3^{2-}$). Essentially more acid, in the form of CO_2 than base in the form of CO_3^{2-} is added to the water during the solubilization of particulate matter. These trends are borne out in Table 4.4, where the actual carbonate species changes are calculated by using the complete carbonate equilibrium equations.

In the above discussion of the response of the carbonate system to changes caused by OM degradation (addition of CO_2) or $CaCO_3$ dissolution (addition of CO_3^{2-}) we relied almost exclusively on changes in the total quantities DIC and A_T (or A_C) to gain insight into how the system responds. The reason for this is that it is possible to predict exactly how the total quantities will change due to organic carbon degradation or $CaCO_3$ dissolution, whereas it is not clear how the equilibria will react without solving the entire set of equations (Fig. 4.2).

As an example, let us start with surface seawater and add 20 μmol kg^{-1} of CO_2 only. DIC in the solution increases by 20 μmol kg^{-1} but A_C does not change, which is roughly analogous to organic matter degradation with no $CaCO_3$ dissolution. We will approximate the change in CO_3^{2-} and HCO_3^- and then see how correct this turns out to be. From the carbonate equilibrium program we find that the distribution of carbonate species is that in Table 4.8(a) for $A_T = 2300$ μeq kg^{-1} and DIC $= 2188$ μmol kg^{-1} at 25 °C and $S = 35$. To predict the change in HCO_3^- and CO_3^{2-} in response to the addition of CO_2 we could take two different routes. First, by the laws of mass action we would predict from the CO_2 hydration equation that bicarbonate would be formed

$$CO_2(aq) + H_2O \rightleftharpoons HCO_3^- + H^+. \tag{4.11}$$

However, there is no way to know how much this would affect the CO_3^{2-} concentration formed by the second carbonate dissociation reaction

$$HCO_3^- \rightleftharpoons CO_3^{2-} + H^+. \tag{4.9}$$

We are stuck unless we do the entire equilibrium – mass balance calculation or refer to Fig. 4.2 to find the answer. The lines in the figure indicate that as CO_2 increases CO_3^{2-} decreases, but we obtain very little information about the fate of HCO_3^-.

Table 4.8. *The degree of approximation involved in calculations using Eq. (4.38)*

(1) Distribution of carbonate species in surface seawater at chemical equilibrium (25 °C, $S = 35$). (2) After the addition of 20 μmol kg^{-1} of CO_2: (a) guess using Eq. (4.38), (b) assuming $A_T = A_C$ only, (c) carbonate equilibrium equations assuming $A_T = A_{C\&B}$. Note differences in the changes in HCO_3^- and CO_3^{2-}. (3) The same as (2) except for dissolution of the equivalent of 20 μmol kg^{-1} $CaCO_3$. All concentrations are in μmol kg^{-1} except $A_{C\&B}$ and A_C, which are μeq kg^{-1}.

	$A_{C\&B}$	A_C	DIC	HCO_3^-	CO_3^{2-}	CO_2
(1) Surf. SW	2300	2188	1950	1696	246	9
(2) +20 μmol kg^{-1} of CO_2						
(a) Eq. (4.38)		2188	1970	1736	226	9
Δ(1 – 2a)		**0**	**+20**	**+40**	**−20**	**0**
(b) A_C, DIC		2188	1970	1732	228	10
Δ(1 – 2b)		**0**	**+20**	**+36**	**−18**	**+1**
(c) $A_{C\&B}$, DIC	2300	2194	1970	1728	233	10
Δ(1 – 2c)	**0**	**+6**	**+20**	**+32**	**−13**	**+1**
(3) +20 μmol kg^{-1} of CO_3^{2-}						
(a) Eq. (4.38)		2228	1970	1696	266	9
Δ(1 – 3a)		**+40**	**+20**	**0**	**+20**	**0**
(b) A_C, DIC		2228	1970	1694	267	9
Δ(1 – 3b)		**+40**	**+20**	**−2**	**+21**	**0**
(c) $A_{C\&B}$, DIC	2340	2221	1970	1701	260	9
Δ(1 – 3c)	**+40**	**+33**	**+20**	**+5**	**+14**	**0**

The other route is to think in terms of mass and charge balance. By subtracting the change in dissolved organic matter, ΔDIC, from the change in carbonate alkalinity, ΔA_C, and realizing that CO_2 is a very small component of DIC and can be neglected in the DIC formula (again, this is true for surface waters but not for those in the deep ocean), the CO_3^{2-} concentration must decrease by about 20 μmol kg^{-1} (Eq. (4.38)):

$$\Delta A_C - \Delta \mathrm{DIC} = \Delta\left[CO_3^{2-}\right] - \Delta[CO_2] \cong \Delta\left[CO_3^{2-}\right] \cong -20\ \mu\mathrm{mol\ kg}^{-1}.$$

Since the only carbonate species added was CO_2, it is reasonable to assume A_C cannot have changed much (we are going to check this below). Thus, any change in CO_3^{2-} will require an opposite change in HCO_3^- of twice the magnitude to maintain a neutral solution. The only way both of these can happen is if HCO_3^- grows by 40 μmol kg^{-1} as CO_3^{2-} decreases by 20 μmol kg^{-1} (Table 4.8(a)).

Calculated changes in HCO_3^- and CO_3^{2-} after the addition of 20 μmol kg^{-1} of CO_2 using the full set of carbonate equilibrium equations (see Appendix 5.1) with the assumption that A_C remains constant are presented in line 2b of Table 4.8. We see that the mass balance calculation is approximately correct (compare the changes under (2a) and (2b) of Table 4.8). Taking the final step towards reality by stipulating that it is carbonate + borate alkalinity that does not change rather than the carbonate alkalinity ($\Delta A_{C\&B} = 0$, 2c in Table 4.8) reveals that the bicarbonate and carbonate changes are somewhat smaller than predicted by the simple calculation represented by Eq. (4.37). The reason for the differences between the changes in 2b and 2c is that addition of the acid CO_2 caused the borate equilibrium in Eq. (4.19) to shift to the left, decreasing the borate concentration, which required an increase in carbonate alkalinity, A_C, for $A_{C\&B}$ to remain constant. The bottom line is that the approximation in Eq. (4.38) overestimates the HCO_3^- and CO_3^{2-} changes by *c*.25 and 50%, respectively.

We can try this again by estimating the HCO_3^- and CO_3^{2-} changes from the addition of 20 μmol kg^{-1} of $CaCO_3^{2-}$ to the same surface water (Table 4.8(3)). In this case the carbonate alkalinity increases by 40 μeq kg^{-1} and the DIC by 20 μmol kg^{-1}. Using the approximation in Eq. (4.38), leads to a change in CO_3^{2-} of + 20 μmol kg^{-1} ($\Delta A_C - \Delta DIC = \Delta CO_3^{2+} = +20$ μmol $kg^{-1} = 40$ μeq kg^{-1}). Since the change in A_C is 40 μeq kg^{-1} there can be virtually no change in HCO_3^-. We see that removing the successive approximations in Table 4.8(3b and c) reveals errors that are of the same magnitude as when we did this for the CO_2 addition in section (2) of the table.

Generally, when estimating the changes to be expected in the carbonate system by organic matter degradation, $CaCO_3$ dissolution or exchange with the atmosphere, it is much safer to deal with changes in the total properties A_C and DIC rather than trying to guess the response of the carbonate equilibrium equations. One can predict precisely how the total quantities change, and then it is possible to show the change in direction and approximate concentration of both CO_3^{2-} and HCO_3^-. Absolute values of the carbonate species change, however, must wait till you consult the simultaneous solution of the carbonate equilibrium equations.

Appendix 4.1 Carbonate system equilibrium equations in seawater

Appendix 4A1.1 describes the equations necessary for determining the concentrations of carbonate species in seawater for the three different definitions of alkalinity given in the text. Appendix 4A1.2 is a listing of the Matlab program for determining carbonate buffer species by using the equations for the case where $A_T = A_{C\&B}$.

4A1.1 Equations

Equation numbers refer to equations in text.

(a) Using carbonate alkalinity, A_C

Five equations, seven unknown chemical concentrations: A_C, DIC, HCO_3^-, CO_3^{2-}, CO_2, H^+, f_{CO_2}.

$$A_C = [HCO_3^-] + 2[CO_3^{2-}] \tag{4.26}$$

$$DIC = [HCO_3^-] + [CO_3^{2-}] + [CO_2] \tag{4.15}$$

$$K_1' = \frac{[HCO_3^-] \times [H^+]}{[CO_2]} \tag{4.11}$$

$$K_2' = \frac{[CO_3^{2-}] \times [H^+]}{[HCO_3^-]} \tag{4.9}$$

$$K_H = \frac{[CO_2]}{f_{CO_2,a}}. \tag{4.14}$$

(b) Using carbonate and borate alkalinity, $A_{C\&B}$

(Seven equations and ten unknown chemical concentrations.)

New unknown concentrations: B_T, $B(OH)_4^-$, $B(OH)_3$.

Substitute Eq. (4.25) for Eq. (4.26):

$$A_{C\&B} = [HCO_3^-] + 2[CO_3^{2-}] + [B(OH)_4^-]. \tag{4.25}$$

Include borate-related equations (4.19, 4.20):

$$[B]_T = [B(OH)_4^-] + [B(OH)_3] \tag{4.20}$$

$$K_B' = \frac{[B(OH)_4^-] \times [H^+]}{[B(OH)_3]}. \tag{4.19}$$

(c) Using the total alkalinity, A_T **but without the acids** HSO_4^-**, HF,** H_3PO_4 **and** H^+

Fourteen equations and 19 unknown concentrations.

New unknown concentrations: Si_T, P_T, $H_3SiO_4^-$, H_4SiO_4, PO_4^{3-}, HPO_4^{2-}, $H_2PO_4^-$, OH^-.

Substitute Eq. (4.23) for Eq. (4.25):

$$\begin{aligned} A_T = &[HCO_3^-] + 2[CO^{3-}] + [B(OH)_4^-] + [H_3SiO_4^-] + [HPO_4^2-] \\ &+ 2[PO_4^{3-}] + [OH^-]. \end{aligned} \tag{4.24}$$

Include new species-related mass balance and equilibrium equations:

$$[Si]_T = [H_3SiO_4^-] + [H_4SiO_4] \tag{4A1.1}$$

$$[P]_T = [H_3PO_4] + [H_2PO_4^{2-}] + [HPO_4^-] + [PO_4^{3-}] \tag{4A1.2}$$

$$K_{Si} = \frac{[H_3SiO_4^-] \times [H^+]}{[H_4SiO_4]} \tag{4A1.3}$$

$$K_{P,1} = \frac{[H_2PO_4^-] \times [H^+]}{[H_3PO_4]} \tag{4A1.4}$$

$$K_{P,2} = \frac{[HPO_4^{2-}] \times [H^+]}{[H_2PO_4^-]} \tag{4A1.5}$$

$$K_{P,3} = \frac{[PO_4^{3-}] \times [H^+]}{[HPO_4^{2-}]} \tag{4A1.6}$$

$$K_W = [OH^-][H^+] \tag{4A1.7}$$

4A1.2

The following Matlab function program finds the root of the cubic equation for $[H^+]$ in terms of $A_{C\&B}$ and DIC resulting from the combination of the equations in 4A1.1 (a) and (b) above (Zeebe and Wolf-Gladrow, 2000). Input values are temperature, salinity, depth, $A_{C\&B}$ and DIC and the outputs are f_{CO_2}, pH, $[CO_2]$, $[HCO_3^-]$ and $[CO_3^{2-}]$. Units and equilibrium constants used are indicated in the comment statements, which are preceded by a % sign.

```
    function [fco2, pH, co2, hco3, co3] = co3eq (temp, s, z, alk, dic)
% Function to calculate fCO2, HCO3, and CO3 from ALK and DIC as a
% f(temp,sal,Z)
% temp = temp(deg C),
% sal = salinity(ppt),depth = z(m),alk = ALK(microeq/kg),
% dic = DIC(micromol kg^-1)
% HCO3, CO3, and CO2 are returned in mol kg^-1, fCO2 in atm
% This program uses the equations in Zeebe and Wolf-Gladrow (2000) and
% Matlab's root finding routine
% checked for fCO2 against Luecker et al. (2000), May 2002;
% Depth dependence has not been checked
    t = temp + 273.15;
    Pr = z/10;
    alk = alk * .000001;
    dic = dic * .000001;
    R = 83.131;
% Calculate total borate (tbor) from chlorinity
    tbor = .000416 * s / 35.0;
% Calculate Henry's Law coeff, KH (Weiss, 1974)
    U1 = -60.2409 + 93.4517 * (100/t) + 23.3585 * log(t/100);
    U2 = s * (.023517 - .023656 * (t/100) + .0047036 * (t/100) ^ 2);
    KH = exp(U1 + U2);
% Calculate KB from temp & sal (Dickson, 1990)
    KB = exp((-8966.9 -2890.53 * s ^0.5 -77.942 * s + 1.728 * s^1.5
       - 0.0996*s^2)/t ... + 148.0248 + 137.1942 * s^0.5 + 1.62142 * s
       - (24.4344 + 25.085 * s^0.5 + ...0.2474 * s) * log(t) + 0.053105 * s^0.5 * t);
% Calculate K1 and K2 (Luecker et al., 2000)
    K1 = 10^(-(3633.86/t - 61.2172 + 9.67770 *
    log(t) - 0.011555*s + 0.0001152 * s^2));
    K2 = 10^(-(471.78/t + 25.9290 -3.16967 *
    log(t) - 0.01781*s + 0.0001122 * s^2));
% Pressure variation of K1, K2, and KB (Millero, 1995)
    dvB = -29.48 + 0.1622 * temp - .002608 * (temp)^2;
    dv1 = -25.50 + 0.1271 * temp;
    dv2 = -15.82 - 0.0219 * temp;
    dkB = -.00284;
    dk1 = -.00308 + 0.0000877 * temp;
```

```
    dk2 = +.00113 -.0001475 * temp;
    KB = (exp( - (dvB/(R*t))*Pr + (0.5 * dkB/(R*t))*Pr^2)) * KB;
    K1 = (exp( - (dv1/(R*t))*Pr + (0.5 * dk1/(R*t))*Pr^2)) * K1;
    K2 = (exp( - (dv2/(R*t))*Pr + (0.5 * dk2/(R*t))*Pr^2)) * K2;
% temperature dependence of Kw (DoE, 1994)
    KW1 = 148.96502-13847.26/t-23.65218*log(t);
    KW2 = (118.67/t-5.977 + 1.0495*log(t))*s ^ .5-0.01615*s;
    KW = exp(KW1 + KW2);
% solve for H ion (Zeebe and Wolf-Gladrow, 2000)
    a1 = 1;
    a2 = (alk + KB + K1);
    a3 = (alk*KB-KB*tbor-KW + alk*K1 + K1*KB + K1*K2-dic*K1);
    a4 = (-KW*KB + alk*KB*K1-KB*tbor*K1-KW*K1 + alk*K1*K2
        + KB*K1*K2-dic*KB*K1-2*dic*K1*K2);
    a5 = (-KW*KB*K1 + alk*KB*K1*K2-KW*K1*K2-KB*tbor*K1*K2 -2*dic*KB*
        K1*K2);
    a6 = -KB*KW*K1*K2;
    p = [a1 a2 a3 a4 a5 a6 ];
    r = roots(p);
    h = max(real(r));
% calculate the HCO3, CO3 and CO2aq using DIC, AlK and H+
    format short g;
    hco3 = dic/(1 + h/K1 + K2/h);
    co3 = dic/(1 + h/K2 + h*h/(K1*K2));
    co2 = dic/(1 + K1/h + K1*K2/(h*h));
    fco2 = co2 / KH;
    pH = -log10(h);
% calculate B(OH)4- and OH
    BOH4 = KB*tbor/(h + KB);OH = KW/h;
% recalculate DIC and Alk to check calculations
    Ct = (hco3 + co3 + co2)*1e6;
    At = (hco3 + 2*co3 + BOH4 + OH-h)*1e6;
```

Appendix 4.2 Equations for calculating the equilibrium constants of the carbonate and borate buffer system

Constants are based on the "total" pH scale, pH_T (Dickson, 1984, 1993). Values are first presented at 1 atm pressure and then equations are given for calculating the pressure effect on K. T is temperature in either degrees Kelvin (T), or degrees centigrade (T_C). Salinities are on the practical salinity scale. Equilibrium constants for the equilibria other than K_H, K_1', K_2', K_B', and K_W' given in Appendix 4A1.1(c) can be found in DoE (1994) and in Zeebe and Wolf-Gladrow (2000).

4A2.1. Values at 1 atmosphere

(a) The Henry's Law constant for CO_2 in seawater (mol kg^{-1}atm^{-1}), Eq. (4.14)

Source: From Weiss (1974) as reported in DoE (1994).

$$\ln K_{\mathrm{H}} = \frac{9345.17}{T} - 60.2409 + 23.3585 \ln\left(\frac{T}{100}\right) + S\left[0.023\,517 - 0.000\,236\,56T + 0.004\,7036\left(\frac{T}{100}\right)^2\right] = -3.5617\ (T_{\mathrm{C}} = 25\ (\mathrm{T} = 298.15), \mathrm{S} = 35). \tag{4A2.1}$$

(b) The first (Eq. (4.11)) and second (Eq. (4.9)) dissociation constants for carbonic acid in seawater (mol kg^{-1})

Mehrbach's constants are given on the total pH scale (Luecker *et al.*, 2000).

$$\mathrm{p}K_1' = \frac{3633.86}{T} - 61.2172 + 9.6777 \ln(T) - 0.011\,555S + 0.000\,1152S^2 = 5.847\ (T_{\mathrm{C}} = 25,\ S = 35) \tag{4A2.2}$$

$$\mathrm{p}K_2' = \frac{471.78}{T} + 25.9290 - 3.169\,67 \ln(T) - 0.017\,81S + 0.000\,1122S^2 = 8.966\ (T_{\mathrm{C}} = 25,\ S = 35). \tag{4A2.3}$$

(c) Boric acid in seawater, mol kg^{-1} (Eqs. (4.19 and 4.20))

Based on Dickson (1990) as reported in DoE (1994).

$$B_{\mathrm{T}} = [\mathrm{B(OH)_3}] + [\mathrm{B(OH)_4^-}] = 4.16 \times 10^{-4}\ (S = 35) \tag{4A2.4}$$

$$\ln K_{\mathrm{B}} = \frac{-8966.90 - 2890.53S^{1/2} - 77.942S + 1.728S^{3/2} - 0.0996S^2}{T} + 148.0248 + 137.1942S^{1/2} + 1.621\,42S - \left(24.4344 + 25.085S^{1/2} + 0.2474S\right)\ln(T) + 0.053\,105S^{1/2}T = -19.7964\ (T_{\mathrm{C}} = 25,\ S = 35). \tag{4A2.5}$$

(d) The dissociation constant of water, $mol^2\ kg^{-1}$

From Dickson and Riley (1979) as reported in DoE (1994).

$$\ln K_{\mathrm{W}} = 148.965\,02 - \frac{13\,847.26}{T} - 23.652\,1 \ln(T) + \left(\frac{118.67}{T} - 5.977 + 1.0495 \ln(T)\right)S^{1/2} - 0.016\,15\,S = -30.434\ (T_{\mathrm{C}} = 25,\ S = 35). \tag{4A2.6}$$

4A2.2 The pressure dependence (from Millero, 1995)

The effect of pressure can be calculated from the molal volume, ΔV, and compressibility, $\Delta\kappa$, changes for any given reaction

$$\ln\frac{K_{\mathrm{P}}}{K_0} = -\left(\frac{\Delta V}{RT}\right)\mathrm{P} + \left(\frac{0.5\Delta\kappa}{RT}\right)\mathrm{P}^2, \tag{4A2.7}$$

Table 4A2.1. *Parameters for calculating the effect of pressure change on carbonate buffer system reactions and values of equilibrium constants at P = 0 and 300 bar*

Constant	a_0	a_1	$a_2 \times 10^3$	$b_0 \times 10^3$	$b_1 \times 10^3$	$P=0$	$P=300$	K^{300}/K^0
pK_1'	25.50	−0.1271	0.0	3.08	−0.0877	5.847	5.726	1.32
pK_2'	15.82	0.0219	0.0	−1.13	0.1475	8.966	8.883	1.21
pK_B'	29.48	−0.1622	2.608	2.84	0.0	8.598	8.455	1.39
pK_W'	25.60	0.2324	−3.6246	5.13	0.0794	13.217	13.106	1.43

where K_P and K_0 are equilibrium constants for the reaction of interest at pressure P and at 0 bars (1 atm), respectively. P is pressure in bars, $R = 83.131$ (cm^3 bar mol^{-1} K^{-1}) and T is in degrees Kelvin. The molar volume (cm^3 mol) and compressibility can be fit to equations of the form ($S = 35$)

$$-\Delta V = a_0 + a_1 T_C + a_2 T_C^2; \quad (4A2.8)$$

$$-\Delta\kappa = b_0 + b_1 T_C, \quad (4A2.9)$$

where T_C is now temperature in degrees C. Values for the coefficients a and b are presented in Table 4A2.1 along with calculated differences in pK' and K' at two different pressures ($T_C = 25$ °C, $S = 35$).

References

Broecker, W. S. and T.-H. Peng (1982) *Tracers in the Sea*. Palisades, NY: ElDIGIO Press.

Dickson, A. G. (1981) An exact definition of total alkalinity and a procedure for estimation of alkalinity and total inorganic carbon from titration data. *Deep-Sea Res.* **28A**(6), 609–23.

Dickson, A. G. (1984) pH scales and proton-transfer reactions in saline media such as sea water. *Geochim. Cosmochim. Acta* **48**, 2299–308.

Dickson, A. G. (1990) Thermodynamics of the dissociation of boric acid in synthetic seawater from 273.15 to 298.15 K. *Deep-Sea Res.* **37**, 755–66.

Dickson, A. G. (1993) pH buffers for sea water media based on the total hydrogen ion concentration scale. *Deep-Sea Res.* **40**, 107–18.

Dickson, A. G. and F. J. Millero (1987) A comparison of the equilibrium constants for the dissociation of carbonic acid in seawater media. *Deep-Sea Res.* **34**, 1733–43.

Dickson, A. G. and J. P. Riley (1979) The estimation of acid dissociation constants in seawater media from potentiometric titrations with strong base I. The ionic product of water (K_w). *Mar. Chem.* **7**, 89–99.

DoE (1994) *Handbook of Methods for the Analysis of the Various Parameters of the Carbon Dioxide System in Sea Water*, version 2 (ed. A. G. Dickson and C. Goyet). ORNL/CDIAC-74.

Emerson, S. (1995) Enhanced transport of carbon dioxide during gas exchange. In *Air-Water Gas Transfer. Selected papers from the Third International Symposium on Air-Water Gas Transfer July 24–27, 1995 Heidelberg University* (ed. B. Jahne and E. C. Monahan), pp. 23–36. Hanau: AEON Verlag.

Harned, H. S. and R. Davis (1943) The ionization constant of carbonic acid in water and the solubility of CO_2 in water and aqueous salt solution from 0 to 50 °C. *J. Am. Chem. Soc.* **65**, 2030–7.

Harned, H. S. and B. B. Owen (1958) *The Physical Chemistry of Electrolyte Solutions*. New York, NY: Reinhold.

Jahnke, R. J. and G. A. Jackson (1987) Role of sea floor organisms in oxygen consumption in the deep North Pacific Ocean. *Nature* **329**, 621–3.

Johnson, K. S. (1982) Carbon dioxide hydration and dehydration kinetics in seawater. *Limnol. Oceanogr.* **27**, 849–55.

Keir, R. S. (1980) The dissolution kinetics of biogenic calcium carbonates in seawater. *Geochim. Cosmochim. Acta* **44**, 241–52.

Key, R. M., A. Kozar, C. L. Sabine *et al.* (2004) A global ocean carbon climatology: results from Global Data Analysis Project (GLODAP). *Global. Biogeochem. Cycles* **18**, GB4031, doi: 10.1029/2004GB002247.

Lewis, E. and D. Wallace (1998) Program developed for CO_2 system calculations. ORNL/CDIAC - 105. Carbon Dioxide Information Analysis Center, Oak Ridge National Laboratory, U.S. Department of Energy, Oak Ridge, TN.

Luecker, T. J., A. G. Dickson and C. D. Keeling (2000) Ocean pCO_2 calculated from dissolved inorganic carbon, alkalinity, and the equations for K1 and K2: validation based on laboratory measurements of CO_2 in gas and seawater at equilibrium. *Mar. Chem.* **70**, 105–19.

Mehrbach, C., C. H. Culberson, J. E. Hawley and R. M. Pytkowicz (1973) Measurements of the apparent dissociation constants of carbonic acid in seawater at atmospheric pressure. *Limnol. Oceanogr.* **18**, 897–907.

Millero, F. J. (1995) Thermodynamics of the carbon dioxide system in the oceans. *Geochim. Cosmochim. Acta* **59**, 661–77.

Sarmiento, J. L., J. Dunne, A. Gnanadesikan *et al.* (2002) A new estimate of the $CaCO_3$ to organic carbon export ratio. *Global Biogeochem. Cycles* **16**, doi: 10.1029/2002GB001010.

Stumm, W. and J. J. Morgan (1996) *Aquatic Chemistry*. New York, NY: Wiley Interscience.

Weiss, R. F. (1974) Carbon dioxide in water and seawater: the solubility of a non-ideal gas. *Mar. Chem.* **2**, 203–15.

Zeebe, R. E. and D. A. Wolf-Gladrow (2000) *CO_2 in Seawater: Equilibrium, Kinetics, Isotopes*. Amsterdam: Elsevier.

5

Stable and radioactive isotopes

Analyses of stable and radioactive isotope compositions have become a mainstay of the chemical perspective of oceanography, owing in large part to their value as tracers of important oceanographic processes. The utility of isotopes as tracers of biological, physical and geological ocean processes is perhaps the main reason that chemical oceanography has become a strongly interdisciplinary science. Small contrasts in stable isotope compositions can carry geographic information for discriminating sources such as different ocean water masses, and marine versus terrestrially derived organic matter. Within fossils, isotope distributions afford information about the temperatures, geographic settings, transport mechanisms, and ecology (e.g. who ate whom) of ancient environments. Stable isotope compositions also integrate the cumulative results of ongoing processes such as the passage of organic elements up trophic levels, climate change, marine productivity, and the formation and melting of continental glaciers.

Stable isotopic signatures can persist over geologic time, even through severe changes in chemical composition. Radioactive isotopes have the additional property of being useful as nuclear clocks that, regardless of environmental conditions, dependably tick away to indicate the age of an object or the dynamics (e.g. turnover time) of a pool of materials. In addition, nuclear decay events often involve conversions of parent isotopes to daughter elements with very different physical and chemical properties, which then can be sensitively traced as they seek new chemical forms and locations in the ocean.

About 15 billion years ago, immediately following the Big Bang, the light elements of H (≈99%), He (≈1%), and trace amounts of Li formed as the universe cooled. Subsequent nuclear reactions during star formation and collapse created (and are still creating) all the remaining elements. On Earth today, there are 92 naturally occurring elements, each of which has a unique number of protons (atomic number, Z) in its nucleus; however, the number of accompanying neutrons (N) often varies. The masses of protons and neutrons are essentially the same, making their numeric sum equivalent to the relative atomic mass (A) of the atom. Atoms of the same element with different numbers of neutrons, and hence different atomic masses, are referred to as *isotopes* of that element (Table 5.1). Because different isotopes of the same element have the same number of electrons, they exhibit almost identical chemical properties. The small differences are that heavier isotopes of an element typically form slightly stronger bonds to other atoms, and molecules containing heavier isotopes move somewhat more slowly at a given temperature owing to their greater mass.

In general, all isotopes can be categorized into stable and radioactive forms, based on whether they spontaneously convert into other nuclei at a discernable rate. Naturally occurring isotopes are either stable on the time scale of the history of the Earth (≈4.5 billion years), or are continuously formed from long-lived parents or cosmic rays. For example, oxygen occurs in three stable isotopic forms as ^{16}O, ^{17}O and ^{18}O, with 8, 9, or 10 neutrons, respectively (Table 5.1). The ^{19}O isotope with 11 neutrons, however, is radioactive. This isotope has zero natural abundance because its half life (the time to decrease its activity by a factor of two) of 29.4 s is extremely short, and it has no continuous formation process outside the

Table 5.1. *The isotopic composition of oxygen*

Isotope	Protons (Z)	Neutrons (N)	Atomic mass (A)	Atom (%)
^{16}O	8	8	16	99.8
^{17}O	8	9	17	0.04
^{18}O	8	10	18	0.2
$^{19}O^{a}$	8	11	19	0

[a] $t_{1/2} = 29.4$ s.

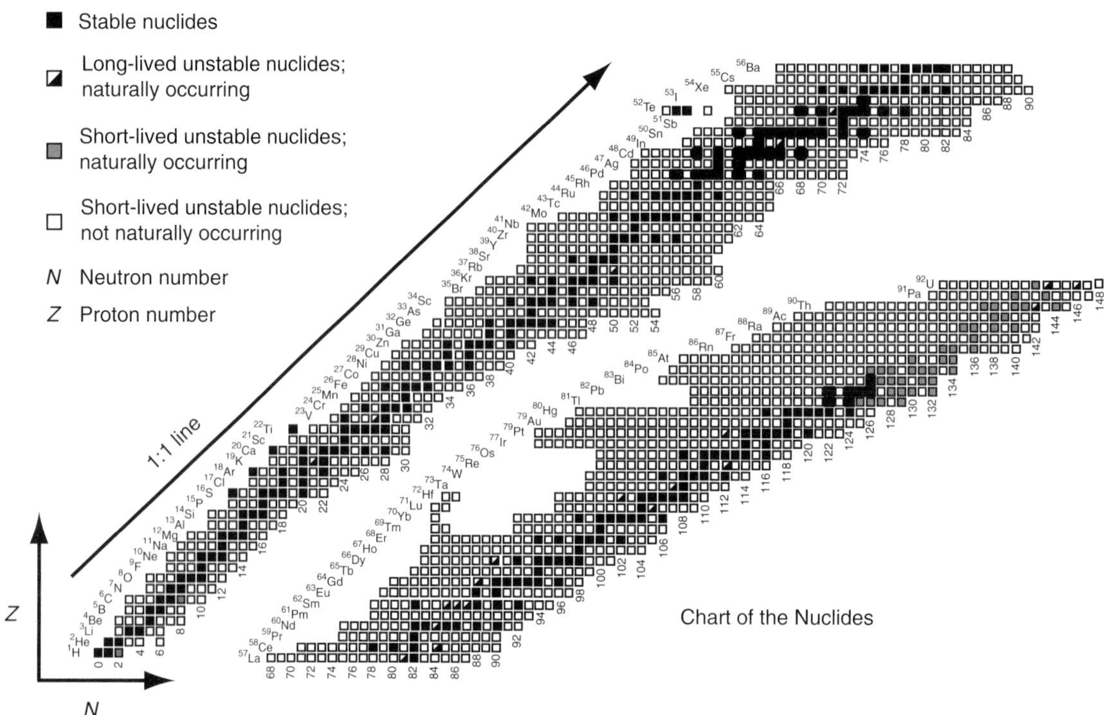

Figure 5.1. Distribution of stable and radioactive isotopes in a plot of atomic number (*Z*) versus number of neutrons (*N*). The horizontal portion of the "stair steps" formed by the stable nuclides (filled squares) represents the number of stable isotopes for any given element. Notice that the number of stable isotopes increases with atomic number and that nearly all naturally occurring unstable nuclides are at the high end of the atomic numbers.

laboratory. The listed percentages of the three stable oxygen isotopes are approximate because the isotopic compositions of many natural substances vary in subtle ways.

One of the most distinctive trends among isotopes is that nuclei with even numbers of both protons and neutrons are more stable, and hence more abundant. For example, of the 264 naturally occurring stable isotopes, 157 contain even numbers of protons and neutrons, and hence an even atomic mass. Approximately a third this number of isotopes contain an odd number of either protons or neutrons, and therefore an odd atomic mass. In sharp contrast, only four stable isotopes (^{2}H, ^{6}Li, ^{10}B, and ^{14}N) contain odd numbers of both protons and neutrons. The "stair step" pattern in Fig. 5.1 illustrates the proton and neutron distributions among the elements. A wide step at a given atomic number (*Z*) indicates a number of different isotopes for the element. The second general trend in the figure is that the number of neutrons in naturally occurring isotopes tends to be equal or greater than the number of protons. The excess of neutrons over protons increases at higher atomic numbers, causing the shift of the isotope distribution in Fig. 5.1 to the right of the 1:1 line. Finally, most nuclei beyond Hg, with *Z* in excess of 80 and *A* greater than 210, are unstable. The naturally occurring nuclides in this super-heavy region are relatively short-lived daughters that ultimately cascade from decay of three extremely long-lived radioactive parents, ^{238}U, ^{235}U, and ^{232}Th.

5.1 Stable isotopes

Most elements have more than one stable isotope (Fig. 5.1). In general, the Earth's crust exhibits relatively homogeneous distributions of the isotopes of each naturally occurring element, with variations typically being on the order of a percent or less. Nevertheless, these small differences in relative abundance can be informative tracers of material sources and processes. Although stable isotopic studies are now being carried out for a wide range of elements because of continuous improvements in technology, this chapter will focus primarily on the most common applications to low mass ($Z < 50$) species, many of which occur in both organic and inorganic form.

5.1.1 Measuring stable isotopic abundances

The keys to measuring small differences in isotope abundances are instrument stability and signal comparability. To obtain the latter, the relative amounts of at least two isotopes are measured at one time, and the resulting abundance ratios are alternately analyzed in sample and standard materials. Isotopes of virtually any mass can now be measured by using many different forms of sample introduction into modern mass spectrometers. The lighter isotopes are typically measured in gaseous molecules that contain the target element as a major component. The analyte gas must also be chemically unreactive with metal surfaces and not prone to stick to surfaces (leaving out H_2O). Examples of the most commonly measured light isotopes and their corresponding analyte gases, abundances and standard materials are given in Table 5.2. Ideally, a standard material

Table 5.2. *Analyte gases and standard materials for light isotope analysis*

Element (analyte)	Stable isotopes	Atom (%)	Standard material
Hydrogen (H_2)	^{1}H	99.99	SMOW[a]
	^{2}H	0.01	
Carbon (CO_2)	^{12}C	98.9	PDB $CaCO_3$
	^{13}C	1.1	
Nitrogen (N_2)	^{14}N	99.6	Air
	^{15}N	0.4	
Oxygen (CO_2)	^{16}O	99.8	SMOW
	^{17}O	0.04	
	^{18}O	0.2	
Sulfur (SO_2)	^{32}S	95.0	Canyon Diablo triolite (FeS)
	^{33}S	0.8	
	^{34}S	4.2	
	^{36}S	0.02	

[a] SMOW, Standard Mean Ocean Water.

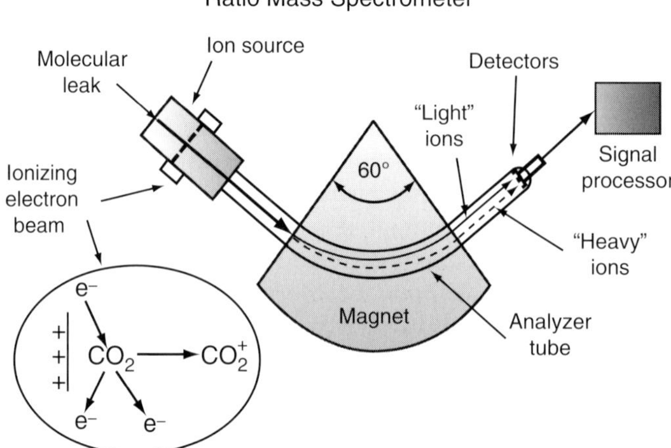

Figure 5.2. Schematic diagram of the main functions of a ratio mass spectrometer. The inset illustrates the reactions that occur during ionization of a CO_2 molecule in the ion source (see text for further explanation).

against which all sample isotope ratios are measured and expressed should be readily available in a homogeneous form. An ideal example is atmospheric N_2. Another widely available reference material (for H and O) is standard mean ocean water (SMOW), although surface ocean waters vary slightly from this reference composition depending on the local balance of evaporation versus precipitation. At the other availability extreme is the PeeDee Belemnite, or PDB standard, a small calcium carbonate fossil that was exhausted in the early days of stable isotope analysis. Although the PDB-based scale is still used for reporting stable carbon isotope compositions, other secondary reference materials that have been linked to PDB now serve as working standards in stable isotope laboratories.

Stable isotope abundances are measured in ratio mass spectrometers (Fig. 5.2). These instruments operate as continuous ion separators. Molecules of the analyte gas prepared from a sample are introduced into the ion source of the instrument through a non-fractionating molecular leak. The entry path to the mass spectrometer must be tiny to maintain a high vacuum within the mass analyzer. The incoming gas molecules are bombarded with electrons that are "boiled" off of a heated filament. Some of the high-energy electrons impact the neutral gas molecules and knock out an electron, producing singly charged positive ions (see insert in Fig. 5.2). These ions are repelled by positively charged plates in the ion source and are accelerated to constant velocity down the analyzer tube. The analyzer tube is evacuated so that the ions do not collide with other particles as they pass through a magnetic field in a continuous beam. Because any charged particle moving within a magnetic field is acted upon by a force oriented perpendicularly to its path, the trajectories of individual ions are deflected. The extent of deflection is greater for lighter ions (e.g. $^{12}CO_2$, $M/Z = 44$) so that they are separated from more massive counterparts (e.g. $^{13}CO_2$, $M/Z = 45$) on the way down the analyzer tube. The process is similar to separating chaff from grain by lofting the mixture in a wind that carries the lighter chaff further away from its initial position. The net result in a mass spectrometer is that the initial mixture of ions is resolved through space into separate beams containing ions of different *M/Z*. The "heavy" and "light" ions are focused onto different detectors that simultaneously create electrical currents proportional to the number of ions striking them. This procedure is carried out repeatedly for gases of the same type that are generated from sample and standard materials and alternately introduced into the ratio mass spectrometer every minute or so. The advantage of sequential analysis of sample and standard gases is that fluctuations in instrument response are largely cancelled out over time when the isotope abundances are expressed as ratios of sample and standard.

The isotope composition of the sample (spl) is measured as the per mil (‰) relative deviation, δ (pronounced "del") of its isotope ratio from the ratio of the simultaneously measured standard (std) material,

$$\delta H = \left[\frac{(H/L)_{spl} - (H/L)_{std}}{(H/L)_{std}}\right] \times 1000 = \left[\frac{R_{spl} - R_{std}}{R_{std}}\right] \times 1000, \tag{5.1}$$

where *H* and *L* represent the "heavy" and "light" isotopes of the target element in the analyte gas and *R* (sometimes written with a lower case *r*) is the *H/L* quotient. In this formulation the sample composition is related to that of the standard gas by both difference and normalization. The results are expressed in parts per thousand (‰) because the absolute relative differences are typically small. In the early days of stable isotope analysis *H/L* values were calculated as absolute values (e.g. 0.0110 for $^{13}C/^{12}C$), a format that proved to be too cumbersome for practical use.

The state of unequal stable isotope composition within different materials linked by a reaction or process is called "isotope fractionation." Isotope fractionation is an observable quantity (or phenomenon) that results from a process called an "isotope effect." One common cause for isotope effects is that molecules or atoms with the same elements, but different numbers of neutrons, have slightly different free energies. If these molecules are able to spontaneously exchange isotopes, they will exhibit slightly different isotopic abundances among their component elements at thermodynamic equilibrium (their lowest energy state; see Chapter 3). This phenomenon is called an "equilibrium isotope effect." In contrast, a "kinetic isotope effect" results from either differential bond breaking rates or diffusion rates of molecules containing the same elements, but different atomic masses (different numbers of neutrons). For the kinetic isotope effect, products containing the lighter isotope will form or diffuse faster, leaving behind a pool of reactants that is enriched in the "heavier" isotope. Kinetic isotope effects caused by bond breaking occur only for atoms directly involved in the bond that is being broken (or formed). This is particularly relevant for large organic molecules, where there can be a pronounced bulk isotope fractionation only when atoms at the cleavage point represent a large fraction of the total element in the parent ion or molecule.

5.1.2 Equilibrium isotope effects

Equilibrium isotope effects are most likely to be exhibited by inorganic chemical species that rapidly interconvert between forms containing the same elements. Organic molecules almost never directly exhibit equilibrium isotope effects because the covalent bonds linking the component atoms do not continuously break and reform. The carbonate buffer system, involving gaseous, $CO_2(g)$, and aqueous, $CO_2(aq)$, carbon dioxide, aqueous bicarbonate, $HCO_3^-(aq)$, and carbonate, $CO_3^{2-}(aq)$, ions, provides an oceanographically important example of a chemical network that exhibits equilibrium isotope effects for both carbon and oxygen. (See Chapter 4 for a discussion of carbonate chemistry.) In such systems, chemical interconversion and attending isotopic exchange continue even when the system is at chemical and isotopic equilibrium, although under this state (of zero free energy) net changes in chemical and isotopic distribution are not observed. It is also possible for a system to be at chemical equilibrium, but not isotopic equilibrium. For example, if one replaced $^{12}CO_2(aq)$ with $^{13}CO_2(aq)$ in a seawater sample, it would take several minutes for the heaver isotope to distribute itself among all the inorganic carbonate species.

An illustration of an isotope exchange reaction is the distribution of stable carbon isotopes between equilibrated $CO_2(g)$ and aqueous HCO_3^- species of the carbonate buffer system. The hydration reaction for CO_2, Eq. (4.11)

$$CO_2 + H_2O \rightleftarrows HCO_3^- + H^+ \qquad K_1' = \frac{[HCO_3^-][H^+]}{[CO_2]} \tag{4.11}$$

can be written for both carbon-12 and carbon-13:

$$^{12}CO_2 + H_2O \rightleftarrows H^{12}CO_3^- + H^+ \qquad {}^{12}K_1' = \frac{[H^{12}CO_3^-][H^+]}{[^{12}CO_2]} \tag{5.2}$$

$$^{13}CO_2 + H_2O \rightleftarrows H^{13}CO_3^- + H^+ \qquad {}^{13}K_1' = \frac{[H^{13}CO_3^-][H^+]}{[^{13}CO_2]}.$$

Rearranging the carbon-12 equation and combining these two gives the isotope exchange reaction involving no net change in chemical species.

$$\begin{gathered} ^{13}CO_2(g) + H^{12}CO_3^- \rightleftarrows {}^{12}CO_2(g) + H^{13}CO_3^- \\ K = \frac{^{13}K}{^{12}K} = \frac{[^{12}CO_2(g)][H^{13}CO_3^-]}{[^{13}CO_2(g)][H^{12}CO_3^-]}. \end{gathered} \tag{5.3}$$

The equilibrium constant, like all K values, is a function of temperature and pressure. In this case, K is equal to 1.0092 at 0 °C and 1.0068 at 30 °C. As is almost always seen for equilibrium isotope effects, the extent of isotopic fractionation becomes less as the temperature of the system increases. Reaction (5.3) is also typical in that the heavier isotope is concentrated in the chemical compound that has the strongest (or most numerous) bonds; in this example carbon-13 is concentrated within $[HCO_3^-]$(aq) as opposed to CO_2(g).

By convention, the magnitude of an equilibrium isotope effect is expressed as a fractionation factor (α). If the product is enriched in the heavy isotope relative to the reactant, α is greater than unity. For reaction (5.3), α has the form

$$\alpha_{H/L} = \frac{(H/L)_{\text{product}}}{(H/L)_{\text{reactant}}} = \frac{(H^{13}CO_3^-/H^{12}CO_3^-)}{(^{13}CO_2/^{12}CO_2)}, \tag{5.4}$$

where H and L again represent heavy and light isotopes in the chemical species exchanging isotopes. For this example, where all stoichiometric coefficients in the balanced equation are equal to one, α has the same value as K in Eq. (5.3). This will not be the case for more complicated isotope equations with non-unit coefficients, where K will have different exponents. A related expression is the "difference fractionation factor" (ε), which is the difference between the δ values of a product and its precursor (the reactant). For carbon this definition becomes:

$$\varepsilon^{13}C = \delta^{13}C_{\text{product}} - \delta^{13}C_{\text{reactant}}. \tag{5.5}$$

The fractionation factor and difference fractionation factor are related by the approximation that

$$\varepsilon \approx 1000 \times \ln\alpha \approx 1000 \times (\alpha - 1). \tag{5.6}$$

For a derivation of the numeric relationships between K, α, δ, and ε for the exchange of ^{18}O between water and carbonate ion, see Appendix 5.1. Difference fractionation factors for many of the important reactions among molecules containing the elements H, C, N, and O are presented in Table 5.3.

A classic example of the application of an equilibrium isotope effect in oceanographic research is the use of stable oxygen isotope methods to estimate the temperature of an ancient environment in which a carbonate shell formed. The reaction on which this application is based is the equilibrium exchange of ^{18}O between $CaCO_3$(s) and water:

$$CaCO_3(s) + H_2{}^{18}O \rightleftarrows CaC^{18}OO_2 + H_2O, \tag{5.7}$$

where ^{16}O atoms are unlabeled for simplicity. Note that the equilibrium between $CaCO_3$ and H_2O involves both dissolved equilibria among the carbonate species and water (Eqs. (4.9) and (4.11)) and between CO_3^{2-} and $CaCO_3$ (Eq. (4.36)). The equilibrium distribution of isotopes among the dissolved carbonate species, water and $CaCO_3$ is temperature-dependent. When $CaCO_3$ precipitates, the carbonate ion is incorporated into the calcite or aragonite ($CaCO_3$) shell of a plant or animal growing in the water. Thus, changes in $\delta^{18}O$ of the $CaCO_3$ shell can be used as a record of the changes in temperature of the surroundings while the shell formed. This ^{18}O paleotemperature method involves several assumptions. The first of these is either that

Table 5.3. *Difference fractionation factors, $\varepsilon = \delta_{product} - \delta_{reactant}$, for important equilibrium (equations with two-way arrows) and kinetic (one-way arrows) reactions among the elements H, C, O, and N*

Equilibrium fractionation factors are for 20 °C. Kinetic fractionation factors are approximate as they vary in the marine environment.

	Reaction	ε (‰)	Reference
H	Evaporation / condensation	+78	Dansgaard, 1965
	$H_2O(g) \rightleftharpoons H_2O(l)$		
C	CO_2 solubility	−1.1	Knox *et al.*, 1992
	$CO_2(g) \rightleftharpoons CO_2(l)$		
	Carbonate equilibria	+8.4	Zhang *et al.*, 1995
	$CO_2 \rightleftharpoons$ DIC	pH = 8.15	
	Photosynthesis	−14 to −19	O'Leary, 1981
	$CO_2 + H_2O \rightarrow CH_2O_{OM} + O_2$		
	Respiration	~0	
	$CH_2O_{OM} + O_2 \rightarrow CO_2 + H_2O$		
	$CaCO_3$ precipitation	+1 (calcite)	Romanek *et al.*, 1992
	$Ca^{2+} + HCO_3^- \rightleftharpoons CaCO_3(s) + H^+$		
	Methane formation	−40 to −90	Lansdown *et al.*, 1992
	$4H_2 + CO_2 \rightarrow CH_4 + H_2O$		
O	Evaporation / Precipitation	+9	Daansgard, 1965
	$H_2O(g) \rightleftharpoons H_2O(l)$		
	O_2 solubility	+0.7	Knox *et al.*, 1992
	$O_2(g) \rightleftharpoons O_2(aq)$		
	Photosynthesis	~0	Guy *et al.*, 1993
	$CO_2 + H_2O \rightarrow CH_2O_{OM} + O_2$		
	Respiration	−20	Kiddon *et al.*, 1993
	$CH_2O_{OM} + O_2 \rightarrow CO_2 + H_2O$		
N	Solubility	+0.7	Knox *et al.*, 1992
	$N_2(g) \rightleftharpoons N_2(aq)$		
	Nitrogen fixation	−1.3 to	Carpenter *et al.*, 1997
	$4H^+ + 6H_2O + 2N_2 \rightarrow 4NH_4^+ + 3O_2$	−3.6	
	NO_3^- uptake by photosynthesis	−5 to −9	Altabet and Francois, 1994
	$2H^+ + NO_3^- + H_2O \rightarrow NH_4^+ + 2O_2$		
	Denitrification	−20 to −40	Wada, 1980
	$4NO_3^- + 4H^+ + 5CH_2O_{OM} \rightarrow 2N_2 + 5CO_2 + 7H_2O$		

the organism precipitated $CaCO_3(s)$ (calcite or aragonite) in isotopic equilibrium with dissolved CO_3^{2-}, or that any non-equilibrium isotope effect is known. If there is a non-equilibrium (vital) effect whereby a specific organism perturbs the expected equilibrium isotope effect, the extent of this offset must be systematically known as a function of temperature. This can be achieved from laboratory simulation experiments or by measuring the oxygen isotope ratio of present-day organisms that are known to have grown in environments of different temperature. The second assumption is that the $\delta^{18}O$ of the original water is known, or can be accurately estimated.

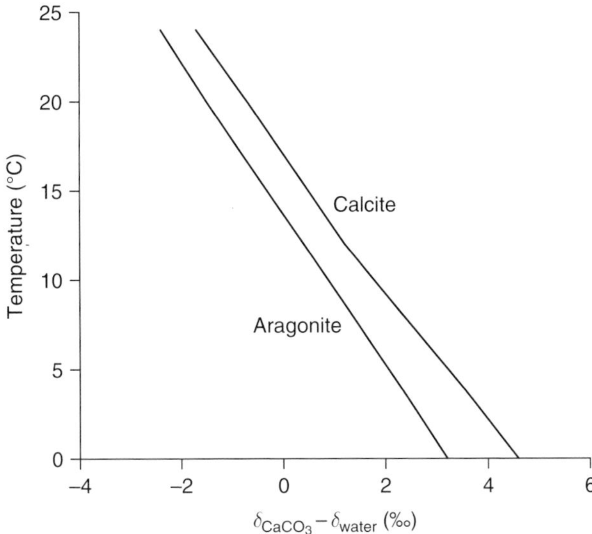

Figure 5.3. Laboratory-determined temperature dependence of the oxygen isotope difference between biologically produced $CaCO_3$ shells (both calcite and aragonite) and the water in which they grew. These curves are drawn by using Eqs. (5.8) and (5.9). There are many different versions of these curves for both inorganically and organically produced $CaCO_3$ (see, for example, Bemis *et al.*, 1998).

Finally, the $\delta^{18}O$ of the shell is assumed to have remained unchanged since the time it was precipitated by the organism.

$\delta^{18}O$ versus temperature curves obtained empirically by growing calcite- and aragonite-secreting organisms in water of known $\delta^{18}O$ are illustrated in Fig. 5.3. The different carbonates secreted at different temperatures, in this case over a range of 0–24 °C, were then individually isolated and analyzed for their $\delta^{18}O$ composition. The isotopic compositions of each carbonate mineral are given as the difference between the $\delta^{18}O$ of the carbonate, δ_{caco_3}, and the $\delta^{18}O$ of the ambient water, δ_{water}. This formulation allows temperature to be determined from δ_{caco_3} for waters of different known δ_{water}. The stable oxygen isotope compositions of the carbonate samples are determined by using a ratio mass spectrometer to measure the δ_{caco_3} of CO_2 released by treatment of the $CaCO_3$ with anhydrous phosphoric acid. The H_3PO_4 must not contain water, which would rapidly equilibrate with the generated CO_2 and change its $\delta^{18}O$. The value of δ_{water} cannot be determined directly because water is a "sticky" molecule that does not behave well in mass spectrometers. This problem is circumvented by equilibrating a small amount of CO_2 at a well-known temperature with the water in which the carbonates were formed. The $\delta^{18}O$ of the equilibrated CO_2 is then measured. To relate this measurement to the $\delta^{18}O$ of the water, the equilibrium fractionation that occurs between CO_2 and H_2O at that temperature must be known and subtracted. The equations for the two calibration lines in Fig. 5.3 are the polynomials:

$$T_{aragonite} = 13.85 - 4.54(\delta_{aragonite} - \delta_{water}) + 0.04 \times (\delta_{aragonite} - \delta_{water})^2 \qquad (5.8)$$

and

$$T_{calcite} = 17.04 - 4.34(\delta_{calcite} - \delta_{water}) + 0.16 \times (\delta_{calcite} - \delta_{water})^2. \qquad (5.9)$$

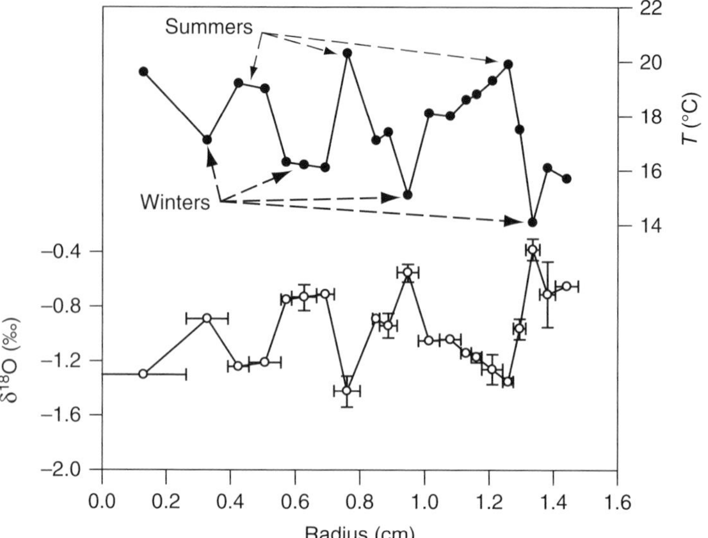

Figure 5.4. The original application of oxygen isotope thermometry. The thermal history (top) of the surroundings in South Carolina, USA, during the Cretaceous period as recorded by oxygen isotope changes (bottom) as a function of shell radius in the PeeDee belemnite (Urey *et al.*, 1951).

There have been a number of empirical equations of this form with slightly different numerical coefficients derived for pure $CaCO_3$ and for different versions of biologically produced $CaCO_3$. A review of these expressions is presented in Bemis *et al.* (1998).

One of the first applications of the paleotemperature method was undertaken by Harold Urey *et al.* (1951), who analyzed incremental sections along the radius of the shell of a fossil belemnite (a type of cephalopod). This bullet-shaped shell, from the Cretaceous PeeDee formation of South Carolina, was the original standard material for $\delta^{13}C$ analysis (Table 5.2). The isotopic "diary" of the PeeDee belemnite (Fig. 5.4) records four cool extremes separated by three warm periods, in what appears to be 3.5 y of life history laid down approximately sixty million years ago. The absolute temperature scale is uncertain, because the $\delta^{18}O$ of the ancient sea where the belemnite lived is unknown.

A more sweeping subsequent application of the paleotemperature method has been by paleoceanographers who measure the $\delta^{18}O$ of Foraminifera (carbonate-secreting animals) from long sediment cores representing several million years of Earth history. The past 700 000 y record (Fig. 5.5) indicates approximately eight glacial–interglacial cycles, with total $\delta^{18}O$ offsets of approximately 1.8 ‰. The main difficulty in interpreting such fluctuations in terms of absolute paleotemperatures is that two separate processes contribute to more positive $\delta^{18}O$ values in carbonates precipitated during glacial times. The first of these is equilibrium fractionation, with its well-known temperature effect (e.g. Fig. 5.3). The second process involves estimating the $\delta^{18}O$ composition of seawater during ice ages. As will be discussed later, net transfer of water from the ocean to continental ice sheets discriminates against isotopically heavier water molecules, leaving continental ice depleted in ^{18}O ($\delta^{18}O \approx -30‰$) and remnant

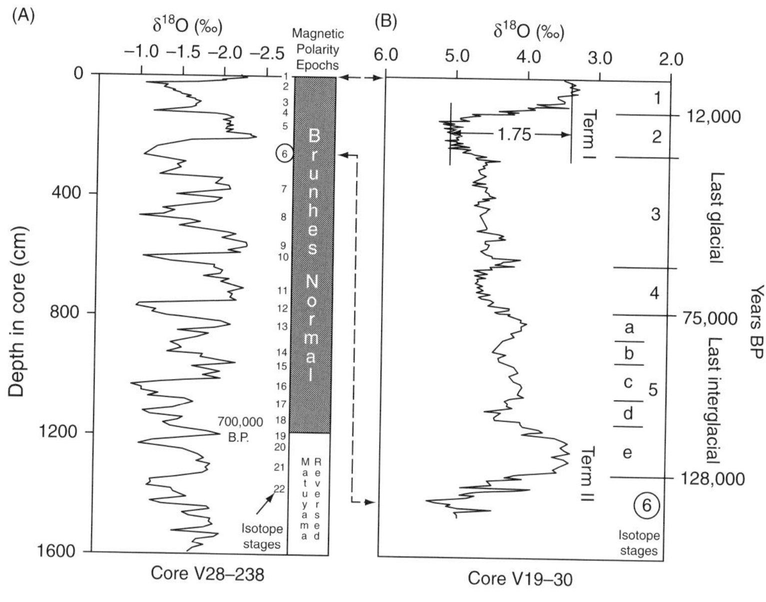

Figure 5.5. The oxygen isotope record recorded in foraminiferan tests in deep sea sediments. The changes are caused by temperature changes and the waxing and waning of glacial ice during the past c.1 million years. The core on the left (A) is a record of planktonic Foraminifera V28–238 from the Equatorial Pacific (Shackleton and Opdyke, 1973). The core on the right (B) is also from the Equatorial Pacific but the isotope data are on benthic foraminiferan tests from V19–30 (Chappell and Shackleton, 1986). The $\delta^{18}O$ scales are different because the planktonic–benthic difference is c.5‰. Redrawn from Crowley (1983) and Broecker (2002).

glacial oceans enriched. Details of this problem and its evolution are covered in Chapter 7.

5.1.3 Kinetic isotope effects

Kinetic isotope effects take place under non-equilibrium physical or chemical conditions. Physically based isotope effects often occur because molecules of the same compound, but containing different stable isotopes, move at different rates (owing to unequal masses). For example, carbon dioxide gas occurs in molecules containing $^{12}CO_2$ (molecular mass, $m_{wt} = 44$) and $^{13}CO_2$ ($m_{wt} = 45$), which must have the same kinetic energy ($E_k = ½\, Mv^2$, where M is mass and v is velocity). Therefore the relation between velocity and mass at the same temperature is:

$$1/2 \times M_{44} \times v_{44}^2 = 1/2 \times M_{45} \times v_{45}^2 \qquad (5.10)$$

$$\frac{v_{44}}{v_{45}} = \frac{\sqrt{45}}{\sqrt{44}} = 1.012. \qquad (5.11)$$

Thus $^{12}CO_2$ molecules diffusing in pure CO_2 gas travel 1.2 % (12 ‰) faster than $^{13}CO_2$ molecules, with the net result that $^{13}CO_2$ will trail in a "pack" of diffusing carbon dioxide molecules. Calculating the diffusion isotope effect of gases in air requires accounting for the molecular mass of both the diffusing gas and the medium in which it diffuses. Molecules of the same compound containing more massive isotopes also diffuse more slowly in liquids, although the isotope effect is much less than in the gas phase and impossible to predict theoretically.

Chemically based kinetic isotope effects occur because molecules of the same compound, but containing different isotopes, react at different rates. In general, more massive isotopes of the same element form stronger chemical bonds that break more slowly during chemical reactions. Overall, molecules of the same compound containing lighter isotopes will move and react faster than molecules containing heavier isotopes.

Essentially all isotope effects involved with the formation and destruction of organic matter are kinetic. Most terrestrial land plants and marine phytoplankton exhibit an enzymatic isotope effect leading to a fractionation of carbon during photosynthesis. The magnitude of this effect depends on the f_{CO_2} and growth rate. Enzymatic fractionation during photosynthesis results in a difference fractionation factor of about − 30‰ (O'Leary, 1981). This entire effect is rarely observed because the isotope ratio of the internal reservoir of CO_2 is controlled by the rate of CO_2 flux across the membrane wall and the rate of depletion of the CO_2 reservoir by carbon fixation. In terrestrial plants the fractionation factor is about − 19‰ with a range from − 26‰ to − 7‰. The isotope fractionation factor for marine plants is smaller: *c.*− 14‰ with a range of − 22‰ to − 8‰ (Table 5.3). The equilibrium and kinetic fractionation factors among organic matter, CO_2 and HCO_3^- control the carbon isotope ratios in the marine DIC, atmospheric carbon dioxide and organic matter reservoirs (Fig. 5.6).

During respiration, carbon isotopes are fractionated very little. Thus, for animals, the adage "you are what you eat" generally holds. A consequence of significant isotopic fractionation during photosynthesis, but little during respiration, is that ^{13}C-depleted organic carbon sinks out of the euphotic zone in the form of dead plants and animals, thereby leaving the surface water DIC enriched in ^{13}C

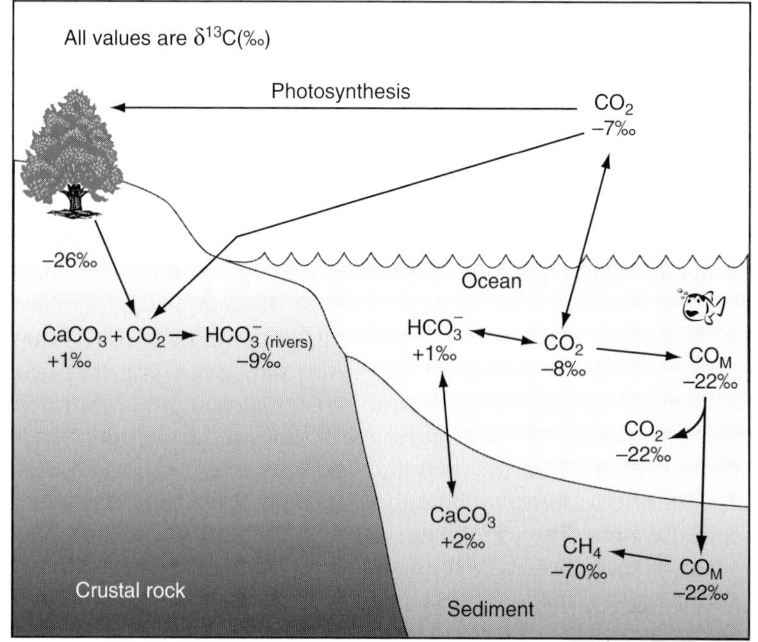

Figure 5.6. Schematic diagram of the exchanges and stable carbon isotope ratios among the reservoirs of atmospheric CO_2, ocean HCO_3^- and CO_2, solid organic carbon (C_{OM}) and $CaCO_3$ and CH_4. Numbers under the chemical symbols represent the isotopic ratio in ‰. See Table 5.3 for an estimate of the fractionation factors.

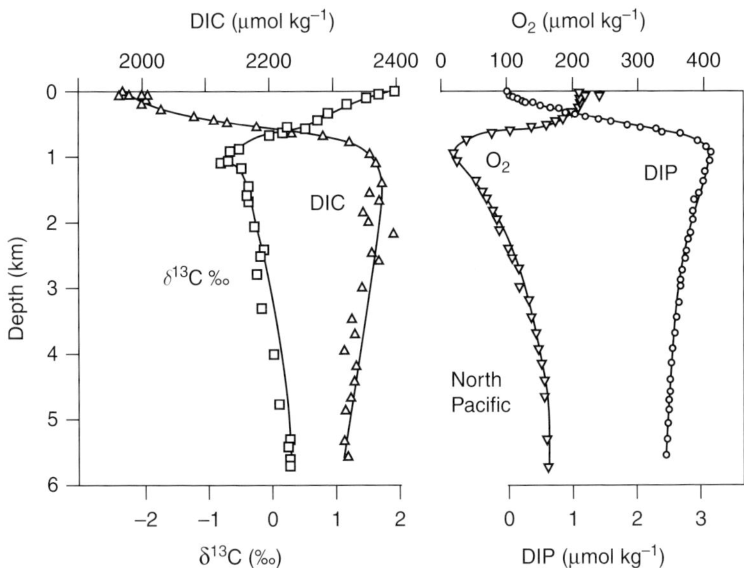

Figure 5.7. A profile of the $\delta^{13}C$ of DIC in the Pacific Ocean showing the surface enrichment and the correlation with DIC, O_2 and dissolved inorganic phosphorus (DIP). Redrawn from Broecker and Peng (1982).

(Fig. 5.7). The organic matter that exits the euphotic zone degrades in the deep sea, slightly depleting the deep water DIC in ^{13}C. The isotope effect is amplified in the surface waters because of the vast difference in the reservoir volumes. (The euphotic zone depth is *c.*100 m versus the aphotic zone depth of *c.*3700 m.) Because of these fractionations and dynamics, carbon isotopes in surface waters of the ocean are useful tracers of the photosynthetically driven flux of organic matter to the deep ocean. Applications of this tracer to determine the rate of carbon export in both modern (Chapter 6) and paleoceanographic (Chapter 7) settings are discussed later in the book.

The fractionation patterns for isotopes of molecular oxygen, O_2, during photosynthesis and respiration are opposite that of carbon in that there is very little kinetic fractionation between H_2O and O_2 during photosynthesis, but a rather large kinetic fractionation ($\varepsilon \approx -20‰$) during respiration (Table 5.3). As a result the $\delta^{18}O$ of oxygen concentrations in the ocean become progressively heavier as the oxygen concentration decreases (Fig. 5.8). The distribution of ^{18}O in molecular O_2 in the ocean is used as a tracer for respiration. One would expect that oxygen isotopes in O_2 would be an ideal paleoceanographic tracer for the extent of O_2 depletion in the deep ocean, given the results shown in Fig. 5.8; however, to date it has not been possible to identify a solid material that faithfully preserves the dissolved O_2 isotope ratio.

Stable nitrogen isotope compositions are useful tracers of both the source of N to the biomass and the history of the organic nitrogen. The $\delta^{15}N$ of dissolved N_2 in surface waters (*c.*+0.6 ‰) is slightly enriched compared with atmospheric N_2 (0‰) because the heavier gas molecules have a lower vapor pressure and are more soluble in water. In contrast, dissolved NO_3^- (by far the most abundant form of oxidized inorganic nitrogen in the sea) exhibits $\delta^{15}N$ values that

Figure 5.8. The relation between the $\delta^{18}O$ of molecular O_2 and the O_2 concentration in the Pacific Ocean. Numbers indicate water column depth in meters. As the oxygen concentration is consumed by respiration the stable isotope ratio becomes progressively heavier because of the kinetic fractionation that accompanies this process. Reproduced from Kroopnick and Craig (1976).

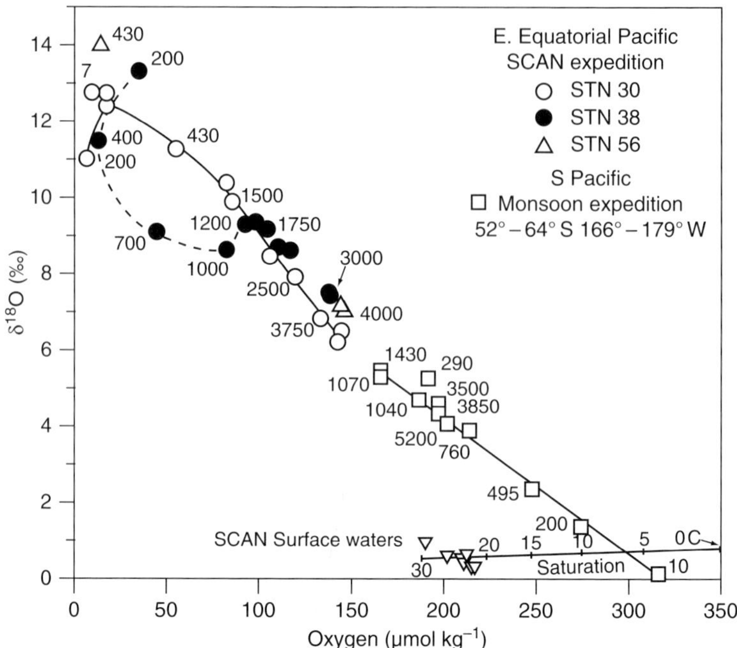

Figure 5.9. Schematic diagram of the stable nitrogen isotope ratios of different nitrogen reservoirs in the sea. The general range of stable isotope ratios (with respect to the atmosphere) found in nature is given in the boxes and the difference fractionation factors ε (in ‰) accompany arrows between the boxes. Many of the values are approximations because of the wide variations of observations. See Table 5.3 for more details of some of the reactions and the text for explanation. Values are based on data presented by Altabet and Small (1990), Altabet and Francois (1994) and Sigman and Casciotti (2001).

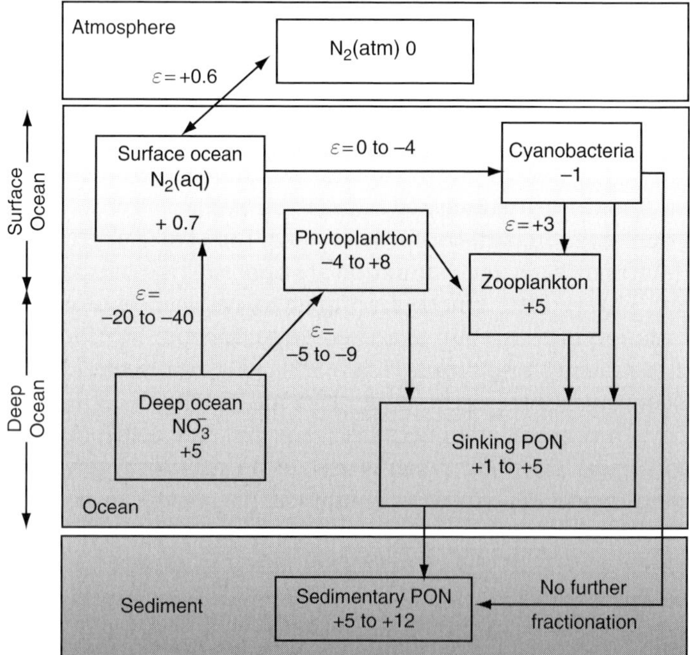

range from + 1 to + 20 with a deep ocean value near + 5‰ (Fig. 5.9). Nitrate is enriched with respect to N_2 because of kinetic fractionation (– 20 to – 40‰) during denitrification, which reduces NO_3^- to N_2 in oxygen-deficient regions of the ocean. Thus, $\delta^{15}N$ of the organic matter formed during photosynthesis is a useful tracer for the nitrogen source. For example, cyanobacteria fix dissolved N_2 from seawater ($\delta^{15}N = +1‰$) into biochemicals with a very small negative isotope effect, such that their biomass exhibits a $\delta^{15}N$ similar to that of N_2 (Fig. 5.9). Most other marine phytoplankton utilize NH_4^+ and NO_3^-, resulting in a $\delta^{15}N$ more like +5‰. Although uptake of NH_4^+ and NO_3^- by phytoplankton can involve a kinetic isotope effect leading to fractionations of −5 to −9‰, the actual amount of isotope discrimination that occurs is a function of nutrient availability versus uptake rates. Mass balance dictates that complete uptake and incorporation of any dissolved species must occur without any attending fractionation because the product contains all the isotopes that were in the reactant. At very high uptake : availability ratios, phytoplankton use up essentially all dissolved NO_3^- and NH_4^+, resulting in minimal fractionation. Under such conditions the biomass of phytoplankton will have a $\delta^{15}N$ similar to that of the nitrogen substrate: $\delta^{15}N \approx +5‰$ from dissolved NO_3^- versus $\delta^{15}N = +1$ for N_2 fixation in cyanobacteria (Fig. 5.9). Nitrogen stable isotope tracers have been used to show that photosynthesis in the subtropical Pacific Ocean, where nitrate is nearly totally depleted in the euphotic zone, is fueled by roughly equal parts of nitrogen fixation and NO_3^- from deeper waters (Karl *et al.*, 1997). In situations where the NO_3^- concentration in the euphotic zone is incompletely utilized by phytoplankton (where surface ocean NO_3^- concentrations are greater than about 1 μmol kg^{-1}) the $\delta^{15}N$ of the phytoplankton will be lighter than the substrate NO_3^- because of the kinetic fractionation factor.

Because of the fractionations described above, measurements of nitrogen isotope ratios in marine sediments have found a utility in paleoceanography. In regions where it can be assumed that there has been constant total NO_3^- depletion in the surface waters through time, the $\delta^{15}N$ in organic matter in sediments depends on the $\delta^{15}N$ of NO_3^- supplied to the euphotic zone. Changes observed in the $\delta^{15}N$ of the organic matter in the sediments through time in this case indicate changes in the $\delta^{15}N$ of the nitrate supply, which is controlled primarily by denitrification in waters below the euphotic zone. By contrast, in regions where one can be sure that surface water NO_3^- has remained in excess through time, an observed change in $\delta^{15}N$ of the organic matter in the sediments would depend on the $\delta^{15}N$ of the surface water NO_3^- pool, which is controlled by the rate of upwelling and removal by photosynthesis. Many authors have interpreted $\delta^{15}N$ changes in situations like this to be the result of variations in the rate of carbon export from the surface ocean (see, for example, Altabet and Francois, 1994).

During respiration, nitrogen isotopes are fractionated by about +3‰ for each trophic level: nitrogen isotopes in animal biomass are on average 3‰ more positive than the $\delta^{15}N$ of their diets. This contrasts with the situation for carbon isotopes, in which there is little fractionation during respiration. If animals are known to have a common dietary source the comparative nitrogen content of individual organisms can be used to roughly discern their average trophic level. Theoretically, the $\delta^{15}N$ of a herbivore should be on average 3‰ greater than its plant diet, whereas a carnivore eating exclusively that herbivore should contain nitrogen exhibiting $\delta^{15}N$ that is again 3‰ greater, and thus 6‰ more positive than the plant source. An advantage of using $\delta^{15}N$ to trace trophic status is that biomass isotopic composition integrates diets over long time scales, thereby providing a record of dietary habits that can be more comprehensive than observations of feeding behavior or gut contents. However, in natural environments, diets are mixed and animals may feed at more than one trophic level, making the $\delta^{15}N$-based interpretations of trophic status complicated.

5.1.4 Rayleigh distillation

The hydrologic cycle provides an interesting combination of equilibrium and kinetic isotope effects accompanying water evaporation, transport and precipitation. The kinetic isotope effect is observed in this process when water molecules evaporate from the ocean surface and the equilibrium effect is imparted when water molecules condense from vapor back into liquid form. At a temperature of 20 °C, there is an equilibrium isotope effect of about 9‰ for $\delta^{18}O$ (Fig. 5.10), which occurs because H_2O molecules containing ^{18}O have a slightly higher boiling point than those containing ^{16}O. Although most of our discussion focuses on oxygen isotope fractionation, the systematics are also true for hydrogen, in which equilibrium isotope effects between liquid and gaseous water are about 10 times larger than those for oxygen.

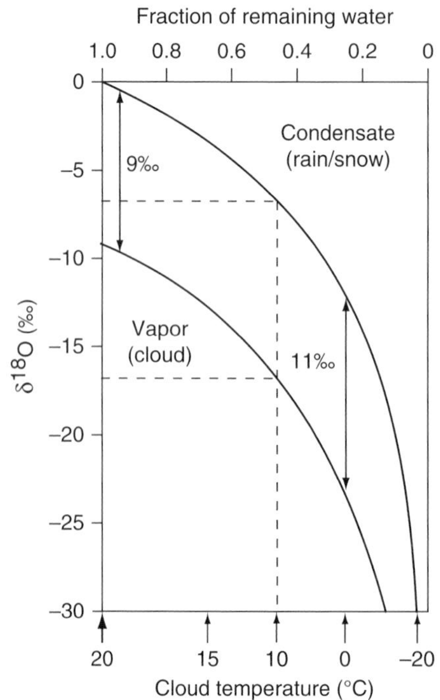

Figure 5.10. An idealized illustration of the differences between the $\delta^{18}O$ of condensate and vapor as a function of the fraction of the remaining water during the Rayleigh distillation process. Envision a cloud that forms at 20 °C and remains a closed system except for water that rains out as it cools from 20 °C to − 20 °C. The equilibrium fractionation factor is temperature dependent, 9‰ at 20 °C and 11‰ at 0 °C. Modified from Dansgaard (1965).

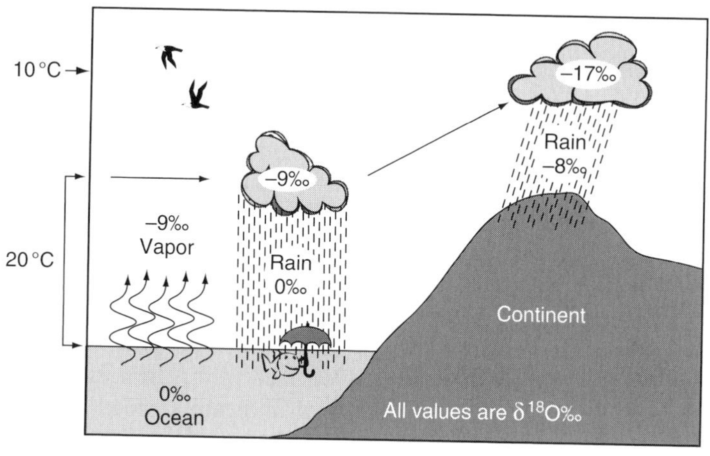

Figure 5.11. Schematic diagram of the oxygen isotopic fractionation among seawater, the atmosphere and rain on land. See Fig. 5.10 also. Modified from Siegenthaler (1976).

When air masses containing water vapor evaporated from the ocean surface are transported to colder regions via global wind patterns, there is a loss of water vapor because the vapor pressure of water decreases progressively with lower temperature. Because the condensate (rain) is approximately 9‰ more enriched than the water vapor, by mass balance the water remaining in the cloud must become progressively more depleted in ^{18}O (and D). Thus, an ocean-derived cloud cooled from 20 to 10 °C would lose approximately half its initial vapor content (Fig. 5.10) and in the process decrease its isotopic composition from −9 to −17‰. The rain falling from the cloud at 10 °C would follow this depletion in ^{18}O and thus would have a $\delta^{18}O$ near −8‰. This "milking" process continues as long as the temperature decreases, as would generally occur as the cloud moved to higher latitudes or to greater elevations, the latter exemplified by moving inland over mountains (Fig. 5.11). At 0 °C the cloud would contain only a quarter of its initial (20 °C) water, and it generates meteoric water (falling rain or snow) with a $\delta^{18}O$ near −12‰ (Fig. 5.10). At sub-zero temperatures like those on the Antarctica or Greenland ice sheets, the cloud and the snow falling from it can both have extremely negative $\delta^{18}O$ values of less than −30‰. Thus the $\delta^{18}O$ of the precipitation falling at any location will reflect both the +9‰ equilibrium fractionation effect between the water vapor and condensate, plus the cumulative $\delta^{18}O$ depletion over time of the cloud's remnant vapor.

The Rayleigh distillation equation was developed to mathematically describe this type of cumulative isotope effect. It relates the initial (R_0) and transient (R_t) stable isotope ratios of a reservoir to the fraction (f) of the initial material that remains (often expressed at the concentration ratio, C_t/C_0 of the more abundant isotope) when product is removed with a constant fractionation factor, α, over a time, t

$$\frac{R_t}{R_0} = f^{(\alpha-1)} = \left(\frac{C_t}{C_0}\right)^{(\alpha-1)}. \tag{5.12}$$

For $\delta^{18}O$, the above equation can be recast as

$$\delta^{18}O_t = (\delta^{18}O_0 + 1000) \times f^{(\alpha-1)} - 1000. \tag{5.13}$$

Equation (5.13) has the advantage that all isotope compositions are given directly in del notation and can be applied to any cumulative isotope fractionation process involving any isotope pair for which α remains constant. See Appendix 5.2 for derivation of both Eq. (5.12) and Eq. (5.13).

The isotope effects (both kinetic and equilibrium) that deplete deuterium, D, and ^{18}O in evaporated water vapor must enrich these isotopes in the surface seawater left behind. This isotope effect can be seen in longitudinal transects of surface waters in both the Atlantic and Pacific Oceans (Fig. 5.12), which indicate a broad maximum of $\delta^{18}O$ between roughly 30° north and south latitudes. Salinity maxima are evident over the same latitude band, with a slight minimum near the Equator. The parallel maxima in $\delta^{18}O$ and salinity result because net evaporation from warm surface waters at lower latitudes preferentially leaves behind both $H_2{}^{18}O$ molecules and salt. Rainfall in excess of evaporation from atmospheric convection cells rising near the Equator is recorded by both a salinity and a $\delta^{18}O$ minimum.

The previous examples are a minute sampling of the many different types of stable isotope studies now being applied to inorganic and organic marine samples. Solid-source mass spectrometers are available for highly sensitive and precise determinations of the stable

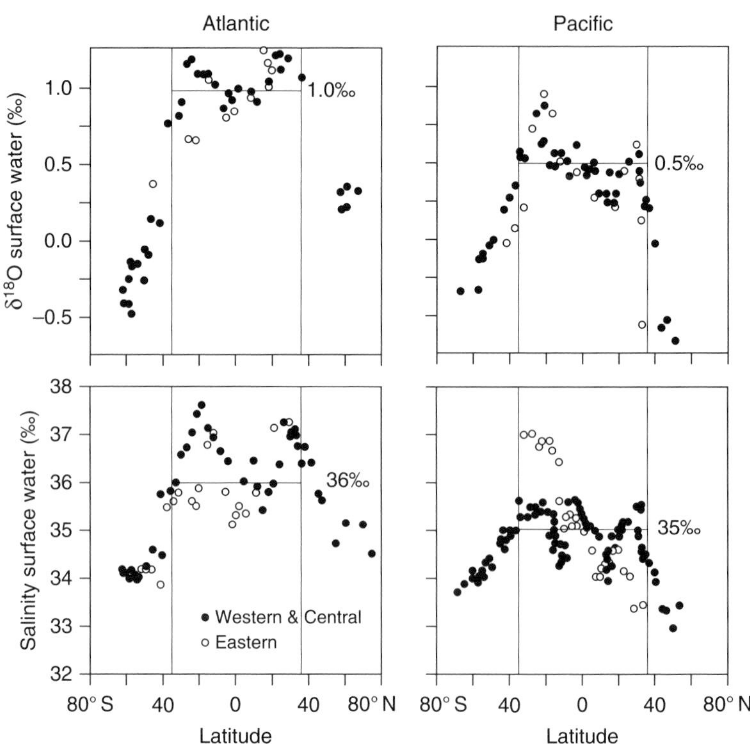

Figure 5.12. The relation between $\delta^{18}O$ of surface water (top two graphs) and salinity (bottom two graphs) of the surface waters of the Atlantic and Pacific Oceans. Variations with respect to latitude are similar for these two indicating that the rain water that dilutes seawater at high latitudes is highly depleted in $\delta^{18}O$. Redrawn from Broecker (2002).

isotope abundances of essentially any element, including transition metals and massive Pb decay products of U and Th. Corresponding applications include sample dating, water mass tracing and studies of paleoceanographic process in sedimentary records. Stable carbon, nitrogen and hydrogen isotope measurements also are now possible on nano- to picogram amounts of individual types of organic molecules separated by gas chromatography (GC) and analyzed "on the fly" by a downstream ratio mass spectrometer (MS). Such isotope ratio monitoring GCMS methods combine the power of using specific organic molecules as biomarkers for particular biological sources with the added dimension that the embedded stable isotope compositions provide complementary source and process indicators.

5.2 Radioactive isotopes

Stable isotope compositions are useful tracers of the sources and transformations of marine materials; however, they carry no direct information about the rates and dates of the associated processes. Such temporal distinctions are possible, however, with the many different naturally occurring radioactive isotopes (Fig. 5.1) and their wide range of elemental forms and decay rates. These highly dependable atomic clocks decay by nuclear processes that allow them to be detected at very low concentrations. Long-lived ^{238}U, ^{235}U, and ^{232}Th radioisotopes are sources of three separate decay series, each of which involves dozens of radioactive daughters with remarkably diverse decay rates and chemical behaviors in the environment. These nuclear cascades continuously replace parent isotopes with daughters at predictable rates. Both the parent and daughter isotopes bear unique chemistries that trace environmental reactions or fluxes that are otherwise difficult to measure. A physical analogy would be a table full of alarm clocks that individually become weightless and float away in the wind for a preset period of time, after which each regains mass and plops to the ground with a characteristic thump. In this fanciful example the distribution of clock carcasses across the landscape would record past directions and speeds of local winds over the time interval they were weightless.

5.2.1 Radioactive decay processes

Radioactivity is characterized by the emission of energy (electromagnetic or in the form of a particle) from the nucleus of an atom, usually with associated elemental conversion. There are four basic types of radioactive decay (Table 5.4), of which alpha (α) and beta (β^-) decay are most common in nature. Alpha emission is the only type of decay that causes a net mass change in the parent nuclide by loss of two protons plus two neutrons. Because two essentially weightless orbiting electrons are also lost when the equivalent of a helium nucleus is emitted, the parent nuclide transmutes into a daughter element two positions to the left on the periodic table. Thus ^{238}U decays by α

Table 5.4. *Different types of radioactive decay process*

Type	Emitted particle	Δ protons	Δ neutrons	Comments
α	He^{2+} (helium nucleus)	−2	−2	loss of 4 atomic mass units
β^-	e^- (electron)	+1	−1	no mass loss
K capture	None	−1	+1	no mass loss, X-ray emission
β^+	e^+ (positron)	−1	+1	no mass loss, X-ray emission

emission to ^{234}Th, skipping direct conversion to protactinium, Pa. Because of their high mass and emission energy, α particles have the potential to damage materials they impact, and to cause recoil of the nucleus from which they emanate (see later discussion). β^- decay involves emission of a negatively charged particle (an electron) from the nucleus, and thus does not lead to a change in atomic mass. Because both mass and charge must be conserved in any nuclear transformation, it is necessary that a positive charge also be generated in the course of β^- decay. This balance is accomplished because the electron is emitted by a neutron and in the process $(n^0 \rightarrow e^- + p^+)$ is converted to a proton of essentially equal mass. This transition moves the element one step to the right on the periodic table: for example, ^{14}C is converted to ^{14}N.

The two other decay processes in Table 5.4 are less common in nature. In K-capture, any orbiting electron (usually in an inner shell) combines with a proton in the nucleus to form a neutron. This relatively rare nuclear transformation process $(e^- + p^+ \rightarrow n^0)$ is just the opposite of that for β^- decay, meaning that the formed nucleus also has the same mass but is displaced one element to the left on the periodic table. Conversion of ^{40}K to ^{40}Ar by K-capture is an example of the chemical conversion that can attend radioactive decay, in this case leading to transformation of a non-volatile alkali metal into the inert gas Ar, the third most abundant gas in the atmosphere. Although no nuclear particle is emitted by K-capture, the attending cascade of electrons into lower orbitals leads to X-ray emission of characteristic energy that can be measured by the appropriate detectors. The last decay process (also rare) involves emission of a positron (β^+), a positively charged electron. The nuclear process $(p^+ \rightarrow n^0 + \beta^+)$ has the same net effect as K-capture and is also characterized by X-ray emission.

5.2.2 Radioactive decay equations

Conversion of a radioactive parent to a single stable daughter, as occurs for ^{14}C, is the simplest form of nuclear decay. Although it is impossible to determine when an individual radioactive nucleus will convert, decay rates become predictable for large populations of radionuclides of a given type. As discussed in Chapter 9, radioactive decay is a perfect example of a first-order irreversible reaction (Eq. (9.27), Fig. 9.5). The general equation describing decay of a parent isotope to a stable daughter is

$$-\frac{dN}{dt} = \lambda N, \tag{5.14}$$

where $-dN/dt$ is the nuclear decay rate (or nuclear activity), N the total number of radioactive atoms present in the system, and λ the first-order decay constant in units of inverse time (e.g. d^{-1}). Integration of (5.14) results in the classic first-order decay equation

$$N = N_0 e^{-\lambda t} \tag{5.15}$$

that allows the number of radioisotope atoms remaining after a given time, t, to be calculated from the number that was present initially, N_0, and the decay constant for that particular nuclide. It is often difficult to directly measure the total number of radioactive atoms in a sample, although such enumerations are now possible by mass spectrometry. In most cases the activity, A (disintegrations per time) is measured rather than the concentration of atoms. Since activity is the rate of decay, it is defined as

$$A = -\frac{dN}{dt} = \lambda N. \tag{5.16}$$

Thus, the number of radioactive isotopes of a given type can always be related to a measurable corresponding activity and known decay constant. This relation also allows the fundamental decay equation to be rewritten in terms of more readily measured activities by simply multiplying the concentrations on both sides of Eq. (5.15) by the decay constant

$$A = A_0 e^{-\lambda t}. \tag{5.17}$$

The standard unit by which radioactivity is measured is the *curie*, which is equal to 3.70×10^{10} disintegrations s^{-1} (dps) ($= 2.22 \times 10^{12}$ disintegrations min^{-1}, dpm). The curie is the amount of radioactivity exhibited by 1.00 g of pure ^{226}Ra and derives its name from Madame Curie, who was a pioneer in the study of radioactivity and Ra. By convention, the relative rate at which a radionuclide decays is expressed in terms of its half life, $t_{1/2}$, which is related to the decay constant by

$$\ln\left(\frac{A}{A_0}\right) = \ln(1/2) = -0.693 = -\lambda t_{1/2}, \tag{5.18}$$

thus,

$$t_{1/2} = \frac{0.693}{\lambda}. \tag{5.19}$$

A related expression called the mean life, τ, is defined as the average time that a radioisotopic nucleus exists before decay.

Mathematically it is the integral of all lives of the atoms in a particular nuclide divided by the initial quantity.

$$\tau = \frac{1}{N_0}\int_0^\infty t\mathrm{d}N = \frac{1}{N_0}\int_0^\infty t\lambda N\mathrm{d}t = \lambda\int_0^\infty t\mathrm{e}^{-\lambda t}\mathrm{d}t = \left[\frac{\lambda t+1}{\lambda}\mathrm{e}^{-\lambda t}\right]_0^\infty = \frac{1}{\lambda} = \frac{t_{1/2}}{0.693}. \tag{5.20}$$

The definitions of half life and mean life are illustrated graphically in Fig. 9.5 of Chapter 9. It can be seen from this figure and Eq. (5.20) that τ is greater than $t_{1/2}$ because some nuclides persist for unusually long life times, dragging out the mean.

Another useful concept is the mathematical expression for the cumulative amount of stable daughter, D, that has been formed from a radioactive parent, P, at any given time. In this case the rate of daughter production is equal to the rate of parent decay

$$\frac{\mathrm{d}N_\mathrm{D}}{\mathrm{d}t} = -\frac{\mathrm{d}N_\mathrm{P}}{\mathrm{d}t} = \lambda_\mathrm{P}N_\mathrm{P}. \tag{5.21}$$

The amount of daughter, N_D, at any time therefore can be expressed as

$$N_\mathrm{D} = N_{\mathrm{P}_0} - N_\mathrm{P} = N_{\mathrm{P}_0} - N_{\mathrm{P}_0}\mathrm{e}^{-\lambda_\mathrm{P}t} = N_{\mathrm{P}_0}\left(1-\mathrm{e}^{-\lambda_\mathrm{P}t}\right). \tag{5.22}$$

This expression is equivalent to the total number of parent atoms initially present multiplied by the fraction of the total parent atoms that remains at any later time.

Conversion of a radioactive parent to radioactive daughter, as occurs within the ^{238}U, ^{235}U, and ^{232}Th decay series, is conceptually and numerically more complex. In this case the rate of change in the number of daughter atoms equals the rate of parent decay minus the rate of daughter decay:

$$\mathrm{d}N_\mathrm{D}/\mathrm{d}t = \lambda_\mathrm{P}N_\mathrm{P} - \lambda_\mathrm{D}N_\mathrm{D}. \tag{5.23}$$

The solution of this differential equation (using the previous P and D notation) for daughter activity is

$$A_\mathrm{D} = \frac{\lambda_\mathrm{D}}{\lambda_\mathrm{D}-\lambda_\mathrm{P}}A_{\mathrm{P},0}\left(\mathrm{e}^{-\lambda_\mathrm{P}t} - \mathrm{e}^{-\lambda_\mathrm{D}t}\right). \tag{5.24}$$

This generally applicable equation simplifies considerably when the half life of the parent is longer than the half life of its radioactive daughter. When $t_{1/2\mathrm{P}} \gg t_{1/2\mathrm{D}}$ then $\lambda_\mathrm{P} \ll \lambda_\mathrm{D}$ and this allows the above equation to be simplified:

$$A_\mathrm{D} \sim A_{\mathrm{P},0}\left(\mathrm{e}^{-\lambda_\mathrm{P}t} - \mathrm{e}^{-\lambda_\mathrm{D}t}\right) = A_\mathrm{p} - A_{\mathrm{p},0}\mathrm{e}^{-\lambda_\mathrm{D}t}. \tag{5.25}$$

For time periods that are on the order of the daughter half life but much shorter than the parent half life, $A_P \sim A_{P,0}$ and the daughter isotope grows into equilibrium with a time constant of its own half life:

$$A_D \sim A_{\mathrm{P},0}\left(1-\mathrm{e}^{-\lambda_\mathrm{D}t}\right). \tag{5.26}$$

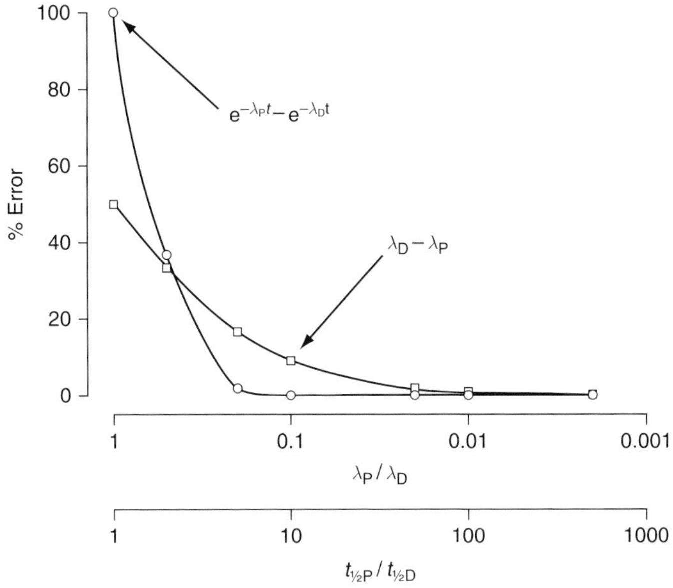

Figure 5.13. The error caused by the assumption that the activities of parent and daughter radioactive isotopes are in secular equilibrium in a closed system (Eq. (5.26)) as a function of the ratio of the parent daughter (P:D) half lives. Very small errors exist as $t_{1/2,P}/t_{1/2,D}$ becomes greater than 50.

For time periods that are long compared to the daughter half life the second term on the right hand side of Eq. 5.25 aproaches zero, and the activities of the parent and daughter are nearly equal, $A_D \sim A_P$.

This condition of essentially equal radioactivity between a long-lived parent and a coexisting short-lived daughter is referred to as "secular equilibrium" and represents one of the most useful relations in radiochemistry.

At what ratios of parent to daughter decay rates do the above approximations become accurate? If we assume that anything smaller than a 2% calculation error in A_D is acceptable, then the λ_P/λ_D ratios for which Eqs. (5.25) and (5.26) are accurate are shown in Fig. 5.13 to be approximately 0.2 and 0.02, respectively. These thresholds correspond to $t_{½P}/t_{½D}$ ratios of 5 and 50, respectively. Comparison of the latter ratios to corresponding values for the ^{238}U, ^{235}U, and ^{232}Th decay series (see later, Fig. 5.19) shows that the half life ratios of the ultimate parents to their longest-lived daughters are always in excess of 10^4. (The longest-lived daughters, ^{234}U and ^{231}Pa, with half lives of 2.5×10^5 and 3.2×10^5 y, respectively, have $t_{½P}/t_{½D}$ ratios of 2×10^4.) Thus an assumption of secular equilibrium between these three ultimate parents and any of their corresponding daughters is in theory acceptable with tiny accompanying mathematical error.

A conceptual analogy of secular equilibrium is two linked water tanks of equal bottom area, each being drained by siphons of different internal diameter (Fig. 5.14). In this analogy, the higher tank contains a height of water, H_P, representing total number of parent radionuclides, N_P. The inner diameter of the siphon draining the upper tank represents the decay constant of the parent, λ_P. The

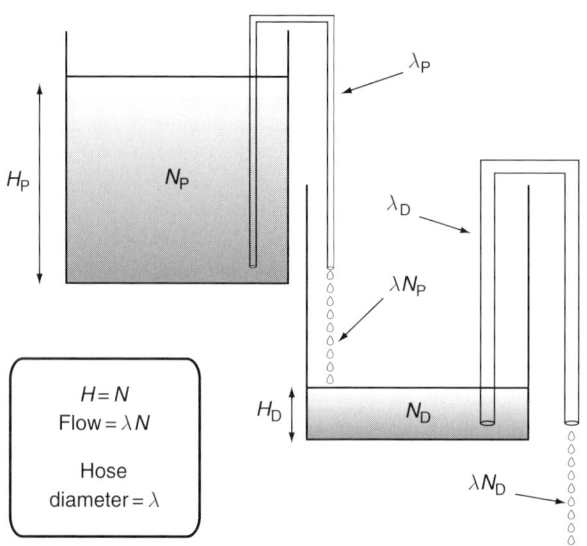

Figure 5.14. A tank and siphon analogy for secular equilibrium. In this analogy the number of atoms (N) of parent (P) and daughter (D) isotopes is proportional to the amount of water in the two reservoirs. The decay half lives of the parent and daughter isotopes are proportional to the hose diameter. If N_P remains constant or nearly so, the flow rate, which is proportional to the reservoir volume times the hose diameter, $N\lambda$, must be the same from both tanks. (See text for further explanation.)

combined effect of this inner diameter and the hydrostatic head determines the rate of water outflow, which is analogous to the activity of A_P. The lower tank is drained by a siphon of greater inner diameter (e.g. $t_{1/2}$ of the daughter is shorter than that of the parent), thus it requires a smaller hydrostatic head to generate a flow that matches the output of the elevated tank. Matched flows from the two tanks eventually will be obtained even if the lower tank is initially empty. In the radiochemical counterpart, the number of atoms of the radioactive daughter supported by the longer-lived parent will be proportionately smaller at secular equilibrium. An additional insight from this analogy is that the time to steady state for the water volume in the lower tank depends on the diameter of the outlet siphon. For example, if the lower tank is empty when the water flow from above started, it will reach its final volume very quickly if the outlet siphon is large and more slowly if it is smaller. For the radioactive counterpart, the time to secular equilibrium is determined by the half life of the daughter, not the parent. This point has important consequences for the utility of parent/daughter isotopes in marine geochemistry.

Secular equilibrium will be obtained only if the system in question remains physically closed, such that radioactive decay is the only process by which a daughter is effectively removed from its parent. Any additional non-radiochemical process that physically removes the daughter will destroy secular equilibrium. We shall see that chemical separation mechanisms prevent a closed system, and this is necessary for useful rate information to be obtained from the radioactive decay series.

5.2.3 Radiochemical sources and distributions: carbon-14

Radiocarbon is one of the most widely applied radionuclides in geochemical studies and one of the most interesting. Like many low mass radionuclides, ^{14}C is produced by cosmic rays. The production process first involves shattering (spalation) of a nucleus (most likely N or O) in the upper atmosphere by a cosmic ray from space. Among the released fragments are neutrons, some of which are slowed by subsequent collisions and then penetrate the nucleus of a ^{14}N atom (in N_2) resulting in the following neutron activation reaction

$$^{14}_{7}N + n^0 \rightarrow {}^{14}_{6}C + p^+ + e^-, \tag{5.27}$$

which releases a proton and an electron from the ^{14}N atom (conserving mass and charge), thereby converting ^{14}N to ^{14}C that largely occurs in the atmosphere as CO_2 gas. ^{14}C, with eight neutrons and six protons, is an unstable nucleus and converts by β^- decay back to ^{14}N with a half life of 5730 y. ^{14}C production is limited largely to the upper atmosphere near the poles where the magnetic lines of force, which shield the atmosphere from cosmic rays, dip into the Earth. Once $^{14}CO_2$ is mixed into the lower atmosphere, and then into the ocean or biomass, it is effectively separated from its source and decays with essentially no *in situ* replenishment. Carbonaceous materials that do not exchange C atoms with the lower atmosphere thus undergo simple first-order decay that is amenable to a variety of dating and isotope balance calculations. A complication in such applications, however, is that the global production rate of ^{14}C varies, in relation to sunspot activities and fluctuations in the strength of the Earth's magnetic field. The history of these variations in ^{14}C production rate, however, can be empirically determined by measuring $^{14}C/^{12}C$ ratios in tree rings and carbonate materials of a known age. The former dating method, dendrochronology, involves matching tree ring patterns in overlapping wood borings taken from different trees in an area to form a continuous record of ^{14}C production (often over periods of thousands of years). Dedrochronological records obtained for a restricted region can often be applied globally because of the relatively fast mixing rate of the atmosphere.

Measurements of ^{14}C in organic samples are made by combustion to CO_2, which is then purified and converted to a suitable form for analysis. Before the widespread availability of accelerator mass spectrometers (AMS), the β^- emission activities of ^{14}C samples were "counted" directly in CO_2 or after carbon was concentrated in a liquid (e.g. benzene) or solid (e.g. graphite) form. Unfortunately direct counting of this type requires more than 100 mg of C and is not feasible for small amounts of sample. Presently, almost all radiocarbon measurements are made by direct counting of individual ^{14}C atoms in a graphite "target" using the AMS methodology that requires less than 100 μg of C. Whatever the method of analysis, the final abundance results are usually calculated and reported in "delta" ^{14}C format

$$\Delta^{14}C = \delta^{14}C - \left[\left(2\,\delta^{13}C + 50\right)\left(1 + \delta^{14}C/1000\right)\right], \tag{5.28}$$

where

$$\delta^{14}C = \left(\frac{A^{14}_{spl}}{0.95 A^{14}_{ox}} - 1 \right) \times 1000 \quad (5.29)$$

and A_{spl} and A_{ox} ^{14}C are activities of the sample and standard (oxalic acid) carbon. The term $[(2\,\delta^{13}C + 50)\,(1 + \delta^{14}C/1000)]$ in Eq. (5.28) corrects the measured $\delta^{14}C$ for isotope fractionation between the sample and standard, as determined by the $\delta^{13}C$ of the same sample. This term is formulated to be equal to zero for a sample having a $\delta^{13}C$ of -25.0‰, as is typical of the woods that were the most common early sample types. The coefficient 2 for $\delta^{13}C$ accounts for the fact that isotope fractionation of ^{14}C for a given sample will be twice as large as for ^{13}C. This is because the magnitudes of carbon isotope effects are proportional to the mass difference from ^{12}C. $\Delta^{14}C$ terminology thus has a built-in correction for stable isotope fractionation so that the observed changes are due only to radioactive decay. Equation (5.28) is set up to give a $\Delta^{14}C$ of zero for a wood formed in 1850, whose ^{14}C activity (or content) would be 95% that of the oxalic acid radiocarbon standard. Normalization to a date approximately 150 y ago is to avoid complications resulting from recent anthropogenic effects on the ^{14}C content of atmospheric CO_2. One of these perturbations is the "Suess Effect," which refers to the net decrease in the ^{14}C content of atmospheric CO_2 that has resulted since 1850 due to the burning of radiocarbon "dead" fossil fuels such as coal and petroleum.

A second, much larger anthropogenic effect results from ^{14}C production during atmospheric testing of thermonuclear devices. The resulting "bomb carbon" nearly doubled the natural level of ^{14}C in atmospheric CO_2 within the Northern Hemisphere between 1945 and 1964, with much of the increase occurring after the late 1950s. The spike in atmospheric ^{14}C in the mid-1960s (Fig. 5.15) resulted from a flurry of thermonuclear detonations in anticipation

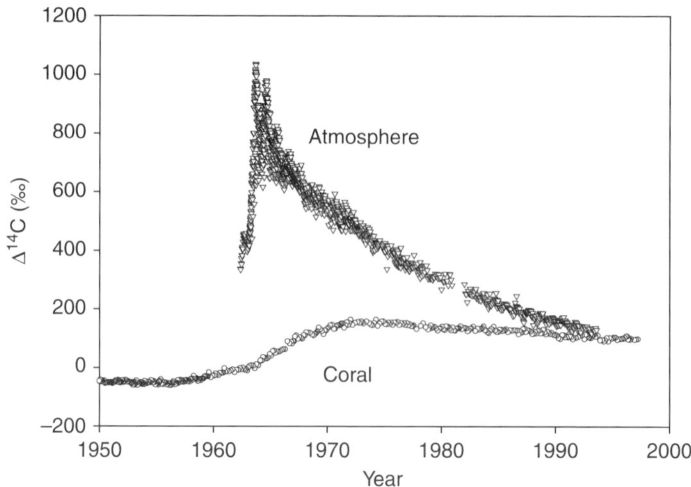

Figure 5.15. The activity of radiocarbon ($\Delta^{14}C$ ‰) in the atmosphere and surface ocean in response to atmospheric nuclear weapons testing. Atmospheric values are from the Carbon Dioxide Information and Analysis Center (CDIAC), Oakridge, TN. Surface ocean data are from measurement of the ^{14}C activity of growth bands of corals (Druffel and Linick, 1978).

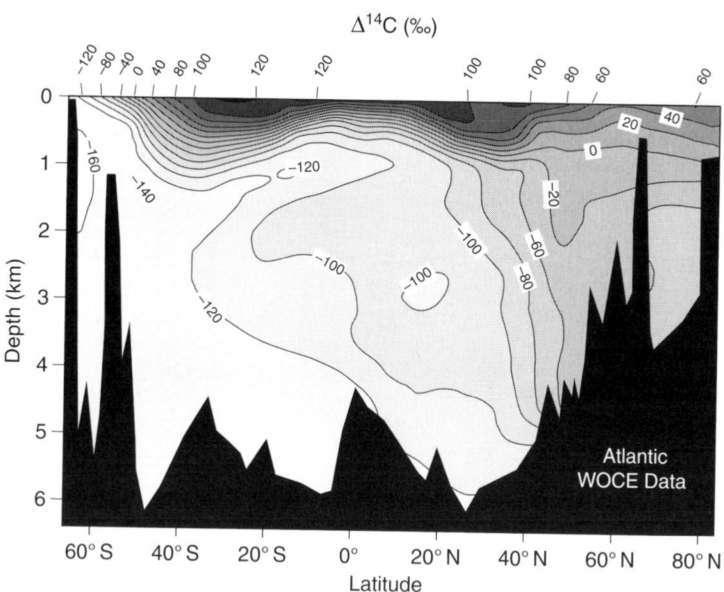

Figure 5.16. A cross section of ^{14}C-DIC in the North Atlantic Ocean, showing the penetration of bomb-produced ^{14}C ($\Delta^{14}C$ values greater than -50‰) to great depth in the northernmost waters. Figure courtesy of Robert Key, Princeton University; Key *et al.* (2004).

of an international moratorium on the atmospheric testing. Since that time the $\Delta^{14}C$ of atmospheric CO_2 has steadily decreased at a rate much faster than to be expected from radioactive decay because of continuous active exchange of carbon between atmospheric CO_2 and other major reservoirs, primarily oceanic dissolved inorganic carbon and carbon in land plants and soil organic matter. The corresponding increase in the radiocarbon content of DIC in the surface ocean is readily measurable but because there were few DIC-^{14}C samples in the 1950s and 1960s, the time course of surface-ocean ^{14}C is recorded with the best detail in corals that grew in mid-latitude surface waters (Fig. 5.15). Approximately five years after the maximum of $\Delta^{14}C$ in atmospheric CO_2, $\Delta^{14}C$ values at mid latitudes increased from approximately -50 (the pre-bomb condition) to +150‰. By the mid-1970s, when the GEOSECS survey took place, measurable vertical penetration of bomb ^{14}C into the surface western Atlantic reached approximately 1 km near 40° north and south latitudes. Carbon-14 measurements during the more recent WOCE program (1990s) clearly show penetration of bomb ^{14}C to the ocean floor in the far North Atlantic due to downwelling in the vicinity of the Greenland Sea (Fig. 5.16). Although injection of "excess" bomb ^{14}C into the contemporary environment has in many cases greatly compromised conventional age determination methods, it has provided a sensitive tracer for the passage of atmosphere CO_2 into the ocean via gas exchange and through the ocean by mixing.

Because water in the deep ocean mixes on time scales of 500–1000 y, the ^{14}C content of DIC in the vast majority of the deep sea waters is still largely unaffected by bomb carbon. In fact, it is via the determination of the ^{14}C age of deep sea DIC that we know the approximate circulation time of the deep ocean. This is illustrated

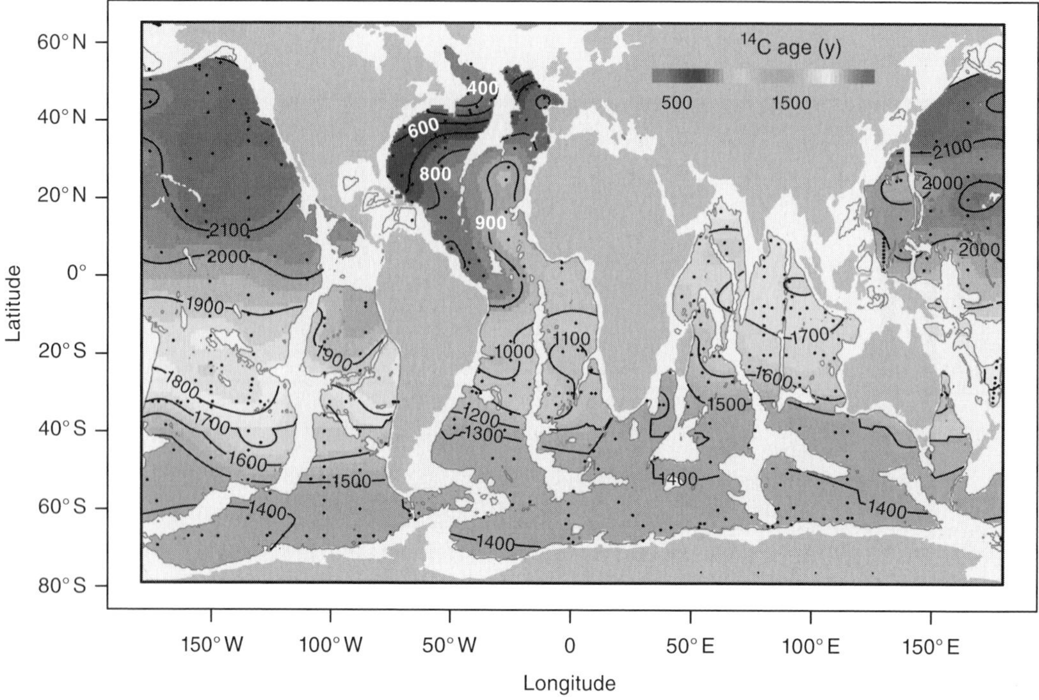

Figure 5.17. The ^{14}C age of DIC in the world's ocean at a depth of 3000 m, determined during the WOCE program in the 1990s. Courtesy of Robert Key, Princeton University; Key *et al.* (2004). (See Plate 3.)

qualitatively in Fig. 5.17 by the distribution of DIC-^{14}C age of water in the deep world's ocean at a depth of 3000 m. A rule of thumb for comparing Δ^{14}C and age is that a 10‰ decrease in Δ^{14}C is roughly equivalent to an increase in age of 80 years. As can be seen, deep circulation rates are sufficiently slow that the entrained ^{14}C decays measurably along flow paths. The younger ages occur in DIC of the North Atlantic Ocean, where active downwelling in the vicinity of the Greenland Sea carries recently vented surface waters to near the ocean bottom. (All samples in this figure are believed to be free of bomb ^{14}C or corrected for the bomb contaminate.)

The DIC ages in Fig. 5.17 represent the decay time with respect to a standard that is calibrated to 1850 wood, which is a close approximation to the pre-bomb modern atmospheric value. The ages in the figure do not strictly represent the time since sinking of deep waters for two reasons: the starting values at locations where surface waters sink are not zero age and the deep sea DIC-^{14}C ages are strongly influenced by mixing between water masses of different ages. Surface water ages in the North Atlantic and Antarctic where deep waters originate are the same as those in Fig. 5.17, *c.* 400 y and *c.* 1400 y, respectively. In both cases the "preformed" ages are greater than atmospheric modern values because it takes a decade or so for surface water DIC-^{14}C to be reset to atmospheric values by air–water gas exchange (see Chapter 10). Surface waters in these regions cool and sink before they have time to lose the carbon-14 age gained from their previous deep ocean journey. One of the ramifications of this is that the deep waters in the Atlantic appear to age from *c.* 400 y to 1400 y from the northern to the southern extreme. In fact we know that the time required for North Atlantic Deep Water to flow from Greenland to Antarctica is only on the order of a few hundreds of years. The "aging" of the water in the figure is primarily a result of mixing of waters that start out with ages of 400 and 1400 y.

Waters in this rapidly circulating Antarctic "hub" exhibit relatively uniform ^{14}C ages of 1300–1400 years, which represents a mixture of ages from all the deep basins. There are no northern end members in the Indian and Pacific Oceans, so the aging that occurs there is representative of the transit time: 200–300 years to flow from the Antarctic to the northern reaches of the Indian Ocean and 500–600 years transit time between Antarctica and the deep waters south of Alaska. Assessments of this process by using both box models and global circulation models indicate a mean "ventilation time" for the ocean of between 500 and 1000 y (Stuiver *et al.*, 1983; Toggweiler *et al.*, 1989). The seemingly simple pattern of conveyor belt circulation of deep water from the Atlantic to the Pacific Ocean (see Chapter 1) is difficult to directly convert into meaningful water velocities and basin flushing rates because ocean waters do not flow like a freight train but recirculate and intermix in complex patterns.

5.2.4 Radiochemical sources and distributions: uranium decay series

Most applications of radiochemical methods to oceanographic studies involve isotopes in the decay series deriving from the three long-lived radioactive parents ^{238}U, ^{232}Th, and ^{235}U, which are all located within the extreme high mass range ($Z > 84$, $N > 126$, $A > 210$) of Fig. 5.1. All three parents have half lives near or in excess of 10^9 y, as is required for their survival in appreciable amounts since the formation of the elements by nucleosynthesis. The three decay series (Fig. 5.18) comprise ten elements and 36 radioactive isotopes and terminate in three stable Pb daughters (^{206}Pb, ^{208}Pb, and ^{207}Pb, respectively). The ten elements involved represent a huge diversity of chemical characteristics, ranging from relatively soluble elements (Rn, U, and Ra) that include a gas (Rn), to highly surface active nuclides (Th, Pa, Po, and Pb) that are readily adsorbed (scavenged) onto particle surfaces. All isotopes in these decay chains interconvert by α or β^- decay, which correspond respectively to two steps down, or one step to the upper right in Fig. 5.18. The half lives of the U and Th series daughters range from fractions of a second (e.g. ^{212}Po, ^{214}Po, ^{215}Po, ^{215}Po, and ^{216}Po) to hundreds of thousands of years (^{230}Th and ^{231}Pa). No daughter has a half life greater than 0.01% of that of its ultimate parent, making secular equilibrium applicable in physically closed systems. In general, all daughters in the ^{232}Th series exhibit relatively short half lives, whereas decay in the ^{238}U and ^{235}U series is rapid only toward the end of the series.

Given the wide range of isotopes (and corresponding half lives) for many of the individual elements (e.g. Th, Ra, Po, and Pb) in the U–Th decay series, some guidelines are useful as to which radioisotope is

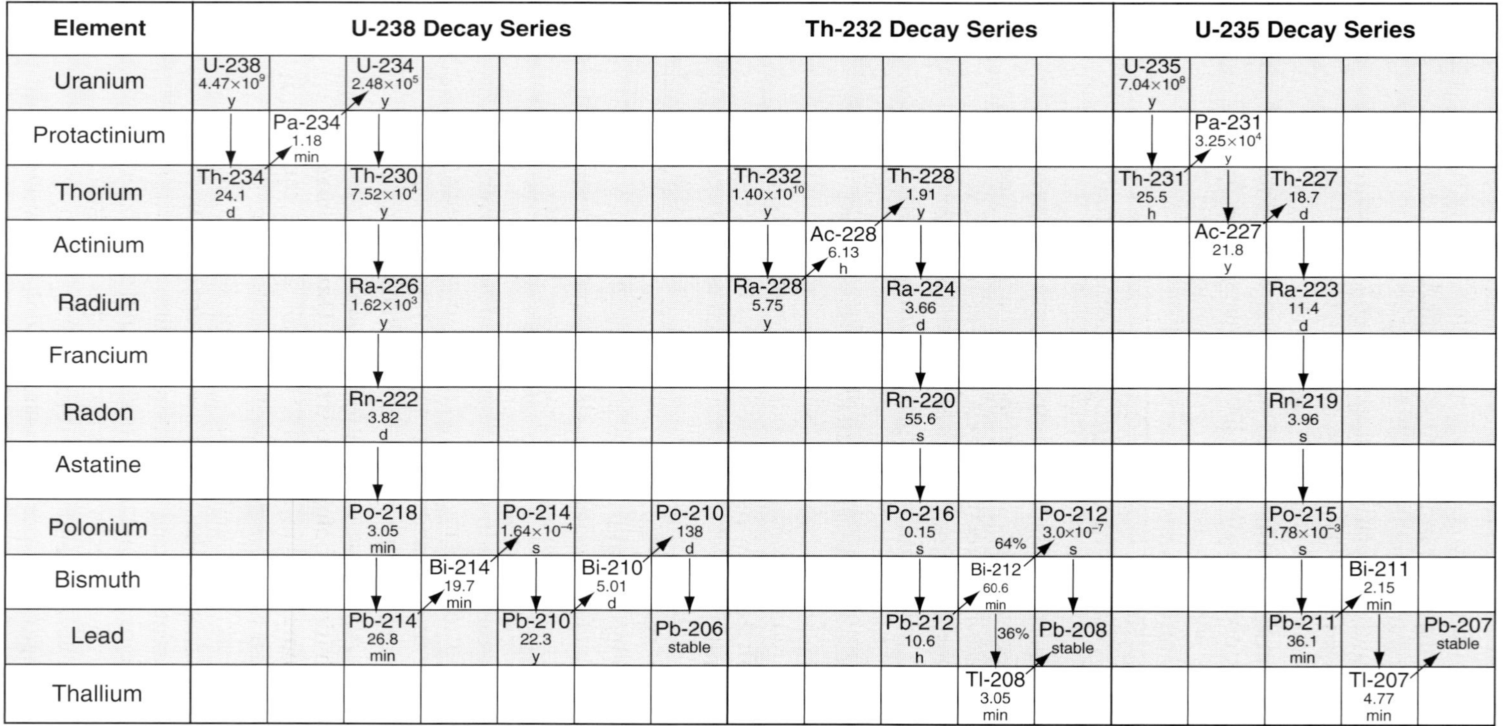

Figure 5.18. A chart of the nuclides showing the decay pathways and half lives of isotopes in the three naturally occurring decay chains. Arrows that point downward indicate α decay, in which a nucleus loses two neutrons and two protons, thus decreasing in atomic number by 2 and atomic mass by 4. The arrows that slant upward to the right indicate β decay, in which a neutron in the nucleus becomes a proton (one negative charge is lost from the nucleus), causing an increase in atomic number but little change in atomic mass.

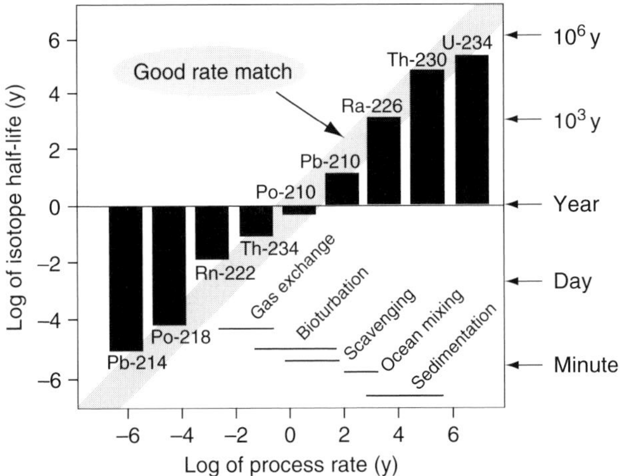

Figure 5.19. The relation between the half life of a radioisotope (ordinate) and the characteristic time scale for marine processes (abscissa). The shaded area indicates the range where the two life times are a good match.

best suited to measure the rate of a particular natural process. Of course, the first consideration is the chemical behavior of the elements in question, which must undergo the process to be investigated (e.g. mixing in dissolved form or particle transfer). An additional necessity in most applications is that at least a fraction of the radioisotope of interest not be continuously supported by a local source such as an "upstream" parent within the U and Th decay series. Specifically, rate determinations are feasible only when the chain of secular equilibrium is physically broken between a parent and its daughter. A final consideration is that the rate of the process to be investigated and the half life of the timing isotope should be matched in magnitude (Fig. 5.19). Thus, rapidly decaying isotopes (e.g. ^{234}Th and ^{222}Rn) are used to measure the rate of fast processes such as particle fluxes or gas exchange in the surface ocean, whereas radioisotopes with long half lives (e.g. ^{230}Th or ^{231}Pa) are useful for determining the rate of slow processes such as the accumulation rates of deep sea sediments. This constraint is largely analytical. An isotope that decays much faster than the characteristic rate of a targeted process will be essentially gone before the process fully expresses itself, whereas an isotope that decays much more slowly will not be measurably changed over the observation time.

Key nuclear and physical transformations that ^{238}U and its longer-lived daughters undergo in the ocean and atmosphere are illustrated in Fig. 5.20. This cartoon is a simplification that excludes some physical sources and sinks, does not specify the chemical forms of the elements, and ignores daughters with half lives less than a day. Like most seawater elements, ^{238}U is weathered out of continental rocks and carried by rivers to the ocean, where it occurs in a highly soluble dissolved form or in detrital sedimentary minerals. Because uranium is strongly complexed by CO_3^{2-} ions, it is relatively inert to particle adsorption, is not readily used by marine biota, and behaves conservatively in seawater. Dissolved ^{238}U, which does decay in the ocean, initiates an interesting series of reactions (Fig. 5.18)

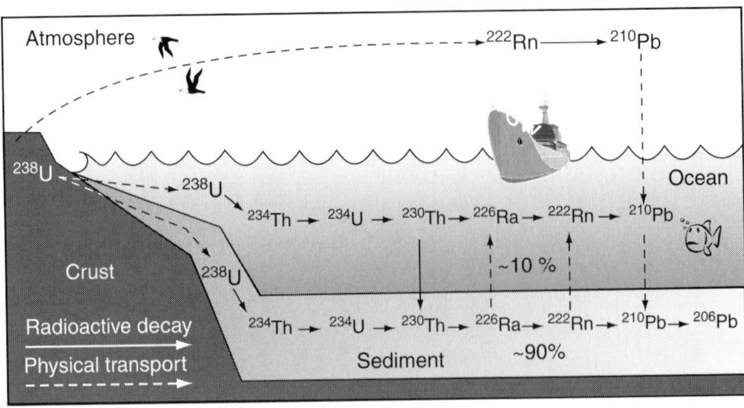

Figure 5.20. A schematic diagram of the pathways of ^{238}U and its daughter products in the ocean, indicating primarily which of the isotopes are soluble in seawater and which are adsorbed on particles and end up in the sediments. Modified from Broecker (1974). See also Table 5.5.

that begins with conversions to ^{234}Th ($t_{1/2} = 24.1$ d) and then to ^{234}U ($t_{1/2} = 2.48 \times 10^5$ y). The bulk of both isotopes remain in seawater, because ^{234}Th is short-lived compared with its physical removal rate and because ^{234}U is conservative in seawater. Although ^{234}U is sufficiently long-lived to survive chemical weathering and transport to the ocean by rivers, the predominant source of this daughter is *in situ* decay of ^{234}Th. By far the most striking aspect of the ^{238}U decay series is the essentially complete removal of ^{230}Th ($t_{1/2} = 7.52 \times 10^4$ y) from the water column following its generation by α decay from dissolved ^{234}U (Table 5.5). This quantitative transfer represents the combined effect of the high affinity of Th for the surfaces of sinking particles and its long half life during which scavenging can occur.

Because ^{230}Th scavenged from the water is concentrated on, rather than within, particles accumulating on the ocean floor, any soluble daughters it forms are in a position to escape to ocean bottom waters. The first daughter of ^{230}Th, ^{226}Ra ($t_{1/2} = 1620$ y), is sparingly surface-active, with the result that about 10% of this nuclide escapes to water in contact with the sea floor. The half life of ^{226}Ra is on the order of the mixing rate of the ocean, allowing this isotope to mix throughout the sea. In turn, ^{226}Ra decays to ^{222}Rn ($t_{1/2} = 3.82$ d), an inert gas with no affinity for mineral surfaces. Because it is a gas with a short half life, ^{222}Rn is an ideal tracer for determining rates of gas exchange (discussed in Chapter 10). Being a gas also makes it possible for ^{222}Rn to be transported from ^{226}Ra sources in rocks and soils on land, through the atmosphere to the open ocean (Fig. 5.20). The daughter of ^{222}Rn is ^{210}Pb ($t_{1/2} = 22.3$ y), a non-volatile metal with a high affinity for particle surfaces. Soon after formation ^{210}Pb is removed from the atmosphere by rain and dust, and falls to the surface of the continents and oceans. In marine systems, excess ^{210}Pb is scavenged from the water column into sediments, where it can serve as a clock for relatively fast sediment accumulation and/or mixing rates. The ^{238}U decay series ends with decay of ^{210}Pb to stable ^{206}Pb, which accumulates in sediments.

A cartoon similar to that in Fig. 5.20 for the ^{235}U decay series would be much simpler, with essentially complete removal of dissolved activity following formation of particle-active ^{231}Pa ($t_{1/2} = 3.25 \times 10^4$ y)

Table 5.5. *Activities in seawater of selective radioisotopes in the ^{238}U, ^{235}U, and ^{232}Th decay series* Modified from Broecker and Peng (1982).

Isotope	Half life (y)	Warm Surface Water	N Atlantic Bottom Water	N Pacific Bottom Water
		dpm $(100\ kg)^{-1}$		
^{238}U (parent)	4.47×10^9	240	240	240
^{234}Th	0.066	230	240	240
^{234}U	248 000	280	280	280
^{230}Th	75 200	c.0	c.0	0.15
^{226}Ra	1620	7	13	34
^{222}Rn	0.010	5	>13	>34
^{210}Pb	22.3	20	8	16
^{210}Po	0.38	10	8	16
^{235}U (parent)	0.7×10^9	13	13	13
^{231}Pa	32 500	c.0	c.0	0.05
^{232}Th (parent)	14×10^9	c.0	c.0	c.0
^{228}Ra	5.8	3	c.0	0.4
^{228}Th	1.9	0.4	c.0	0.3

via the short-lived intermediate ^{231}Th ($t_{1/2} = 25.5$ h). Essentially all the resultant excess radioactivity would remain in the sediments, because the subsequent daughters are either surface-active (e.g. ^{227}Ac, $t_{1/2} = 21.8$ y) or short-lived (e.g. ^{219}Rn, $t_{1/2} = 3.96$ s). The major application deriving from this decay series is the use of ^{231}Pa to clock processes such as particle scavenging and sediment accumulation. Because the ^{235}U decay series begins with a parent having an odd atomic mass and only involves mass decreases of −4 atomic mass units due to alpha emission, all its daughters down to stable ^{207}Pb can be recognized by their odd atomic masses. The ^{232}Th decay series uniquely begins in the sediments because this parent is insoluble. The only radioactivity to escape is in the form of ^{228}Ra ($t_{1/2} = 5.75$ y), which is partly released into bottom waters. This unsupported daughter decays via ^{228}Ac ($t_{1/2} = 6.13$ h) to ^{228}Th ($t_{1/2} = 1.91$ y), which could be used for measuring fast scavenging rates. All the rest of the lower-mass daughters, including stable ^{208}Pb, have very short half lives and limited geochemical application.

As a result of the interplay of physical and radiochemical processes illustrated in Fig. 5.20, isotopes in the ^{238}U, ^{235}U, and ^{232}Th decay series exhibit different activity patterns in seawater (Table 5.5). For example, although the chemical reactivities of ^{238}U and ^{234}U are the same throughout the ocean, the activity of the daughter isotope is about 15% higher than that of its direct parent. This observation is inconsistent with secular equilibrium. The reason for this offset is that the high-energy α particle emitted by ^{238}U shatters a portion of the parent crystal within the rock, which then dissolves on average more rapidly than intact crystals, releasing more ^{234}U to the environment. Thus rivers carry ^{234}U-enriched solutions from the continents to the ocean, where this daughter is sufficiently long-lived to maintain its excess activity. A second major pattern is that the seawater radioactivities of all daughter isotopes except ^{234}U are very small compared with those of their ultimate parents. This is because ^{232}Th is initially sedimentary and early daughters of the ^{238}U and ^{235}U decay series (^{230}Th and ^{231}Pa, respectively) are both removed to the sediments. An additional trend is that four of the ^{238}U daughters (^{226}Ra, ^{222}Rn, ^{210}Pb, and ^{210}Po) increase "downstream" between North Atlantic and North Pacific bottom waters. The reason for this is that ^{226}Ra behaves, at least partly, like a nutrient element in the sea and is associated with the cycles of silica and $CaCO_3$. The daughters of ^{226}Ra (^{222}Rn, ^{210}Pb, and ^{210}Po) follow the trend of their parent. The somewhat lower activities of ^{210}Pb and ^{210}Po versus their ^{226}Ra parent indicate that these surface-active elements are removed by sinking particles. Finally, only ^{210}Pb and ^{228}Ra exhibit pronounced activity maxima in warm surface ocean waters, reflecting their steady inputs from the atmosphere as windblown ^{210}Pb on particles and ^{228}Ra in coastal surface waters because of the source from its parent ^{232}Th in sediments.

Measuring the removal rates of organic matter from the surface ocean via particles is of great interest to oceanographers because it is correlated to the biologically driven export of organic matter. The isotope pair with the most utility as a tracer of this process is ^{234}Th–^{238}U. Because thorium is particle-reactive and has a relatively short half life (24.1 d) it is deficient in the surface ocean, where scavenging is rapid, with respect to its ingrowth by ^{238}U decay. Below the surface (*c.*100 m), where biological particles are less abundant, ^{234}Th and ^{238}U are near secular equilibrium. The difference in the activities of these two isotopes from secular equilibrium in the surface ocean is used as a tracer of particle export, which is correlated with net organic matter export from the oceans. Details of how this tracer is used to determine net carbon export from the surface ocean are described in Chapter 6.

In deeper waters the deficiency of ^{230}Th from its uranium precursor ^{234}U is dramatic (Table 5.5) because this thorium isotope has a very long half life (*c.* 75 000 y) and thus particle scavenging is much more effective at removal than the ingrowth toward secular equilibrium with ^{234}U. Bacon and Anderson (1982) showed that depth profiles of dissolved and particulate ^{230}Th could be used to demonstrate the dynamic relationship of metal exchange between particulate and dissolve forms. They argued that the thorium–uranium isotope pair could be used as a tracer of particle removal rates for those metals that fall in the category of adsorbed in Fig. 1.3.

The same isotopes that are useful tracers of adsorption in the water column are ideal for determining sediment accumulation rates because they are chemically separated from their radioactive parents in the overlying water column. Radionuclides commonly used for sediment dating include ^{230}Th ($t_{1/2} = 75\,200$ y), ^{231}Pa ($t_{1/2} = 32\,500$ y) and ^{210}Pb ($t_{1/2} = 22.3$ y). Given the need to match magnitudes of the daughter's half life with the time scale represented by the sediment core to be dated (Fig. 5.19), ^{230}Th and ^{231}Pa are most useful for long cores of slowly depositing offshore sediments, whereas ^{210}Pb often works well for rapidly depositing near shore sediments. The longer-lived nuclides will not decay appreciably within young sequences from near shore deposits. Cosmogenic ^{14}C ($t_{1/2} = 5700$ y) in sedimentary carbonates or organic remains provides a useful intermediate time scale for dating sediment. The shorter-lived isotopes of ^{210}Pb and ^{234}Th are useful in all sediments as indicators of bioturbation by animals. Application of some of these tracers to determining the age of sediments is discussed later in the chapter on paleoceanography (Chapter 7).

Appendix 5.1 Relating K, α, δ, and ε in stable isotope terminology

The following is a derivation of the relations between K, α, δ and ε for the oxygen isotope exchange reaction between CO_3^{2-} and H_2O that is the basis of the ^{18}O paleotemperature method. For the reaction

$$CO_3^{2-} + H_2{}^{18}O \rightleftharpoons C^{18}OO_2^{2-} + H_2O, \tag{5A1.1}$$

where the superscript for ^{16}O has been dropped for simplicity, K is related to α as follows.

$$\begin{aligned} K_{1-2} &= \frac{[C^{18}O_3^{2-}]\,[H_2O]}{[CO_3^{2-}]\,[H_2{}^{18}O]} \\ &= \frac{[C^{18}O_3^{2-}]}{[CO_3^{2-}]} \times \left(\frac{[H_2{}^{18}O]}{[H_2O]}\right)^{-1} = R_1 R_2^{-1} = \frac{R_1}{R_2} = \alpha_{1-2}, \end{aligned} \tag{5A1.2}$$

where R is the ^{18}O/^{16}O ratio and subscripts 1 and 2 refer to CO_3^{2-} and H_2O, respectively. The subscript 1–2 represents the reaction direction as written. Now, α_{1-2} can be related to ε $(\delta_1 - \delta_2)$ by generating a relative difference expression in R:

$$\alpha_{1-2} - 1 = \frac{R_1}{R_2} - 1 = \frac{R_1 - R_2}{R_2}. \tag{5A1.3}$$

However, for sample i by definition

$$\delta^{18}O \equiv \frac{R_i - R_{std}}{R_{std}} \times 1000, \tag{5A1.4}$$

which can be rearranged to give

$$R_i = \frac{\delta^{18}O_i \times R_{std}}{1000} + R_{std} = \left(\frac{\delta^{18}O_i}{1000} + 1\right) R_{std}. \tag{5A1.5}$$

Substitution of (5A1.5) into (5A1.3) (with R_{std} canceling out), results in

$$\frac{R_1 - R_2}{R_2} = \frac{\left(\frac{\delta^{18}O_1}{1000} + 1\right) - \left(\frac{\delta^{18}O_2}{1000} + 1\right)}{\left(\frac{\delta^{18}O_2}{1000} + 1\right)} \tag{5A1.6}$$

and

$$\alpha_{1-2} - 1 = \frac{\delta^{18}O_1 - \delta^{18}O_2}{\delta^{18}O_2 + 1000}, \tag{5A1.7}$$

$$\varepsilon = (\alpha_{1-2} - 1) \times 1000 \cong \delta^{18}O_1 - \delta^{18}O_2. \tag{5A1.8}$$

Appendix 5.2 Derivation of the Rayleigh distillation equation

Changes in the concentration of light [L] and heavy [H] isotopes with respect to time as a result of reaction in a closed system are proportional to the first-order rate constants, k, and the concentrations of the isotopes

$$\frac{d[L]}{dt} = k_L[L]; \tag{5A2.1}$$

$$\frac{d[H]}{dt} = k_H[H]. \tag{5A2.2}$$

Dividing (5A2.1) by (5A2.2) gives

$$\frac{d[H]}{d[L]} = \frac{k_H}{k_L}\frac{[H]}{[L]} = \alpha\frac{[H]}{[L]}, \tag{5A2.3}$$

where the ratio of the rate constants is the isotope fractionation factor, α:

$$\alpha = \frac{k_H}{k_L}. \tag{5A2.4}$$

Rearranging (5A2.3)

$$\frac{d[H]}{[H]} = \frac{d[L]}{[L]}\alpha \tag{5A2.5}$$

and integrating from the initial values ($t = 0$, $[H] = [H^0]$, $[L] = [L^0]$) to the values [H] and [L] at time t,

$$\int_{[H^0]}^{[H]} \frac{d[H]}{[H]} = \alpha \int_{[L^0]}^{[L]} \frac{d[L]}{[L]}, \tag{5A2.6}$$

gives

$$\ln\left(\frac{[H]}{[H^0]}\right) = \alpha \ln\left(\frac{[L]}{[L^0]}\right), \tag{5A2.7}$$

which equals

$$\frac{[\mathrm{H}]}{[\mathrm{H}^0]} = \left(\frac{[\mathrm{L}]}{[\mathrm{L}^0]}\right)^{\alpha}. \qquad (5A2.8)$$

Dividing by [L]/[L^0] gives

$$\frac{[\mathrm{H}]/[\mathrm{L}]}{[\mathrm{H}^0]/[\mathrm{L}^0]} = \left(\frac{[\mathrm{L}]}{[\mathrm{L}^0]}\right)^{\alpha-1} = f^{(\alpha-1)}, \qquad (5A2.9)$$

where f is the fraction of the light isotope remaining at time t. If you substitute R for the concentration ratios, then you have the Rayleigh distillation equation:

$$\frac{R_t}{R_0} = f^{(\alpha-1)}. \qquad (5.12)$$

Now, this equation can be transformed to δ notation by rearranging the definition of the del notation (Eq. (5.1)):

$$R_t = \left(\frac{\delta_t}{1000} + 1\right)R_{\mathrm{std}};\ R_0 = \left(\frac{\delta_0}{1000} + 1\right)R_{\mathrm{std}}, \qquad (5A2.10)$$

and substituting these values into Eq. (5.12) results in

$$\delta_t = (\delta_0 + 1000)f^{(\alpha-1)} - 1000. \qquad (5.13)$$

References

Altabet, M. A. and R. Francois (1994) Sedimentary nitrogen isotopic ratio as a recorder for surface ocean nitrate utilization. *Glob. Biogeochem. Cycles* **8**, 103–16.

Altabet, M. A. and L. F. Small (1990) Nitrogen isotopic ratios in fecal pellets produced by marine zooplankton. *Geochim. Cosmochim. Acta* **54**, 155–63.

Bacon, M. P. and R. F. Anderson (1982) Distribution of thorium isotopes between dissolved and particulate forms in the deep sea. *J. Geophys. Res.* **87**, 2045–56.

Bemis, B. E., H. J. Spero, J. Bijma and D. W. Lea (1998) Reevaluation of the oxygen isotopic composition of planktonic foraminifera: experimental results and revised paleotemperature equations. *Paleoceanography* **13**, 150–60.

Broecker, W. S. (1974) *Chemical Oceanography*. New York, NY: Harcourt Brace Jovanovich.

Broecker, W. S. (2002) *The Glacial World According to Wally*. Palisades, NY: Eldigio Press, Lamont-Doherty Earth Observatory.

Broecker, W. S. and T. H. Peng (1982) *Tracers in the Sea*. Palisades, NY: Eldigio Press, Lamont-Doherty Earth Observatory.

Carpenter, E. J., N. R. Harvey, B. Fry and D. G. Capone (1997) Biogeochemical tracers of the marine cyanobacterium *Trichodesmium*. *Deep-Sea Res.* **44**, 27–38.

Chappell, J. and N. J. Shackleton (1986) Oxygen isotopes and sea level. *Nature* **324**, 137–40.

Crowley, T. J. (1983) The geologic record of climate change. *Rev. Geophys. Space Phys.* **21**, 828–77.

Dansgaard, W. (1965) Stable isotopes in precipitation. *Tellus* **16**, 436–68.

Druffel, E.M. and T.W. Linick (1978) Radiocarbon in annual coral rings of Florida. *Geophys. Res. Lett.* **5**, 913–16.

Guy, R.D., M.L. Fogel and J.A. Berry (1993) Photosynthetic fractionation of stable isotopes of oxygen and carbon. *Plant Physiol.* **101**, 37–47.

Karl, D., R. Letelier, L. Tubas *et al.* (1997) The role of nitrogen fixation in biogeochemical cycling in the subtropical North Pacific Ocean. *Nature* **388**, 533–8.

Key, R.M., A. Kozyr, C.L. Sabine *et al.* (2004) A global ocean carbon climatology: results from Global Data Analysis Project (GLODAP). *Global Biogeochem. Cycles* **18**, GB4031, doi: 10.1029/2004GB002247.

Kiddon, J.M., L. Bender and J. Orchardo (1993) Isotopic fractionation of oxygen by respiring marine organisms. *Global Biogeochem. Cycles* **7**, 679–94.

Knox, M., P.D. Quay and D. Wilber (1992) Kinetic isotopic fractionation during air-water gas transfer of O_2, N_2, CH_4 and H_2. *J. Geophys. Res.* **97**, 20335–43.

Kroopnick, P. and H. Craig (1976) Oxygen isotope fractionation in dissolved oxygen in the deep sea. *Earth Planet. Sci. Let.* **32**, 375–88.

Lansdown, J.M., P.D. Quay and S.L. King (1992) CH_4 production via CO_2 reduction in a temperate bog: a source of ^{13}C depleted CH_4. *Geochim. Cosmochim. Acta* **56**, 3493–503.

O'Leary, M.H. (1981) Carbon isotope fractionation in plants. *Phytochemistry* **20**, 553–67.

Romanek, C.S., E.L. Grossman and J.W. Morse (1992) Carbon isotope fractionation in synthetic aragonite and calcite: effects of temperature and precipitation rate. *Geochim. Cosmochim. Acta* **56**, 419–30.

Shackleton, N.J. and N.D. Updyke (1973) Oxygen isotope and a paleomagnetic stratigraphy of equatorial Pacific core V28–338: oxygen isotope temperatures and ice volume on a 105 and 106 year scale. *Quat. Res.* **3**, 39–55.

Siegenthaler, U. (1976) *Lectures in Isotope Geology* (ed. E. Jager and J.C. Hunziker). Heidelberg: Springer-Verlag.

Sigman, D.M. and K.L. Casciotti (2001) Nitrogen isotopes in the ocean. In *Encyclopedia of Ocean Sciences* (ed. J.H. Steele, K.K. Turekian and S.A. Thorpe), pp. 1884–94. London: Academic Press.

Stuiver, M., P.D. Quay and H.G. Ostlund (1983) Abyssal water carbon-14 distribution and the age of the world oceans. *Science* **219**, 849–51.

Toggweiler, J.R., K. Dixon and K. Bryan (1989) Simulations of radiocarbon in a coarse-resolution world ocean model 1. Steady state prebomb distributions. *J. Geophys. Res.* **94**, 8217–42.

Urey, H.C., H.A. Lowenstam, S. Epstein and C.R. McKinney (1951) Measurement of paleotemperatures and temperatures of the upper Cretaceous of England, Denmark and the Southeastern United States. *Geol. Soc. Am. Bull.* **62**, 399–416.

Wada, E. (1980) Nitrogen isotope fractionation and its significance in biogeochemical processes occurring in marine environments. In *Isotope Marine Chemistry* (ed. D.E. Goldberg and Y. Horibe), pp. 375–98. Tokyo: Uchida Rokakuho.

Zhang, J., P.D. Quay and D.O. Wilbur (1995) Carbon isotope fractionation during gas-water exchange and dissolution of CO_2. *Geochim. Cosmochim. Acta* **59**, 107–14.

6

Life processes in the ocean

Patterns of chemical distributions within the ocean are primarily controlled by biological processes and ocean circulation. Major features of this biogeochemical mosaic include removal of nutrients from warm surface ocean waters, concentration of these same nutrients in deep-ocean waters, and depletion of dissolved oxygen at intermediate water depths. These patterns are imprinted as mixing and advection carry nutrient-laden water from ocean depths into the sunlit upper water. These nutrients are used during photosynthesis to generate particulate and dissolved products that sink or are mixed into the interior ocean, where they are respired back into dissolved metabolites. Interactions of these physical and biological processes occur on time scales of days to hundreds of years and are expressed by the vertical concentration profiles of a variety of dissolved chemical species throughout the ocean. The chemical perspective of oceanography involves using the distributions of metabolic products to derive information about the rates and mechanisms of ocean processes in this largely unobserved sphere.

The effects of life processes are felt in every chapter of this book. In this chapter we introduce the methods by which chemical tracers have been used to determine biological fluxes. We begin with a whole-ocean point of view in which chemical differences between the sunlit upper ocean and the dark deep waters are interpreted by using a two-layer model. The chapter progresses to a description of the chemical signatures of the two main biological processes, photosynthesis and respiration, starting at the top in the euphotic zone where organic matter is produced and progressing to deeper regions where it is respired.

6.1 A simple model of ocean circulation and biological processes

Descriptive insight into biological processes can be derived from details of concentration distributions for various chemical species. However, quantitative inferences require a model that describes the relations among concentrations, biological fluxes and circulation rates. A basic construct historically often used by chemical oceanographers to model large-scale circulation and fluxes involves dividing the ocean into well-mixed reservoirs (or boxes). The simplest of these is the two-layer ocean that was employed effectively by W. S. Broecker and others, primarily in the 1970s and 1980s (see, for example, Broecker, 1971; Broecker and Peng, 1982). In this model (Fig. 6.1) the upper

Figure 6.1. A schematic representation of the two-layer ocean model including the equations for the surface and deep-ocean mass balance of dissolved constituent C (mol m^{-3}). *J*, particle flux (mol y^{-1}); V_D, deep ocean volume (1.35×10^{18} m^3); V_S, surface ocean volume (3.62×10^{16} m^3); *R*, river water flow (3.5×10^{13} m^3 y^{-1}); v_M, water exchange rate (m^3 y^{-1}); *B*, burial flux (mol y^{-1}).

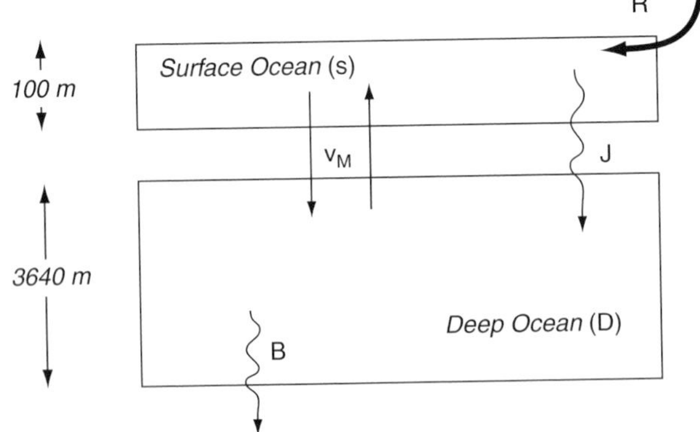

Surface layer

$$V_S \times \frac{d[C_S]}{dt} = R \times [C_R] - v_M \times ([C_S] - [C_D]) - J$$

Deep Ocean

$$V_D \times \frac{d[C_D]}{dt} = v_M \times ([C_S] - [C_D]) + J - B$$

ocean is assumed to consist of a surface layer that is *c*.100 m deep, which is the mean depth of the winter mixed layer in most of the ocean and the approximate depth of the 1% light level above which most photosynthesis occurs. The rest of the ocean, including the thermocline, is in the deep layer, which is on average 3700 m thick.

Although one can derive some fabulous first-order insight by using the two-layer model, one also has to be careful not to apply it blindly. Because the model assumes that the surface and deep ocean are well mixed and homogeneous, it can give very misleading results about processes that occur on time scales that are shorter than whole-ocean mixing or space scales smaller than indicated by the boxes. A few examples of problems that cannot be treated in with this construct are: (1) the uptake by the ocean of anthropogenic CO_2, because it is a transient process that has occurred over only the past several hundred years and the deep reservoir of the model ocean has a much longer residence time; (2) the influence of changing circulation on the biological pump, because low- and high-latitude surface waters are very different in their nutrient concentrations, but the model has only one global surface ocean reservoir; and (3) burial of $CaCO_3$, because the saturation state changes with depth and the deep reservoir of the model has no bathymetry. The solution to interpreting the chemical distributions caused by these processes is to create models with more layers and more realistic physics. The two-reservoir model is introduced here to demonstrate the first-order relation between nutrient distribution and biological export from the euphotic zone, and later in Chapter 10 to demonstrate the first-order response of atmospheric f_{CO_2} to changes in ocean circulation and biological carbon export.

The only nutrient input to the ocean in this simple model is via rivers, with a flow rate of R ($m^3\ y^{-1}$) and the only output is burial in sediments, B ($mol\ y^{-1}$). The surface and deep reservoirs communicate by upward and downward advection or mixing, v_M ($m^3\ y^{-1}$) and by particle transport, J ($mol\ y^{-1}$). The changes in concentration of component C with respect to time for the surface (S) and deep (D) reservoirs are:

$$V_S\left(\frac{d[C]_S}{dt}\right) = R\,[C]_R - v_M\left([C]_S - [C]_D\right) - J \qquad (mol\ y^{-1}); \tag{6.1}$$

$$V_D\left(\frac{d[C]_D}{dt}\right) = v_M\left([C]_S - [C]_D\right) + J - B, \tag{6.2}$$

where V_S and V_D are the volumes of the surface and deep reservoirs, respectively. At steady state the changes with respect to time are zero and the flow of species C from rivers is equal to its burial rate in sediments:

$$\frac{d[C]}{dt} = 0 \text{ and } R[C]_R = B. \tag{6.3}$$

Since there are only two reservoirs, Eqs. (6.1) and (6.2) are not independent.

In order to gain quantitative insight about fluxes the mixing rate, v_M, must be evaluated. This requires knowledge of the concentration distribution of a chemical tracer that has a known time history or a built-in radioactive clock. To achieve this we use the steady-state equation, and the mass balance of dissolved inorganic carbon (DIC) and the naturally occurring (i.e. not bomb-produced) radioactive isotope of carbon, ^{14}C, with a half life of 5730 y.

The mass balance for DIC in the deep box (Eq. (6.2)) at steady state is:

$$-v_M([DIC]_S - [DIC]_D) = J_C - B_C. \tag{6.4}$$

The flux of DIC from the deep reservoir to the surface ocean is equal to the particle rain rate of carbon from the surface reservoir minus the burial rate of organic carbon. The equation for $DI^{14}C$ is exactly identical, with the exception that there is an additional loss term on the right side for radioactive decay:

$$-v_M([DI^{14}C]_S - [DI^{14}C]_D) = J_{^{14}C} - B_{^{14}C} - [DI^{14}C]_D \cdot \lambda_{^{14}C} \cdot V_D. \tag{6.5}$$

The last term on the right-hand side represents the decay of ^{14}C, where $\lambda_{^{14}C}$ is the decay constant ($\lambda_{^{14}C} = 0.693/5730\ y = 1.21 \times 10^{-4}\ y^{-1}$; see Chapter 5 for details about radioactive decay). ^{14}C values are presented in units of $\Delta^{14}C$, which are related to the ratio of ^{14}C and ^{12}C in DIC. For convenience here, we define the term $r^* = DI^{14}C / DIC$ and rewrite the above equation as:

$$-v_M([DIC]_S \cdot r_S^* - [DIC]_D \cdot r_D^*) = J_C r_S^* - B_{^{14}C} - [DIC]_D \cdot r_D^* \cdot \lambda_{^{14}C} \cdot V_D. \tag{6.6}$$

Let us assume that the burial flux on the right side is much smaller than the particle transport and solve the equations; we will check this assumption later to see whether it is correct. Substituting Eq. (6.4) into (6.6) to eliminate the biological flux term gives:

$$v_M \cdot (r_S^* - r_D^*) = r_D^* \cdot \lambda_{^{14}C} \cdot V_D. \tag{6.7}$$

The turnover time or residence time of the deep ocean with respect to mixing in the model is $\tau_M = V_D / v_M$ ($m^3 / m^3\ y^{-1} = y$) so the above equation can be rewritten as:

$$\tau_M = \frac{V_D}{v_M} = \left(\frac{r_S^*}{r_D^*} - 1\right)\left(\frac{1}{\lambda_{^{14}C}}\right). \tag{6.8}$$

The relation between $\Delta^{14}C$ and r^* is approximately:

$$\Delta^{14}C \cong \left(\frac{r^*_{sample}}{r^*_{standard}} - 1\right) \times 1000. \tag{6.9}$$

Here we have deleted the corrections for stable isotope fractionation because we are comparing samples with the same source: seawater (see Chapter 5, Eq. (5.28)). Thus,

$$\frac{r_S^*}{r_D^*} \cong \left(1 + \frac{\Delta^{14}C_S}{1000}\right) \Big/ \left(1 + \frac{\Delta^{14}C_D}{1000}\right). \tag{6.10}$$

The residence time, τ, can now be written in terms of the natural $\Delta^{14}C$ of the surface and deep water of the ocean.

Presently ocean surface waters are contaminated with bomb-produced ^{14}C (Fig. 5.16); however, the few measurements from pre-atmospheric weapons testing and more recently on corals that grew at that time suggest that surface waters had values of about – 50‰ (Fig. 5.16). Since the Antarctic circumpolar water is a mixing region for the whole ocean, we take its value ($\Delta^{14}C = -160‰$) as an estimate of the mean value for deep waters. (This value is similar to the mean derived from more rigorous methods of volume averaging the deep water values (Broecker and Peng, 1982).) Thus, the average deep water $\Delta^{14}C$ is 110‰ lower than that for the surface, giving a value for r_S^*/r_D^* of 1.13. Now, from Eq. (6.8) the residence time of water in the deep ocean becomes

$$\tau_M = \frac{V_D}{v_M} = (1.13 - 1)\,(8268\,\mathrm{y}) = 1073\ \mathrm{y}. \tag{6.11}$$

We are not quite finished because it is necessary to evaluate our assumption at the beginning that carbon burial is negligible compared with ocean circulation in the deep ocean DIC balance. At steady state the burial rate and river inflow rate are the same. Using the global river flow rate (Table 1.1) and DIC concentration (Table 2.3) to calculate the burial rate gives

$$\begin{aligned} B = R \cdot [\mathrm{DIC}]_R &= (3.5 \times 10^{13}\ \mathrm{m^3\ y^{-1}})\,(0.96\ \mathrm{mol\ m^{-3}}) \\ &= 3.3 \times 10^{13}\ \mathrm{mol\ y^{-1}}. \end{aligned} \tag{6.12}$$

The mixing flux can be calculated from Eq. (6.4) by using the value of v_M from Eq. (6.11) and mean values for the DIC in the surface and deep ocean of 2.269 – 1.941 mmol kg^{-1} (Toggweiler and Sarmiento, 1985):

$$\begin{aligned} v_M \cdot ([\mathrm{DIC}]_D - [\mathrm{DIC}]_S) &= (1.26 \times 10^{15}\ \mathrm{m^3\ y^{-1}})\,(0.33\ \mathrm{mol\,m^{-3}}) \\ &= 41 \times 10^{13}\ \mathrm{mol\ y^{-1}}. \end{aligned} \tag{6.13}$$

Comparison of these fluxes indicates that mixing moves about 12 times more DIC from deep waters to the surface than burial removes to the sediments. The assumption that the particulate carbon rain rate is the dominant of these two fluxes is justified for our rough calculation.

Earlier in Chapter 5 a map of the ^{14}C age of DIC in the ocean's deep waters (Fig. 5.17) revealed that the age difference between the northern North Atlantic Deep Water and that in the Northeast Pacific is *c.*1700 y. This value compares the most recent and most ancient ventilation ages of the deep ocean, whereas the box model compares the mean deep water age of the entire ocean, *c.* – 160‰ (*c.* 1480 y) with that of the surface ocean, *c.* – 50‰ (400 y): (1480 – 400 = 1080 y). In some ways, the largest task of the two-layer-ocean calculation is determining representative $\Delta^{14}C$ values for the mean surface and deep ocean. More complicated models with more reservoirs (see, for example, Stuiver *et al.*, 1983; Toggweiler *et al.*, 1989) better illustrate the role of mixing among different ocean waters in determining the $DI^{14}C$ distribution. The more sophisticated calculations indicate that most of the aging of the deep ocean occurs in the Pacific and the residence time of water in this basin is about 500 y. The importance of the above calculation is that it returns a first-order value for circulation time in the two-layer ocean and creates a time scale for the model.

It is now possible to make a few general statements about biogeochemical dynamics in the ocean. First, the mixing rate, v_M, is about 40 times greater than the inflow rate from rivers:

$$\frac{v_M}{R} = \frac{1.26 \times 10^{15}}{3.5 \times 10^{13}} = 36. \tag{6.14}$$

Water circulates on average about 40 times through the surface ocean before it evaporates. This is consistent with the mean residence time of water with respect to river inflow of 40 000 y calculated in Chapter 2 (Table 2.3).

We also can evaluate the importance of the mixing (upwelling) rate and river inflow to the delivery of nutrients to the surface ocean, and the fraction of the particle flux that is buried. (See Broecker and Peng (1982) for more details.) Phosphorus will be used rather than nitrate as limiting nutrient in this illustration to avoid the complexities of nitrogen fixation and denitrification; however, phosphate and nitrate are related stoichiometrically in most of the ocean, so nitrate could also be used (Fig. 6.2). The average dissolved inorganic phosphorus concentrations in rivers, $[DIP]_R$, is 1.3 μmol kg^{-1} (Meybeck, 1979). The mean concentrations in the deep sea and surface ocean are $[DIP]_D = 2.3$ μmol kg^{-1} and $[DIP]_S = 0.0–1.3$ μmol kg^{-1}, respectively. (The latter value is the range for average subtropical ocean to

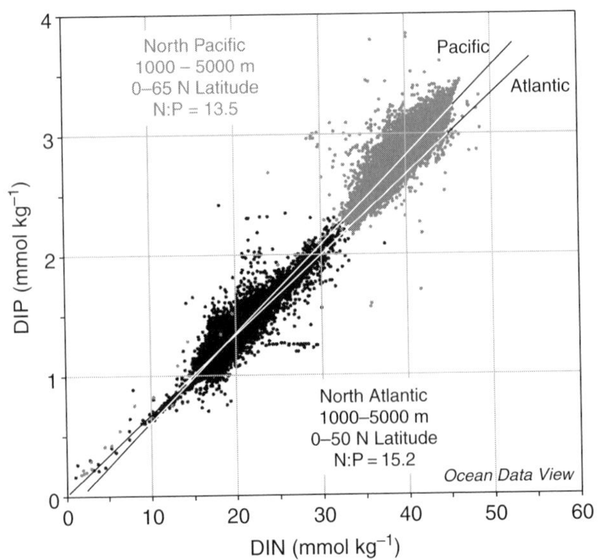

Figure 6.2. Dissolved inorganic phosphorus, DIP, versus dissolved inorganic nitrate (DIN, which is mostly NO_3^-) concentrations in the world's oceans between 1000 and 5000 m (plotted using Ocean Data View). Dark points represent data from the North Atlantic and lighter points are data from the North Pacific. Lines indicate linear regressions through the data from the Atlantic and Pacific.

high-latitude surface waters (Toggweiler and Sarmiento, 1985).) The model phosphorus fluxes are thus:

sediment burial:

$$B = R\,.\,[\mathrm{DIP}]_R = (3.5\times10^{13}\ \mathrm{m^3\,y^{-1}})\,(1.3\times10^{-3}\ \mathrm{mol\,m^{-3}}) \\ = 4.6\times10^{10}\ \mathrm{mol\,y^{-1}}; \qquad (6.15)$$

upwelling to the surface layer:

$$v_M\,.\,[\mathrm{DIP}]_D = (1.26\times10^{15}\ \mathrm{m^3\,y^{-1}})\,(2.3\ \times10^{-3}\mathrm{mol\,m^{-3}}) \\ = 2.9\times10^{12}\ \mathrm{mol\,y^{-1}}; \qquad (6.16)$$

and particulate rain rate to the deep layer:

$$J_P = v_M\cdot(\,[\mathrm{DIP}]_D - [\mathrm{DIP}]_S\,) = (1.26\times10^{15}\ \mathrm{m^3\,y^{-1}})\,(1.0\text{–}2.3\ \times10^{-3}\mathrm{mol\,m^{-3}}) \\ = 1.3\text{–}3.0\times10^{12}\ \mathrm{mol\,y^{-1}}. \qquad (6.17)$$

Implications of these results are that phosphorus removed from the surface waters as biological flux is 30–65 times more likely to come from upwelling than from rivers ($1.3\text{–}3.0\times10^{12}/4.6\times10^{10}$), indicating that ocean circulation is far more important in regulating biological productivity than river inflow. Also, only 1 in 30–65 atoms of P that rains to the deep ocean is actually buried; the rest are degraded in the deep and recycled back to surface waters. This results in a residence time for phosphorus with respect to burial of 30 000–65 000 y: 30–65 times the ocean circulation rate.

One of the most important fluxes that we will be discussing in this chapter and the following chapter on the carbon cycle is the biological carbon flux from the surface to the deep ocean. The flux is sometimes called the *biological pump* because it represents the force that "pumps" carbon from the surface ocean and atmosphere to the deep sea. The magnitude of this flux calculated in Eq. (6.13) is *c*.$5\times10^{15}\,\mathrm{g\,C\,y^{-1}}$ ($5\,\mathrm{Pg\,y^{-1}}$). This is an upper limit for the *organic carbon* flux as calculated from this model because the surface to deep DIC gradient is controlled in reality by both organic C and $CaCO_3$ carbon fluxes and we assumed in this calculation that it was entirely due to the OM flux. The biologically driven carbon flux from the euphotic zone is an important quantity because of its implications for the global carbon cycle and climate. It has been estimated by several independent models and experimental methods. We shall see later in this chapter how the value calculated here compares with those that are directly measured, but first we provide some detail of the chemical reactions that describe photosynthesis and respiration.

6.2 The euphotic zone

6.2.1 Photosynthesis

The average total mass of living organisms in surface seawater is approximately $10\,\mu\mathrm{g\,l^{-1}}$ (10 ppb). Thus, the organic matter concentration in the sea is much smaller than the comparable land surface values, which include large standing crops of trees and grasses. However, the fluxes of chemical species through the marine biomass by photosynthesis are about the same as those on land, so organic matter turnover times in the ocean must be relatively short. Overall, most marine plants and animals are small, on the order of micrometers, and are either passively drifting (planktonic) or weakly swimming. The average life times of the bacteria and phytoplankton are on the order of hours to days. Zooplankton have a much broader size range and life span, but still have life times of only days to months. The combined result of this limited mobility and brief life span is that most marine organisms are captive to, and characteristic of, the seawater in which they occur. The most common exceptions to this generalization are larger animals that can swim appreciable distances moving between water types, and sinking particles that can penetrate the thermocline into the interior ocean. Because planktonic ocean life is so dynamic, diverse, and endemic, its net effect on chemical fluxes is often better understood by the chemical patterns it creates and responds to, rather than by direct observation. A brief review of the main algal types and the methods of determining rates of net community production was presented near the end of Chapter 1, which is a good background to the following section.

One of the most useful findings that oceanographers have made in support of a chemical perspective on ocean processes has been that the C:N:P ratios of mixed marine plankton (zooplankton and phytoplankton) collected by towing nets (>64 μm mesh) through the surface ocean occur at relatively constant values, near 106:16:1. This observation was published by Redfield (1958) and then later elaborated by Redfield, Ketchum, and Richards (RKR, 1963), who "fleshed out" the ratios in the form of an equation for photosynthesis:

$$106CO_2 + 16HNO_3 + H_3PO_4 + 122H_2O \rightarrow (CH_2O)_{106}(NH_3)_{16}H_3PO_4 + 138O_2. \quad (6.18)$$

The left side of this RKR equation gives the moles of various inorganic nutrients (in their uncharged forms) that are converted by photosynthesis to plankton biomass and molecular oxygen. This reaction is highly endothermic (energy-requiring) and produces, in the form of O_2 and organic matter, the strongest oxidizing agent (electron acceptor) and reducing agent (electron donor) that occur in appreciable amounts on Earth. (See redox process in Chapter 3.)

The number of moles of oxygen produced in Eq. (6.18) was estimated theoretically, assuming that one mole of O_2 is released for every atom of carbon converted into biomass, and two moles for every atom of nitrogen. The reduction half-reactions for Eq. (6.18) are:

$$CO_2 + 4H^+ + 4e^- \longrightarrow CH_2O + H_2O; \quad (6.19)$$

$$NO_3^- + 9H^+ + 8e^- \longrightarrow NH_3 + 3H_2O; \tag{6.20}$$

and the oxidation half-reaction is:

$$2H_2O \longrightarrow O_2 + 4H^+ + 4e^-. \tag{6.21}$$

Equations (6.19) and (6.20) are combined in a ratio of 106:16 and the result is added to Eq. (6.21) by adjusting the stoichiometry so that the electrons cancel. This accounts for all the hydrogens that show up in $(CH_2O)_{106}(NH_3)_{16}H_3PO_4$ and the production of 138 mol of O_2. Whereas all the dissolved reactants for photosynthesis must travel with ambient seawater, O_2 can escape to the atmosphere and organic matter can sink in particles or be mixed in dissolved form out of the upper ocean. The differential mobilities of the two chemically extreme products of photosynthesis largely control the distribution of life and redox-sensitive compounds in the atmosphere and oceans.

As useful as the RKR equation has proven to be, it is not perfect. For example, any organic chemist would recognize that the formula given for biomass is impossibly hydrogen-rich and suspiciously overpacked with oxygen as well. These excesses result in part because the formula $(CH_2O)_{106}(NH_3)_{16}H_3PO_4$ assumes that organic carbon in plankton occurs exclusively in the form of carbohydrate (CH_2O), which is the most hydrogen- and oxygen-rich of all biochemicals (see Table 8.4). In addition, NH_3 carries roughly three times the amount of hydrogen that occurs in protein, the major form of nitrogen in living organisms. A plot of the H/C versus C/O quotients, sometimes called a "van Krevelen plot," of phytoplankton from five different regions of the ocean (Hedges *et al.*, 2002) is illustrated in Fig. 6.3A. As can be seen the RKR formula $(C_{106}H_{263}O_{110}N_{16}P)$, is richer in hydrogen (H/C = 2.45) and oxygen (O/C = 1) than the proteins, polysaccharides and lipids from which plankton are composed, and is thus impossible. Based on biochemical analysis, average marine plankton contain roughly 65% protein, 19% lipid and 16% carbohydrate, with the rest comparably divided between lipid and polysaccharide. The corresponding nominal formula $(C_{106}H_{177}O_{37}N_{17})$ corresponds to atomic H/C and O/C quotients near 1.60 and 0.38, respectively. To convert the RKR formula to the biochemical counterpart, it is necessary to remove approximately 46 water molecules. This correction is required because the CH_2O and NH_3 units in the RKR equation lose water when coupled into polysaccharides and proteins. RKR plankton also contain too much oxygen because polysaccharides are more O-rich than protein and lipid. The N/C quotient versus the RQ (O_2/C) for these plankton samples (Fig. 6.3B) illustrates that the respiration quotient of 1.30 for RKR plankton is the minimum possible value compared with 1.44, which best fits the plankton data.

So far we have considered only carbon and nitrogen, which dominate the redox chemistry during photosynthesis and respiration.

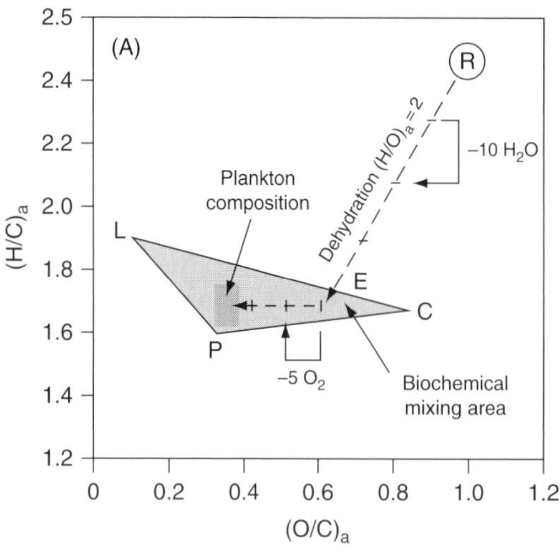

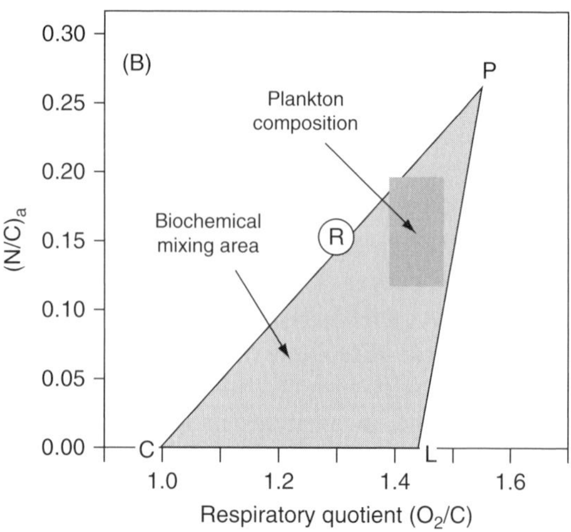

Figure 6.3. Quotients of H, C, N and O in four plankton samples from different locations in the ocean. (A) A van Krevelen plot of the H/C and O/C quotients on an atomic (subscript a) basis. The triangle has values for proteins, P, carbohydrates, C, and lipids, L, at the apices. The dark rectangle represents the plankton values and R is the RKR value. The trend from R to the triangle is the result of removing H_2O from the formula; the horizontal trend is for O removal alone. (B) The nitrogen/carbon quotient in organic matter, N/C, versus the molecular oxygen utilization to organic carbon degradation rate, the respiratory quotient ($\Delta O_2/\Delta C$). R is the RKR formula and the dark box the results of the plankton analysis. Redrawn from Hedges *et al.* (2002).

The ratio of these components to phosphorus is difficult to determine by measuring the P content of organic components because it is in much lower concentration. The ratios of C, N and O to phosphorus have been investigated by measuring changes of DIC, DIN and DIP on constant density and neutral surfaces in the aphotic zone of the ocean. These results conclude that a N:P ratio of 16:1 best fits the data in areas where denitrification is unimportant (Anderson and Sarmiento, 1994; Schaffer *et al.*, 1999). The nitrogen concentration relative to carbon is somewhat higher (106:17) in analysis by Hedges *et al.* (2002), probably because several of the samples were from the Southern Ocean where high nitrate assimilation may make these data more N-rich than the global average. We assume that the long-standing Redfield stoichiometry of N:P = 16:1 is appropriate for organic matter respiration in the absence of denitrification or nitrogen fixation.

The C:P ratio of marine plankton is more difficult to determine by analysis of DIC:DIP changes along constant density surfaces because: DIP concentrations are lower and therefore the DIP data must be more precise; DIC changes by organic matter degradation are small compared with a relatively large background value so these measurements must be very accurate; and thermocline density surfaces are contaminated by fossil fuel CO_2. Studies in which contamination by anthropogenic CO_2 sources is removed by the methods described in Chapter 11 indicate a C/P quotient that is in the vicinity of 120 (Anderson and Sarmiento, 1994; Schaffer *et al.*, 1999; Körtzinger *et al.*, 2001). Data used to determine C/P values of around 120, however, come from ocean depths greater than 400 m. In those studies where the value was determined as a function of depth (Schaffer *et al.*, 1999; Kortzinger *et al.*, 2001) a quotient of about 100 in the shallower regions of the thermocline was observed to increase to values between 120 and 130 with depth. For our purposes we stay with the C/P stoichiometric value of 106 because it is more likely to reflect the ratio that enters and exits the euphotic zone; however, it is clearly acknowledged that this number is still relatively uncertain.

To calculate O_2 consumption during respiration from the new Redfield stoichiometry, use the formula from Anderson (1995):

$$C_\alpha H_\beta N_\delta O_\gamma P_\phi + \omega O_2 \rightarrow \alpha CO_2 + 0.5(\beta - \delta - 3\phi)H_2O + \delta HNO_3 + \phi H_3PO_4; \tag{6.22}$$

and

$$\omega = \alpha + 0.25\beta + 1.25\,\delta - 0.5\,\gamma + 1.25\,\phi. \tag{6.23}$$

For the analysis here we use the stoichiometry of the phytoplankton analyzed by Hedges *et al.* (2002), but corrected for N by increasing the C:N ratio from 106:17 to 106:16 and including phosphorus in the C:P ratio of 106:1. The coefficients for Eqs. (6.22) and (6.23) are α, β, δ, γ, and $\phi = 106$, 179, 16, 38 and 1, respectively. Using these values in the above equation results in an oxygen : phosphate stoichiometry, $O_2 : P = 153$, and thus a molar ratio of oxygen change to carbon change ($\Delta O_2:\Delta OC$, sometimes called a respiration quotient, RQ) of 1.44. The RKR equation corresponding to this analysis becomes:

$$106CO_2 + 16HNO_3 + H_3PO_4 + 80H_2O \rightarrow C_{106}H_{179}O_{38}N_{16}P(OM) + 153O_2. \tag{6.24}$$

Clearly, there remain a range of possible stoichiometries and this "mean" value will probably continue to be revised. Based on the plankton study of Hedges *et al.* (2002) and the analysis of Anderson (1995), formulas in the range $C_{106-120}H_{170-180}O_{35-45}N_{14-18}P$, which require an O_2/C quotient of 1.42–1.46, appear to be realistic.

A final caveat regarding the RKR equation is that it includes only the *macro*nutrients (P, N and C) necessary for life. In Chapter 1 it was pointed out that some trace element profiles in the ocean are so well correlated with dissolved inorganic phosphorus and nitrogen in seawater that they suggest these metal concentrations are controlled by metabolic processes. Indeed, a number of first-row transition metals (manganese, iron, cobalt, nickel, copper, and zinc) in addition to the second-row metal cadmium are known to be essential for the growth of organisms (Morel *et al.*, 2003). In addition to being depleted in surface waters (except for Mn), most of these essential trace metals are chelated (complexed) by organic ligands (dissolved molecules that bind to the metals) so that the unchelated metal concentrations (either free metal or complexed by inorganic species in seawater) are present in extremely small amounts, between 10^{-15} mol kg^{-1} for Co and 10^{-11} mol kg^{-1} for Zn (Fig. 6.4). The exact nature of the organic ligands is not certain, but they are believed to be created by biological processes, and usually it is the unchelated metal concentration that is biologically available (Bruland and Lohan, 2003).

Figure 6.4. Fe, Mn, Co, Cu, Zn, Ni and Cd profiles as a function of depth in the oceans. Profiles in the left graph for each element are total metal concentration (the horizontal axis is linear). The profiles on the right are concentrations that are unchelated by organic metal complexes. This includes the free metal ion and all inorganic complexes. Notice that the horizontal axis for the graphs of unchelated metals is logarithmic (except for Mn). Redrawn from Morel *et al.* (2003).

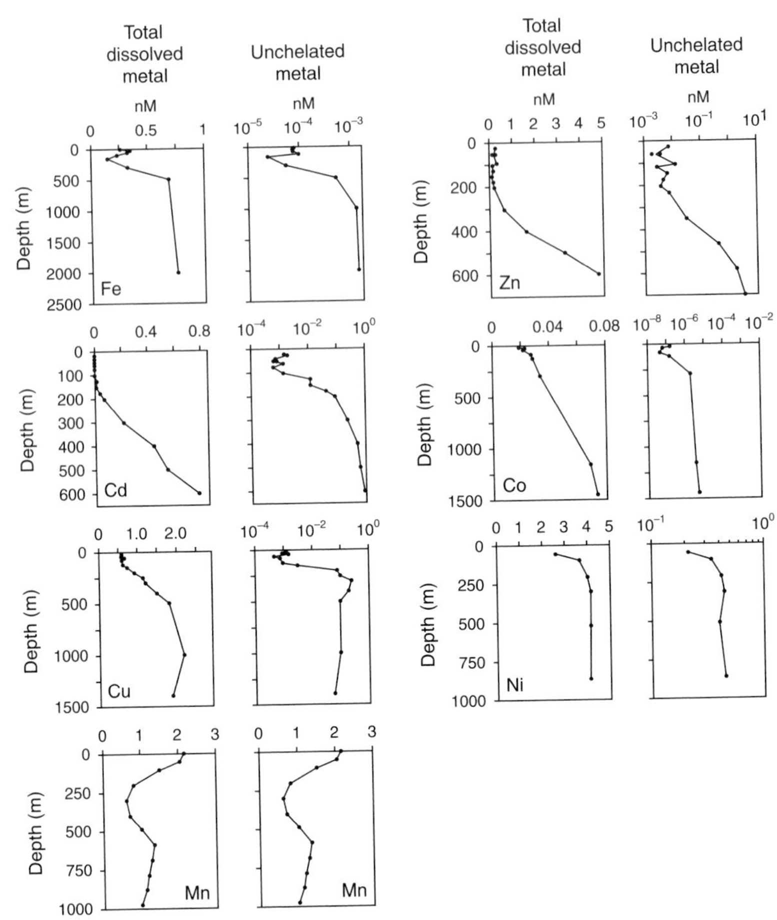

Measurements of the essential trace metals in some eukaryotic marine plankton grown in culture resulted in a Redfield-like stoichiometry for the essential trace metals (Morel *et al.*, 2003):

$$(C_{106}N_{16}P)_{1000}\,Fe_8Mn_4Zn_{0.8}Cu_{0.4}Co_{0.2}Cd_{0.2}. \tag{6.25}$$

This stoichiometry cannot be taken very seriously at this point as it is variable from species to species, and it is sure to evolve as more data become available. None the less, it clearly demonstrates the order of magnitude of the concentration of trace metals necessary for plankton growth. Iron and manganese have concentrations ten times higher than the other trace metals, but still 100 times lower than the macronutrient phosphorus.

Both Fe and Mn play important roles in electron transport in photosynthetic systems I and II, and this is the reason for their high concentrations relative to other trace metals in organisms (Morel *et al.*, 2003). Iron is the only trace metal that has, to date, been definitively shown to limit photosynthesis in the sea, and some estimate the area of iron limitation to be as high as 40% of the ocean's surface waters (Moore *et al.*, 2002). The reason for limitation by Fe and not by Mn appears to be the differences in their chemistry in seawater. Iron is strongly chelated, leaving surface waters extremely depleted in free iron (0.01–0.1 pmol) that can be used by phytoplankton. Manganese, on the other hand, is apparently not chelated by organic compounds to any great extent and is available in concentrations that are at least 1000 times those of Fe (Fig. 6.4). The Mn water column profile and its role in photosynthesis are examples of how our elemental classification scheme (Chapter 1) can, at times, be misleading. Mn is considered to be an adsorbed element because of its vertical profile shape in the sea; however, it is also one of the most important trace metal nutrients. The Mn profile shape implies something about the dominant oceanographic mechanism controlling its concentration, but does not rule out involvement in other important processes.

Iron is also a constituent of enzymes that are necessary for many of the transformations among the nitrogen containing compounds. Ammonium ion, NH_4^+, is the form most readily taken up by phytoplankton, but concentrations of NH_4^+ in seawater are very low because it is thermodynamically unstable and oxidizes to NO_3^- in oxygenated waters. A consistent observation in Fe enrichment experiments, in which the surface ocean was seeded with dissolved Fe and the response monitored, has been that larger diatoms grew much more readily than the other soup of phytoplankton after the Fe addition. The reason for this is believed to stem from the physics of nutrient supply to the cell and the availability of NH_4^+. Because large diatoms have a relatively low surface area to volume ratio, they require higher nitrogen concentrations to maintain the supply necessary for growth. Nitrate is unavailable unless reduced to NH_4^+, which requires iron-rich enzymes. In this scenario the limiting factor to large diatom growth is the availability of iron to aid the reduction of NO_3^- to NH_4^+ so it is available in concentrations high enough to support growth. Perhaps the most extreme example of Fe requirement is during nitrogen fixation (the transformation of N_2 gas to NH_4^+ in surface waters depleted in other dissolved nitrogen compounds). Enzymes responsible for N_2 fixation are very iron-rich, and growth on N_2 can require as much as ten times more Fe than growth on NH_4^+.

The trace metals Ni, Zn, Co, and Cd are essential elements for enzymes that carry out various functions of metabolism. Zinc is the most predominant metal in the enzyme carbonic anhydrase (CA), which catalyzes the transformation of HCO_3^- to CO_2. This mechanism is important in the sea because the pool of HCO_3^- contains 100 times more carbon than the pool of CO_2 at the pH of seawater, and it is usually CO_2 that is reduced enzymatically to organic carbon. Diatoms and some cyanobacteria also use CA to concentrate CO_2, and it has been observed that in some cases both Co and Cd can substitute for Zn in the carbonic anhydrase enzyme.

Zn is also an important component of the enzyme alkaline phosphatase, which is necessary for phytoplankton to be able to use dissolved organic phosphorus (DOP) during growth. Ni is necessary for the enzyme urease, which is required for phytoplankton to utilize urea as a nitrogen source. Cu has been labeled a "Goldilocks" metal because it is necessary for growth of cyanobacteria at the concentrations available in seawater, but at higher concentrations it is toxic. The organic chelators present in seawater maintain free Cu at a concentration that is "not too low" and "not too high," but "just right" (Bruland and Lohan, 2003).

Both macro- and micronutrient limitation and light availability conspire to create a complex distribution of marine photosynthesis in the sunlit surface of the ocean that generates net primary productivity of about 50 $Pg\,C\,y^{-1}$ (see Table 11.1). This flux is derived from compilations of ^{14}C productivity measurements and from global satellite color estimates of chlorophyll distributions. Because of the uncertainty of these methods the error in this estimate is large, probably at least $\pm 30\%$. Most of the organic matter produced during photosynthesis is respired in the euphotic zone by grazers, but some escapes to the deeper ocean, where it has a profound effect on the chemistry of the deep sea. Before moving on to biological processes that occur in the vast region of the ocean below the euphotic zone, we present a short discussion of the rates of respiration that accompany photosynthesis in the upper ocean.

6.2.2 Respiration in the upper ocean

For the chemist, respiration is simply photosynthesis (Eq. (6.24)) run backwards, with net destruction of organic matter and O_2 by flow of electrons from the former to the latter. This highly exothermic process fuels metabolic activity for both plants and animals and regenerates dissolved nutrients. (Sometimes this process is called "remineralization," but we refrain from using this term because of the mismatch between what the word says and the meaning of respiration.) Heterotrophic respiration (that is, excluding the respiration that occurs in phytoplankton) involves enzymatically controlled, thermodynamically favored, kinetic processes carried out by a variety of organisms that range in size from bacteria to whales at rates that are not readily predicted. The importance of microbial respiration was fully appreciated only when it became possible to measure the growth rates of marine bacteria by determining the incorporation of dissolved radioactive molecules (e.g. thymidine and leucine) into bacterial biomass. Such measurements throughout the surface ocean indicated that up to half of the carbon formed by photosynthesis was shunted via dissolved organic-molecule intermediates into bacteria. Bacteria supported by dissolved organic matter (DOM) are grazed by single-cell zooplankton (heterotrophic protozoa) that in turn are the food source for larger zooplankton and fishes (Fig. 6.5). This pathway for the flow of nutrients and energy up marine trophic levels has become known as the *microbial loop*.

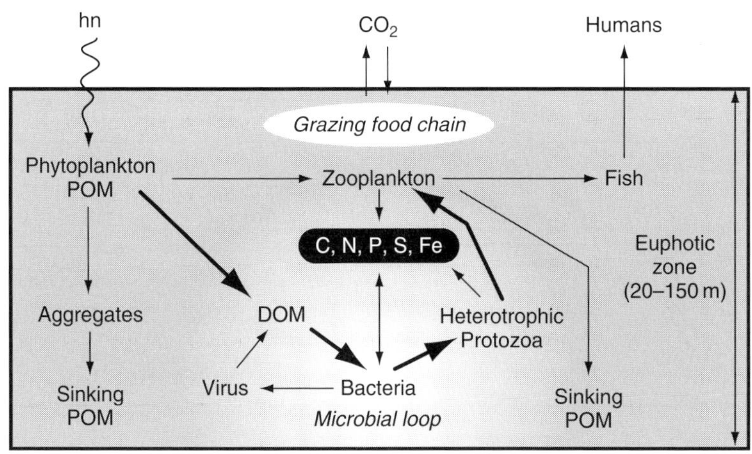

Figure 6.5. A schematic illustration of the food chain in the ocean indicating pathways of particulate organic matter (POM) and dissolved organic matter (DOM). The latter, indicated by the bold arrows in the figure, is called the microbial loop.

The key link between larger phytoplankton and zooplankton and bacteria is dissolved organic matter (DOM), which can be metabolized by single-celled heterotrophs because it can pass through microbial membranes. From an analytical chemical point of view DOM must be operationally defined because there is a continuum of sizes of organic matter particles in the ocean. The distinction between particulate organic matter (POM) and DOM is defined based on filter pore size and the ability of particles to gravitationally settle (Fig. 6.6). The general cutoff between particles that settle and those that are "dissolved" in seawater is 0.5 μm, which is the pore size of most frequently used filters. Thus, small particles and colloids are included as DOM. These particles are small enough and their density difference to water low enough that they are transported largely by water flow along with truly dissolved organic matter.

Because much of the respiration in the upper ocean is caused by degradation of DOM and small particles, a rough approximation of the rate of community respiration can be evaluated by measuring oxygen depletion in samples of seawater incubated at *in situ* temperature and covered in black material to block out the light. Data from such experiments (Fig. 6.7) carried out in the open ocean indicate that respiration rates in the euphotic zone are of the order of $1\,\mu mol\,kg^{-1}\,d^{-1}$ and decrease dramatically with depth. As the oxygen concentration in the surface ocean is on the order of $200\text{–}300\,\mu mol\,kg^{-1}$, this represents a change of $<0.5\%$ in a day, which is difficult to determine accurately by the incubation methods. At steady state over long periods of time the difference between photosynthesis and respiration in the euphotic zone (the 1% light level is at 100 m in most of the open ocean) is equal to the net flux of organic carbon from the upper ocean: the biological pump. This flux is of great importance to oceanography, but is generally much smaller than photosynthesis or respiration and cannot be determined by subtracting measurements of two large numbers that are

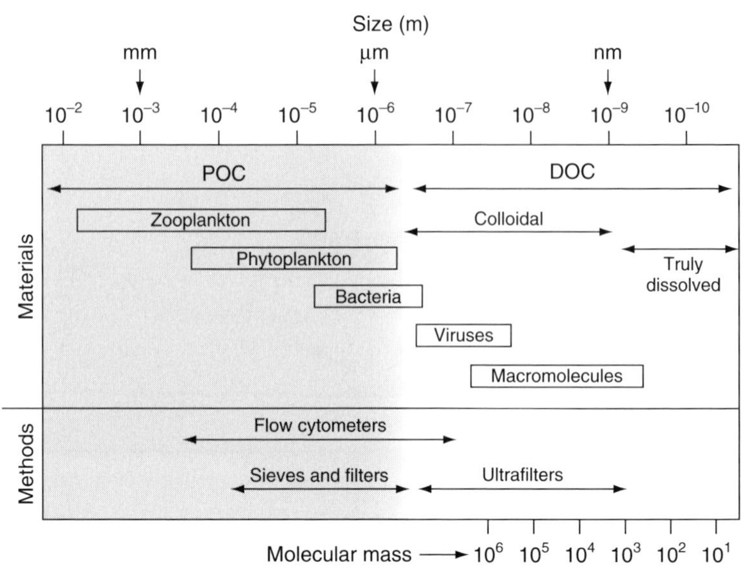

Figure 6.6. A schematic depiction of the sizes of aggregates containing organic carbon in the ocean. The distinction between particulate organic carbon (POC) and dissolved organic carbon (DOC) is operationally determined by the pore size of filters (usually 0.45 μm) used to separate "dissolved" from "particulate" material. Constituents of the POM and DOM are indicated, along with methods used to distinguish them.

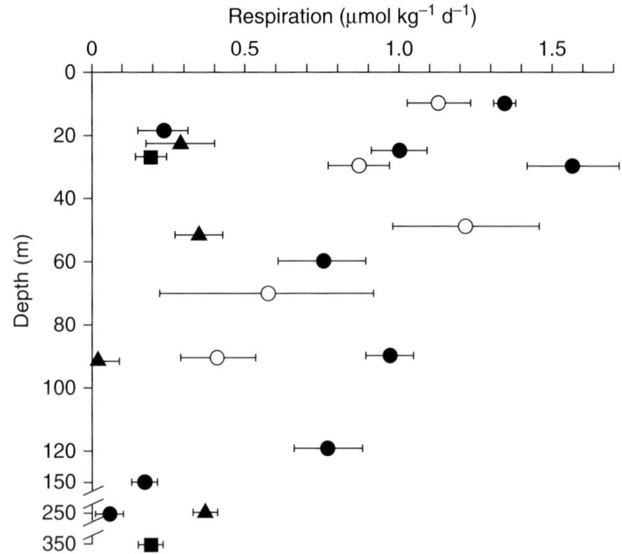

Figure 6.7. Measurements of *in vitro* changes in oxygen concentration in dark bottles containing samples taken from different depths in the ocean from the subtropical North Pacific. Samples from different depths were incubated in bottles either on deck or *in situ*. Symbols indicate different casts. Redrawn from Williams and Purdie (1991).

relatively uncertain. Methods for determining the flux of organic matter from the euphotic zone are the next topic, but we will have more to say about ocean respiration below the euphotic zone later in the chapter.

6.3 Biologically driven export from the euphotic zone

Biologically driven organic matter export from the euphotic zone regulates the p_{CO_2} of the atmosphere and provides the substrate for most of the chemical changes observed in the deep sea. Metabolic products of photosynthesis leave the ocean surface primarily as particulate matter, dissolved organic matter and oxygen gas. Biologically produced particles that sink out of the surface ocean are made mostly of organic carbon and mineral shells. This is the material that is caught during filtration of seawater and in sediment traps. DOM is transported with the water and thus must have a concentration gradient from surface to depth to drive a carbon flux from the euphotic zone. These gradients are observed almost everywhere in the ocean (see Fig. 6.8), and they are variable in different productivity regimes. It has been estimated based on DOC profiles that the DOM flux out of the euphotic zone is roughly 15%–20% of the total carbon export globally (Hansell and Carlson, 1998).

Since there are no standards for flux determinations, as there are for chemical concentration measurements in water or sediments, evaluating their accuracy is difficult. One is forced to rely on comparison of different analytical approaches and modeling of nutrient and oxygen distributions to achieve a consensus. Of the different methods used to determine the organic carbon export presented in the following section, the first two (sediment traps and thorium isotope disequilibria) are used to evaluate the POC flux only. They must be combined with estimates of the DOC flux to achieve the total. The second two methods, molecular oxygen and carbon isotope mass balance, determine the sum of DOC and POC export.

6.3.1 Particle flux

Sediment traps are rain-gauge-like collectors of particles (Fig. 6.9) that have been deployed in the ocean hundreds of times. Although these deployments have revealed a wealth of information about the nature of the sinking particles, it is generally agreed that sediment

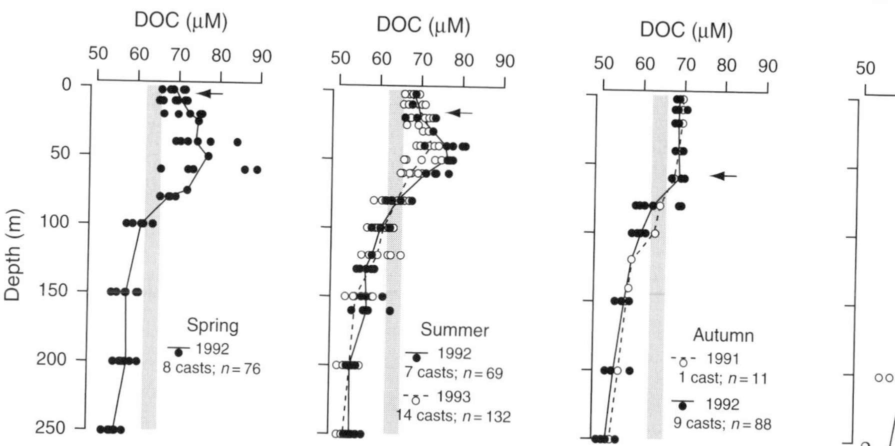

Figure 6.8. Profiles of dissolved organic carbon (DOC) for four different times of the year from the Bermuda Atlantic Time-series Station (BATS). The arrow indicates the approximate mixed layer depth for these times of the year. The vertical shaded area represents the wintertime values. Redrawn from Carlson *et al.* (1994).

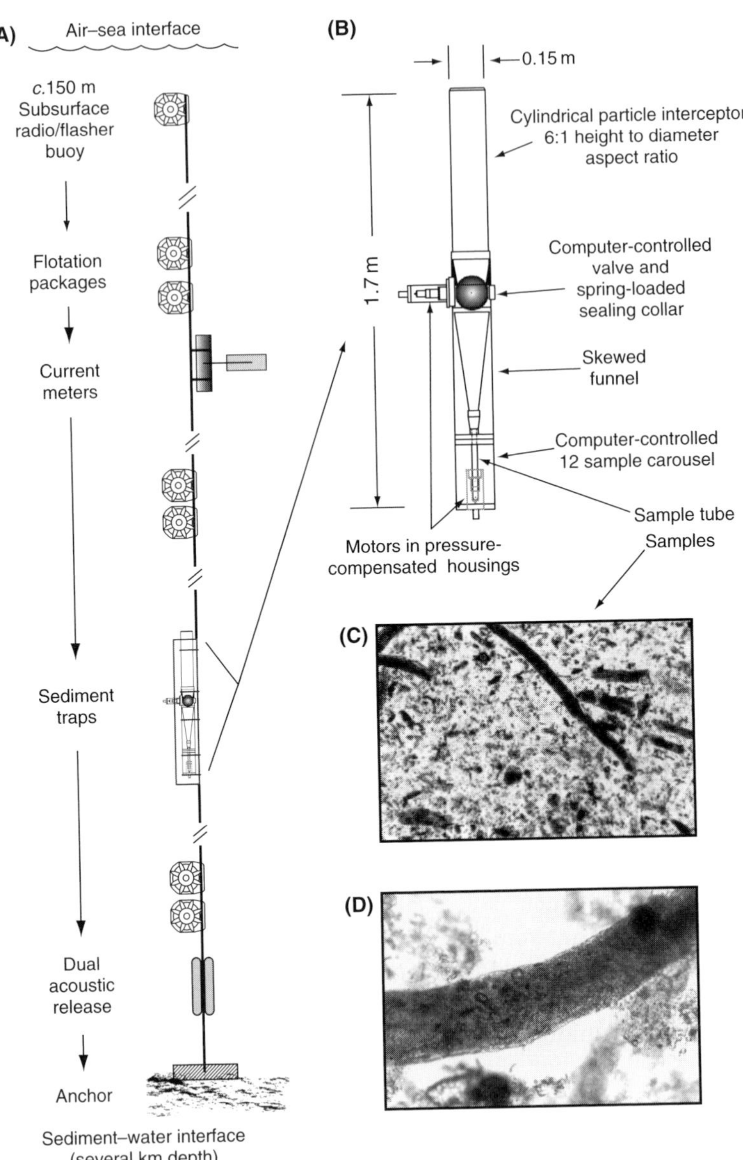

Figure 6.9. (A) A schematic drawing of a moored sediment trap indicating the anchor and flotation. (B) A close-up of the sediment trap. Particles enter the cylinder at the top and are deposited on the surface of the ball. The ball rotates after a short period and dumps the particles in successive sample compartments. The rotating ball prevents swimmers from entering the sediment trap to feed on the particles in the carousel. (C) A microscopic photograph of particle aggregates taken from sediment traps; the shorter fecal pellets are several millimeters long. (D) An enlargement of the fecal pellet in (C), showing that it contains small particles of phytodetritus. Photo courtesy of Michael Peterson, University of Washington.

traps (especially those deployed within several hundreds of meters of the surface) are not very good at quantifying particle fluxes. There are good reasons, based on flume experiments in the laboratory, to expect problems in determining the true falling flux of particles with different sizes and densities by using a hollow cylinder in a moving fluid. Practically, though, the main reason that fluxes measured by sediment traps are suspect is that over an annual cycle they are usually lower than particle fluxes estimated by other methods.

The most important contribution of sediment trap studies to the chemical perspective of oceanography is the vast amount that has been learned about the nature of particles that leave the upper ocean and the rate at which they degrade and dissolve as they pass through the abyss. Shells of $CaCO_3$ and opal have a major, and often overlooked, effect on the vertical transport of organic matter. One can estimate the weight percent (wt%) of the three main constituents of the particulate flux by using estimates of their mole fraction leaving the euphotic zone. Based on worldwide gradients of DIC and alkalinity in the top of the permanent thermocline, it has been suggested that about 6% of the carbon that leaves the upper ocean is $CaCO_3$ (Sarmiento *et al.*, 2002) and the $SiO_2 : CaCO_3$ ratio found in sediment traps is about 2:1 (Broecker and Peng, 1982). Thus, to a first approximation, for each 100 atoms of carbon sinking there are 6 carbonate carbons and 12 silicon atoms. Given approximately 50 wt% C in organic matter, 12 wt% C in $CaCO_3$ and 42 wt% Si in SiO_2 (assuming 10% water), average marine particles (AMPs) contain roughly 60, 15, and 25 wt% of organic matter, calcium carbonate and opal, respectively. AMPs represent a mixture of raining plankton, fecal pellets, and flocks. Given that most larger particles sink from the ocean in aggregates, the AMP concept should be a useful collective model for dynamic studies.

At first thought, marine plankton carrying around heavy shells of calcium carbonate and opal seems to make as much sense as birds with lead feathers. However, sinking can be advantageous for small organisms because it minimizes the diffusive boundary layer around cells, thereby increasing exchange rates of nutrients and waste products. The downside, so to speak, is that sinking acts to remove cells from the lighted upper ocean if local rates of vertical water mixing are insufficient for resuspension.

The fact that particles sinking out of the surface ocean already contain on average 30–40 wt% of mineral matter has at least two major implications for the transport efficiency of organic carbon (and other bioactive elements) into the interior ocean. The first of these is that the organic component of AMP should sink much faster than organic particles not accompanied by mineral matter. According to Stokes' law, the sinking rate of a particle is directly proportional to its density difference from the ambient fluid. Assuming that the average densities ($g\,cm^{-3}$) of calcite and opal are *c.* 2.5, and that organic matter and seawater have densities of *c.* 1.1 and *c.* 1.0, mineral and organic matter will have relative density differences from water of *c.* 1.5 and *c.* 0.1, respectively. Under these conditions, a pure inorganic particle will sink 15 times faster than an organic particle of the same dimensions. An illustration of the effect on sinking organic matter using a simple Stokes' law settling model (Fig. 6.10) indicates that ballast minerals proportionately deepen the respiration profile in the water column. In fact, it is predicted in this calculation that there would be little organic matter reaching below the euphotic zone without ballast. A detailed analysis of this effect and comparison to field data is presented in Armstrong *et al.* (2002).

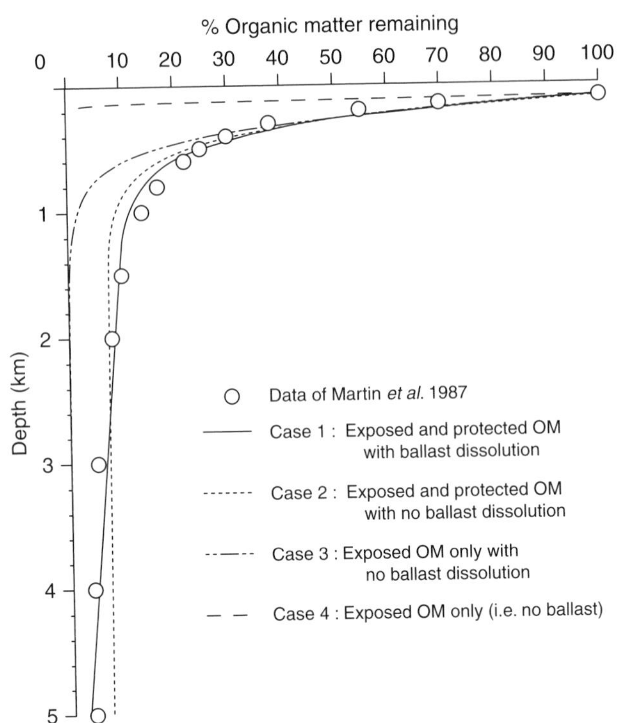

Figure 6.10. Comparison of the organic matter remaining as a function of depth in sediment traps (circles, from Martin *et al.*, 1987) with predictions using a Stokes settling model for several different cases. Particles were 60% organic matter mass and 40% mineral mass with an initial diameter of 125 μm. Densities were: $\rho_{seawater} = 1.0$, $\rho_{OM} = 1.1$, $\rho_{min} = 2.5$ g cm^{-3}. Organic matter degradation rates were 0.87 d^{-1} and mineral dissolution rates were 0.044 d^{-1}. Sinking particulate organic matter can fit the measured values only if it contains mineral ballast and some of it is protected from degradation. From Michael Peterson, unpublished results.

The second implication of the mixture of organic and inorganic matter in marine particles is that a fraction of the organic matter in AMPs may be physically protected by mineral armor, and thus only be respired when the associated mineral is dissolved. Because minerals dissolve more slowly than unprotected organic matter is respired, mineral armor should be especially effective in transporting organic matter through the deep ocean. The curve in Fig. 6.10 that matches observations of particles collected in sediment traps as a function of depth, the "Martin curve" (Martin *et al.*, 1987), was determined by arbitrarily setting the amount of organic matter to be protected from degradation to 15% of the total. Although there is as yet little independent evidence to predict what fraction of the organic matter is protected, the important information from the model is that without this protection organic matter would be oxidized in the upper 2000 m of the water column. Thus, physical protection may explain the characteristic hyperbolic form of POC fluxes and its asymptotic approach to non-zero values at depth.

An oceanographically reasonable combination of mineral ballasting and armoring effects can give a good fit to the respiration curve presently used to model the penetration of organic matter into the deep ocean (Fig. 6.10). There may be other mechanisms that could accomplish the observed result, but one must entertain the possibility that mineral ballast and protection may be a prerequisite for penetration of organic matter into the deep ocean.

6.3.2 The ^{234}Th method of determining particle fluxes

A direct geochemical check on the particle flux of organic matter from the euphotic zone determined by sediment traps is achieved by using the mass balance of thorium isotopes in surface waters. Since decay systematics and chemistry of the uranium series isotopes were introduced in Chapter 5, we will only briefly reiterate them here. ^{238}U is relatively unreactive in oxic seawater and exists in the ocean as a conservative element, i.e. the concentration normalized to salinity is everywhere the same to within measurement error. ^{238}U decays to ^{234}Th, which is very reactive to particles and has a relatively short radioactive half life of 24.1 d:

$$^{238}\mathrm{U} \xrightarrow{t_{1/2} = 4.5\times 10^9\,\mathrm{y}} {}^{234}\mathrm{Th} \xrightarrow{t_{1/2} = 24.1\,\mathrm{d}} {}^{234}\mathrm{U}. \quad (6.26)$$

Most of the atoms of ^{234}Th that are produced from ^{238}U decay readily attach to particles and are removed from solution. In the surface ocean there are enough particles formed to create a deficiency in ^{234}Th activity from that to be expected at secular equilibrium with ^{238}U. At steady state, the depth-integrated deficiency of the activity concentrations of ^{234}Th in the euphotic zone is equal to its flux from the surface ocean on particles. If one then knows the ^{234}Th : C ratio in the particles, the flux of particulate carbon can be calculated. An example of ^{234}Th measurements in the surface waters of the subtropical Pacific (see Fig. 6.11) indicates that difference in ^{234}Th activity from that expected at secular equilibrium (equal to the activity of ^{238}U) is small but readily measurable.

Mathematically, the above explanation is expressed as a mass balance for ^{234}Th, in which the change of the concentration of dissolved thorium-234, $[^{234}\mathrm{Th}^{\mathrm{d}}]$ (atoms m^{-3}), is equal to the production rate by uranium-238 decay, $[^{238}\mathrm{U}]\cdot\lambda_{238}$ (atoms m^{-3} d^{-1}), minus the decay rate of dissolved thorium-234, $[^{234}\mathrm{Th}^{\mathrm{d}}]\cdot\lambda_{234}$, and adsorption of dissolved ^{234}Th onto particles, Ψ_{234}.

$$\frac{\mathrm{d}\left[^{234}\mathrm{Th}^{\mathrm{d}}\right]}{\mathrm{d}t} = \left[^{238}\mathrm{U}\right]\lambda_{238} - \left[^{234}\mathrm{Th}^{\mathrm{d}}\right]\lambda_{234} - \Psi_{234}\,(\mathrm{atoms\ m^{-3}d^{-1}}). \quad (6.27)$$

Multiplying by the decay constant for ^{234}Th changes the concentrations to activity concentrations ($A = [\,]\,\lambda$, with units of disintegrations per minute per m^{-3}, dpm m^{-3}), which is convenient because usually it is the activity of radioisotopes that is measured:

$$\begin{aligned}\frac{\mathrm{d}A^{\mathrm{d}}_{234}}{\mathrm{d}t} &= \lambda_{234}\left[^{238}\mathrm{U}\right]\lambda_{238} - \lambda_{234}\left[^{234}\mathrm{Th}^{\mathrm{d}}\right]\lambda_{234} - \Psi^{*}_{234}\\ &= \lambda_{234}\left(A_{238} - A^{\mathrm{d}}_{234}\right) - \Psi^{*}_{234}\,(\mathrm{dpm\ m^{-3}d^{-1}}).\end{aligned} \quad (6.28)$$

(The * on the Ψ indicates activity concentration instead of chemical concentration.) A similar mass balance for particulate thorium states that the change in particulate thorium activity in the water, $A^{\mathrm{p}}_{\mathrm{Th}}$, with time is equal to the gain from adsorption of dissolved thorium minus the decay of particulate thorium and the vertical flux of particulate thorium, J^{*}_{234}.

$$\frac{dA^{p}_{Th}}{dt} = \Psi^{*}_{234} - A^{p}_{Th}\lambda_{Th} - J^{*}_{234}(\text{dpm m}^{-3}\text{d}^{-1}). \tag{6.29}$$

Substituting Eq. (6.29) into (6.28) to eliminate the adsorption term and assuming steady state gives:

$$0 = \lambda_{234}(A_{238} - A^{d}_{234}) - A^{p}_{Th}\lambda_{234} - J^{*}_{234}. \tag{6.30}$$

Combining both particulate and dissolved thorium activities, $A_{Th} = A^{d}_{Th} + A^{p}_{Th}$, and integrating over the depth of the euphotic zone results in a relation between the flux of thorium activity at the base of the euphotic zone, $z = h$, and the integrated activities of uranium-238 and thorium-234 in units of dpm m^{-2} d^{-1}:

$$0 = \lambda_{234}\int_{z=0}^{z=h}(A_{238} - A_{234})\,dz - \int_{z=0}^{z=h} J^{*}_{234}\,dz, \tag{6.31}$$

which at steady state indicates that the flux of particulate ^{234}Th out of the euphotic zone, F^{*}_{234}, is equal to the integrated deficiency of dissolved and particulate ^{234}Th.

$$F^{*}_{234}\big|_{z=h} = \int_{z=0}^{z=h} J^{*}_{234}\,dz = \lambda_{234}\int_{z=0}^{z=h}(A_{238} - A_{234})\,dz. \tag{6.32}$$

Many ^{234}Th measurements in the upper ocean indicate that it is mostly dissolved even though thorium is relatively particle-reactive because dissolved thorium grows into secular equilibrium faster than it is depleted by adsorption to particles. To determine an annual particle flux at a given location one would like to sample the ocean at the frequency of the ^{234}Th half life: about one month. This has been done at the Hawaii Ocean Time series (HOT) near Hawaii, and a few of the monthly profiles are presented in Fig. 6.11 (Benitez-Nelson *et al.*, 2001). In this experiment, monthly estimates of the particulate ^{234}Th flux determined by water column profiles were transformed to carbon fluxes by using measured particulate C : ^{234}Th ratios and compared with the carbon flux from sediment traps. The two different estimates varied by almost a factor of two, with the trap samples being lower. The largest differences were observed during times

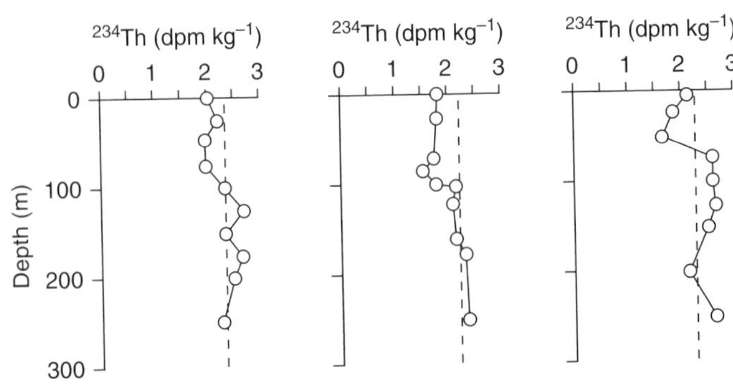

Figure 6.11. ^{234}Th activity concentration (disintegrations per minute per kg seawater, dpm kg^{-1}) as a function of depth for three different months in the Hawaii Ocean Time series (HOT). The vertical line indicates the activity of ^{238}U. Redrawn from Benitez-Nelson *et al.* (2001).

Table 6.1. *The annual organic carbon export from the surface ocean determined at three time series locations by different methods*

These locations are the time series stations at BATS (near Bermuda), HOT (near Hawaii), and Station P (in the subarctic Pacific). Total organic C flux at BATS is the sum of the sediment trap and DOC flux. At HOT it is the sum of the ^{234}Th particle flux, DOC flux, DOC accumulation rate (0.3 mol C m^{-2} y^{-1}) and zooplankton migration flux (0.2 mol C m^{-2} y^{-1}).

	Organic C (mol m^{-2} y^{-1})		
	Subtropical Atlantic (BATS)	Subtropical Pacific (HOT)	Subarctic Pacific (Station P)
^{14}C primary productivity	12.7[a]	14.6[b]	17.9[c]
Estimates of organic C export			
Sediment traps	0.7[d]	0.8 ± 0.1[e,f]	
^{234}Th particle flux		1.5 ± 1.0[f]	
DOC flux	1.1 ± 0.1[d]	0.4 ± 0.2[e,f]	
Total organic C flux	1.8 ± 0.1[d]	2.4 ± 0.9[f]	
Oxygen mass balance	3.6 ± 0.6[g]	2.7 ± 1.7[e], 1.1–1.7[l]	2.0 ± 1.0[h]
DIC and δ^{13}C DIC	3.5 ± 0.5[i]	2.7 ± 1.3[j]	
^{3}H–^{3}He (OUR)	2.8[k]		

[a] Michaels and Knap (1996)
[b] Karl *et al.* (1996)
[c] Varela and Harrison (1999)
[d] Carlson *et al.* (1994)
[e] Emerson *et al.* (1997)
[f] Benitez-Nelson *et al.* (2001)
[g] Spitzer and Jenkins (1989)
[h] Emerson *et al.* (1991)
[i] Gruber *et al.* (1998)
[j] Quay and Stutzman (2003)
[k] Jenkins and Wallace (1992)
[l] Hamme and Emerson (2006)

of high flux, when the sediment traps collected much less than the flux indicated by the thorium–uranium method. Values for the ^{234}Th-determined POC export at HOT are presented in Table 6.1. After augmenting the thorium-based POC flux with an estimate of the DOC flux, the total carbon flux is (within error) the same as those determined by other methods.

6.3.3 Dissolved O_2 mass balance

During net photosynthesis approximately 153 mol of O_2 are produced for every 106 mol of organic carbon (Eq. (6.24)). Since some fraction of the organic matter produced escapes the euphotic zone

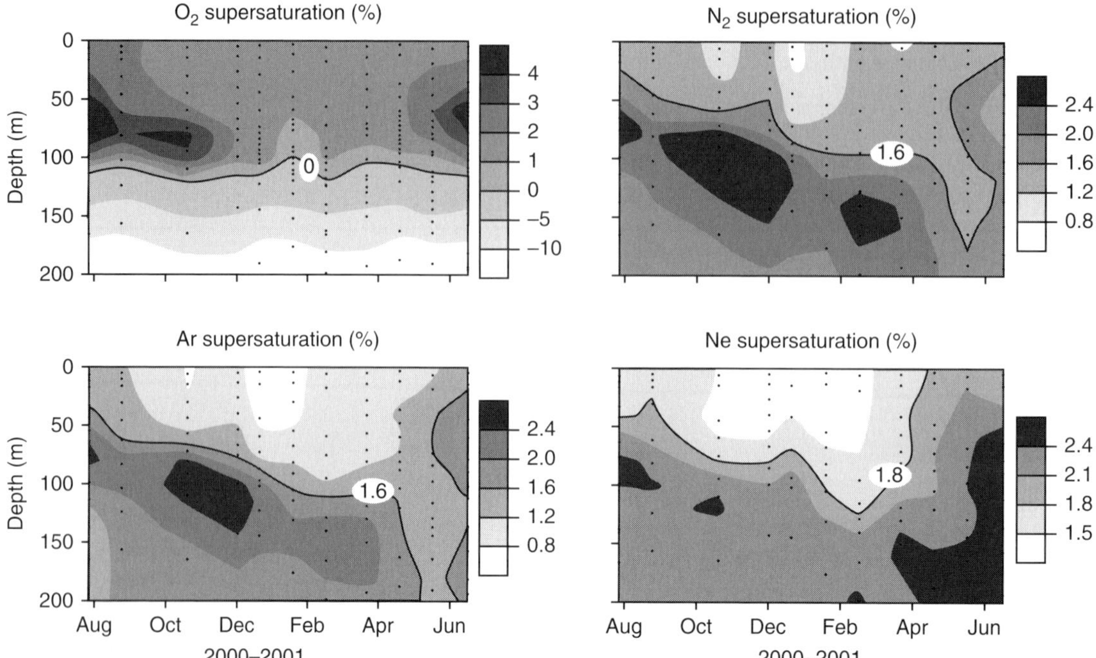

Figure 6.12. The degree of supersaturation (in %) for the gases O_2, N_2, Ar and Ne as a function of time at the Hawaii Ocean Time series (HOT). The shading for the gas supersaturations is different except for Ar and N_2, because of the different range of the data. Above 100 m all the gases are supersaturated year-round. From Hamme and Emerson (2006).

before it is respired, at steady state a corresponding stoichiometric amount of O_2 also escapes. If one can determine the net annual biological O_2 production from an upper ocean O_2 mass balance, then this value is also a measure of the annual carbon export.

O_2 concentrations in ocean surface waters are greater than expected at saturation equilibrium with the atmosphere and a subsurface O_2 maximum forms in the upper 100 m during summer in most regions of the ocean (Fig. 6.12), suggesting a measurable biological component to the upper ocean O_2 mass balance. The flux of O_2 to the atmosphere can be determined by measuring the concentration difference between the value in the mixed layer and that expected at atmospheric equilibrium and multiplying this value by the gas exchange mass transfer coefficient (see Chapter 10). The complicating factor in this calculation is that there are other physical processes that also create gas supersaturation. This can be seen in an annual survey of the degrees of supersaturation of O_2, N_2, Ar, and Ne in the subtropical North Pacific Ocean (Fig. 6.12). There are no biological factors affecting the inert gases Ar and Ne and this is also true for N_2, because rates of nitrogen fixation are not great enough to change the concentration by more than *c.*0.1%. Physical factors that affect the degree of saturation are: (1) temperature changes (as waters warm in summer the saturation concentrations fall and water becomes supersaturated) and (2) bubble processes (breaking waves inject small bubbles that collapse as the hydrostatic pressure increases below the surface) (see Chapter 10).

To determine the net biological O_2 production from measurements of O_2, one must evaluate the physical factors causing gas supersaturation. This is where determination of the inert gases comes in handy. Since these gases are not influenced by biological processes the concentration and saturation state change with time can be attributed entirely to physical processes that can then be evaluated. To do this it is useful to have a number of inert gases that have different physical characteristics and respond differently to temperature change and bubble processes. For example, among the gases N_2, Ar, and Ne in Fig. 6.12, the Henry's Law saturation equilibrium for Ne is lowest, and it is much less temperature-dependent than the other two gases (Fig. 3.11). Thus, the saturation state of Ne is mostly dependent on bubble processes. The saturation states of both N_2 and Ar are affected equally by temperature change and bubbles, but Ar is about twice as soluble as N_2. Once the physical processes are determined in a model that reliably reproduces the inert gas concentrations and saturation states, these processes are used to determine their influence on the oxygen saturation state. If the gas measurements are made accurately enough, the biological and physical processes affecting the O_2 concentration can be separated in this way.

An estimation of the portion of the O_2 supersaturation that is due to biological processes is achieved by using the difference in saturation state between O_2 and Ar (Fig. 6.13). This simple difference illustrates the biological component of the O_2 supersaturation because the solubilities and diffusion coefficients of these two gases are very similar. Thus, the degree of supersaturation caused by bubble injection and temperature change is nearly the same for both gases, and subtraction of O_2 supersaturation from that of Ar supersaturation leaves the biological component. The data in Fig. 6.13 indicate that biologically induced supersaturation in the surface waters of the subtropical North Pacific Ocean is between 0.5% and 1.5%. In order to derive an estimate of the fluxes, a model of the upper ocean is required.

Mathematically, one can represent the upper ocean O_2 mass balance by writing equations for dominant fluxes of each gas. We will assume at the outset that horizontal fluxes are not important,

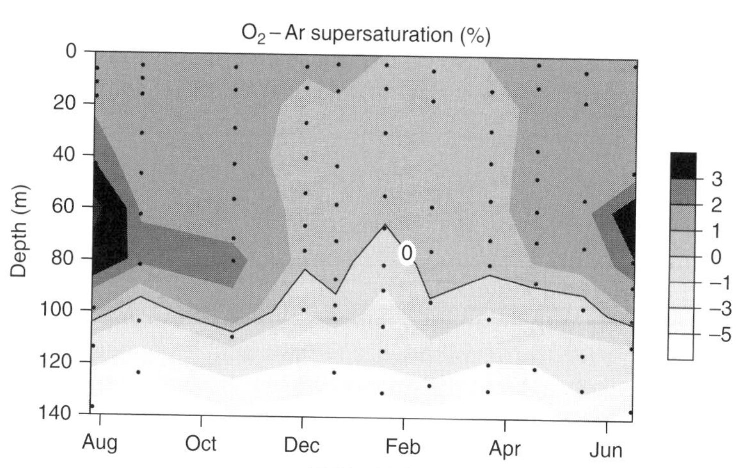

Figure 6.13. The difference in oxygen and argon supersaturation (%) in the upper ocean at HOT. (The individual data are in Fig. 6.12.) This index isolates the biologically produced component of the gas data. Redrawn from Hamme and Emerson (2006).

which is reasonable for gases that are near saturation equilibrium and have air–ocean exchange times of about one month. (A surface current of 10 cm s^{-1} will move water about 300 km in one month; thus the biological O_2 supersaturation will be reset by gas exchange over distances of several degrees of latitude or longitude. The surface ocean signal averages biological processes over distances of this magnitude.) The integrated concentration of gas, [A], over the depth of the euphotic zone is equal to the flux across the air–water interface, F_{atm}, minus the flux to the ocean interior, F_z, plus the production by biological processes, J_A:

$$\frac{\mathrm{d}\left(\int_{z=0}^{z=h} [\mathrm{A}]\,\mathrm{d}z\right)}{\mathrm{d}t} = F_{atm} - F_z + J_A. \tag{6.33}$$

The air–water flux model (Chapter 10, Eq. (10.24)) has a term for molecular exchange across the interface, F_{awi}, and one for bubble processes, F_{Bub}:

$$F_{atm} = F_{awi} + F_{Bub}. \tag{6.34}$$

The first term on the right side represents the interface exchange that is the product of a gas transfer velocity, G, and the concentration difference between that measured in the ocean's surface, $[A^{surf}]$, and that expected at saturation equilibrium, $[A^{sat}]$:

$$F_{atm,A} = G_A \cdot \left(\left[A^{sat}\right] - \left[A^{surf}\right] \right) + F_{Bub}. \tag{6.35}$$

The gas transfer velocity is proportional to wind speed and can be estimated if this is known. The bubble terms, however, must be determined from a model of bubble processes (explained in Chapter 10) and the concentrations of the inert gases Ar, N_2 and Ne.

The transfer of gases at the base of the euphotic zone is by a combination of advection and diffusion processes and in reality is probably dependent on mechanisms that are intermittent rather than constant in time. Here we write the flux as a simple one-dimensional diffusion process dependent on the concentration gradient at the base of the euphotic zone and a parameter that is assumed to be analogous to molecular diffusion, an "eddy diffusion coefficient," K_z:

$$F_{z=h} = K_z \left(\frac{\mathrm{d}[\mathrm{A}]}{\mathrm{d}z}\right)_{z=h}. \tag{6.36}$$

This parameterization is probably mechanistically incorrect, but it incorporates the likelihood that exchange at this boundary depends on the concentration gradient.

In principle there are enough equations and tracers (Ar, N_2, He, O_2) to solve for all the unknowns: two for bubbles imbedded in F_{Bub} (Eq. (10.35)), K_z and J. At the time of writing this book it has not been possible to do this in practice at the few locations where it has been tried (the location of the data in Fig. 6.12 is the Hawaii Ocean Time series, HOT), because the inert gas signals are small and can be explained by a range of bubble mechanisms without constraining the eddy diffusion coefficient. For this reason researchers have attempted to independently determine the value of K_z by using heat flux balance and seasonal temperature changes in one-dimensional models. This has also had problems because of the large errors in the air–water heat flux terms (Hamme and Emerson, 2006). At present, the best one can do is place limits on the value of K_z that do not violate the observations. Once this is done, it is possible to determine a range of possible rates of net O_2 production that satisfy the measured inert gas and O_2 distribution.

Estimates of carbon fluxes determined by O_2 mass balances at marine time series stations are presented in Table 6.1, where it is demonstrated that they are consistent with the fluxes determined by the ^{234}Th method, but higher than those estimated from sediment trap particle flux. Presently, the main weaknesses in determining the upper ocean mass balance for O_2 are a roughly $\pm 30\%$ error in estimating the air–water transfer velocity from wind speeds and the inability to constrain mixing at the base of the euphotic zone. The latter problem can result in errors of up to 50% (Hamme and Emerson, 2006). As more inert gases are used and models take into consideration lateral processes in addition to vertical ones to constrain K_z, it may be possible to overcome some of these problems and make a more tightly constrained estimate of the net biological oxygen production.

6.3.4 Carbon isotopes of dissolved inorganic carbon in surface waters

During photosynthesis the light isotope of carbon, ^{12}C, is preferentially transformed to organic carbon, causing an isotope fractionation of $-15‰$ to $-20‰$ (Table 5.3), but there is very little isotope discrimination during respiration. Thus, organic matter created in the euphotic zone is lighter than the dissolved inorganic carbon (DIC) of the surface waters and this causes DIC surface values to be heavier than those below the euphotic zone. Carbon isotope measurements of DIC in the ocean (Fig. 6.14) indicate that net organic carbon export has a readily measurable effect on the δ^{13}C of DIC. Seasonal trends indicate an increase of δ^{13}C-DIC accompanying the summertime DIC decrease. The processes constraining the mass balance of DIC and its trace isotope of carbon, DI^{13}C, are different because of fractionation during photosynthesis and air–water exchange processes. This difference is the reason that equations describing the mass balances of DIC and DI^{13}C are independent and can be used along with gradients like those in Fig. 6.14 to determine the rate of organic carbon export (Quay and Stutzman, 2003; Gruber *et al.*, 1998).

For the upper ocean, equations for the mass balance of DIC and DI^{13}C are similar to those used for the O_2 mass balance; however, there are two important differences. First, bubble processes can be neglected for gas transfer of CO_2 because it is a much more soluble gas. At solubility equilibrium CO_2 partitions about equally between the atmosphere and seawater whereas O_2, Ar, N_2, and Ne remain primarily

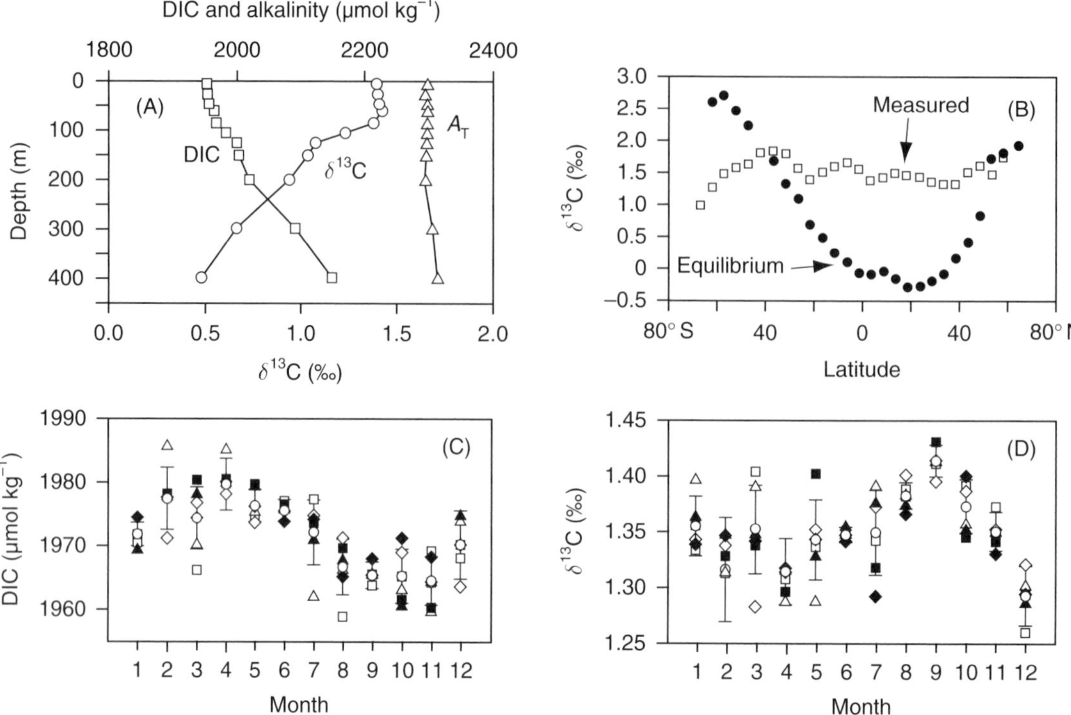

Figure 6.14. DIC and carbon isotope ratio of DIC, δ^{13}C-DIC, in the North Pacific Ocean. The depth and time series data are from the Hawaii Ocean Time series (HOT) between 1994 and 1999. (A) Depth profile of DIC (squares), total alkalinity, A_T (triangles) and δ^{13}C-DIC (circles). (B) Zonally averaged mean values of the surface water δ^{13}C-DIC in the Pacific Ocean (squares) and the δ^{13}C-DIC expected if the surface ocean DIC were in chemical equilibrium with the atmosphere (circles). (C) Monthly values of DIC in the mixed layer at HOT. (D) Monthly values of δ^{13}C-DIC in the mixed layer at HOT. Redrawn from Quay and Stutzman (2003).

(>95%) in the atmosphere (Table 3.6). Bubble processes have much less of an effect on more soluble gases. This is because injecting a small piece of atmosphere into the water increases the concentration above that predicted by solubility equilibrium (the Henry's Law coefficient). If the gas is insoluble there is relatively little dissolved in the water and the bubble process increases the saturation state by a lot, but if the gas is already very soluble there is already a lot of gas in the water and the bubble process affects the saturation state very little. This is a major advantage for computing the carbon mass balance via the DIC and DI^{13}C method because bubble processes are difficult to characterize.

The second difference in the DIC versus O_2 mass balance equations is a disadvantage due to the very long residence time of DIC in the surface ocean with respect to gas exchange. Since only about 1% of the surface ocean DIC is a gas, it takes about one hundred times longer to renew the upper ocean DIC and DI^{13}C reservoirs via gas exchange with the atmosphere than the analogous residence time for the inert gases or O_2. With a residence time of about 10 y with respect to gas exchange, the DIC and DI^{13}C concentrations at any location in the surface ocean are influenced by processes that have occurred over distances that are of basin scale rather than within several degrees of longitude or latitude as indicated for oxygen concentrations. Thus the horizontal transport terms in the equations cannot be neglected and it is necessary to evaluate surface ocean DIC and DI^{13}C gradients.

Mathematically, the DIC and DI^{13}C tracers can be described in two independent equations. First, the integrated DIC concentration in the euphotic zone is depicted as the mean value in the euphotic zone, $[\overline{\mathrm{DIC}}]$, multiplied by the euphotic zone depth, h. This value depends on the fluxes at the air–water interface, F_{atm}, and across the upper ocean thermocline boundary, F_z, in addition to horizontal advection, F_{H}, and organic carbon export, J_{C}.

$$h\frac{\mathrm{d}[\overline{\mathrm{DIC}}]}{\mathrm{d}t} = F_{\mathrm{atm}} + F_z + F_{\mathrm{H}} - J_{\mathrm{C}}\,(\mathrm{mol\,C\,m^{-2}\,d^{-1}}). \tag{6.37}$$

Each of these terms can be expanded by using the same parameters as in the O_2 equations, with the exceptions that the term for bubbles is deleted, and the horizontal transport, F_{H}, depends on the horizontal mean velocity, U (m d^{-1}) and the horizontal concentration gradient. For simplicity, these values are indicated here as horizontal using the subscript x: U_x, and $\mathrm{d}(h[\overline{\mathrm{DIC}}])/\mathrm{d}x$ (mol m^{-2}), where in reality it must be considered in two dimensions, x and y.

$$h\frac{\mathrm{d}[\overline{\mathrm{DIC}}]}{\mathrm{d}t} = -G_{\mathrm{CO_2}}\cdot K_{\mathrm{H}}\cdot\left(f^{\mathrm{ml}}_{\mathrm{CO_2}} - f^{\mathrm{atm}}_{\mathrm{CO_2}}\right) + K_z\cdot\left(\frac{\mathrm{d}[\mathrm{DIC}]}{\mathrm{d}z}\right)_{z=h} + U_x\frac{\mathrm{d}(h[\overline{\mathrm{DIC}}])}{\mathrm{d}x} - P_{\mathrm{C}} + R_{\mathrm{C}}. \tag{6.38}$$

The atmospheric gas exchange rate depends on the CO_2 fugacity difference between the air, $f^{\mathrm{atm}}_{\mathrm{CO_2}}$, and water $f^{\mathrm{ml}}_{\mathrm{CO_2}}$, where superscript ml indicates CO_2 fugacity in the ocean mixed layer. Fugacity and concentration are related through the Henry's Law coefficient, K_{H} (mol m^{-3} atm^{-1}). The net organic carbon flux from the euphotic zone, J_{C}, has been expanded in the above equation to show that it is the difference between photosynthesis, P_{C}, and respiration, R_{C} ($J_{\mathrm{C}} = P_{\mathrm{C}} - R_{\mathrm{C}}$).

The equation for DI^{13}C is exactly analogous to that for DIC except that there are fractionation factors, α (Eq. (5.4)), that account for the difference in rates of some processes for the two different isotopes. It is convenient to write the DI^{13}C equation in terms of the concentrations of the ^{12}C terms times the ratio of the isotopes, $r = {}^{13}\mathrm{C}/{}^{12}\mathrm{C}$. For example, the relation between DIC and DI^{13}C is:

$$\mathrm{DI^{13}C} = \mathrm{DIC}\cdot r^{\mathrm{DIC}}. \tag{6.39}$$

The ratios, r, are directly related to δ^{13}C values via the equation for stable isotope notation (Eq. (5.1)). With this formalism the DI^{13}C mass balance can be written:

$$h\frac{\mathrm{d}[\mathrm{DI^{13}C}]}{\mathrm{d}t} = -G_{\mathrm{CO_2}}\cdot\alpha_{\mathrm{ge}}\cdot K_{\mathrm{H}}\cdot\alpha_{\mathrm{s}}\cdot\left(f^{\mathrm{ml}}_{\mathrm{CO_2}}\cdot r^{\mathrm{DIC}}\cdot\alpha_{\mathrm{DIC}} - f^{\mathrm{atm}}_{\mathrm{CO_2}}\cdot r^{\mathrm{atm}}\right) + K_z\cdot\left(\frac{\mathrm{d}[\mathrm{DIC}]\cdot r^{\mathrm{DIC}}}{\mathrm{d}z}\right)_{z=h} + U_x\frac{\mathrm{d}(h[\overline{\mathrm{DIC}}]\,r^{\mathrm{DIC}})}{\mathrm{d}x} - P_{\mathrm{C}}\cdot r^{\mathrm{DIC}}\cdot\alpha_{\mathrm{P}} + R_{\mathrm{C}}\,r^{\mathrm{OM}}. \tag{6.40}$$

Fractionation factors are the key to the difference between the DI^{12}C and DI^{13}C equations, and they must be known experimentally. The fractionation factors in Eq. (6.40) represent the kinetic and equilibrium fractionations during air–seawater gas exchange, α_{ge} and α_{s}, respectively; the equilibrium isotope difference between DIC and f_{CO_2}, α_{DIC} (this is necessary because the isotope ratios are measured on DIC and not CO_2); and the kinetic fractionation during photosynthesis, α_P (Table 5.3). With vertical, horizontal, and time-dependent measurements of DIC and δ^{13}C-DIC and surface measurements of f_{CO_2}, the unknowns in Eqs. (6.38) and (6.40) are the gas exchange mass transfer coefficient, G_{CO_2}, the carbon export rate, J_C, and the physical transport terms K_z and U_x. The easiest of these to estimate independently are the gas exchange rate, via estimates of G_{CO_2} using correlation to wind speed and the horizontal velocity from Ekman transport calculations. This has been done at the two time series stations near Hawaii and Bermuda; the resulting carbon export rate (Table 6.1) agrees well with those determined by ^{234}Th and O_2 mass balance methods. The largest errors in this approach are in the determination of the gas exchange rate by correlation to wind speed and in evaluating the horizontal and vertical flux components.

6.3.5 Comparison of methods for determining organic carbon export

There have been only a few places in the ocean where different methods of estimating the organic carbon export have been compared. The only locations where there have been sufficient observations to determine annual fluxes are those that have been designated and funded as time-series sites. Results at the Bermuda Atlantic Time-series Study (BATS) and the Hawaii Ocean Time series (HOT) using mass balances of O_2 and inert gases, and DIC and carbon isotopes, agree to within the error estimates (Table 6.1), as do fluxes determined by ^{234}Th isotopes when augmented by DOC fluxes at HOT. In general, however, sediment trap fluxes by themselves are significantly smaller than those determined by other methods. This may be the reason that the total organic

Table 6.2. *Estimates of the carbon export from the upper ocean by the biological pump*

Method	Flux ($Gt\ r^{-1}$)	Explanation
Two-layer ocean model	5	Globally average nutrients and circulation
Sediment traps	3.4 – 4.7[a], 6[b]	Global extrapolation from local estimates
O_2 mass balance	13–15[c]	Global extrapolation from four local estimates
Global surface ocean O_2	4.5–5.6[d]	Extra-tropical and Spring–Summer only
Satellite color	11[e]	Assumes knowledge of color–C export relation
Global circulation models	13–17[f]	Mean of 11 models evaluated at 133 and 75 m
GCM inverse model	11[g]	Uses nutrient distribution to calculate model fluxes

[a] Eppley and Peterson (1979)
[b] Martin *et al.* (1987)
[c] Emerson (1997)
[d] Najjar and Keeling (1999)
[e] Laws *et al.* (2000)
[f] Najjar *et al.* (2007)
[g] Schlitzer (2000)

flux at BATS is less than that determined by O_2 and DIC–$DI^{13}C$ mass balances.

Between 10% and 30% of the carbon synthesized at these locations, as measured by ^{14}C primary production, escapes the euphotic zone. There is about one chance in five that an atom of carbon fixed into the organic phase will escape to the upper thermocline of the ocean. Extrapolation of these observations to the entire subtropical oceans and augmenting them with other mass balance estimates from the Equatorial Pacific result in a global carbon export rate from the ocean's euphotic zone of 13–15 $Pg\,y^{-1}$ (Table 6.2). This is about twice the values predicted from particle sediment traps and the simple two-layer model we used at the beginning of this chapter. The experimentally determined global fluxes are, however, in the same range as values estimated from satellite-driven calculations and those derived from global circulation models. From these estimates one might suggest that the global organic carbon export from the euphotic zone is 10–15 $Pg\,y^{-1}$. The largest uncertainty at the time of writing this book is in the unknown contribution of the continental margin regions, where advection makes it difficult to use the mass balance techniques.

6.4 Respiration below the euphotic zone

Oxygen depletion below the mixed layer of the ocean results from a combination of degradation of organic matter that escapes the euphotic zone and renewal of oxygen in the water by contact with the atmosphere. Since we know the ratio of metabolic products P, N, C and O_2 (Eq. (6.24)), it is possible to compare the O_2 demand in the ocean below the euphotic zone with the supply of organic matter and oxygen from above. For this comparison (Table 6.3), imagine bringing a cubic meter of average deep water to the surface ocean and

Table 6.3. *Relative availability of the major bioactive elements and O_2 versus their use by average marine plankton*

Seawater concentrations are the mean values of dissolved P, N and C in the deep ocean and the atmospheric saturation value in the atmosphere. *Availability ratio* refers to the concentrations of dissolved N, C and O_2 relative to P. *Use ratio* refers to the relative stoichiometry of P, N, C, and O_2 during respiration.

Dissolved bioactive element	Seawater concentration ($\mu mol\,kg^{-1}$)[a]	Availability ratio	Use ratio	Availability/use
P	2.3	1	1	1
N	34.5	15	16	0.94
C	2340	1017	106	9.6
O_2	360	156	154	1.0

[a] For average (deep) seawater, $S‰ = 35$ ppt. O_2 at saturation with the atmosphere at 0 °C.

following the fate of the P, N, C and O_2 through a photosynthesis, export and respiration cycle. When the water arrives at the surface P, N, and C are in the relative ratios found in the deep sea water, these are the "availability ratios" in Table 6.3. During photosynthesis OM is created and returned to the aphotic zone with the concentration of oxygen equal to the atmospheric saturation value at the surface temperature. During respiration P, N and C are respired back to the water and O_2 is depleted in the ratios indicated as the "use" ratios in the table. From the ratios of availability versus use on the right of Table 6.3 it is evident that the availability and use of both dissolved N and P are balanced. Either of these macronutrients has the potential to be limiting for primary production, with fixed N (NO_3^-) being slightly more in deficit. There is a comparatively large surplus of inorganic carbon, such that when all N and P have been stripped from the water to form biomass in the euphotic zone, only about 10% of the total DIC is removed.

The somewhat unexpected result from this thought experiment is that cold seawater contains at saturation very little excess of dissolved O_2 compared with the amount required to completely oxidize the biomass that can be made from its dissolved nutrients. Thus, when seawater downwells at a cold (*c.*0 °C) high latitude it carries just enough O_2 to meet the demand from respiring organic matter. There is clearly a mismatch between this calculation and the real ocean in that the average O_2 concentration in the deep sea is *c.* 150 μmol kg^{-1} instead of near zero. This inconsistency was first noted by Redfield (1958). The reason for it is that surface ocean nutrient concentrations are not uniformly near the detection limit. In polar waters near the Arctic and Antarctic Oceans, nutrients that upwell are not totally removed by biological processes (see, for example, Fig. 6.15), but have a roughly equal probability of being downwelled by mixing or advection. Essentially, not all of the 2.2 μmol kg^{-1} of PO_4^{3-} in the deep water arrived there from biological processes; some was mixed or advected into the deep ocean from surface waters in areas that have high nutrient concentrations. This is also why it is not possible to match both the nutrient and O_2 distributions between the warm surface and deep ocean by using a simple two-layer ocean like the one in Fig. 6.1 if one assumes the surface nutrients are near zero. Using the mean ocean phosphate concentration in the two-layer model equations, Redfield ratios for the organic matter flux, J, and *no surface water nutrients*, causes the O_2 content of the deep ocean to be completely depleted. (Try it!) The model is too simple to reproduce actual ocean measurements, because it does not take into consideration elevated nutrient concentrations in high-latitude waters. Slightly more complicated models with another surface water reservoir are able to avoid this problem at the expense of creating a more complicated paradigm that requires more mixing terms among the different boxes that must be evaluated.

The other major trend in ocean chemistry that results from deep respiration and large-scale ocean circulation is the general

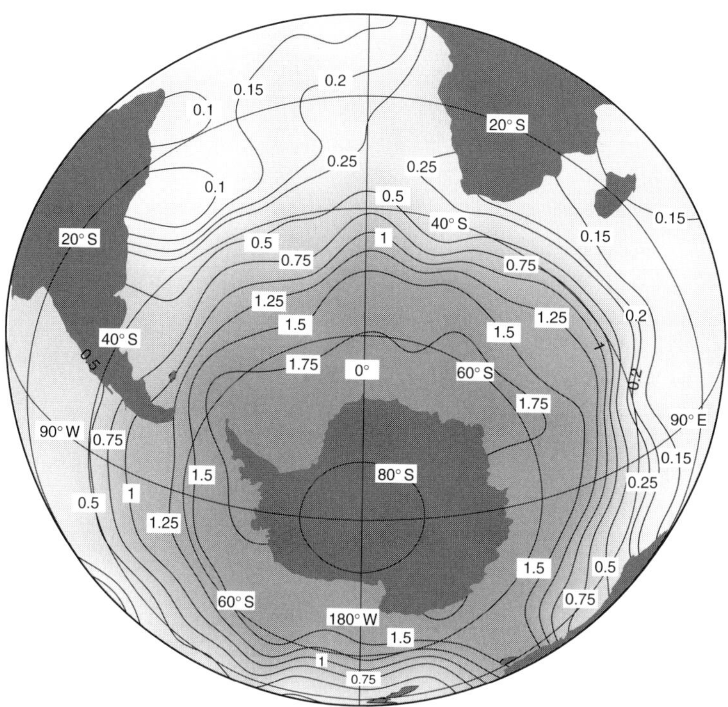

Figure 6.15. Surface water dissolved phosphate distribution, illustrating the relatively high concentrations (in μmol kg^{-1}) in the Southern Ocean. From Ocean Data View (Schlitzer, 2002).

downstream increase of nutrients and other metabolic products that leads to greater concentrations in the deep Pacific and Indian Oceans compared with the Atlantic Ocean. This general trend has already been demonstrated in Chapter 1 (e.g. Fig. 1.4), and illustrated by the ^{14}C distribution at 3000 m depth in Chapter 5 (Fig. 5.17). The conveyor belt circulation trend is particularly vivid when comparing the pattern of ^{14}C-DIC and NO_3^- at 3000 m of the world's oceans (Fig. 6.16). The two concentration trends are identical, indicating that the nutrient content increases as the waters are further from the downwelling source in the North Atlantic. To imprint such distinct geographic trends it is not sufficient that sustained deep water flow paths occur. In addition, dissolved nutrients that are upwelled along the global deep water flow path must be rapidly returned to the deep before they can disperse. This reconcentration process is accomplished by rapid and efficient incorporation of upwelled nutrients into particles that then sink and are recycled.

6.4.1 Apparent oxygen utilization (AOU) and preformed nutrients

Oxygen profiles in the ocean do not continually decrease with depth (Fig. 1.4). A typical dissolved O_2 profile exhibits a minimum that is positioned above 1000 m. The main processes that contribute to this profile are the rapid and efficient respiration of settling organic matter (with more than half being degraded between 100 and

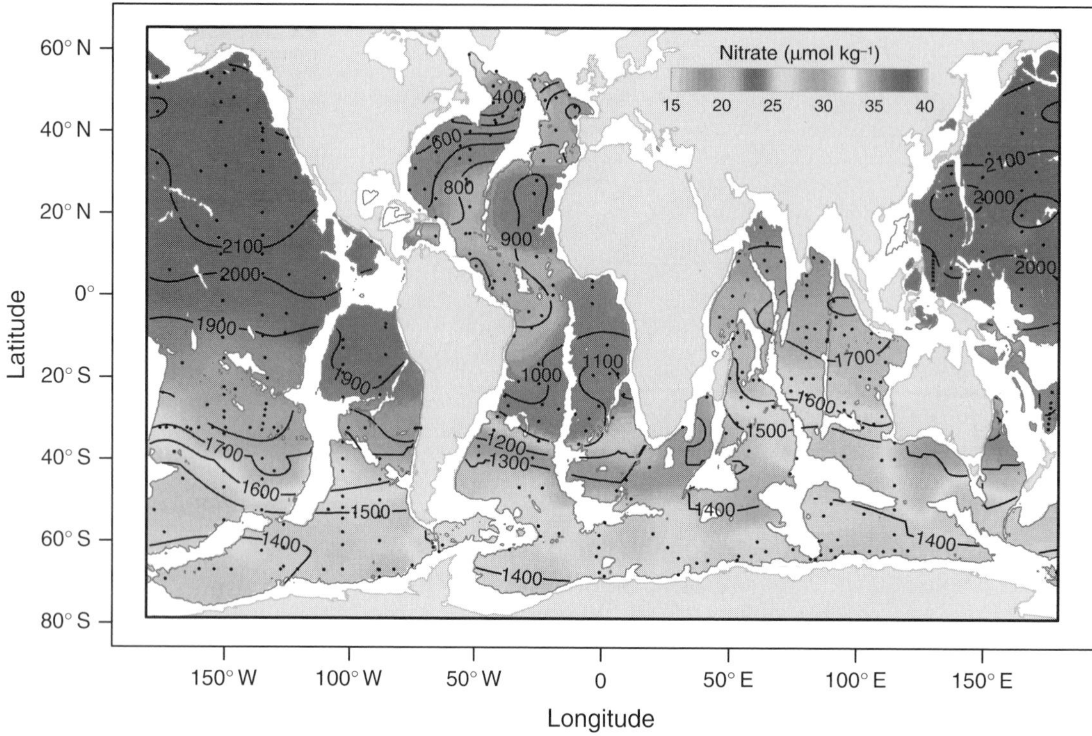

Figure 6.16. The distribution of nitrate (μmol kg^{-1}) and ^{14}C-DIC age (in years) at a depth of 3000 m, illustrating the coincidence between the increase in nutrients and the aging of seawater in the deep ocean. Modified from a figure supplied by Robert Key (Key *et al.*, 2004) (see Plate 4).

300 m) (Fig. 6.10) and the local ventilation of shallow regions of the ocean basins down to several hundred meters in the thermocline. Below this, where O_2 minima form, waters have been out of contact for much longer times. Finally, O_2 concentrations typically rise with depth below 1000 m depth owing to introduction via horizontal advection of relatively O_2-rich waters from high-latitude sources. Being cold, these waters acquire high O_2 concentrations before downwelling and then flow at depths below where most sinking organic-rich particles penetrate.

The amount of O_2 deficit due to organic matter respiration in a water sample can be estimated by knowing the temperature, salinity, and O_2 concentrations. The difference between the O_2 calculated to be at equilibrium, $[O_2^{sat}]$, and the measured O_2 value is called the apparent oxygen utilization or AOU:

$$\mathrm{AOU} = \left[O_2^{sat}\right] - [O_2]. \tag{6.41}$$

With this value it is possible to compare biological O_2 utilization in waters that have very different temperatures and hence O_2 saturation values. Sections of AOU in the major oceans (Fig. 6.17) resemble those for O_2 (Fig. 1.4), but where AOU increases O_2 decreases. There are major differences in magnitude, however, because the saturation concentration of O_2 in surface waters varies by 150 μmol kg^{-1} owing to the temperature dependence.

Using the AOU, one can determine the fractions of dissolved N and P concentrations that are derived from respiration versus those that

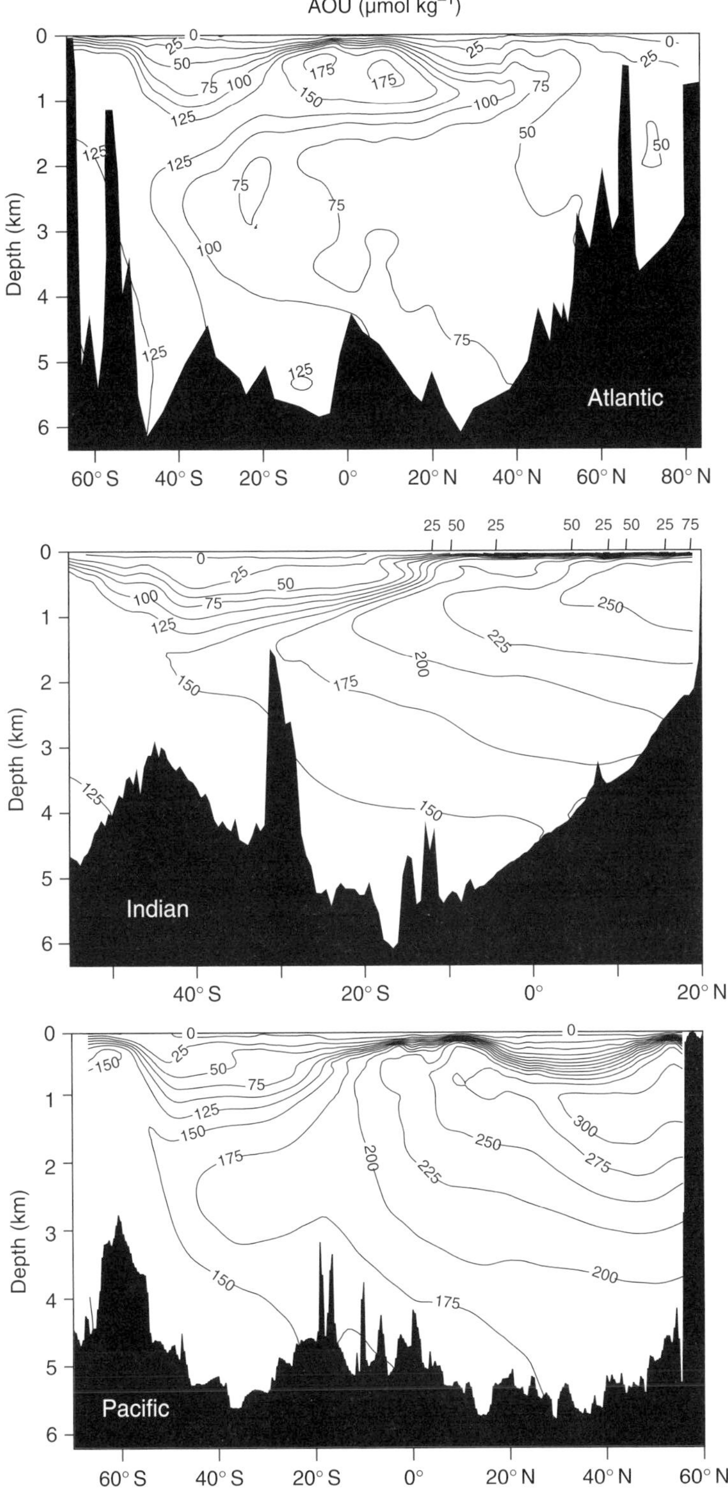

Figure 6.17. Cross sections of apparent oxygen utilization (AOU) in the Atlantic, Indian and Pacific Oceans. Modified from figures supplied by Robert Key (Key *et al.*, 2004).

were present in the water mass when it initially sank from the surface. The latter are termed *preformed nutrients* and are independent of the amount of respiration that has occurred since downwelling. This is the portion of the deep nutrient concentration that was mixed or advected into the deep ocean rather than respired from organic matter. Preformed and respiration-derived nutrient concentrations are separated by evaluating the latter component. The method for estimating respiration-derived nutrient concentrations assumes that the Δnutrient : ΔO_2 ratio during respiration, $r_{\text{Nutrient}:O_2}$, is known (via the modified RKR ratios discussed earlier in this chapter) and that AOU is a measure of the O_2 consumed during respiration. The amounts of dissolved inorganic phosphorus and nitrogen (here indicated as DIP and DIN, though in practice almost all the DIN is in the form of NO_3^-) that have been generated cumulatively in a water body since it sank from active contact with the atmosphere can be estimated by multiplying together the AOU and nutrient respiration ratio. Subtracting this value from the measured value yields the *preformed nutrient* concentrations:

$$\text{DIP}^{\text{pre}} = \text{DIP} - r_{\text{P}:\text{O}_2} \times \text{O}_{2,\text{resp}} = \text{DIP} - r_{\text{P}:\text{O}_2} \times \text{AOU} \tag{6.42}$$

$$\text{DIN}^{\text{pre}} = \text{DIN} - r_{\text{N}:\text{O}_2} \times \text{O}_{2,\text{resp}} = \text{DIN} - r_{\text{N}:\text{O}_2} \times \text{AOU}. \tag{6.43}$$

The concept of nutrient and O_2 change along a surface of constant density in the upper ocean is illustrated in Fig. 6.18. When waters subduct (surface waters flow along density horizons into the thermocline) it is assumed that they carry with them O_2 concentrations near saturation equilibrium with the atmosphere and preformed nutrient concentrations. The assumption of saturation equilibrium is not exactly correct but in most cases this is probably not a serious error because surface oxygen measurements indicate near-saturation equilibrium everywhere except in the Southern Ocean south of the polar front, where concentrations can be up to 10% undersaturated. As a water parcel moves along a constant-density surface into the upper thermocline, respiration consumes the O_2 concentration while creating nutrients and AOU in the water mass. At any point in the ocean interior, preformed nutrient concentrations can be calculated if one knows the temperature, salinity (for determining $[O_2^{\text{sat}}]$), nutrient and O_2 concentrations.

A useful application of preformed nutrient concentrations is that they are intrinsic to different water masses and sometimes can be used as conservative tracers. For example, the main sources of deep water in the Pacific Ocean are North Atlantic Deep Water (NADW), Antarctic Intermediate Water (AAIW) and Antarctic Bottom Water (AABW), all of which are at least partly homogenized in the Antarctic Circumpolar Water (AACW). It is not possible to determine how much of each of these sources contributes to Pacific deep water by using end member mixing of the conservative properties temperature and salinity because salinities of the end members are not sufficiently different. Since concentrations of DIP are well above detection limits in

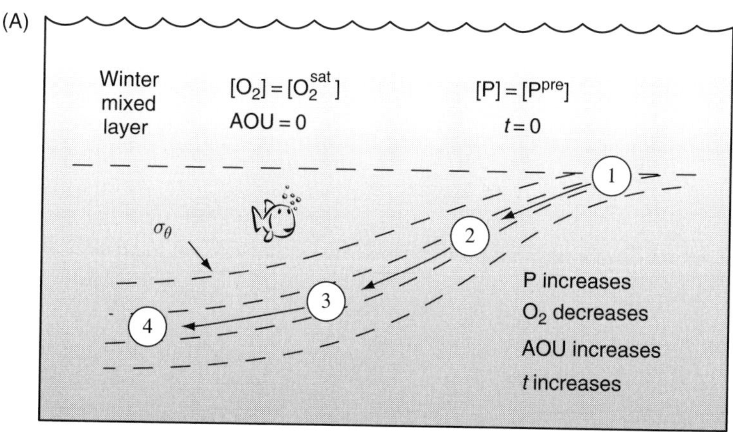

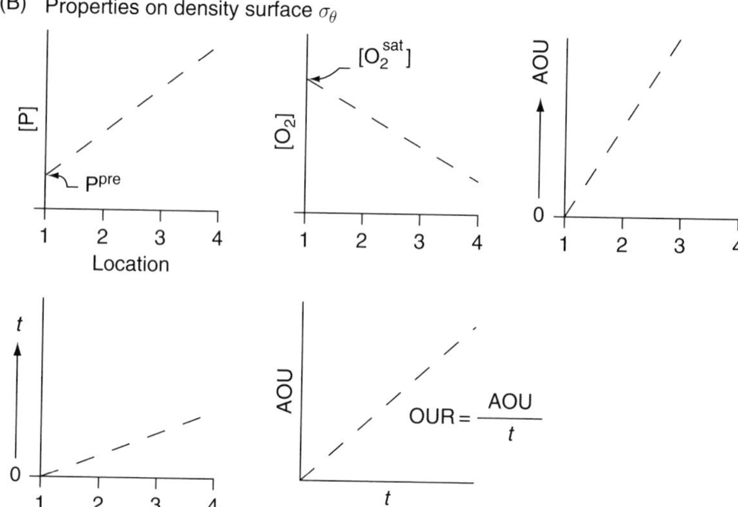

Figure 6.18. (A) A schematic cross section of the upper ocean, illustrating surface water values and trends on subsurface isopycnals for O_2, AOU, preformed nutrients, P^{pre}, and time since the water mass was at the surface, t. (B) Schematic plots of O_2, P, AOU and t versus locations 1, 2, 3 and 4 in (A); and a plot of AOU versus t used to derive oxygen utilization rates (OUR).

high-latitude surface waters, and very different among the three sources, Broecker *et al.* (1985) used the preformed nutrient phosphate (which he called PO or PO_4^*) along with temperature to determine that the Deep Pacific Water is made of roughly equal parts of NADW and the two southern source waters (AAIW and AABW).

Preformed tracers analogous to DIP^{pre} and DIN^{pre} have been developed for nearly all the major macronutrients (P, N, DIC) and O_2 (Gruber and Sarmiento, 2002). In some cases these tracers can be used to identify sources and sinks that are different from traditional photosynthesis and respiration. For example, the tracer N* is the concentration of dissolved inorganic nitrogen (almost all NO_3^-) to be expected in deep waters if the only process occurring is oxic respiration of organic matter:

$$N^* = DIN - 16 \cdot DIP + 2.9 \ \mu mol \ kg^{-1}, \tag{6.44}$$

where DIP indicates the phosphate concentration and the Redfield ratio $r_{N:P} = 16$. The value 2.90 μmol kg^{-1} has no other significance

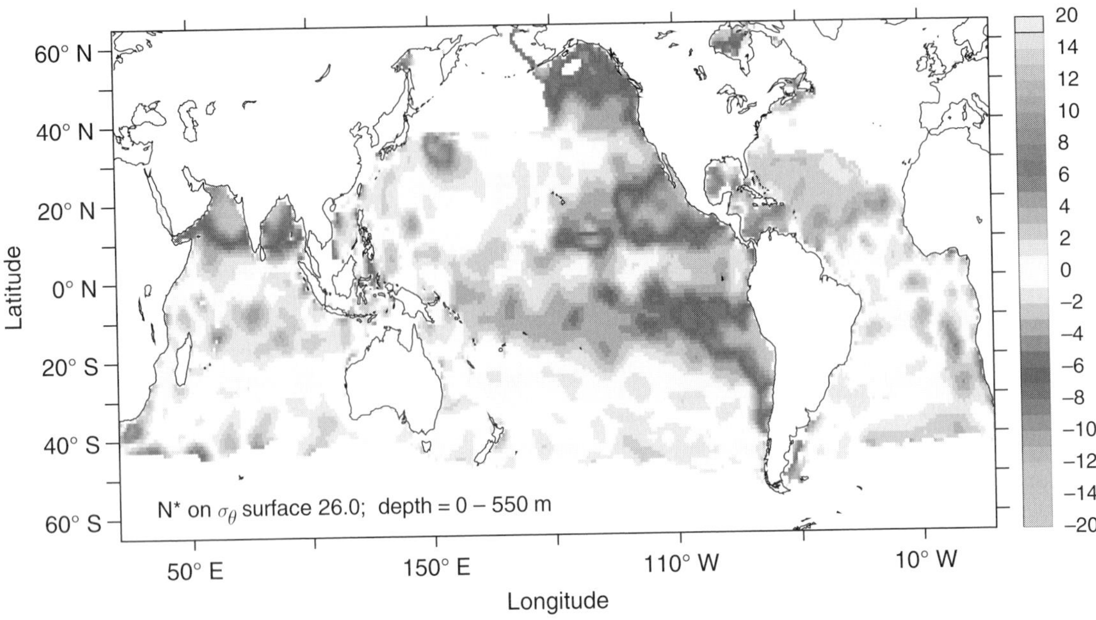

Figure 6.19. N* (NO_3^- – 16 DIP + 2.9 μmol kg^{-1}, see text) on $\sigma_\theta = 26.0$ in the world's ocean basins. Calculated from the Levitus Atlas nutrient data. This map is used to determine locations of denitrification. Modified from a figure supplied by Curtis Deutsch, University of Washington (Deutsch *et al.*, 2007). (See Plate 5.)

than to make the global value of N* zero; i.e. when both N* and DIN = 0 there is a positive value of DIP. Where N* is positive there is production of DIN in the absence of DIP, which indicates nitrogen fixation. In regions where the N* is negative DIN is consumed without DIP being consumed, indicating denitrification. Values of N* on a density surface of the shallow thermocline for the global oceans (Fig. 6.19) indicate very high values in the North Atlantic and low values in the regions of the O_2 minima, the Eastern Equatorial Pacific and Arabian Sea. North Atlantic surface waters are known to be regions of strong nitrogen fixation, and the E Equatorial Pacific and Arabian Sea are regions where oxygen concentrations are low enough that nitrate is used to degrade organic matter rather than oxygen (denitrification).

A similar value has been conceived for DIC, ΔC* (Gruber *et al.*, 1996). In this case the value of preformed DIC is evaluated and its changes due to organic matter degradation and $CaCO_3$ dissolution are calculated from changes in AOU, NO_3^-, and Alk. These values are then compared with measured DIC values and the excess is attributed to the concentration of anthropogenic CO_2 that has penetrated the ocean. Details of this procedure are described in Chapter 11 on the carbon cycle.

6.4.2 Oxygen utilization rate (OUR)

If the amount of time that a water mass has been away from the surface ocean mixed layer can be evaluated, then this "age" can be used along with the AOU to determine the O_2 utilization rate (OUR). The concept is illustrated in Fig. 6.18, where the OUR, in a closed system uncomplicated by advection and concentration gradients, is the slope of AOU versus water parcel age. The OUR is thus the mean rate of respiration that has occurred in the water parcel since it left

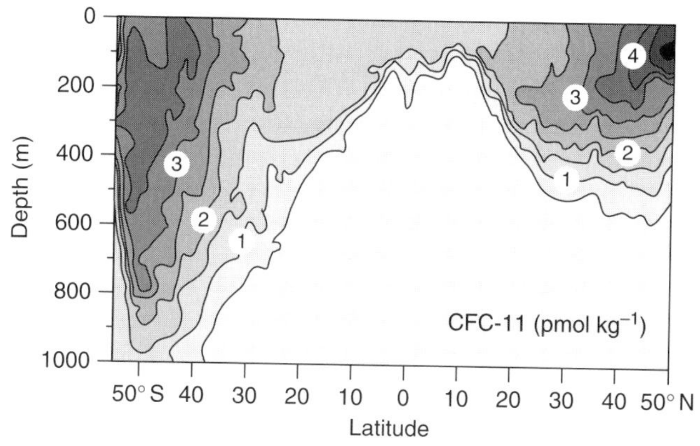

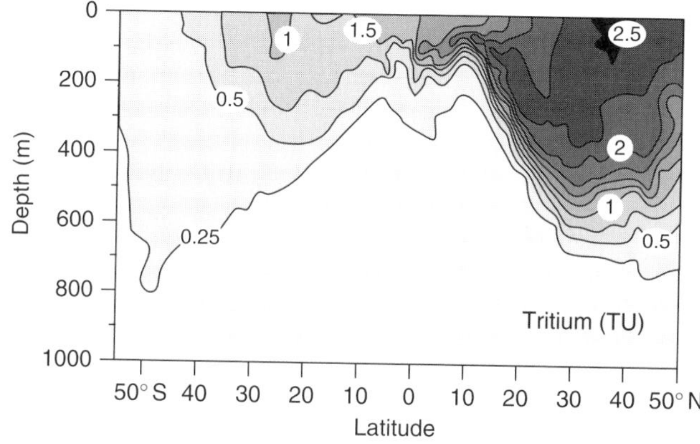

Figure 6.20. CFC and ^3H sections as a function of depth along 135° W in the Pacific Ocean. Both tracers have an anthropogenic origin and were introduced to the atmosphere primarily in the past 50 years. Tritium concentrations are in tritium units (TU) (1 TU = 1 tritium atom per 10^{18} hydrogen atoms). Redrawn from Jenkins (2002).

the surface ocean. There are two tracers of water mass age that are ideal over the decadal-scale ranges appropriate for regions of the top of the thermocline where most organic matter is degraded: the ^3H–^3He pair and chlorofluorocarbons (CFCs). Both tracers are gases and stem from anthropogenic activities, either from nuclear weapons testing, ^3H (tritium), or industrial processes, CFCs. They have entered the upper ocean and are presently penetrating the thermocline (Fig. 6.20). We illustrate the application of these tracers by describing the evolution of tritium and ^3He in the upper ocean.

Because there is a very low level of natural ^3H on the Earth, nearly all that is present in today's ocean derives from nuclear weapons testing in the atmosphere in the 1950s and 1960s. Immediately after it was produced by nuclear explosions in the atmosphere, ^3H was incorporated into water and rained out to the land and ocean surfaces. Thus, it entered the ocean as a spike in the same fashion as bomb-produced ^{14}C (Fig. 5.15), only for tritium the spike was much sharper because it was removed from the atmosphere as rain, which is faster than the process of CO_2 gas exchange at the atmosphere–ocean inter face for ^{14}C. ^3H is radioactive and decays to ^3He with a half life of 12.5 y:

$$^3H \xrightarrow{t_{1/2} \,=\, 12.5\,y} {}^3He. \tag{6.45}$$

Because most ^3He that leaks from the Earth's mantle escapes to the stratosphere, there is very little in the atmosphere. Thus, surface ocean water in equilibrium with the atmosphere is nearly devoid of this tracer. One can therefore write a rather simple equation describing the age, $t_{^3He}$, of an isolated water mass (in a closed system) after it leaves the surface ocean. The rates of change in ^3H and ^3He are determined by the decay constant of ^3H (Chapter 5):

$$-\frac{d[^3He]}{dt} = \frac{d[^3H]}{dt} = -\lambda_{^3H} \cdot [^3H]. \tag{6.46}$$

Since there is almost no initial ^3He in surface waters, the concentration of ^3He at any time t is related to the difference between the values of ^3H at t and $t = 0$:

$$[^3He]_t = [^3H]_{t=0} - [^3H]_t. \tag{6.47}$$

The first-order decay law for ^3H is:

$$[^3H]_t = [^3H]_{t=0} \cdot e^{-\lambda_{^3H} t}. \tag{6.48}$$

Substituting Eq. (6.48) into Eq. (6.47) and solving for t gives:

$$t_{^3He} = \frac{1}{\lambda_{^3H}} \cdot \ln\left(1 + \frac{[^3He]_t}{[^3H]_t}\right). \tag{6.49}$$

If there were no mixing with surrounding waters, this value of the ^3H–^3He age could be plotted against AOU to determine the OUR. In practice mixing is important and the three-dimensional distributions of ^3H, ^3He and O_2 must be considered along with equations that include advection and mixing. William Jenkins solved the problem in both ways for different density levels in an area in the eastern subtropical Atlantic Ocean (Jenkins, 1998) and showed that the more careful solution gives lower values at depth and a more realistic distribution of OUR as a function of depth (Fig. 6.21).

This result is probably the best estimate of the depth dependence of respiration in the upper thermocline of the ocean. An exponential regression through the data indicates a scale height of 165 m. This is the distance required for the rate at the base of the euphotic zone (*c*.100 m) to decay to 1/e of its value, and is the reason that it is often stated that most of the organic carbon that exits the euphotic zone is respired in the upper 200 m of the thermocline.

At steady state, the depth-integrated OUR must equal the flux of organic matter from the upper ocean after converting O_2 to C via the stoichiometric ratio $\Delta O_2:\Delta C = 1.4:1$. The integrated OUR in the eastern subtropical Atlantic is 4.1 ± 0.8 mol O_2 m^{-2} y^{-1}, which implies an organic C consumption rate of 2.8 ± 0.5 mol C m^{-2} y^{-1}. This value is very similar to values determined by the same ^3H–^3He method on the other side of the subtropical Atlantic Ocean at the Bermuda Time Series (BATS) location (Table 6.1). The integrated OUR value provides further support for the organic matter export rates determined by O_2 and carbon isotope mass balance in the surface ocean in this area.

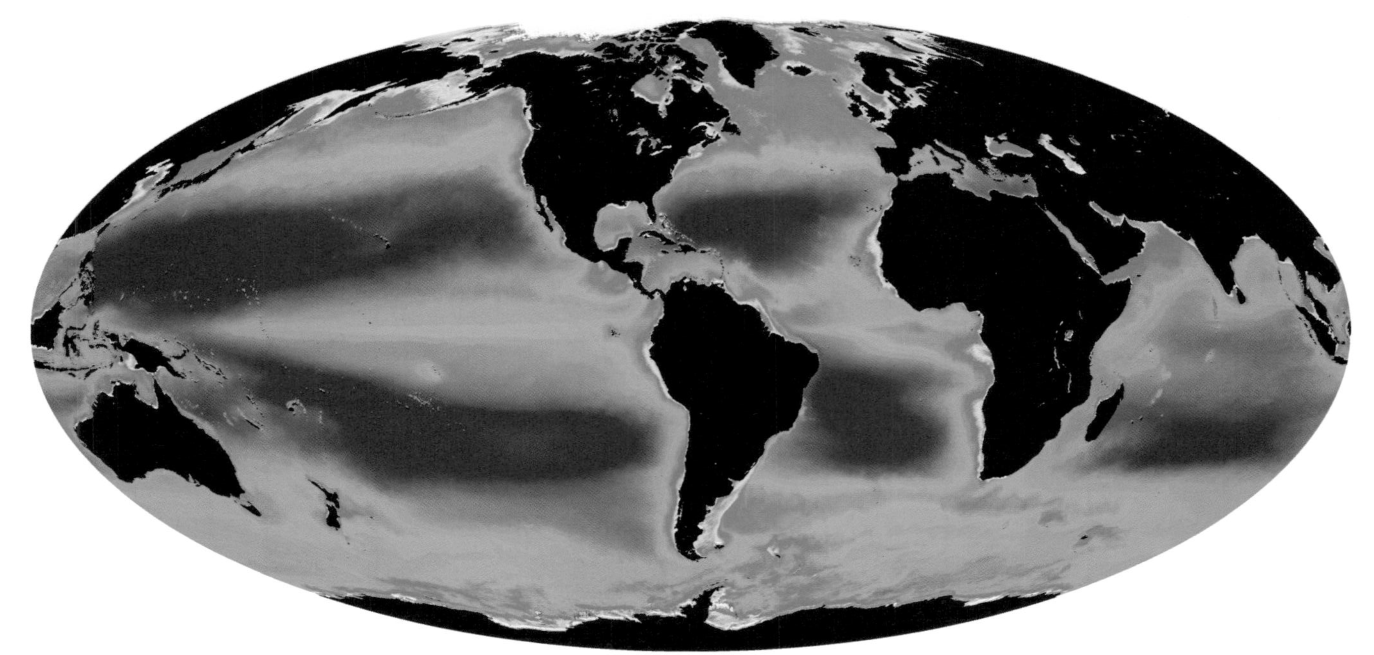

Plate 1 A global map of surface-ocean chlorophyll derived from satellite ocean color imagery. Images like this are used to determine relative distributions of ocean primary production; however, these are approximate because the relation between surface ocean color and primary productivity is variable and in some cases uncertain. The figure is the average of c.10 years of SeaWIFS ocean color data, 1997–2007. (The image is from the NASA/MODIS ocean color web site, http://oceancolor.gsfc.nasa.gov.)

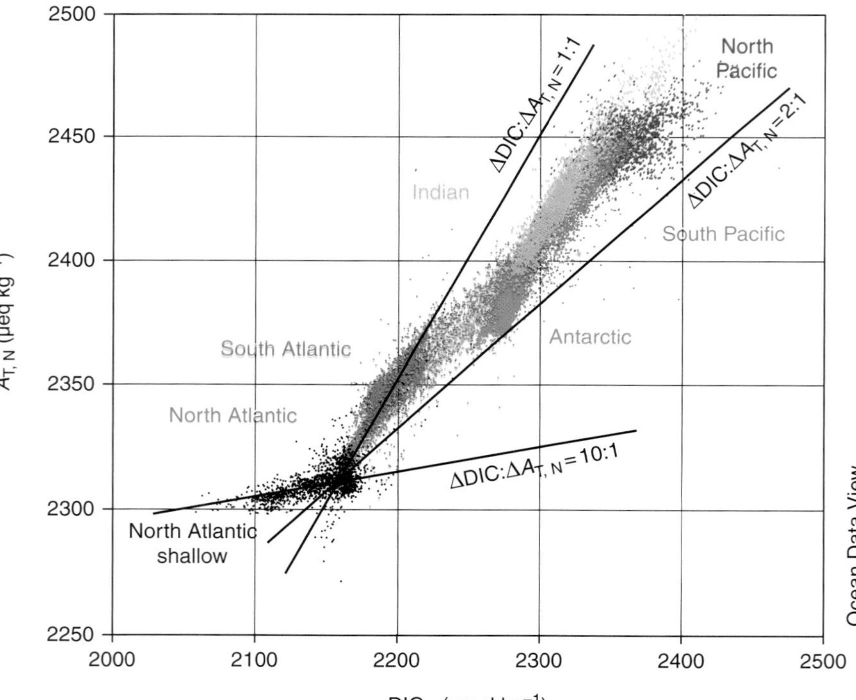

Plate 2 Salinity-normalized ($S = 35$) total alkalinity, $A_{T,N}$, versus salinity-normalized dissolved inorganic carbon, DIC_N, for the world's ocean. Data are for the deep ocean at depths >2.5 km except for the section labeled "North Atlantic Shallow," which is 100–1000 m in the North Atlantic Ocean. Lines indicate different $DIC_N : A_{T,N}$ ratios.

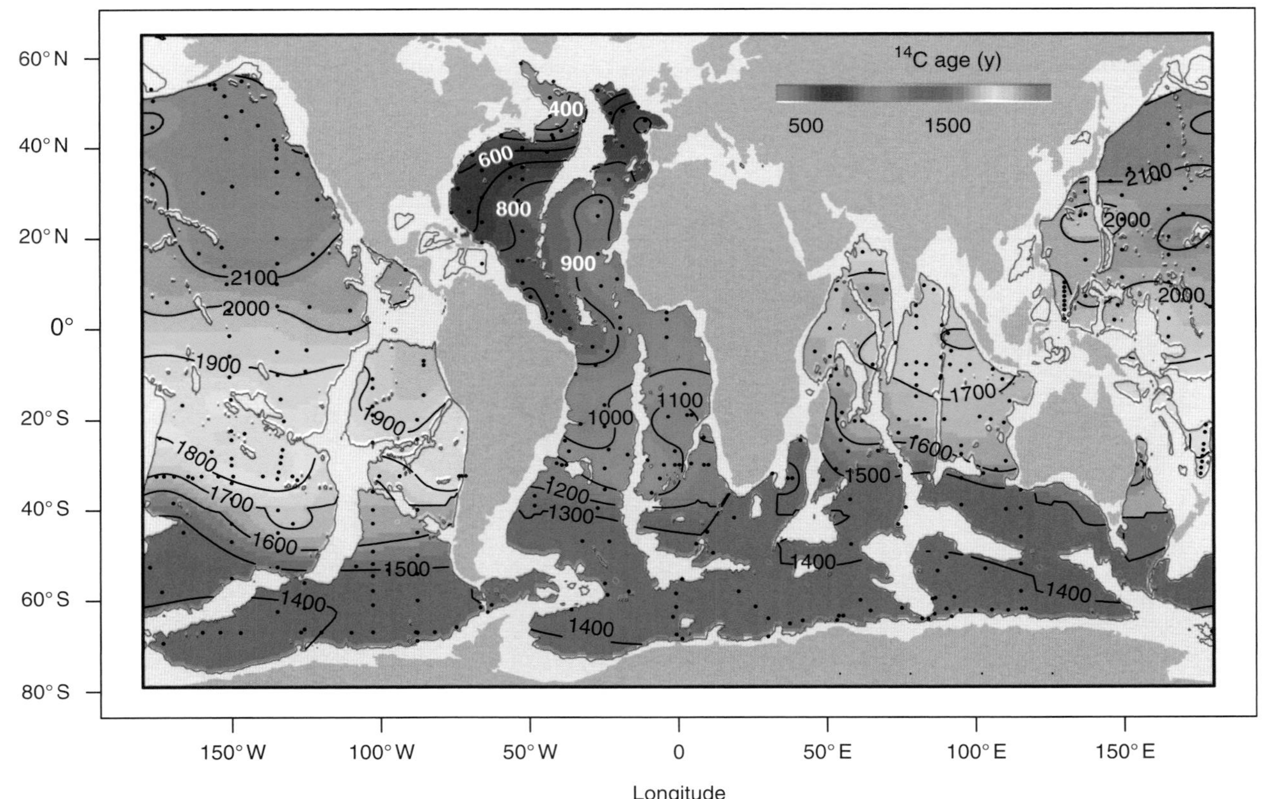

Plate 3 The ^{14}C age of DIC in the world's ocean at a depth of 3000 m, determined during the WOCE program in the 1990s. Courtesy of Robert Key, Princeton University; Key *et al.* (2004).

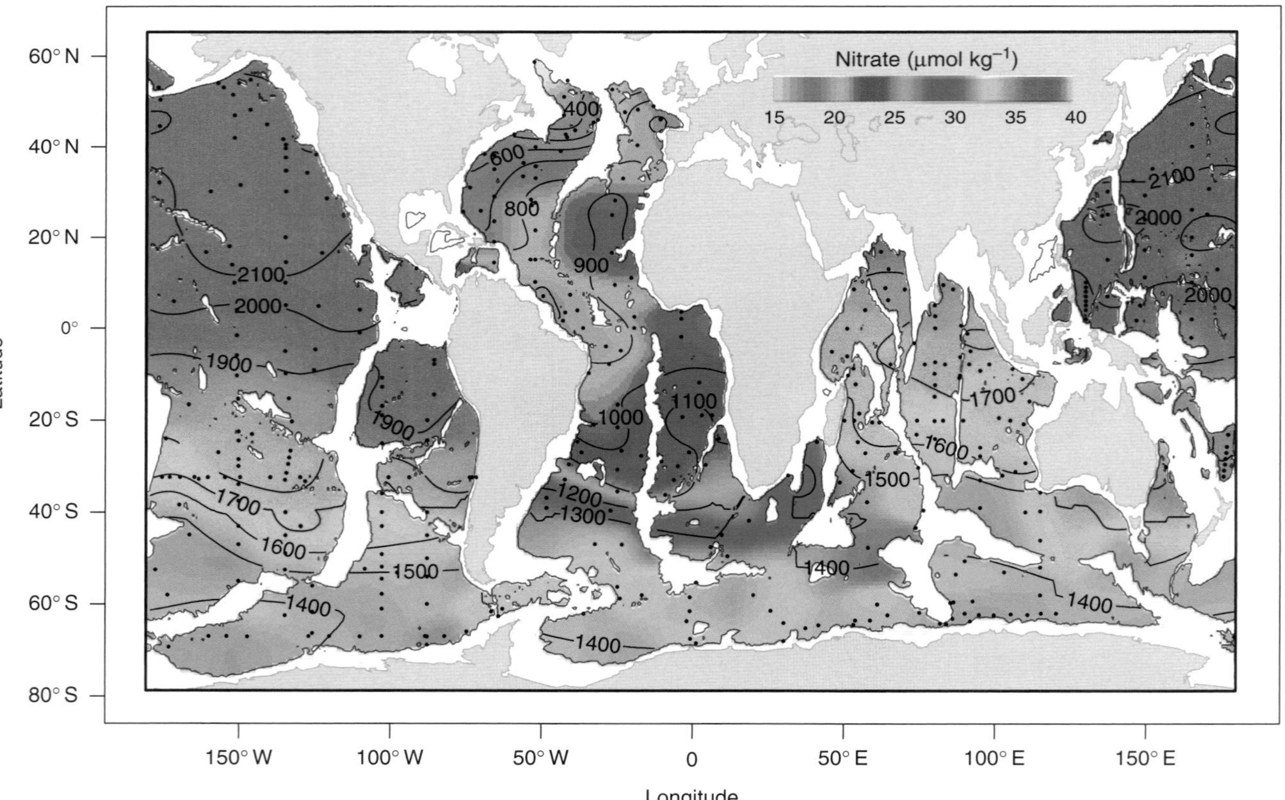

Plate 4 The distribution of nitrate ($\mu mol\ kg^{-1}$) and ^{14}C-DIC age (in years) at a depth of 3000 m, illustrating the coincidence between the increase in nutrients and the aging of seawater in the deep ocean. Modified from a figure supplied by Robert Key (Key *et al.*, 2004).

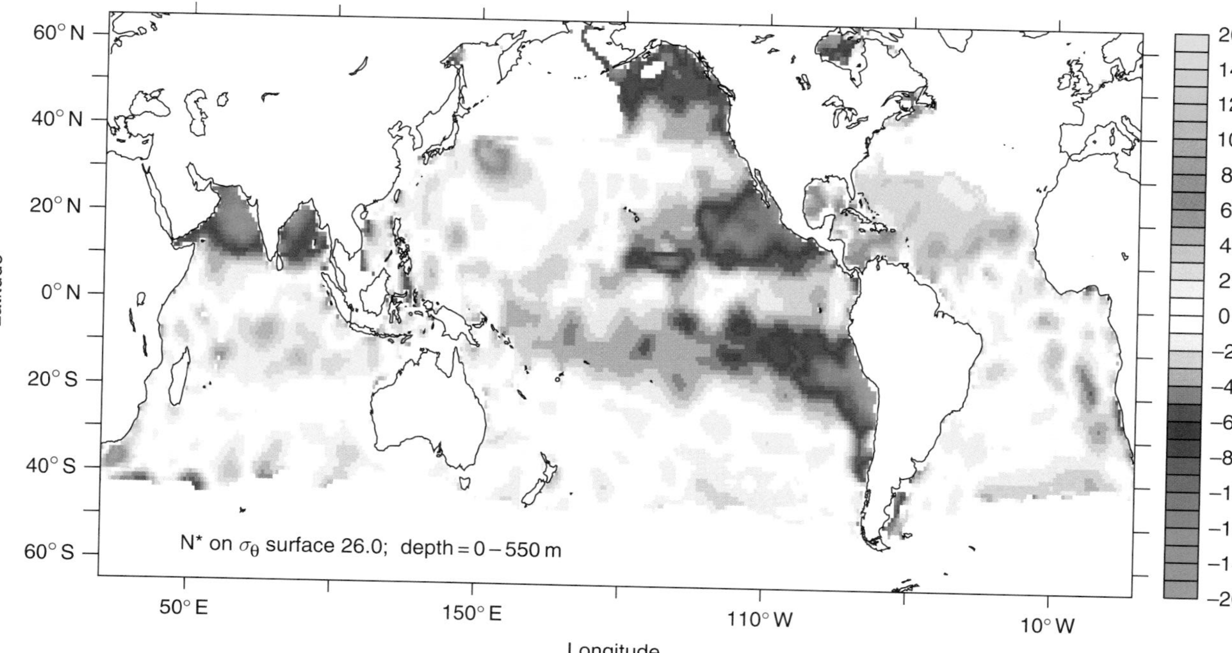

Plate 5 N* (NO_3^- − 16 DIP + 2.9 μmol kg^{-1}, see text) on $\sigma_\theta = 26.0$ in the world's ocean basins. Calculated from the Levitus Atlas nutrient data. This map is used to determine locations of denitrification. Modified from a figure supplied by Curtis Deutsch, University of Washington (Deutsch *et al.*, 2007).

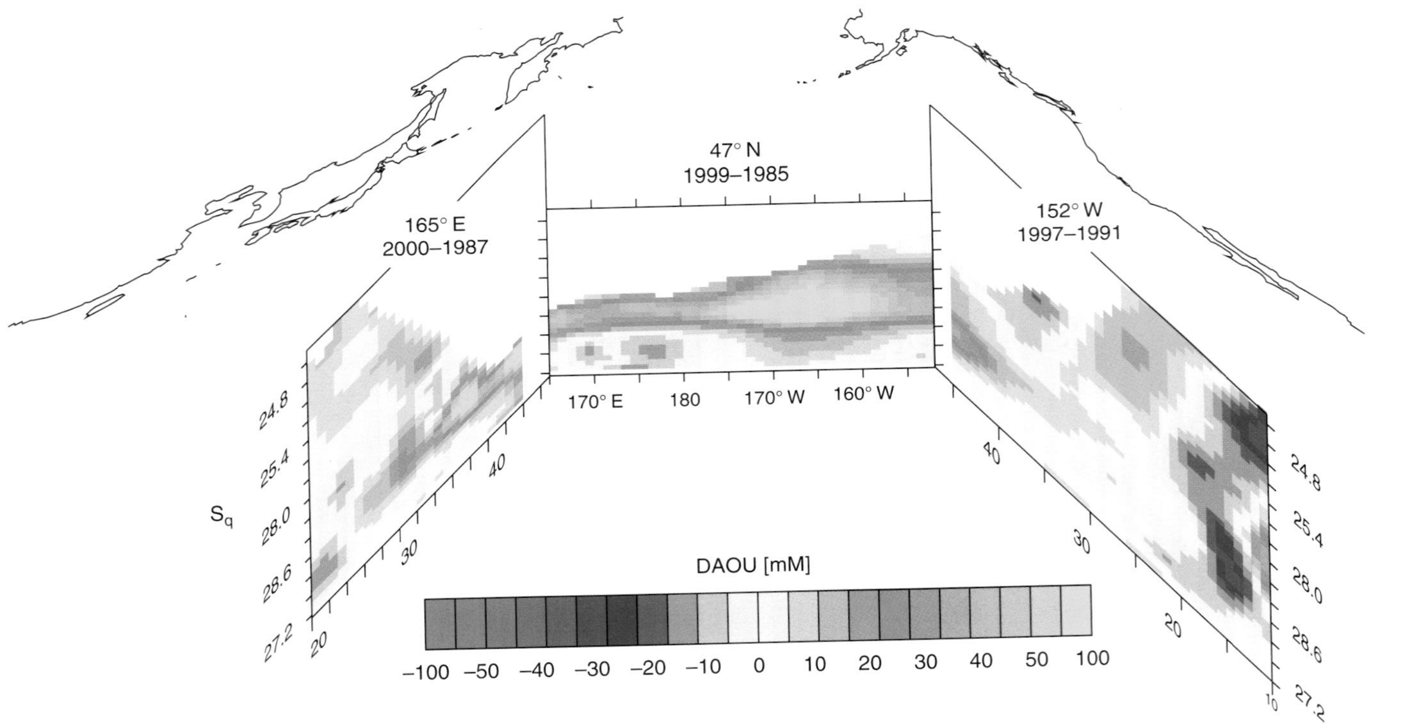

Plate 6 A three-dimensional figure of AOU differences (DAOU) between the 1980s and 1990s in the North Pacific Ocean. Differences were determined from repeat sections in the eastern and western subtropical North Pacific and a zonal section in the subarctic North Pacific. Reproduced from Deutsch *et al.* (2006).

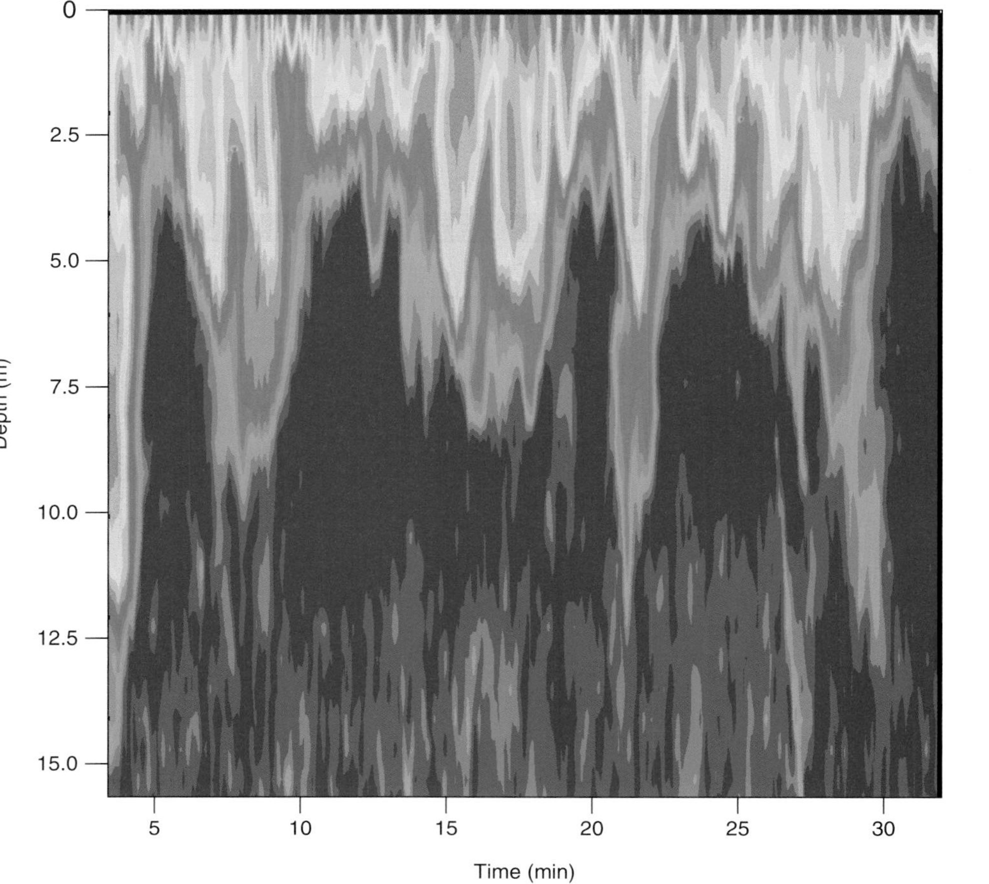

Plate 7 The intensity of acoustic backscatter as a function of depth in the ocean at Stn. P in the subarctic Pacific at a wind speed of $12\ \mathrm{m\ s^{-1}}$. Backscatter intensity is an indication of the depth of penetration of bubbles caused by breaking waves. (Data courtesy of Sven Vegel of the Institute of Ocean Sciences, Sidney, BC.)

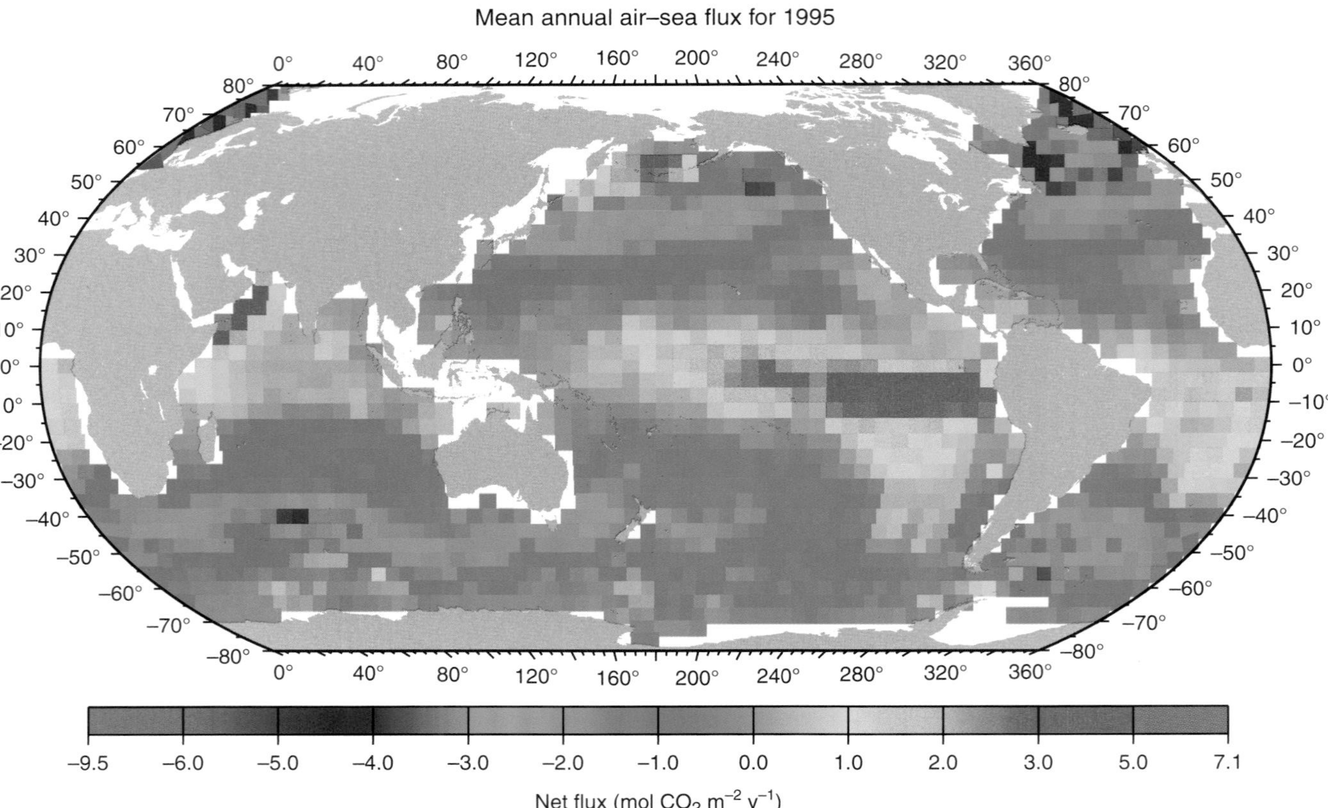

Plate 8 The mean annual flux of CO_2 between the atmosphere and ocean, based on measurements of f_{CO_2} in the ocean and atmosphere, and the gas exchange mass transfer coefficient determined from wind speed. From Takahashi *et al.* (2002).

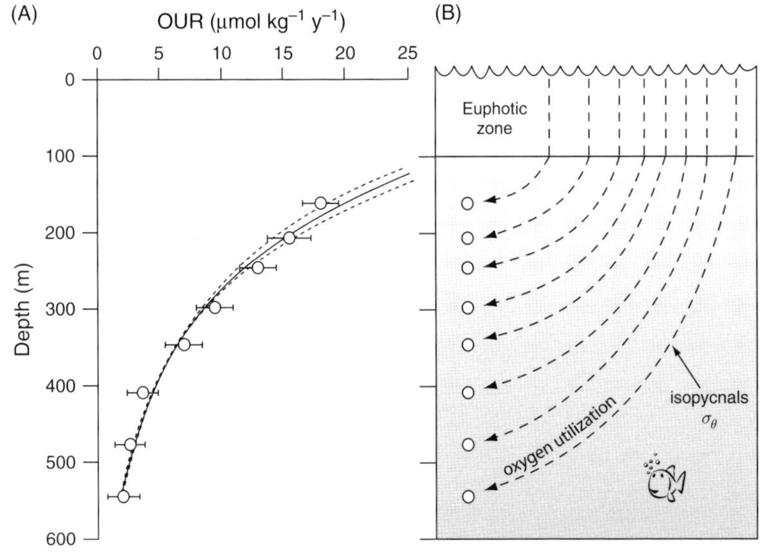

Figure 6.21. The oxygen utilization rate (OUR) versus depth determined by measuring ^3H, ^3He and AOU on isopycnal surfaces in the eastern subtropical North Atlantic Ocean. (A) The OUR results redrawn from Jenkins (1998). (B) A schematic diagram indicating the pathway of constant density surfaces (isopycnals) from the surface ocean into the thermocline. Because fresher organic matter degrades on surfaces closest to the euphotic zone, the OUR rates are greater on the shallower isopycnals even though O_2 concentrations are lower on the deeper ones.

What little is known about rates of respiration in the ocean below 1000 m is inferred from deep sediment traps and by the study of O_2 consumption at the sediment–water interface. The efficiency of deep sediment traps has been calibrated by using the longer-lived thorium isotope ^{230}Th in the same way that shallow traps are calibrated with ^{234}Th. In this case the fluxes agree well with those expected from the decay of ^{234}U, indicating that these devices are more accurate at determining particle fluxes in the deep ocean than in shallow waters. This is to be expected since in the deep-sea currents are slower and the particles that carry the organic matter freight all the way to the deep ocean are more homogeneous. Organic matter fluxes from deep sediment traps gradually decrease with depth between 1000 and 5000 m. Since many of the particles that reach the deep sea are predominantly mineral tests or in fecal pellets of zooplankton they transit the water column fairly quickly (in weeks) so that some material relatively rich in organic matter reaches the sea floor.

Benthic respiration is studied by placing chambers on the sea floor and measuring the decrease in O_2 in the chambers and by determining O_2 gradients in porewaters and calculating the diffusive flux (see also Chapter 12). Rates of respiration by this method have been compiled and compared with the respiration estimated in the deep ocean via sediment trap experiments to determine the amount of O_2 consumption in each cubic meter of seawater that could be attributed to benthic respiration and subsequent lateral exchange. As illustrated in Fig. 6.22, below about 3000 m the amount of O_2 consumed at the sediment–water interface dominates respiration in the deep ocean. The reason for this is that high levels of benthic respiration are concentrated in continental margin regions near locations of upwelling and the hypsometry of the deep sea dictates that the

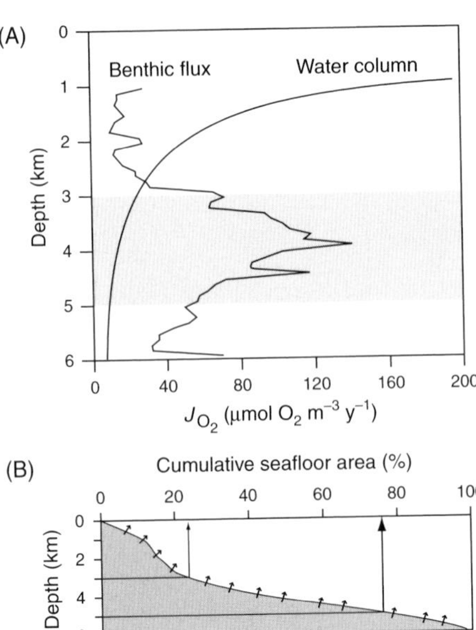

Figure 6.22. (A) Water column oxygen consumption rate (from sediment trap and OUR determinations) and benthic flux (μmol O_2 m^{-3} y^{-1}) as a function of depth in the ocean. Redrawn from Jahnke and Jackson (1987). The benthic fluxes are normalized to the volume of water exposed per unit of sediment area and indicate that below 3000 m the respiration contribution from the sediments is greater than that in the water. The hypsometric curve in (B) indicates that the region between 3000 and 5000 m depth also has the greatest sea floor area to ocean volume ratio, which is indicated by the shaded region in (A).

surface area to volume ratio increases rapidly below about 3000 m (the ocean has a relatively flat bottom).

6.4.3 Aphotic zone respiration summary

Most of the respiration below the euphotic zone of the ocean occurs in the upper 200 m of the thermocline. Oxygen utilization rate (OUR) values determined from dating the ventilation age of water by using CFCs and AOU data in many locations of the ocean confirm results determined by using the ^{3}H–^{3}He isotope pair in the North Atlantic. The concentration of oxygen below the ocean mixed layer (more precisely, the apparent oxygen utilization, AOU) is controlled by a combination of ocean circulation, ventilation (subduction of water from the ocean mixed layer into the upper thermocline), and the strength of the biological pump. Until recently it has been assumed that the values of AOU in the subsurface oceans have been constant in time. At the time of writing this book there is growing evidence that there have been decadal-scale changes, at least in the shallow thermocline (see, for example, Emerson *et al.*, 2004) (Fig. 6.23). Modeling studies of this process in the North Pacific Ocean indicate that the changes are due primarily to decadal-scale variations in circulation (Deutsch *et al.*, 2006). Time will tell whether the observed trends are part of a natural oscillation (for example, driven by the Pacific Decadal Oscillation) or a result of global warming. In any case, shallow thermocline AOU changes are sensitive indicators of changes in their biological and physical forcing and may act as an early indicator, "a canary in the mineshaft," for the effect of anthropogenic changes on the biogeochemistry of the ocean.

The variation of the respiration rate with depth in the ocean depends strongly on factors controlling organic matter degradation and particle sinking rates. The role of mineral content ($CaCO_3$ and

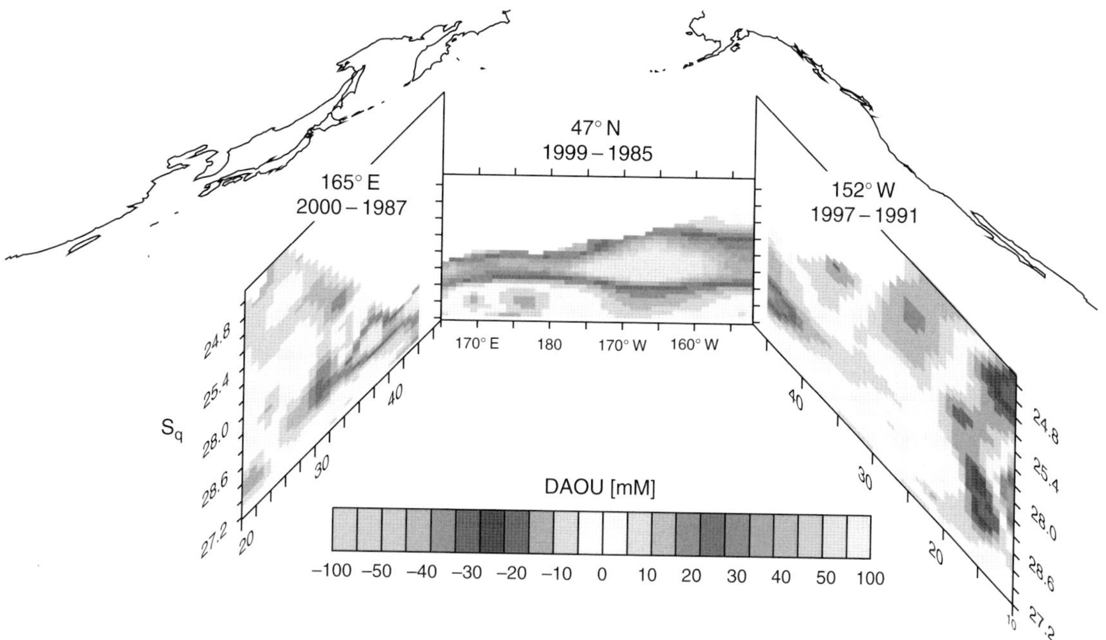

Figure 6.23. A three-dimensional figure of AOU differences (DAOU) between the 1980s and 1990s in the North Pacific Ocean. Differences were determined from repeat sections in the eastern and western subtropical North Pacific and a zonal section in the subarctic North Pacific. Reproduced from Deutsch *et al.* (2006). (See Plate 6.)

SiO_2) in particulate matter in controlling the depth distribution of respiration has not been evaluated experimentally. None the less, the interplay of organic matter degradation and particle sinking velocities allows enough organic matter to reach the sediments so that respiration at this interface is an important factor in global ocean respiration below depths of about 3000 m. Determining the mechanisms that control the depth dependence of organic matter degradation is necessary to be able to predict deep-water metabolite distributions in response to variations in ecology and ocean circulation.

References

Anderson, L. A. (1995) On the hydrogen and oxygen content of marine phytoplankton. *Deep-Sea Res. I* **42**, 1675–80.

Anderson, L. A. and J. L. Sarmiento (1994) Redfield ratios of remineralization determined by nutrient data analysis. *Global Biogeochem. Cycles* **8**, 65–80.

Armstrong, R. A., C. Lee, J. I. Hedges, S. Honjo and S. G. Wakeham (2002) A new mechanistic model for organic carbon fluxes in the ocean based on the quantitative association of POC with ballast minerals. *Deep-Sea Res. II* **49**, 219–36.

Benitez-Nelson, C., K. O. Buesseler, D. M. Karl and J. Andrews (2001) A time-series study of particulate matter export in the North Pacific subtropical gyre based on $^{234}Th : ^{238}U$ disequilibrium. *Deep-Sea Res. I* **48**, 2595–611.

Broecker, W. S. (1971) A kinetic model for the chemical composition of seawater. *Quat. Res.* **1**, 188–207.

Broecker, W. S. and T. H. Peng (1982) *Tracers in the Sea*. Lamont–Doherty Earth Observatory, Palisades, NY: Eldigio Press.

Broecker, W. S., T. Takahashi, H. J. Simpson and T. -H. Peng (1985) Sources of flow patterns of deep ocean waters as deduced from potential temperature, salinity and initial phosphate concentration. *J. Geophys. Res.* **90**, 6925–39.

Bruland, K. W. and M. C. Lohan (2003) Controls of trace metals in seawater. In *The Oceans and Marine Chemistry* (ed. H. Elderfield), vol. 6, *Treatise on Geochemistry* (ed. H. D. Holland and K. K. Turekian), pp. 23–47. Oxford: Elsevier-Pergamon.

Carlson, C. A., H. W. Ducklow and A. F. Michaels (1994) Annual flux of dissolved organic carbon from the euphotic zone in the northwestern Sargasso Sea. *Nature* **371**, 405–8.

Deutsch, C., S. Emerson and L. Thompson (2006) Physical-biological interactions in the North Pacific oxygen variability. *J. Geophys. Res.* C09S90, doi: 10.1029/2005JC003179.

Deutsch, C., D. J. L. Sarmiento, D. A. Sigman, N. Gruber and J. A. Dunne (2007) Spatial coupling of nitrogen inputs and losses in the ocean. *Nature* **445**, 163–7.

Emerson, S. (1997) Net biological oxygen production: a global estimate from oceanic measurements. In *Biogeochemical Processes in the North Pacific* (ed. S. Tsunogai), pp. 143–55. Tokyo: Japan Marine Science Foundation.

Emerson, S. *et al.* (1997) Experimental determination of the organic carbon flux from open-ocean surface waters. *Nature* **389**, 951–4.

Emerson, S., P. D. Quay, C. Stump, D. Wilbur and M. Knox (1991) O_2, Ar, N_2 and ^{222}Rn in surface waters of the subarctic ocean: net biological O_2 production. *Global Biogeochem. Cycles* **5**, 49–69.

Emerson, S., T. W. Watanabe, T. Ono and S. Mecking (2004) Temporal trends in Apparent Oxygen Utilization in the upper pycnocline of the North Pacific: 1980–2000. *J. Oceanogr.* **60**, 139–47.

Eppley, R.W. and B. J. Peterson (1979) Particulate organic flux and planktonic new production in the deep ocean. *Nature* **282**, 677–80.

Gruber, N. and J. L. Sarmiento (2002) Large scale biogeochemical-physical interactions in elemental cycles. In *The Sea* (ed. A. R. Robinson, J. J. McCarthy and B. J. Rothschild), pp. 337–99. New York, NY: John Wiley.

Gruber, N., C. D. Keeling and T. F. Stocker (1998) Carbon-13 constraints on the seasonal inorganic carbon budget at the BATS site in the northwestern Sargasso Sea. *Deep-Sea Res. I* **45**, 673–717.

Gruber, N., J. L. Sarmiento and T. F. Stocker (1996) An improved method for detecting anthropogenic CO_2 in the oceans. *Global Biogeochem. Cycles* **10**, 809–37.

Hamme, R. C. and S. R. Emerson (2006) Constraining bubble dynamics and mixing with dissolved gases: implications for productivity measurements by oxygen mass balance. *J. Mar. Res.* **64**, 73–95.

Hansell, D. A. and C. A. Carlson (1998) Net community production of dissolved organic carbon. *Global Biogeochem. Cycles* **12**, 443–53.

Hedges, J. I. J. A. Baldock, Y. Gelinas *et al.* (2002) The biochemical and elemental compositions of marine plankton: a NMR perspective. *Mar. Chem.* **78**, 47–63.

Jahnke, R. J. and G. A. Jackson (1987) Role of sea floor organisms in oxygen consumption in the deep North Pacific Ocean. *Nature* **329**, 621–3.

Jenkins, W. J. (1998) Studying subtropical thermocline ventilation and circulation using tritium and ^{3}He. *J. Geophys. Res.* **103**, 15 817–31.

Jenkins, W. J. (2002) Tracers of ocean mixing. In *The Oceans and Marine Chemistry* (ed. H. Elderfield), vol. 6, *Treatise on Geochemistry* (ed. H. D. Holland and K. K. Turekian), pp. 223–46. New York, NY: Elsevier.

Jenkins, W.J. and D.W.R. Wallace (1992) Tracer based inferences of new primary production in the sea. In *Primary Production and Biogeochemical Cycles in the Sea* (ed. P.G. Falkowski and A.D. Woodhead), pp. 299–316. New York, NY: Plenum.

Karl, D.M., J.R. Christian, J.E. Dore *et al.* (1996) Seasonal and interannual variability in primary production and particle flux at Station ALOHA. *Deep-Sea Res.* II, **43**, 539–68.

Key, R.M., A. Kozar, C.L. Sabine *et al.* (2004) A global ocean carbon climatology: results from Global Data Analysis Project (GLODAP). *Global Biogeochem. Cycles* **18**, GB4031, doi: 10.1029/2004GB002247.

Kortzinger, A., J.I. Hedges and P.D. Quay (2001) Redfield ratios revisited: removing the biasing effect of anthropogenic CO_2. *Limnol. Oceanogr.* **46**, 964–70.

Laws, E., P. Falkowski, W.O. Smith, H. Ducklow and J.J. McCarthy (2000) Temperature effects on export production in the open ocean. *Global Biogeochem. Cycles* **14**, 1231–46.

Martin, J.H., G.A. Knauer, D.M. Karl and W.W. Broenkow (1987) VERTEX: carbon cycling in the northeast Pacific. *Deep-Sea Res.* **34**, 267–5.

Meybeck, M. (1979) Concentrations des eaux fluviales en elements majeurs et apports en solution aux oceans. *Rev. Geol. Dny. Geogr. Phys.* **21**(b), 215–46.

Michaels, A. and A. Knap (1996) Overview of the U.S. JGOFS Bermuda Atlantic Time-Series Study and the Hydrostation S program. *Deep-Sea Res.* **43**, 157–8.

Moore, J.K., S.C. Doney, D.M. Glover and I.Y. Fung (2002) Iron cycling and nutrient limitation patterns in the world oceans. *Deep-Sea Res. II* **49**, 463–507.

Morel, F.M.M., A.J. Milligan and M.A. Saito (2003) Marine bioinorganic chemistry: the role of trace metals in oceanic cycles of major nutrients. In *The Oceans and Marine Chemistry* (ed. H. Elderfield), vol. 6, *Treatise on Geochemistry* (ed. H.D. Holland and K.K. Turekian), pp. 113–43. Oxford: Elsevier-Pergamon.

Najjar, R.O. and R.F. Keeling (1999) Mean annual cycle of the air-sea oxygen flux: a global view. *Global Biogeochem. Cycles* **14**, 573–84.

Najjar, R.O. *et al.* (2008) Impact of circulation on export production, dissolved organic matter and dissolved oxygen in the ocean: results from OCMIP-2. *Global Biogeochem. Cycles*, in press.

Quay, P. and J. Stutsman (2003) Surface layer carbon budget for the subtropical N. Pacific: $\delta^{13}C$ constraints at station ALOHA. *Deep-Sea Res. I* **50**, 1045–61.

Redfield, A.C. (1958) The biological control of chemical factors in the environment, *Amer. Sci.* **46**, 205–21.

Redfield, A.C., B.H. Ketchum and F.A. Richards (1963) The influence of organisms on the composition of seawater. In *The Sea*, vol. 2 (ed. M.N. Hill), pp. 26–77. New York, NY: Interscience.

Sarmiento, J.L., J. Dunne, A. Gnanadesikan *et al.* (2002) A new estimate of the $CaCO_3$ to organic carbon export ratio. *Global Biogeochem. Cycles* **16**, doi: 10.1029/2002GB001010.

Schaffer, G., J. Bendtsen and O. Ulloa (1999) Fractionation during remineralization of organic matter in the ocean. *Deep-Sea Res. I* **46**, 185–204.

Schlitzer, R. (2000) Applying the adjoint method for biogeochemical modeling: export of particulate organic matter in the world ocean. In *Inverse Methods in Global Biogeochemical Cycles* (ed. P. Kasibhatla, M. Heinmann, D. Hartley *et al.*) *Geophys. Monogr.* **114**. Washington, D.C.: American Geophysical Union.

Schlitzer, R. (2002) Carbon export from the Southern Ocean: results for inverse modeling and comparison with satellite based estimates. *Deep-Sea Res. II* **49**, 1623–44.

Spitzer, W.S. and W.J. Jenkins (1989) Rates of vertical mixing, gas exchange and new production: estimates from seasonal gas cycles in the upper ocean near Bermuda. *J. Mar. Res.* **47**, 169–96.

Stuiver, M., P.D. Quay and H.G. Ostlund (1983) Abyssal water carbon-14 distribution and the age of the world oceans. *Science* **219**, 849–51.

Toggweiler, R. and J. Sarmiento (1985) Glacial to interglacial changes in atmospheric carbon dioxide: the critical role of the ocean surface water in high latitudes. In *The Carbon Cycle and Anthropogenic CO_2* (ed. E. Sundquist and W. S. Broecker), pp. 163–84. Washington, D.C.: American Geophysical Union.

Toggweiler, R.J., K. Dixon and K. Bryan (1989) Simulations of radiocarbon in a coarse-resolution world ocean model I. Steady state prebomb distributions. *J. Geophys. Res.* **94**, 8217–42.

Varela, D.E. and P.J. Harrison (1999) Seasonal variability in nitrogenous nutrition of the phytoplankton assemblages in the northeastern subarctic Pacific Ocean. *Deep-Sea Res. II* **46**, 2505–38.

Williams, P.J. leB. and D.A. Purdie (1991) *In vitro* and *in situ* derived rates of gross production, net community production and respiration of oxygen in the oligotrophic subtropical gyre of the North Pacific Ocean. *Deep-Sea Res.* **38**, 891–910.

Paleoceanography and paleoclimatology

We end Part I of this book with a study of past changes in the Earth's atmosphere, oceans and ice volume. Interpretation of past climatic conditions from chemical tracers and isotopes preserved in the geological record requires knowledge and intuition developed from the study of present-day oceanography. For this reason descriptions of how paleoceanographic tracers are used to unravel insights about past ocean circulation and biogeochemistry serve as a review of the geochemical perspectives presented in the first six chapters of this book.

Present human activities create chemical sources to the environment that are, in some cases, comparable to those of the natural (pre-industrial) Earth. Since some of these anthropogenic additions may affect the natural order of the Earth's climate system, it is urgent to understand how the natural system functions mechanistically. For example, the effect of rapidly rising anthropogenic atmospheric CO_2 on the climate system of the Earth is a first-order question (see Chapter 11). Even though global models that incorporate physical and biological interactions among the atmosphere, ocean, terrestrial and ice "spheres" now allow scientists to recreate the Earth's system, reasons for even the most first-order observations of climate change during the past million years are still poorly understood. The waxing and waning of glacial ice with roughly a 100 ky cycle is likely triggered by variations in the amount of solar energy reaching the

Earth's surface because of changes in the Earth's orbital motions around the sun. The timing and magnitude of these solar insolation variations are accurately understood (the Milankovitch cycles, see later), but it is widely believed that these variations are too small to cause the major changes in climate characterized by the ice ages. There must be feedbacks that amplify the climate response to relatively small changes in incoming solar radiation. It is critical to understand these feedbacks well enough to insure early awareness of anthropogenic changes that might be detrimental to climate and the habitability of our planet.

At present, interactions that control climate are too complicated to understand even with the most sophisticated models. One seeks to improve the prediction skill of these models by attempting to understand how the Earth has responded to natural changes in the past. To this end, paleoceanographers and paleoclimatologists attempt to understand the complexities of the climate system by reconstructing past climate variability from natural archives such as sediment and ice cores, tree rings, and corals. Although this endeavor covers time scales that span the past one billion years, our goal here is to present an introduction to the milestones of paleoceanography and paleoclimatology research over the most recent ice age cycles with an emphasis on the past glacial–interglacial transition.

Paleoceanography and paleoclimatology have become increasingly focused on the more recent past because the history of these intervals can be resolved on decadal to centennial time scales, which are relevant to human history and environmental change. The bulk of this chapter deals with background information about marine sedimentary and ice core records of climate change. This is followed by a short review of information about millennial-scale changes derived from both the ocean and ice core records. Our attempt to summarize the high points of this increasingly vast field in one chapter necessarily requires that many important discoveries be omitted. For more detail the reader is referred to some excellent books about this subject (Alley, 2000; Broecker, 2002; Kennett *et al.*, 2003; Ruddiman, 2001).

7.1 The marine sedimentary record: 0–800 ky

The Cenozoic Era (*c.*65 million years BP to the present) encompasses the Tertiary and the Quaternary Periods. During the Tertiary the Earth's climate began an overall cooling trend of about 12 °C in the past 40 million years. Over the past two and a half million years the climate has varied from cool to warm periods, accompanied by massive expansions and contractions of the polar ice caps. This period of climate fluctuation is termed the Quaternary Period and spans the geologic time scale from the end of the Pliocene Epoch, roughly 1.8–2.6 million years ago, to the present. The Quaternary Period includes the Pleistocene and Holocene Epochs, with the Holocene beginning a time of relatively uniform climatic conditions at the end of the last glacial excursion (*c.*11 000 years ago).

Reconstruction of the record of climate change preserved in marine sediments began with observations of how the abundance of certain species of planktonic Foraminifera, identified by the morphology of their calcareous tests (Fig. 1.13), waxed and waned with depth (and time) in sediment cores raised from the ocean floor. Since geological oceanographers knew the temperature of the environment in which different foraminiferan species presently live, it was possible to suggest how the surface temperature changed in the past based on relative compositions of planktonic foraminiferan assemblages preserved in sediments. As interest in the evolution of Earth's climate grew, the subject evolved from the study of faunal changes to the more quantitative investigation of isotopes and chemical tracers preserved in microfossils and sediments.

7.1.1 Ice volume and temperature change during the Pleistocene

The single most influential contribution of chemical tracers to the study of past climate change has been the investigation of stable isotope ratios of oxygen and carbon in $CaCO_3$ tests of planktonic and benthic Foraminifera. Caser Emiliani, who learned the study of isotopes from Harold Urey at the University of Chicago, was the first to make measurements of stable oxygen and carbon isotopes in $CaCO_3$ tests (shells) of planktonic Foraminifera preserved in deep ocean sediment cores. Early isotope studies by Emiliani and others led to the identification of eight saw-tooth-like cycles of oxygen isotope change over the past *c.*800 ky, having a temporal periodicity of roughly 100 ky and a magnitude of 1‰–2‰. These cycles were discussed briefly in Chapter 5 on isotopes and are reproduced here (Fig. 7.1). Emiliani attributed the change in the $^{18}O:^{16}O$ ratio to lower temperature in surface waters of glacial times in which the planktonic Foraminifera grew, since it was known that the chemical equilibrium relation between the oxygen isotope ratio of water and $CaCO_3$ is temperature-dependent (Fig. 5.3). Indeed, Urey had used the annual changes in the oxygen isotope ratio of a fossil belemnite (Fig. 5.4) to predict the winter–summer differences in temperature in South Carolina during the Cretaceous Period (135–60 million years ago). Using the calcite–water oxygen isotope chemical equilibrium temperature relations (Eq. (5.9)), a 1.5‰ variation in $\delta^{18}O$ of $CaCO_3$ corresponds to a surface water temperature lowering of about 6 °C in the tropical ocean during glacial periods, which seemed reasonable.

It was soon realized, however, that not only did the temperature of the water in which the Foraminifera grew change, but also the isotope ratio of the water probably changed as well. Through the study of ancient sea level terraces, it became evident that sea level was about 100 m lower when ice sheets were at their maximum volume. Since the $\delta^{18}O$ of polar ice is presently on average 30‰–40‰ lighter than seawater (ice *c.* − 40‰; seawater 0‰), accumulation of

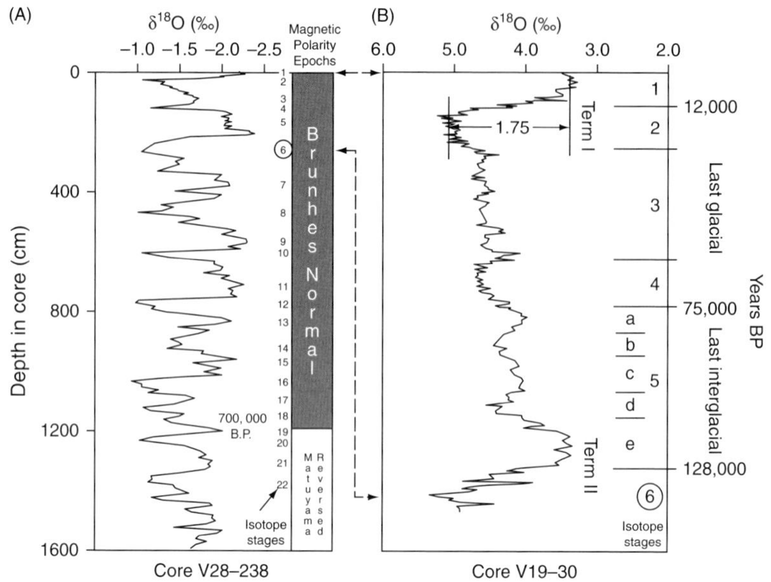

Figure 7.1. The oxygen isotope record recorded in foraminiferan tests in deep sea sediments. The changes are caused by temperature changes and the waxing and waning of glacial ice during the past c.1 million years. The core on the left (A) is a record of planktonic Foraminifera V28–238 from the Equatorial Pacific (Shackleton and Opdyke, 1973). The core on the right (B) is also from the Equatorial Pacific but the isotope data are on benthic foraminiferan tests from V19–30 (Chappell and Shackleton, 1986). The $\delta^{18}O$ scales are different because the planktonic–benthic difference is c. 5‰. Redrawn from Crowley (1983) and Broecker (2002).

glacial ice represents a large enough sequestration of light water to affect the $\delta^{18}O$ of seawater. A rough idea of the magnitude of change can be readily evaluated from the fractional change in ocean volume and the isotope difference between seawater and polar ice. During glacial periods roughly one fortieth of the ocean's water was stored as ice (the mean depth of the ocean is 3800 m, so $100/3800 \approx 1/40$). If we assume this ice was 40‰ lighter than seawater, then this transformation of H_2O from a liquid to a solid state left the ocean *c.*1‰ enriched in $\delta^{18}O$ during glacial periods. This value is in the same direction as the observed $\delta^{18}O$ change in foraminiferal calcite between glacial and interglacial times and is clearly great enough to be potentially important. The problem with our simple calculation is that the mean $\delta^{18}O$ of glacial ice is uncertain.

As mass spectrometers improved, the sample size necessary for the analysis of oxygen isotopes decreased to the point that it was possible to determine the isotope ratio of benthic Foraminifera, which are typically much less abundant in marine sediments than planktonic Foraminifera. In the early 1970s, Nick Shackleton (Shackleton and Opdyke, 1973) showed that the $\delta^{18}O$ changes in planktonic and benthic Foraminifera from the same core did not differ greatly (Fig. 7.2). Of course, it is possible that the isotope changes in the planktonic and benthic foraminiferan tests are equally affected by a temperature change, but this seemed unlikely for the benthic Foraminifera because deep water temperatures are *c.*1 °C, and seawater freezes at *c.* − 2 °C. Since changes in the amount of glacial ice would affect planktonic and benthic foraminiferan $\delta^{18}O$ equally and the magnitude of $\delta^{18}O$ change attributed to ice volume corroborated other studies of sea level change, the analysis in Fig. 7.2 led the community to suspect that much of the observed oxygen isotope

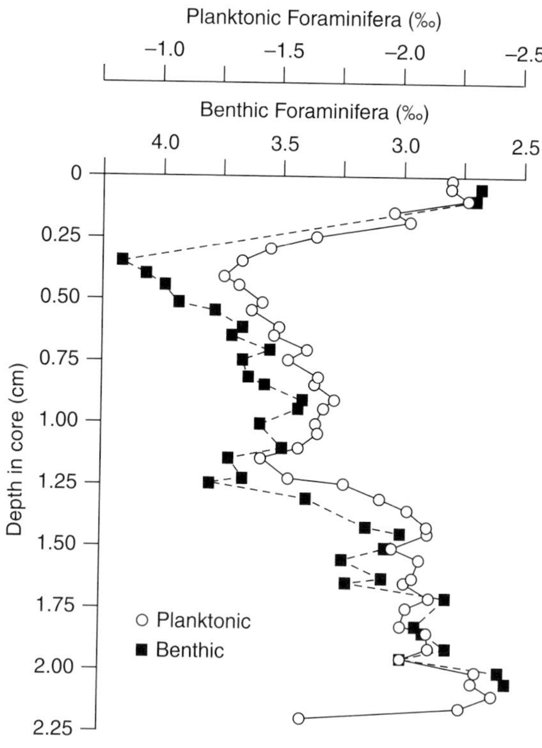

Figure 7.2. The $\delta^{18}O$ in $CaCO_3$ tests of planktonic and benthic Foraminifera as a function of depth in a sediment core from the Equatorial Pacific, core V28–238. The scales are the same but offset by 5.3‰, the present-day planktonic–benthic difference. Modified from Shackleton and Opdyke (1973).

change in foraminiferan $CaCO_3$ was caused by variations in ice volume and that temperature effects were difficult to quantify and probably of second order.

Recently, progress has been made in resolving the mechanisms that produce the $\delta^{18}O$ changes observed in benthic foraminiferan tests by actually measuring the $\delta^{18}O$ change of ocean bottom water over the last glacial–interglacial period. This was accomplished by determining the $\delta^{18}O$ of interstitial waters of marine sediments in very long sediment cores recovered by the Ocean Drilling Project (ODP) (Adkins and Schrag, 2001). As sediments accumulate on the sea floor the ambient water is also incorporated. The water exchanges by molecular diffusion with the overlying bottom water as the sediments accumulate, but since molecular diffusion is a relatively slow process the signature of the $\delta^{18}O$ content of previous times is not totally erased. The observed porewater $\delta^{18}O$ changes are much smaller than at the time the waters were being incorporated into the sediments because of the smoothing by molecular diffusion. Fortunately, the effect of molecular diffusion in changing the signal preserved in the sediment interstitial waters is readily predictable. The $\Delta\delta^{18}O$ values on the curves of Fig. 7.3A indicate the glacial–interglacial differences, which were applied as boundary conditions for the differential equations that generated the model profiles shown in the figure. Clearly, most of the data are adequately fit with a bottom water $\delta^{18}O$ difference between today and the last glacial maximum

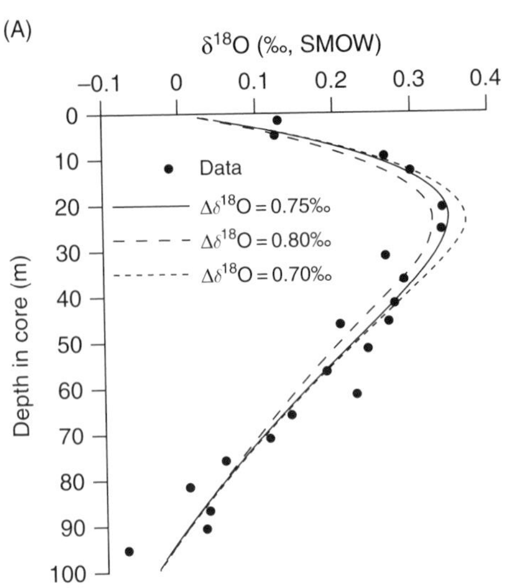

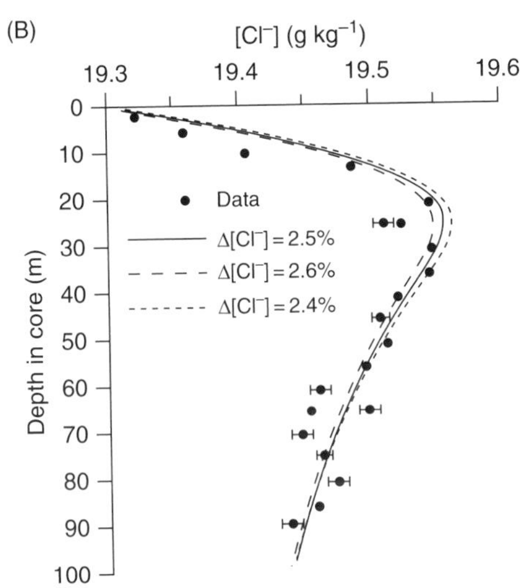

Figure 7.3. δ^{18}O-H_2O (A) and [Cl^-] (B) from interstitial waters of an Ocean Drilling Project core from the Bermuda Rise. Data from core depths of 20–30 m are higher because seawater [Cl^-] and δ^{18}O-H_2O were greater in bottom waters buried with sediments during the last glacial maximum. Different lines indicate model results for various Holocene–Glacial Maximum changes in the bottom waters. Redrawn from Adkins and Schrag (2001).

of 0.75‰. Adkins and Schrag estimate that the average global change in δ^{18}O determined from the porewaters of geographically diverse ODP cores was about 0.95‰. The average global change in δ^{18}O of benthic foraminiferan tests between the Holocene and last glacial maximum from about 20 globally distributed cores is 1.75‰ (Broecker, 2002). Thus the seawater isotope change is 0.95/1.75 = 54% of the glacial–interglacial change observed in benthic Foraminifera, leaving about half of the signal to be accounted for by temperature change. Interpretation of the glacial–interglacial change in the δ^{18}O of foraminiferan calcite has evolved from one in which the origin was perceived to be primarily due to temperature change, to one due primarily to ice volume change, to being a roughly even split between these two forcings.

The porewater and benthic foraminiferan oxygen isotope measurements suggest that deep ocean temperatures were 3–4 °C lower during the last glacial maximum. The results in Fig. 7.3 indicate that, at this location, the temperature in the last glacial period was − 1.8 °C, which is indistinguishable from the freezing temperature of seawater! The porewater measurements also imply that temperature changes in the ocean's surface waters recorded by planktonic Foraminifera are closer to 2–3 °C than the 6 °C originally predicted by Emiliani.

These conclusions have been corroborated by other geochemical tracers archived in marine sediments. The two most prominent independent geochemical temperature tracers are the Mg:Ca ratio of foraminiferan $CaCO_3$ and organic lipid molecules called alkenones that are created primarily by coccolithophorid phytoplankton. It has been shown that the Mg:Ca ratio incorporated into the $CaCO_3$ of foraminiferan shells varies with the temperature in which the shells grew. This ratio is faithfully preserved as long as the $CaCO_3$ shells are not partly dissolved during burial. Similarly, the degree of unsaturation of alkenone molecules has been shown in modern ocean samples to correlate with the temperature in which coccolithophorids grow (see Fig. 8.13). Degradation of organic molecules during burial in sediments has been shown to have little effect on the degree of undersaturation over 10–100 ky time scales. The Mg:Ca ratio in the $CaCO_3$ of the planktonic Foraminifera from the Equatorial Pacific suggests a change of 2–3 °C between the last glacial period and the Holocene (Lea *et al.*, 1999); alkenones preserved in sediments from the western tropical Atlantic Ocean indicate a temperature change of about 3 °C (Ruhlemann *et al.*, 1999).

A direct check on the conclusions of deep water temperature change implied by the porewater and benthic foraminiferan $\delta^{18}O$ measurements comes from measurements of the changes in the Mg:Ca ratio in benthic foraminiferan tests from the deep Pacific and Atlantic Oceans. The changes in the Mg:Ca ratio over glacial–interglacial times also suggest that the deep water temperature changed by *c.*4 °C during this period (Martin *et al.*, 2002).

7.1.2 Dating the marine sedimentary archives

Marine sediments are useful for interpreting climate history because in some areas the accumulation rate is controlled by the relatively constant and quiescent sinking of marine biogenic and terrestrial detritus to make a continuous and smooth historical record. Unfortunately these ideal conditions are often not the case, particularly on continental margins, where accumulation rates are elevated and where the potential to retrieve high-resolution paleoclimate records is greatest. In these regions deposition rates change over time and horizontal movement of sediment over the sea floor can cause some areas to be eroded while other areas accumulate material intermittently and at varying rates. Thus, interpretations of the isotope and faunal records require accurate knowledge of past sediment accumulation rates.

A first-order understanding of the age of sediments is achieved by locating in the sediments signatures of events with known time histories, for example volcanic ash layers or magnetic reversals. The Earth's magnetic field has reversed suddenly and intermittently on geologic time scales. These changes are recorded in marine sediments by the alignment of magnetically susceptible minerals, which can be measured. The Brunhes–Matuyama magnetic reversal, dated radiometrically in lava flows on land at 700 ky BP, has been an important age datum for placing Quaternary $\delta^{18}O$ cycles in deep sea sediments in a temporal perspective (Fig. 7.1). If marine sediment accumulation rates were perfectly uniform this would be all that is necessary to establish the age of the sediments over this period, but of course they are not. Accumulation rates of sediments in some areas varied dramatically between glacial and interglacial periods, so estimates of sedimentation rates on time scales much shorter that 700 ky are necessary. Natural radiotracers with half lives that fit the time scales of interest (0–100 ky) and have the appropriate chemistry are ^{14}C (half life 5730 y) and the uranium-series isotope, ^{230}Th (half life *c*.75 ky). The sources and sinks and chemistry of both of these isotopes are discussed in Chapter 5 (see, for example, Figs. 5.18 and 5.20). Here we describe the application of these two isotopes for determining marine sediment chronology in the Quaternary Period.

If we assume a coordinate system in which the sediment water interface is at zero depth, z, with positive downward (i.e. the boundary is moving with respect to the center of the Earth as the sediment accumulates), the change in property, p, at a given depth, x, below the sediment–water interface is given by:

$$\left(\frac{\partial p}{\partial t}\right)_{z=x} = \frac{\mathrm{d}p}{\mathrm{d}t} - s\left(\frac{\partial p}{\partial z}\right)_t. \tag{7.1}$$

This equation states that the change in p with respect to time at a depth, x, below the sediment–water interface is equal to the total derivative of p with respect to time minus the flux of p transported by sedimentation rate, s (cm y^{-1}) along gradient $(\partial p/\partial z)$ at time, t. The total derivative refers to factors that change p as a function of t in a layer that is stationary with respect to the Earth's center and moves away from the interface ($z = 0$) with velocity equal to the sedimentation rate, s. A steady-state condition is one in which the property p does not change at a given depth below the interface so the left side of Eq. (7.1) is zero. The property p can change as functions of t and z, but must be constant at a given depth below the sediment–water interface. In this case the change of p depicted by the total derivative is exactly balanced by the deposition of p from above:

$$0 = \frac{\mathrm{d}p}{\mathrm{d}t} - s\left(\frac{\partial p}{\partial z}\right)_t. \tag{7.2}$$

This is the general equation for steady-state diagenesis without compaction or bioturbation. Consult Berner (1980), Boudreau (1997) or Burdige (2006) for elaboration and variations on this simple equation. If we substitute the concentration of a radioisotope ([C], atoms cm^{-3})

for p and the decay rate, $\lambda[\mathrm{C}]$ (atoms $\mathrm{cm}^{-3}\,\mathrm{y}^{-1}$) for the total derivative, one arrives at the general equation for determining sediment accumulation rates, s, by using radioisotope tracers.

$$0 = s\left(\frac{\partial[\mathrm{C}]}{\partial z}\right) - \lambda[\mathrm{C}]. \tag{7.3}$$

This relatively uncomplicated equation has the following solution for the initial condition in which the concentration is $[\mathrm{C}_{t=0}]$:

$$\ln\left(\frac{[\mathrm{C}]}{[\mathrm{C}_{t=0}]}\right) = -\left(\frac{\lambda}{s}\right)z. \tag{7.4}$$

The reader may recognize the similarity between this equation and one relating the radioisotope concentration to time (Eq. (5.18)). The left side is the same, but the right side of Eq. (5.18) had time, t, in the place of (z/s). Thus, the age, t, determined by the fraction of the radioisotope that remains is equal to z/s, which is also equal to the left-hand side of Eq. (7.4) divided by the decay constant, λ:

$$\text{age} = \ln([\mathrm{C}]/[\mathrm{C}_{t=0}])\frac{1}{\lambda} = \frac{z}{s}. \tag{7.5}$$

Examples of the application of these equations to the activities of $^{14}\mathrm{C}$ and $^{230}\mathrm{Th}$ in marine sediments are presented in Figs. 7.4 and 7.5. $^{14}\mathrm{C}$ with a half life of 5730 y is appropriate for dating sediments younger than 4–5 half lives or up to c.30 ky BP. This includes the late Quaternary through the Holocene. Profiles of $^{14}\mathrm{C}$ versus depth for the carbonate-rich sediment cores shown in Fig. 7.4 reveal a surface layer of constant age in the top 5–10 cm followed by linear decreases with depth that reflect sedimentation rates of 0.25 and 1 cm ky^{-1}. The uniform top section is mixed by bioturbation (mixing

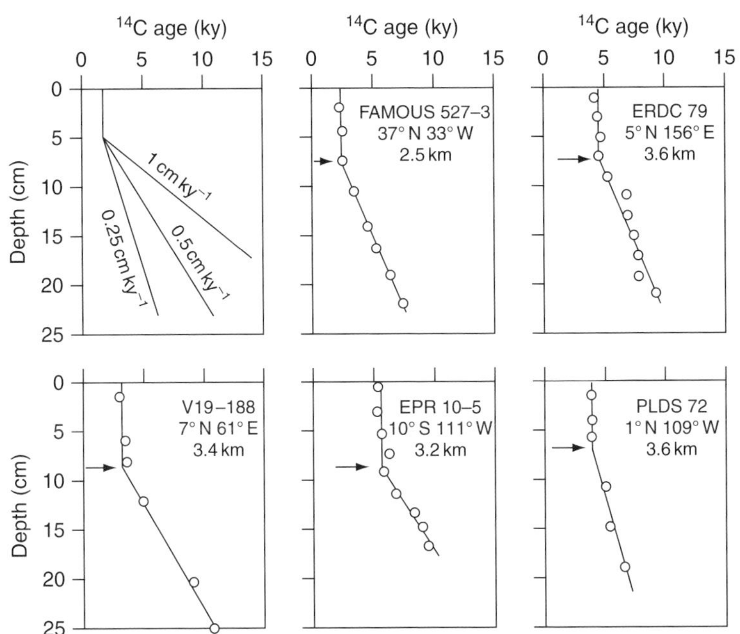

Figure 7.4. The carbon-14 age of calcite-rich deep sea sediments as a function of depth in the sediments at different locations in the ocean. Expedition names and core numbers are indicated along with the core location and water depth. Data (circles) indicate a 5–10 cm $^{14}\mathrm{C}$ mixed layer in the surface sediments everywhere in the deep ocean due to bioturbation by benthic fauna. The figure in the upper left illustrates the idealized relation between profile slope and sedimentation rate. Redrawn from Peng and Broecker (1984).

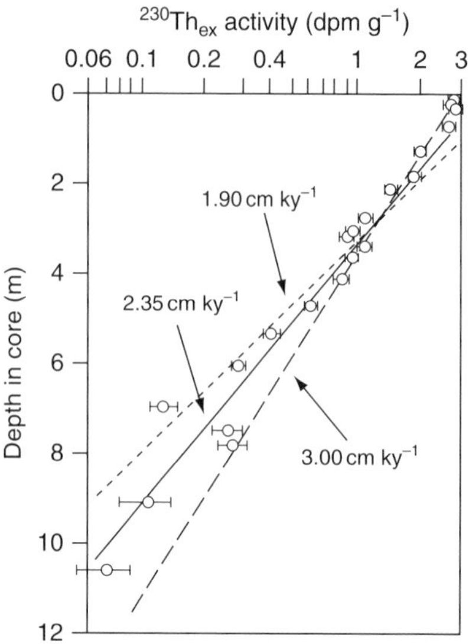

Figure 7.5. The ^{230}Th excess activity as a function of depth in the top 10 m of a sediment core from the Caribbean Sea. Different sedimentation rates are indicated for a model that assumes a continuous and constant sediment accumulation (see text). Redrawn from Ku *et al.* (1972).

by benthic animals) to a depth of 5–10 cm in deep-sea deposits. This mixing depth is deeper in continental-margin sediments, but probably relatively uniform in the deep sea because of the size and burrowing habits of benthic fauna there. Bioturbation has ramifications for the utility of any tracer in marine sediments to resolve temporal changes because it acts as a smoothing filter with a time scale equal to the mean age of the bioturbated layer. For sedimentation rates of 1 cm ky^{-1} the filter provides a running mean of about 7 ky; for sedimentation rates of 0.1 cm ky^{-1}, characteristic of $CaCO_3$-poor regions of the North Pacific, it is a 70 ky filter! Clearly one must sample sediments with accumulation rates significantly faster than 1 cm ky^{-1} in order to investigate millennial-scale climate change.

Another uncertainty for the application of ^{14}C to determining the age of deep-sea deposits is that the initial value $[C]_{t=0}$ in Eqs. (7.4) and (7.5) is not constant. As discussed in Chapter 5, ^{14}C is produced in the upper atmosphere by cosmic ray spallation (bombardment of atmospheric molecules with high-energy cosmic-ray protons). Production rates have changed over the past millennia because of variations in the Earth's magnetic field, which shields the atmosphere from incoming cosmic rays, and because of changes in the ventilation of the deep sea. Calibration of the initial value for ^{14}C has been determined by a number of techniques that allow both ^{14}C and independent dating to be achieved. Among the most successful of these methods are the comparison of the age of very old tree rings with ^{14}C dates of the wood, simultaneous measurements of ^{230}Th and ^{14}C in ancient corals, and measurements of ^{14}C in planktonic foraminiferan tests from varved sediments of the Cariaco Trench, an anoxic basin off Venezuela. In all cases the "calendar" age is determined along with the ^{14}C activity precisely and independently so the Δ^{14}C activity of the sample can be adjusted for radioactive decay to determine the activity of the sample when it was deposited.

Comparison of tree-ring age and ^{14}C age is limited to the time scale of trees (*c*.5000 y), which is much shorter than the useful life of ^{14}C of about 30 ky. Extending the calibration of the initial value for ^{14}C back in time became possible when it became technologically feasible to compare the ^{14}C and ^{230}Th age of corals. Corals are ideal for both ^{14}C and ^{230}Th dating because they are made of aragonite, a form of $CaCO_3$ that has a relatively open structure and incorporates high concentrations of U, but very little Th. Thus, as long as corals do not exchange U or Th with the surrounding seawater after they die, virtually all of the ^{230}Th present in the aragonite is derived from ^{234}U decay. Because of the relatively long half life of ^{230}Th, accurately dating samples with ages of only a few thousand years is not possible by radioactive decay counting and became feasible only when analytical methods were developed for measuring tiny amounts of ^{230}Th in coral samples by using mass spectrometry.

Since the life time of corals is only several hundred years, a long-term record was obtained by sampling submerged corals on the fringes of tectonically stable shorelines. During the last glacial period sea level was approximately 100 m lower, so corals grew on the shoreline between 0 and *c*.100 m during the transition from the last glacial maximum to the Holocene. These fossil corals are now submerged, so they were sampled by a series of cores from the sea surface to more than 100 m depth. Dating of both ^{14}C and ^{230}Th on cores from off the Bahamas Islands revealed ages that are nearly the same in corals less than 10 ky old, but the two ages diverged by up to 3 ky in corals of age 20 ky (Bard *et al.*, 1990). Since it is difficult to suggest a reason why the ^{230}Th ages could be in error, it has been assumed that the production rate of ^{14}C is the reason for the difference.

An independent, continuous and longer record of the initial value for atmospheric $\Delta^{14}C$ has been obtained by dating planktonic Foraminifera from the sediments of the Cariaco Trench (Fig. 7.6) (Hughen *et al.*, 2004). Since the sediments of this anoxic basin are varved, the age filter applied to most sediment cores by bioturbation is not an issue. Calendar ages of the varves in the sediments of this basin were determined by matching the percent reflectance (a measure of the color of the sediments) with $\delta^{18}O$ variations in the ice of a Greenland ice core (described later in Fig. 7.19). Since the latter record is precisely dated back to 40 000 years by actual counting of annual ice layers, and the two records are undeniably correlated, it was possible to determine an accurate "calendar age" for the Cariaco Trench sediment core by using variations in the percent reflectance record. The results in Fig. 7.6 indicate offsets of up to 5 ky between ^{14}C age and calendar age at about 30 ky BP and an abrupt shift at 40 calendar kiloyears (cal. ky) BP in which 7000 ^{14}C years elapsed in only 2000 y. The results have been explained as variations in the source function and the ventilation of the deep sea and are now used to correct ^{14}C dates back to more than 40 cal. ky BP.

Another application of the isotope ^{230}Th has been to date marine sediments that span a full 100 ky glacial cycle. In this case bulk

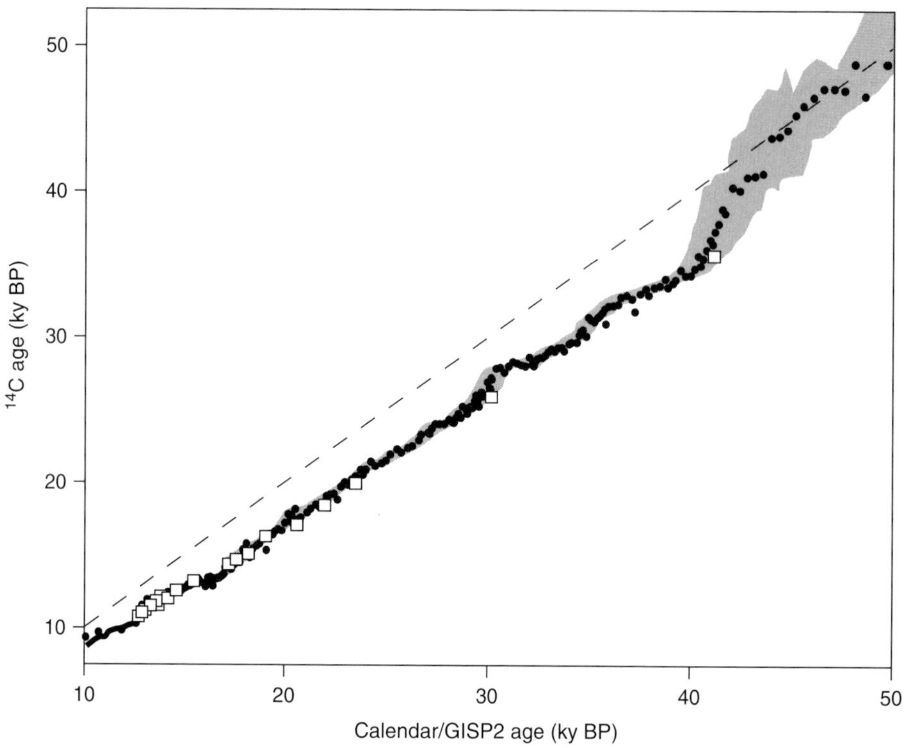

Figure 7.6. $\Delta^{14}C$ activity versus calendar age of sediments from the Cariaco Trench. The carbon-14 dates illustrated by the dots were determined on the shells of planktonic Foraminifera preserved in the sediments. Calendar ages were determined by correlating sediment reflectance measurements from the cores with $\delta^{18}O$ changes in the GISP2 ice core from Greenland (see Fig. 7.19). The shaded region represents an estimate of the error due to calendar age uncertainty. The solid line spanning the calendar age 10–12 ky BP is from tree rings and the open squares are from paired ^{232}Th and ^{14}C ages on corals. Redrawn from Hughen *et al.* (2004).

sediments are used rather than only $CaCO_3$ shells, so there is enough ^{230}Th available to make the measurement by using radioactive decay techniques. As stated earlier, ^{230}Th is a daughter product of ^{234}U, which is relatively soluble in seawater and conservative with respect to chemical reactions so its concentration is nearly the same everywhere in the ocean. Thorium, however, is quite insoluble and adsorbs readily to particles that rain through the water column and come to rest on the sea floor (see Fig. 5.20). The result of these different chemistries is that sediments have very low ^{234}U activity compared with that of ^{230}Th. The ^{230}Th in the sediment is free to decay until it reaches the background activity of ^{234}U in the sediments, at which time it is *supported* by and in secular equilibrium with its parent, ^{234}U (see the discussion of secular equilibrium in Chapter 5). The systematics of Eq. (7.4) have been applied to measurements of *excess* (or *unsupported*) ^{230}Th in long cores from $CaCO_3$-rich sediments. An example is the 10 m long Caribbean core shown in Fig. 7.5, which has a mean sedimentation rate below 10 cm of between 2 and 3 cm ky^{-1}. It may be that there are larger changes in accumulation rate on time scales that are not resolved by this technique, which points out the relatively large errors that are inherent in the method. None the less, it is the only direct method for determining the age of the penultimate interglacial period (100–125 ky BP).

A natural extension of the ^{230}Th method is to use the inventory in sediments as an indication of the source of sediment material

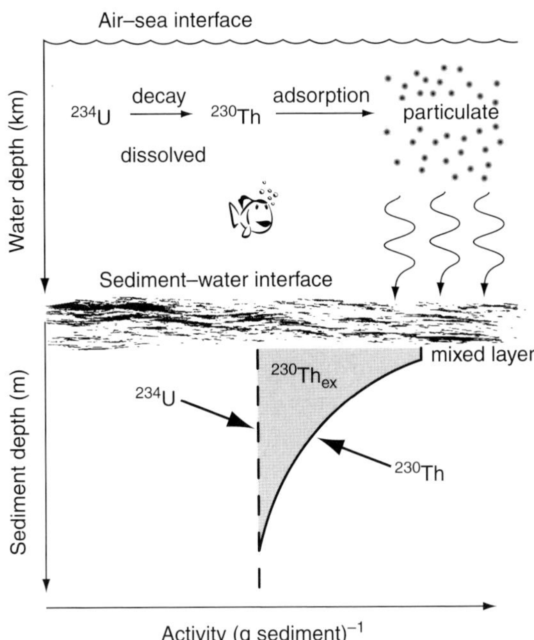

Figure 7.7. A schematic illustration demonstrating the relation between dissolved ^{234}U and its daughter ^{230}Th in the ocean. Uranium is dissolved in the water, but thorium is quantitatively adsorbed to particles and removed to the sediments. At secular equilibrium the integrated activity of excess ^{230}Th in the sediment should equal the depth-integrated activity of ^{234}U on the overlying water.

(Fig. 7.7). As long as one can assume that the source of particulate Th is from the overlying water column, which has been demonstrated to be accurate away from the ocean's margins, then the inventory of the ^{230}Th activity in the sediments must be equal to the ^{234}U activity in the water column overlying the sediments. This is calculated simply by multiplying the ^{234}U activity by the water depth. (Since ^{234}U is conservative in the ocean it holds a constant ratio to salinity and the activity of ^{234}U, A_{234}, is 2.84 dpm kg^{-1} at $S = 35$.) If the ^{230}Th inventory is larger (or less) than the expected inventory, then the location of study is receiving (donating) sediment from (to) locations that are laterally adjacent. This method has allowed climate scientists to correct sediment accumulation rates as a function of time for the influence of changes in horizontal transport. An example of how sediment redistribution might complicate interpretation of the sedimentary record is in the interpretation of changes in mass accumulation rate of $CaCO_3$, opal, or organic carbon between glacial and interglacial times. The difference can be interpreted as either a change in the rain rate from above, which indicates a change in the production rate in the euphotic zone, or a variation in the direction and strength of deep water circulation which redirected or focused sediments into or out of the area. From the point of view of climate, one would like to interpret the sedimentary record in terms of what was happening in surface waters, so any change due to sediment focusing should be normalized. This can be done by assuming that the integrated activity of ^{230}Th in the sediments derived from the above water column must equal the production rate in the overlying water. Correcting accumulation rates to conform to this constraint eliminates falsely interpreting a change in benthic sediment focusing or erosion as a change in input from the overlying euphotic zone.

Although radioisotope methods are the ultimate tracer of sediment ages over the past 100 ky, presently the most commonly used dating method for filling in the details and extending the chronology in marine sediments is an indirect one based on the known timing of changes in solar insolation throughout glacial time. Orbital cycles that influence the amount of heat reaching the surface of the Earth were calculated by the Yugoslavian mathematician Milutin Milankovitch in the 1920s and 1930s and are thus referred to as Milankovitch cycles. Years of study have shown that mechanisms (ice volume and temperature changes) controlling the $\delta^{18}O$ cycles in marine sediments (Figs. 7.1 and 7.2) are correlated with the changes in the amount of solar energy that reaches the surface of the Earth. Naturally there must be some lag between these two, and this lag is assumed to have been constant with time throughout the Quaternary. Spectral analysis of foraminiferan $\delta^{18}O$ records covering many 100 ky cycles from different locations in the ocean has shown that these data are composed of regular cycles with periods corresponding to changes in Earth's orbital parameters, which control the amount of sunlight that reaches the Earth's surface. The timing of these cycles is accurately known over *c.*30 million years. Thus, foraminiferan $\delta^{18}O$ cycles can be dated by matching them with patterns in the change of solar insolation. Orbital-induced variations in solar insolation are believed to be the primary forcing for climate change, but the magnitude is probably insufficient to cause vast ice sheets to grow and recede. Consequently, feedbacks in the Earth's climate system are required.

There are primarily three properties of the Earth's movement around the sun that influence the strength of the seasons and thus affect the amount of the sun's energy that reaches the Earth's surface during different times of the year. Since they are described in great detail in many books about climate change (e.g., Ruddiman, 2001; Broecker, 2002), we will be very brief here. The first repeating temporal variation in the amount of solar energy that reaches the Earth is due to changes in obliquity: the tilt of the spin axis of the Earth with respect to the plane of its orbit. Seasons on Earth exist because the planet is presently 23.5° out of vertical with respect to the orbital plane around the sun. The direction of the tilt does not change perceptibly during the annual circuit, so half of the time the North Pole is tilted toward the sun and more sunlight reaches the Northern Hemisphere, causing summer in the northern half of the Earth and winter in the southern half. The opposite occurs when the South Pole is tilted toward the sun. Temporal variation in the strength of the seasons results from the fact that the tilt of the Earth has varied between 22.2° and 24.5° with a periodicity of 41 ky, primarily because of the gravitational tug of large planets such as Jupiter.

The second cycle that affects the amount of the sun's heat that reaches Earth seasonally is the eccentricity of the Earth's orbit around the sun. Earth–sun distances presently vary annually between 153 and 158 million km, because the orbit around the sun is not a perfect circle. The eccentricity of an orbit has varied in Earth's history again because of the gravitational pull of the larger planets, from nearly perfectly round to much greater distance extremes than today, with the primary periodicity being *c.*100 ky.

The final orbital cycle is caused by the combination of the precession of the Earth's spin axis and the precession of the Earth's orbit around the sun. The tilt of the Earth with respect to the plane of its orbit around the sun precesses like a top owing to the gravitational pull of the sun and moon on the slight bulge of the Earth's diameter at the Equator. The precession cycle is 25.7 ky and affects the timing of the seasons; the dates of the summer and winter solstices (the longest and shortest days of the year, respectively). This precession by itself causes no change in the amount of the sun's radiation that reaches the Earth seasonally. It only changes the timing of the solstices; however, this motion combines with the precession of the Earth's orbit around the sun to cause a change in the strength of the seasons. Not only is the Earth's orbit elliptical, but the orbit itself rotates in space such that the long end of the ellipse makes one full revolution around the Sun in 105 ky. The combination of the precession of the Earth's spin axis and the precession of the Earth's orbit around the Sun causes a variation in the amount of radiation that reaches the Earth seasonally with cycles of between 19 and 23 ky.

A simple combination of sine curves with periods of 23, 41 and 100 ky (Fig. 7.8) illustrates how difficult it is to deconvolute the cycles by eye. This problem is compounded in natural samples because of their imperfect nature. Climate scientists use spectral analysis filtering as a tool to deconvolute the $\delta^{18}O$ records into their predominant cyclic components. When this is done the location of the main 23, 41 and 100 ky cycles can be identified (Fig. 7.9) and assigned ages. Because the 100 ky forcing is weak, climate scientists primarily rely on the 23 and 41 ky cycles to date $\delta^{18}O$ records. With the 41 ky cycle alone, one can match the cycles, but it is difficult to know which cycle is which because the amplitude of each is the same.

This problem is mitigated by the combined cycles caused by eccentricity (100 ky) and precession of the Earth's spin axis and solar orbit (23 ky). These two cycles combine to create 23 ky cycles that vary in amplitude with a 100 ky periodicity (eccentrically modulated precession cycles). This effect is illustrated by the top curve in Fig. 7.8B. The 23 ky cycles and the 100 ky modulations can be extracted from the longer $\delta^{18}O$ records to distinguish where the 23 ky cycles belong in time.

Dating of marine sediments has relied on a combination of radioactive tracers and stable isotope tracers tuned to known cycles in orbital forcing. The latter process involves comparing the peaks and valleys of a standard $\delta^{18}O$ time series generated from averaging $\delta^{18}O$ curves from different areas in the ocean. This "wiggle matching" is

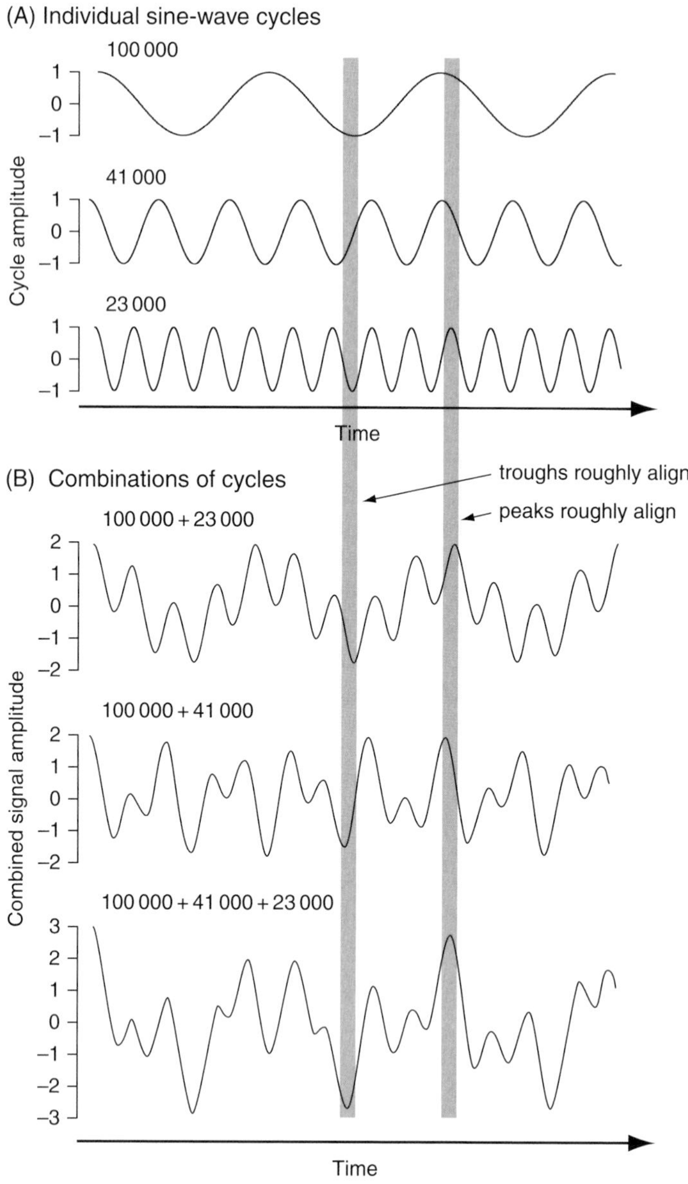

Figure 7.8. Demonstration of the complex curves that result from combination of sine waves with amplitudes of 100, 41 and 23 ky, the three predominant Milankovitch cycles. Combinations of the three sine curves illustrate the difficulty in backing out cycles from the data. The compilation here does not look like $\delta^{18}O$ vs. depth curves in ocean sediments because the amplitudes are the same in the illustration. Redrawn from Ruddiman (2001).

the least expensive and most often the first cut in determining whether a core has a continuous temporal record. Dating by this method is approximate because local changes in sedimentation rate cause the distance between peaks and valleys to expand or contract and become obscured in some cases. Absolute dating over the past *c.*35 ky still relies on ^{14}C, which means it is presently difficult to obtain accurate ages in sediments that do not have $CaCO_3$ fossils or are older than 30 ky. New procedures of determining ^{14}C dates on individual organic compounds of bulk sediment or locked in the shells of diatoms show promise for expanding the ^{14}C method to all areas of ocean sediments.

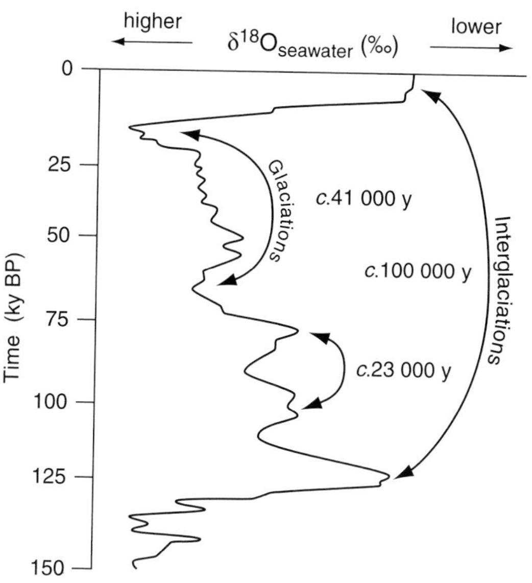

Figure 7.9. An illustration of the 23, 41 and 100 ky cycles imbedded in an idealized curve of $\delta^{18}O$ versus age from deep sea sediments. Compare this curve with data in Fig. 7.2 to be convinced that the cycles are real. Redrawn from Ruddiman (2001).

7.1.3 Changes in ocean chemistry

Now that we have a basis for tracers of temperature and ice volume changes as well as a temporal history of climate cycles, we can turn our attention to sedimentary tracers of changes in ocean chemistry that accompanied the ice ages. This aspect of the story is critical to interpretations of the ocean's role in controlling changes in climate and the f_{CO_2} of the atmosphere (see later). Since the beginning of the systematic study of deep sea sediments, marine geologists have recognized large changes in the carbonate content of sediments associated with climate change (see, for example, Arrhenius, 1952). Although this change may partly result from variations in rain rate of biogenic carbonates from the surface ocean, it is also caused by changes in the chemistry of the deep ocean. Developing tracers that can resolve exactly what changed and how much has been a complicated task.

One of the most quantitative tracers of ocean nutrients has been the $^{13}C/^{12}C$ isotope quotient in foraminiferan tests preserved in sediments. The quotient is measured by mass spectrometry simultaneously with the determination of oxygen isotope ratios, so that one obtains information about both stratigraphy and chemical changes in one analysis. As described in Chapter 5 and illustrated in Figs. 5.7 and 7.10, there is a direct relation between the carbon isotope ratio of DIC and the nutrient content of seawater. The reason for this is the large kinetic isotope fractionation during photosynthesis that discriminates against the heavier isotope, ^{13}C, creating relatively light (depleted in ^{13}C) organic matter and causing the surrounding DIC in surface waters to become heavier (enriched in ^{13}C) (see Chapter 5). During respiration there is very little fractionation, so water in the aphotic zone becomes lighter in $\delta^{13}C$-DIC as the concentration of nutrients increases. The $\delta^{13}C$ content of benthic foraminiferal $CaCO_3$ records the deep water DIC $\delta^{13}C$ because there is an isotopic

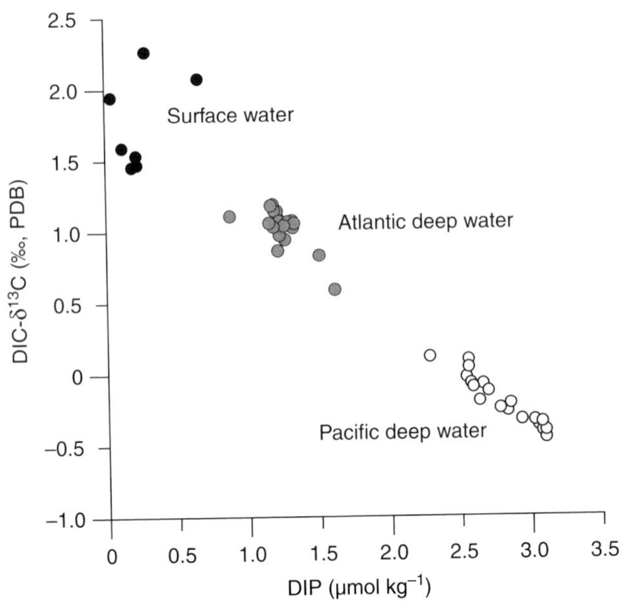

Figure 7.10. DIC-δ^{13}C versus dissolved inorganic phosphate (DIP) for representative seawater samples from the surface ocean, the deep Atlantic and deep Pacific. Redrawn from Boyle (1986).

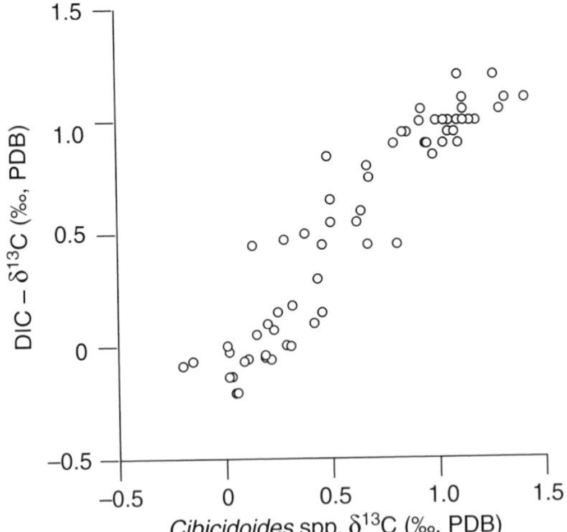

Figure 7.11. DIC-δ^{13}C of $CaCO_3$ from core top shells of the benthic foraminiferan *Cibicidoides* spp. versus the δ^{13}C of bottom waters in the same area of the ocean. Redrawn from Duplessy *et al.* (1984) as reproduced in Curry *et al.* (1988).

equilibrium relation, with a very small fractionation, between the carbon isotope ratio of foraminiferan $CaCO_3$ tests and the DIC of the water in which they grew (Fig. 7.11). In this way variations in the δ^{13}C of $CaCO_3$ in fossil benthic Foraminifera buried in deep sea sediments are a proxy for changes in the nutrient content of deep waters.

Variations in the carbon and oxygen isotope ratio of benthic Foraminifera in cores from the Atlantic and Pacific Oceans (Fig. 7.12) suggest that the δ^{13}C of DIC of the deep water in both oceans was lighter during the last glacial maximum and values in the Atlantic changed at least twice as much as those in the Pacific. When many marine sediment cores from different geographic locations are averaged, the whole-ocean DIC-δ^{13}C increase from glacial to

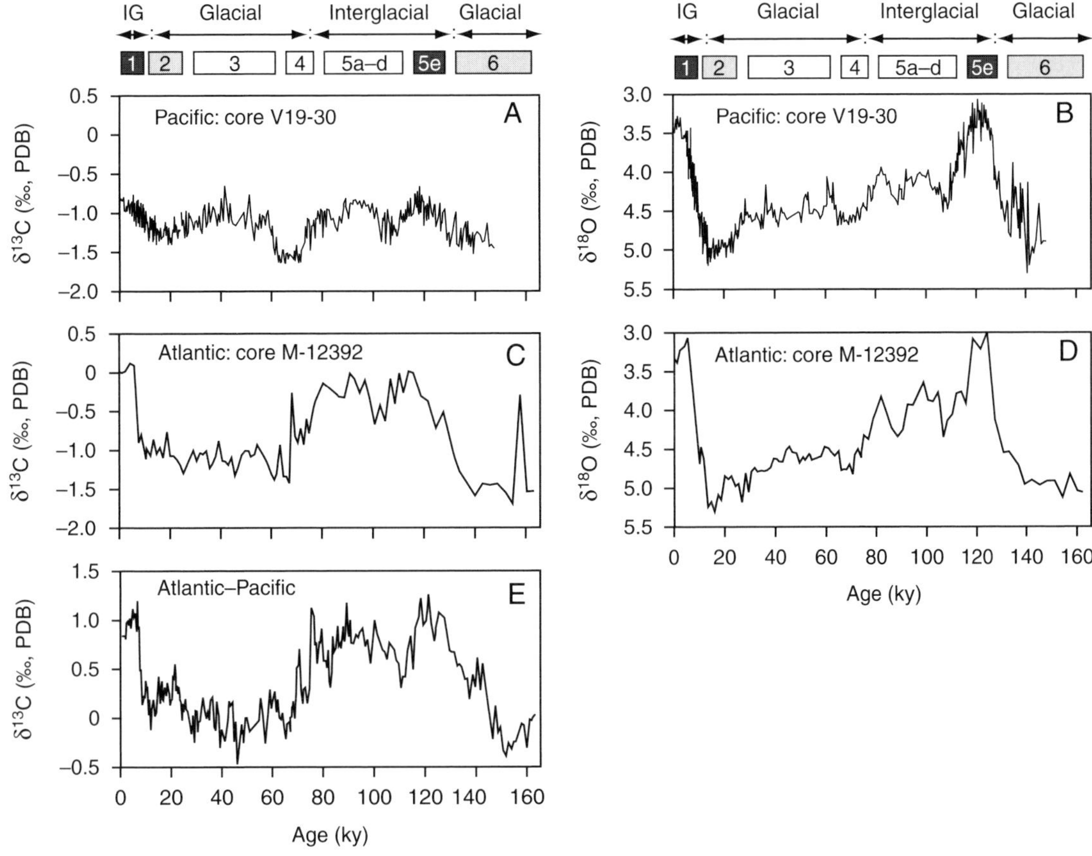

Figure 7.12. $\delta^{18}O$ and $\delta^{13}C$ composition of benthic Foraminifera (*Uvigerina* sp.) in cores from the East Equatorial Atlantic (A, B) and Eastern Equatorial Pacific (C, D). The $\delta^{18}O$ data are presented to indicate the timing of the glacial–interglacial cycle. The difference in $\delta^{13}C$ between the records in (A) and (C) is presented in (E) and suggests how the Atlantic–Pacific carbon isotope ratio of DIC changed between glacial and interglacial times. Redrawn from Shackleton *et al.* (1983).

interglacial time is *c.*0.35‰ (Curry *et al.*, 1988). Because it seems impossible that this change could represent a decrease in the whole-ocean nutrient content on such a short time scale, it is assumed that this global change represents a transfer of CO_2 from terrestrial and marine carbon reservoirs during the glacial period. This change is within constraints of other estimates of terrestrial carbon reservoir variations during glacial times. Because the difference in $\delta^{13}C$ between terrestrial carbon and marine DIC is about -26‰, a 0.35‰ change in whole-ocean DIC-$\delta^{13}C$ requires a transfer of carbon from the terrestrial biosphere to the oceanic–atmospheric reservoir that is about 1.3‰, of the present size of the oceanic–atmospheric reservoir (0.35‰ / 26‰ = 0.013). This amounts to about 500 Pg C (38 000 Pg C $\times$ 0.013 $\approx$ 500 Pg C), roughly equivalent to one quarter the size of the present above-ground and soil terrestrial carbon reservoirs combined (see Table 11.1).

Given a plausible explanation for a whole-ocean $\delta^{13}C$-DIC change between the Holocene and last glacial maximum, it is possible to use the differences in the benthic foraminiferan $\delta^{13}C$ records (Fig. 7.12) in the Atlantic and Pacific Oceans to infer changes in deep water circulation between glacial and Holocene periods. First, benthic foraminiferan $\delta^{13}C$ indicates that the $\delta^{13}C$-DIC of the Atlantic Ocean during the Holocene was *c.*1‰ more positive (heavier) than in the Pacific. If we assume this difference is a proxy for the nutrient distribution (Fig. 7.10), it corresponds to a dissolved inorganic phosphorus (DIP) difference of about 1.0 μmol kg^{-1}, which is equal to the present Atlantic–Pacific deep water difference (Fig. 1.4). It thus seems plausible to interpret the benthic foraminiferan $\delta^{13}C$ differences from the different ocean basins as proxies for nutrient concentrations in those basins. During the last glacial period (15–50 ky BP) the benthic foraminiferan $\delta^{13}C$ values for the Atlantic and Pacific were about the same (Fig. 7.12), implying that the nutrient content of the entire deep ocean was roughly homogeneous at that time. If we view present deep waters of the Atlantic as a mixture between relatively nutrient-poor North Atlantic Deep Water (NADW, DIP $\approx$ 1.0 μmol kg^{-1}) (Fig. 1.4) and relatively nutrient-rich Antarctic Bottom Water (AABW, DIP $\approx$ 2.2 μmol kg^{-1}), the similarity of deep Atlantic and Pacific $\delta^{13}C$ values during glacial time implies that there was much less influence of the NADW in glacial ocean deep waters.

The interpretation of carbon isotope changes in the DIC of seawater is complicated because the $\delta^{13}C$ distribution depends not only on fractionation during photosynthesis, but also on the exchange of CO_2 between the atmosphere and ocean. There is a strong temperature dependence of the air–sea carbon isotope equilibrium such that the difference between equilibrium at the Equator and at high latitudes is about 2.5‰. As illustrated in Fig. 6.14B this difference is not achieved in the $\delta^{13}C$–DIC of today's Pacific Ocean surface waters, which are poorly equilibrated with atmospheric CO_2 $\delta^{13}C$. The reasons for the dramatic disequilibrium are that ocean chemistry in the high latitudes is strongly influenced by mixing with the deep ocean, and that the residence time of water in the surface ocean is too short to exchange enough CO_2 with the atmosphere to reach isotopic equilibrium (consult Chapter 10). If the amount of time that high-latitude surface waters spent in contact with the atmosphere was different in the past, there may have been a different $\delta^{13}C$-DIP relation than we see today.

Indeed, as the body of carbon isotopic data from both planktonic and benthic Foraminifera throughout the world's oceans has increased, it has become evident that, the inventory of $\delta^{13}C$-DIC in deep waters of the ocean was roughly 0.5‰ lighter than in surface waters during glacial time. If we assume that the isotope ratio is a proxy for nutrients, this implies a whole-ocean increase in deep water nutrient concentrations during glacial times. This conclusion, however, does not agree with the other main paleoceanographic nutrient proxy, the foraminiferan Cd/Ca quotient (see later), which indicates that Cd did not change appreciably between glacial and interglacial periods (Boyle, 1992). Toggweiler (1999) has demonstrated by using multi-box models of the ocean's carbon cycle that the $\delta^{13}C$-DIC–DIP relation of the ocean's deepest water can be uncoupled under certain scenarios in which gas exchange with the atmosphere is suppressed. Thus, there are arguments that the $\delta^{13}C$ of DIC may not have been a stable nutrient tracer between glacial and interglacial times.

Other potential problems with the benthic foraminiferan $\delta^{13}C$ exist. For example, some species of benthic Foraminifera have been shown to become lighter in $\delta^{13}C$ as the rain rate of organic matter increases (Mackensen *et al.*, 1993). This is because the shells are so tiny and near the sediment–water interface that the $\delta^{13}C$ of their surroundings is altered by the chemistry in the sediment pore-waters, which become lighter in $\delta^{13}C$ with greater organic matter degradation. Furthermore, cultured planktonic foraminiferan studies suggest that the incorporation of ^{13}C is dependent on the carbonate ion concentration (Spero *et al.*, 1997). Planktonic Foraminifera grown in seawater with identical DIC but different $[CO_3^{2-}]$ have different $\delta^{13}C$ values. A carbonate ion concentration change between 200 and 250 $\mu mol\ kg^{-1}$ caused a decrease in the $\delta^{13}C$ by about 0.5‰ in *G. ruber*. This is about the difference in carbonate ion to be expected if the alkalinity of surface waters remained constant while f_{CO_2} increased *c*.80 ppm from glacial to interglacial times (see later in this chapter). Thus, one might expect a decrease in $\delta^{13}C$ of about this magnitude due to this effect alone without any change in phosphate. It is still uncertain whether this effect is as pronounced in benthic Foraminifera.

Fortunately, the $^{13}C:^{12}C$ ratio is not the only nutrient tracer in the arsenal of chemical oceanographers. It has been shown that the distributions of some trace metals, such as Cd, are proportional to nutrient concentrations (Fig. 1.6) (Boyle, 1988), indicating that they have a nutrient-like biogeochemistry in the ocean. Since Cd has the same charge and is about the same size as Ca, conditions are ideal for it to be incorporated into $CaCO_3$ during foraminiferan test growth. After much careful cleaning of the foraminiferan tests to remove adsorbed trace metals, it was shown that foraminiferan Cd content is proportional to the Cd content of the waters in which they grow (Hester and Boyle, 1982) (Fig. 7.13). Thus, foraminiferan Cd:Ca ratios

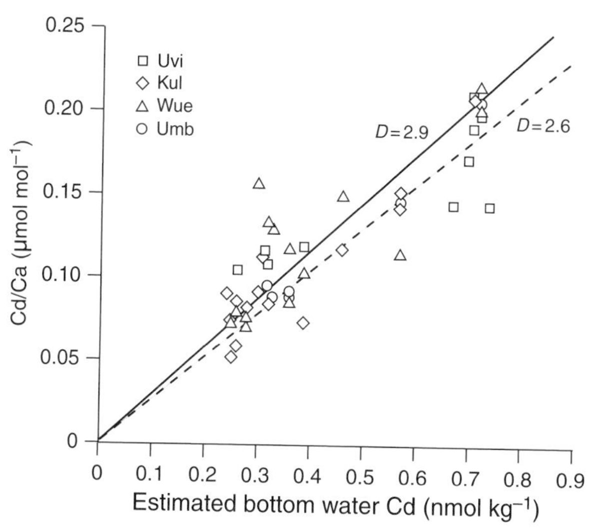

Figure 7.13. Correlation between the Cd:Ca quotient of four different species of benthic foraminiferan (Uvi, *Uvigerina* sp.; Kul, *Cibicidoides kullenbergi*; Wue, *Cibicidoides wuellerstorfi*; Umb, *Nutallides umbonifera*) in core tops from different locations in the world's ocean with the estimated cadmium concentration in these waters. Cd concentrations are estimated from the phosphate concentration and the dissolved Cd:P ratio. *D* is the distribution coefficient, which is equal to the ratio of the Cd/Ca in the foraminiferan test to the Cd/Ca in seawater. Redrawn from Boyle (1992).

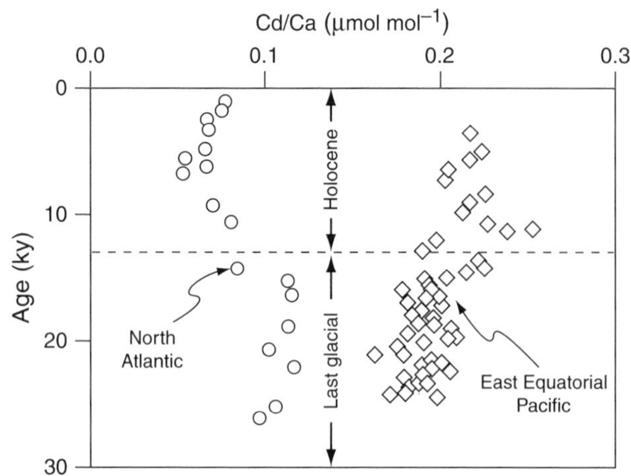

Figure 7.14. The Cd/Ca quotient in benthic foraminiferan tests in cores from the North Atlantic (circles) and East Equatorial Pacific (diamonds) as a function of core age. The data illustrate how the Cd concentration difference in deep waters of the two oceans changed between glacial and interglacial times. Reproduced from Broecker (2002) as derived from the data of E. A. Boyle.

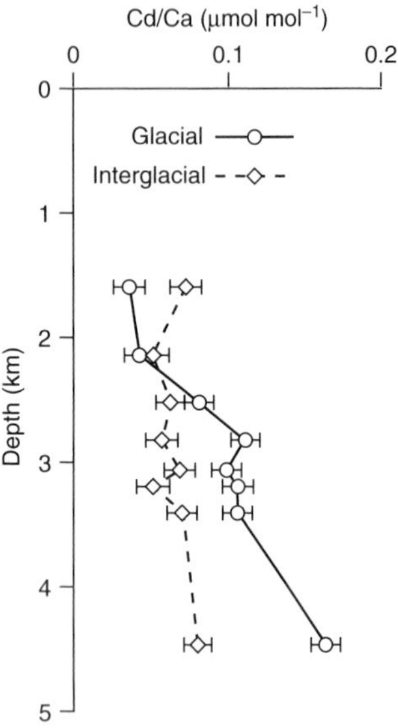

Figure 7.15. The Cd/Ca quotient of benthic Foraminifera sampled from core tops (diamonds) and from the sediment depth of the last glacial maximum (circles) of eight cores from the North Atlantic at different water depths. The difference in depth dependence illustrates the different profiles of dissolved cadmium in the North Atlantic at these times. Reproduced from Broecker (2002) as derived from the data in Boyle and Keigwin (1987).

are a proxy for nutrient concentrations that do not suffer the caveats attributed to foraminiferal $\delta^{13}C$ values.

By measuring benthic foraminiferal Cd:Ca ratios, Ed Boyle and colleagues (Boyle and Keigwin, 1987) were able to demonstrate that the nutrient contents of the Atlantic and Pacific deep waters were more similar during the last glacial maximum than today (Fig. 7.14). We mentioned earlier that the two end member deep and bottom water sources NADW and AABW have very different preformed nutrient contents: NADW is nutrient-poor and AABW nutrient-rich. The data for the Atlantic Ocean in Fig. 7.14 indicate that the deep water of the Atlantic had higher Cd, and thus higher DIP, during glacial times. This result indicates that the deep waters of the Atlantic Ocean were less influenced by NADW during glacial time. Cd concentrations in the Pacific decreased slightly, resulting in concentrations in the Atlantic and Pacific deep waters that were less divergent than today, but still different. This is the same trend as observed in the carbon isotope data but the differences in implied circulation change are less extreme.

Detailed analyses of foraminiferal Cd/Ca in North Atlantic sediment cores indicate that the depth distribution of nutrients was much more stratified below 1 km during the last glacial maximum than it is today (Fig. 7.15). While water below about 2.5 km was much more influenced by high nutrients from Southern sourced waters (AABW) (see above), shallow waters (<2 km) had lower nutrient concentrations than today. This result, coupled with the Pacific–Atlantic differences in nutrient concentrations, suggests that during the last glacial maximum NADW production occupied shallower depths in the North Atlantic Basin.

With the accumulation of a global data set from glacial age sediments, Boyle and colleagues (Boyle, 1992) were able to demonstrate that Cd increases in the Atlantic and decreases in the Pacific resulted in a constant Cd inventory for the entire ocean between glacial and interglacial times (and, by proxy, a constant nutrient inventory). These results represent a major departure from the $\delta^{13}C$ evidence, which suggests a change in whole-ocean nutrients. This seems to imply that we do not completely understand the detailed meaning of the carbon isotope and Cd:Ca ratio tracers or some of the known artifacts are influencing their distributions. None the less, because both tracer records provide similar information regarding the nutrient distribution between the Atlantic and Pacific Oceans, this interpretation is widely accepted within the paleoceanographic community.

Suggestions that NADW played a less prominent role in controlling the properties of the deep Atlantic Ocean during the last glacial maximum are consistent with relatively recent results from investigations of the ^{14}C content of paired planktonic and benthic Foraminifera. The ^{14}C content of deep water is a sensitive tracer of its source, because today the ^{14}C age of Southern source water is *c.*1400 y as opposed to *c.*400 y for NADW (Fig. 5.17). The $CaCO_3$ tests of Foraminifera incorporate the ^{14}C of DIC during growth just as they do ^{13}C. The difference between the age of deep waters and surface waters recorded in the planktonic and benthic foraminiferan tests reveals that glacial deep water (>2.5 km) in the Western Atlantic Ocean was approximately 1000 y as opposed to <500 y today (Keigwin, 2004; Robinson *et al.*, 2005). Thus, the Western Atlantic contained a larger Southern source component during glacial times. Results of this type of analysis in the Pacific Ocean have not yet been accomplished because of the relative lack of preserved benthic Foraminifera. Fortunately, given recent advances in analytical capabilities there should soon be experimental evidence for the ventilation age of Pacific deep waters.

As one might expect, the list of chemical proxies found in foraminiferan $CaCO_3$ continues to grow. Some of the most fruitful results have been from Ba analyses, which in today's ocean correlate with the alkalinity of seawater and Zn analyses, which correlate with the CO_3^{2-} ion concentration. Results from Ba:Ca ratios reveal many of the same conclusions as those for Cd:Ca ratios, $\delta^{13}C$ and $\Delta^{14}C$ of planktonic and benthic Foraminifera regarding the redistribution of nutrient-like tracers between the ocean basins and with depth in the North Atlantic (Lea and Boyle, 1990). Thus, there are at least four indicators of chemical rearrangement in the deep sea. There are also clear differences among these tracers, but because it is uncertain why Ba correlates with alkalinity in today's ocean, other than that they both appear to have a strong sedimentary source, researchers have been reluctant to interpret changes in the Ba:Ca ratio as indicating variations in alkalinity.

One of the critical uncertainties of ocean chemistry that will reveal much about the role of the ocean in influencing atmospheric f_{CO_2} changes through glacial times is the nature of the change in the CO_3^{2-} ion content (and thus pH) in the deep ocean. A tracer for this value is perhaps presently the most sought after "holy grail" in paleochemical studies of the oceans. Two promising tracers of the carbonate system are presently evolving. The boron acid–base system is the second most important acid in seawater with a p*K* very near the pH of seawater (Fig. 4.2). There is a fractionation of *c.*20‰ between borate ($B(OH)_4^-$) and boric acid ($B(OH)_3$) so the $\delta^{12}B$ of borate is very sensitive to the pH of seawater. If it is true that borate is the sole species of boron dissolved in seawater that is incorporated into $CaCO_3$ when foraminiferan tests form and there are not serious vital effects (processes that alter the simple inorganic incorporation of tracers and isotopes into the shells of living organisms), then the boron isotope ratio may be an accurate tracer for seawater pH. The other possibility for an alkalinity proxy is based on the fact that the partition coefficient between the Zn:Ca ratio in seawater and in foraminiferan $CaCO_3$ is correlated with the degree of seawater calcite supersaturation in the ocean (Marchitto *et al.*, 2005). Preliminary measurements of Zn in benthic foraminiferan $CaCO_3$, along with the assumption that the Zn concentration in the deep ocean has not changed through time, suggest that changes in the calcite saturation state have occurred in the past. If both of these proxies can be shown to give geographically consistent information, then paleoceanographers may be a step closer to constraining the carbonate chemistry of the deep ocean through time.

The final ocean chemistry tracer we consider here is salinity. We previously discussed that it has been possible to determine the change in $\delta^{18}O$ of bottom water from glacial to Holocene times by measuring $\delta^{18}O$ in the porewaters of long sediment cores obtained by the Ocean Drilling Project. It is also possible to make very precise measurements of Cl^- on small porewater samples that can be used as a proxy for salinity (Fig. 7.3B). Most of the data suggest an increase in Cl^- associated with the last glacial maximum of about 1 g kg^{-1} (Adkins and Schrag, 2001), which is expected based on a sea level change of 120 m (120 m / 3800 m $\times$ 35 g kg^{-1} = 1.1 g kg^{-1}). In one core from 50° S in the Southern Ocean, however, the change in Cl^- content was much larger, nearly 2 g kg^{-1}. Today we think of Antarctic Bottom Water (AABW) as being relatively fresh, but dense, because it is supercooled beneath Antarctic ice shelves. As we discussed earlier, measurements of $\delta^{18}O$ in the ODP porewaters (Fig. 7.3), along with benthic foraminiferan $\delta^{18}O$ changes, indicate that glacial bottom waters were much colder (*c.* – 1.8 °C) and indistinguishable from freezing. Now, from the Cl^- analyses it appears that these waters were also about 1 g kg^{-1} more saline than today, probably because of the exclusion of salt from ice during sea ice formation. If this result is corroborated, it may indicate that the deep ocean was highly stratified with a very dense abyssal layer during glacial time.

All proxies of the temporal change in the chemical composition of seawater are complicated by artifacts. Some have been discarded because of difficulties in interpretation and may reappear as they are better understood. Taken together, there are a few results about which we may be relatively confident. It is pretty clear that the chemical tracers and age of North Atlantic Deep Water indicate that it was less prominent in the Atlantic relative to Antarctic Bottom Water during the last glacial period. Cd:Ca ratios indicate that the nutrient content of the ocean's deep reservoir did not change appreciably from glacial to interglacial time and the deepest waters of the Atlantic Ocean during glacial times were more nutrient-rich whereas nutrient concentrations in shallower waters (about 2000 m) were less than today. Questions about the reliability of the tracers of pH and carbonate ion content persist, but these may be resolved in the near future. It has been about 30 y since paleoceanographers began interpreting tracers of ocean chemistry; progress has been slow because the work is tedious and leads to many blind alleys. The rewards, however, are great as they promise no less than the keys to understanding ocean feedback in the climate system.

7.2 The ice core record: 0–800 ky

Ice cores in polar regions and mountain glaciers archive past changes in Earth's atmosphere just as marine sediment cores archive paleoceanographic change. As ice accumulates it incorporates a record of the temperature of the air in the vicinity, the local composition of atmospheric dust and the composition of the atmosphere. International scientific teams have succeeded in gaining important insight into Earth's past millennial- to orbital-scale climate change by analyzing the chemistry of long ice cores collected from both Greenland and Antarctica. In this section, we will first present the atmospheric counterpart to the deep-sea record of climate change on orbital time scales and then move into the more recent record of abrupt climate change.

7.2.1 Glacial–interglacial changes in atmospheric chemistry

Long ice cores exist from both Greenland and the Antarctic ice sheets. Ice thicknesses are greater and accumulation rates slower on the Antarctic continent than in Greenland, so the longest ice core records are from the South Pole. Greater detail of short-term changes exists in the Greenland cores. The most recent ice core from Antarctica resulted in a 800 ky record from the Dome C location of Antarctica (EPICA, 2004). As yet, little detailed chemical information has been published about this core, so we present results from the past 400 ky preserved in the Vostok ice core record, cored in central East Antarctica (78° S, 106° E) (Petit *et al.*, 1999). Isotopic results from Vostok indicate variation of the deuterium content of the ice by *c.*70‰ over four regular cycles (Fig. 7.16A) in the 3500 m length of the core. The pattern of change in the $\delta^{18}O$ of ice is similar, only the equilibrium isotope temperature effects are about ten times smaller for $\delta^{18}O$ than they are for δD (see Chapter 5). These changes are much too large to be an indication of the variability of ice volume as is the case for $\delta^{18}O$ of foraminiferan tests in deep-sea cores: rather, the changes indicate temperature variations of the air during snow deposition. The cycles are progressively shorter at depth because ice is elastic: it compresses and flows as the weight of snow accumulates. Decompressing the

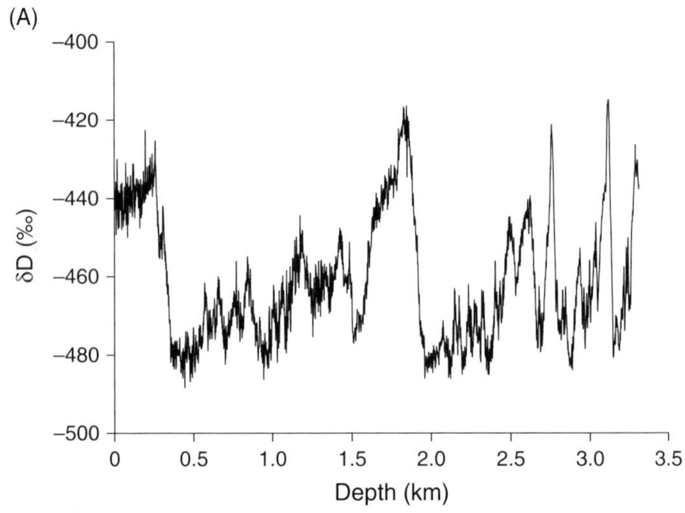

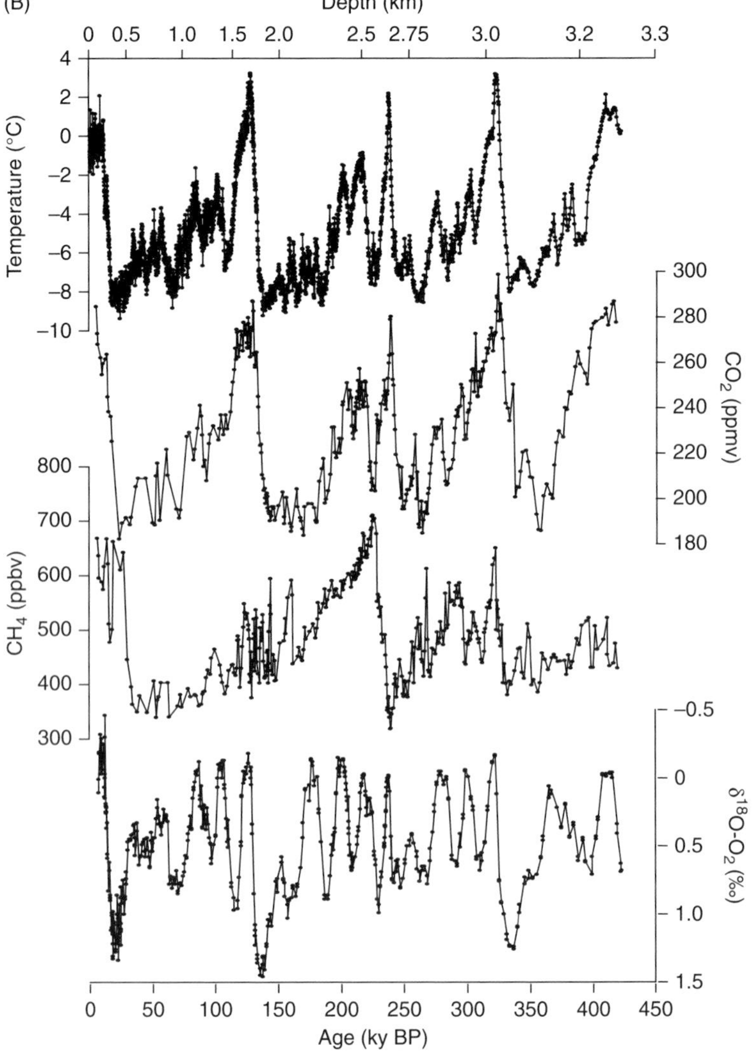

Figure 7.16. Climate indicators in the Vostok Antarctic ice core (78° S, 106° E, elevation 3488 m). (A) The δD of the ice as a function of depth in the 3300 m core. The characteristic *c.*100 ky cycles become closer together with depth owing to ice compaction. (B) Ice core records plotted against age. The temperature of the atmosphere was calculated from the δD data. CO_2, CH_4 and $\delta^{18}O$-O_2 are from air trapped in the ice as a function of age. The age scale is based on an ice accumulation model and control points in the δD record that are assumed to correspond to isotope stages 5.4 and 11.3.4. Redrawn from Petit *et al.* (1999).

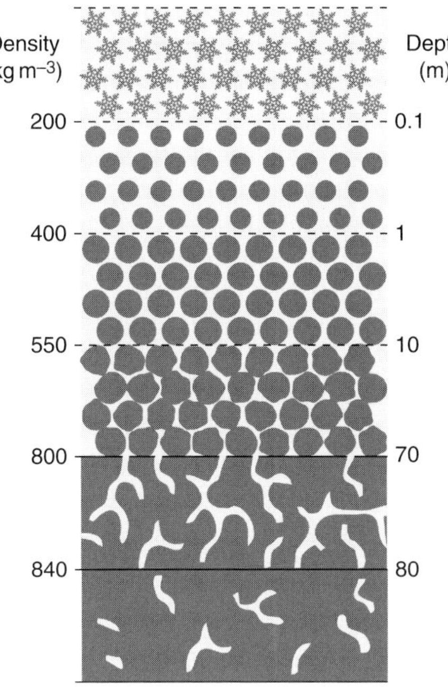

Figure 7.17. A schematic illustration of the closing of bubbles in ice as the density and depth increase. The actual close-off depth depends on snow accumulation rates and other factors. Redrawn from Broecker (2002).

lower regions by using age models and correlation of the changes with deep-sea $\delta^{18}O$ stratigraphy reveals the now classic picture of changes in temperature (from the δD changes) and atmospheric gas concentration over the past 400 ky (Fig. 7.16B).

As snow accumulates and converts to ice it incorporates bubbles of atmospheric gas (Fig. 7.17). These bubbles become isolated from the atmosphere when they close tens to hundreds of meters below the snow surface. At this point a small portion of the atmosphere is sequestered in cold storage in a relatively bacterium-free environment for hundreds of thousands of years. Climate scientists have developed methods for measuring CO_2, CH_4, the oxygen isotope ratio of molecular oxygen, and a host of other trace gases from very small samples. This was accomplished in the late 1970s and early 1980s (see, for example, Berner *et al.*, 1978) when the first systematic measurements of the CO_2 content in bubbles from Greenland and Antarctic ice cores indicated, to the great surprise of the scientific community, that during glacial times the atmosphere had an f_{CO_2} of *c.*190 ppm, about 80 ppm lower than values in pre-industrial times. Subsequently, the very regular changes in f_{CO_2} accompanied by changes in CH_4 and the stable isotope ratio of molecular oxygen associated with 100 ky glacial cycles (Fig. 7.16) have been observed in ice cores from both Antarctica and Greenland.

Because both CO_2 and CH_4 are greenhouse gases and their partial pressures in the atmosphere are lower during glacial periods, they may play an important role in the feedback mechanism necessary to produce climate cycles. Although the processes controlling CO_2 changes are still uncertain approximately 30 y after their discovery, it is fairly certain that the mechanism has to do with variations in the oceanic carbon cycle. The largest global carbon reservoirs with residence times less than millions of years are terrestrial biota and soils (*c.*2000 Pg) and the DIC of the deep sea (38 000 Pg). It is widely believed that the carbon reservoir on land was smaller during glacial periods instead of larger, so the transfer of CO_2 from the land to the atmosphere–ocean reservoir during glacial periods would be in the wrong sense to explain the lower f_{CO_2} values observed at these times. Since the f_{CO_2} of the atmosphere is strongly buffered by exchange with the massive dissolved inorganic carbon reservoir in the ocean, it is this carbon reservoir that is the most likely source of the glacial–interglacial change in f_{CO_2}.

The most probable cause for the observed changes in atmospheric CH_4 is the production rate of methane in tropical wetlands, because this is the largest source of CH_4 to the atmosphere today. It seems likely that during warmer periods, the tropical/subtropical CH_4 sources greatly expanded relative to glacial times. An alternative theory is that the CH_4 cycles might have been caused by periodic release of CH_4 from clathrate reservoirs (Kennett *et al.*, 2003). (Clathrates are CH_4-H_2O solid crystals preserved in marine sediments and in high-latitude wetlands.) Although this is likely to have been an important mechanism of climate change at specific intervals in the distant past, measurements of carbon isotopes in CH_4 from ice core bubbles seem to point toward wetlands as being the dominant source of the 100 ky cycles.

Past changes in atmospheric CO_2 and CH_4 indicate that the two most important greenhouse gases varied along with glacial cycles on 10–100 ky time scales. The third atmospheric gas variable shown in Fig. 7.16 is the stable isotope ratio of atmospheric O_2. We shall see in the next section that changes in this tracer are helpful in correlating the ice-core and deep-sea records.

7.2.2 Correlating atmosphere and ocean changes

Changes in the $\delta^{18}O$ of molecular oxygen have been valuable in the temporal correlation of the ice core records with their deep sea counterparts. (Be careful! We are now discussing the $\delta^{18}O$ of molecular oxygen, O_2, not the $\delta^{18}O$ of H_2O or $CaCO_3$. The similar terminology can be confusing, but the distinction is extremely important.) We demonstrated earlier that the $\delta^{18}O$ of $CaCO_3$ is in equilibrium with CO_3^{2-} in seawater, which is in isotopic equilibrium with water. By contrast, the oxygen isotopes of O_2 are not in isotopic equilibrium with anything because O_2 is produced by photosynthesis, a kinetic process. The isotope ratio of O_2 in the atmosphere is +23.5‰ with respect to the standard SMOW (Standard Mean Ocean Water) and is controlled by the isotope ratio of the oxygen atoms of water because that is the origin of molecular O_2 during photosynthesis. Notice that the $\delta^{18}O$ changes in Fig. 7.16 are about 1‰, the range expected for the variation in the isotope ratio of seawater due to the ice volume effect. (The isotope values presented in Fig. 7.16 vary around 0‰ rather than 23.5‰ because they are presented with respect to the atmosphere which is +23.5‰ greater than the standard SMOW.)

A very brief digression on the origin of the oxygen isotope ratio in the atmosphere will help clarify the reasons for its change over glacial–interglacial cycles. The process controlling the oxygen isotope ratio of the atmosphere is referred to as the Dole effect after its discoverer, Malcolm Dole. Very simply, it represents a mass balance between the processes of photosynthesis and respiration. Globally, the rates of photosynthesis, P, and respiration, R, are about equal because the O_2 concentration of the atmosphere is, as far as we know, stable ($P = R$). Thus, the rate of ^{18}O production by photosynthesis is equal to the rate of ^{18}O destruction by respiration:

$$\left(\frac{^{18}O}{^{16}O}\right)_{H_2O} P\alpha_P = \left(\frac{^{18}O}{^{16}O}\right)_{O_2} R\alpha_R, \tag{7.6}$$

where the alphas are fractionation factors. As reviewed in Chapter 5, there is very little oxygen isotope fractionation during photosynthesis, but during respiration the difference fractionation factor, ε, is about −20‰. Rearranging Eq. (7.6) and incorporating $P = R$ gives:

$$\left(\frac{^{18}O}{^{16}O}\right)_{O_2} \Big/ \left(\frac{^{18}O}{^{16}O}\right)_{H_2O} = \alpha_P/\alpha_R = 1/0.980 = 1.020. \tag{7.7}$$

If we subtract 1 and multiply by 1000 on both sides, the left side of the equation is equal to the definition of $\delta^{18}O$ of O_2 with respect to the standard SMOW, and the right side is +20‰. This value is lower than the measured value of 23.5‰ because roughly half of the oxygen production by photosynthesis occurs on land and about half in the ocean. O_2 from terrestrial photosynthesis is 4‰–8‰ heavier than SMOW because evaporation exceeds precipitation in the water of leaves, where most of the photosynthesis of the terrestrial biosphere occurs. The mean $\delta^{18}O$ from oceanic and terrestrial photosynthesis is thus +22‰ to +24‰. If one assumes that the partitioning of photosynthesis between the land and ocean remained the same and fractionation factors did not change in the past, then the sole reason for changes in the $\delta^{18}O$-O_2 of the atmosphere trapped in bubbles of the ice is due to the variation in the isotope ratio of seawater.

One of the really tricky aspects of comparing the paleorecords of ice cores and marine sediments is the offset between the age of the ice and the age of the atmosphere trapped in the ice (Fig. 7.17). Models of this process indicate that the age difference is dependent on many factors of ice deposition that also change from glacial to interglacial time. Thus, the error in the ice age – gas age difference is several thousand years in old ice (10–100 ky) and several hundreds of years for younger ice (1–10 ky). By comparing the changes in $\delta^{18}O$-O_2 with variations in CO_2 and CH_4, all of which are measured in bubbles and represent the same atmosphere, the timing of the change in ice volume and atmospheric chemistry change can be directly compared. The only correction that must be made is the *c.*2 ky lag between changes in the $\delta^{18}O$ of seawater and atmospheric $\delta^{18}O\text{-}O_2$ because of the residence time of O_2 in the atmosphere, τ_{atm}, which is equal to the moles of O_2 in the atmosphere divided by the global photosynthesis rate:

$$\tau_{atm}(y) = \frac{1.77 \times 10^{20} \text{mol air} \times 0.21 \text{ mol } O_2/\text{mol air}}{1.8 \times 10^{16} \text{mol } O_2 \text{ y}^{-1}} = 2.0 \text{ ky}. \tag{7.8}$$

With this correction it is possible to compare the $\delta^{18}O$ changes in the marine foraminiferan record, which has been systematically dated with radioisotopes and orbital forcing, with the atmospheric $\delta^{18}O$ changes measured in bubbles trapped in ice cores (Sowers *et al.*, 1993).

The three main atmospheric gas tracers in ice core bubbles (f_{CO_2}, CH_4 and $\delta^{18}O\text{-}O_2$) reveal changes in very different processes. Changes in CO_2 are correlated with ocean circulation and biogeochemical processes, CH_4 changes reflect the extent of wetlands, and the $\delta^{18}O$ of oxygen is primarily a tracer of the $\delta^{18}O$ of seawater fractionated by photosynthesis. With these three tracers one can make some judgments about the timing of the physical processes that occurred during glacial–interglacial transitions. The time course of changes in these tracers has been measured during the past four glacial–interglacial transitions in the Vostok ice core (Fig. 7.18). In each case the vertical dashed line in the figure is located at precipitous jumps in CH_4 concentration, which roughly coincide with the beginning of the increase in $\delta^{18}O\text{-}O_2$ as one moves from a glacial to an interglacial period. This line thus marks the beginning of substantial melting of the ice if we are reading the tracers correctly. Because of the resi-

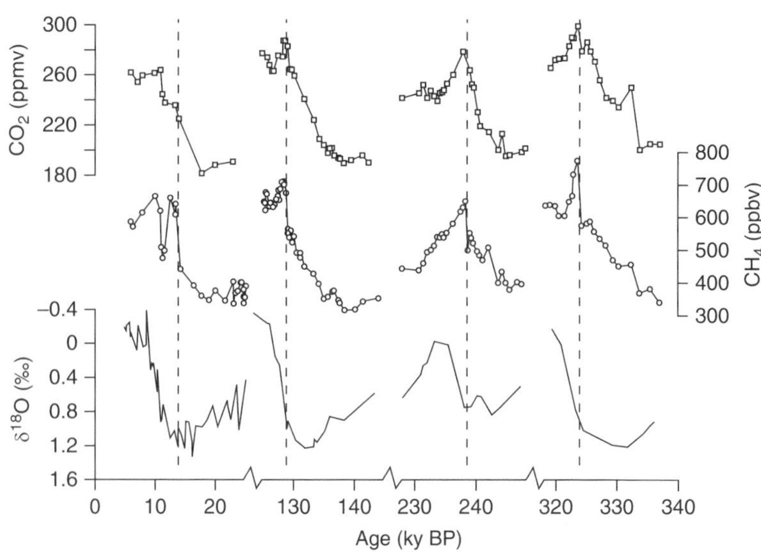

Figure 7.18. The four glacial–interglacial transitions over the past 400 ky in the Vostok ice core, showing the relations among the gases CO_2, CH_4 and $\delta^{18}O\text{-}O_2$ in the bubbles of the ice core. Vertical dashed lines are drawn through the "jumps" in the CH_4 concentrations. These data illustrate the temporal relations between the changes in atmospheric chemistry and ice volume change (see text). Redrawn from Petit *et al.* (1999).

dence time for O_2 in the atmosphere, we would expect the ice melting to have begun 2 ky earlier than the atmospheric $\delta^{18}O$-O_2 change. Close inspection of the CO_2 and CH_4 data indicates that they both begin to rise, in each transition, 4–8 ky before the $\delta^{18}O$-O_2 becomes lighter. Hence, the processes controlling atmospheric gas concentrations led the change in ice volume during the glacial–interglacial transitions. One of the main theories for the cause of glacial–interglacial transitions is that increases in solar insolation initiate ice melting, which is then enhanced by albedo feedback caused by melting glacial ice (Ruddiman, 2001). The data from the ice core would suggest that glacial–interglacial processes in the ocean and in tropical land areas occur before changes in sea level due to ice melting. Leads and lags of physical processes during glacial–interglacial transitions are a subject of much controversy today, and the actual sequence of the onset of different processes is presently uncertain. However, the simplest interpretation of the ice core tracers suggests that changes in atmospheric chemistry occur before changes in tracers of polar ice volume.

7.3 Abrupt (millennial-scale) climate change

Focusing in on the glacial–interglacial transitions has brought to light the rapidity and extreme nature of climate change indicated in the ice core records. This is illustrated most clearly in the $\delta^{18}O$ (or δD) of ice for the past 110 ky in the GISP2 ice core from Greenland (Fig. 7.19) (Grootes *et al.*, 1993). Notice first that the time interval of the Holocene, the past *c.*11 ky, represents an anomalously stable climate period as recorded by the $\delta^{18}O$ in the ice core over the past 100 ky. Before the Holocene the record indicates a series of abrupt warmings in the last glacial period that have been named the Dansgaard–Oeschger (D/O) interstadials (the numbers in the figures) after Danish and Swiss climate scientists who discovered them. Because the snow accumulation rate is relatively fast on Greenland it is possible to count annual seasonal bands in the ice at least as far back as the last glacial–interglacial transition. Changes in CH_4 concentrations in bubbles of the GISP2 ice core faithfully follow temperature changes, with higher CH_4 values in warmer periods, and are believed to indicate the expansion and contraction of wetlands associated with the relatively rapid warming and cooling events. Unfortunately it has not been possible to determine the CO_2 content of bubbles in glacial ice from Greenland because it is contaminated by carbonate in the dust of the ice.

Three key questions we will address here are: what was the geographic extent of the millennial-scale climate events; how fast and how much warming occurred; and what was the cause of these rapid climate changes? It is more difficult to determine short-term climatic events in Antarctic ice because accumulation rates are slower, but it has been addressed by extremely detailed sampling of ice δD and the

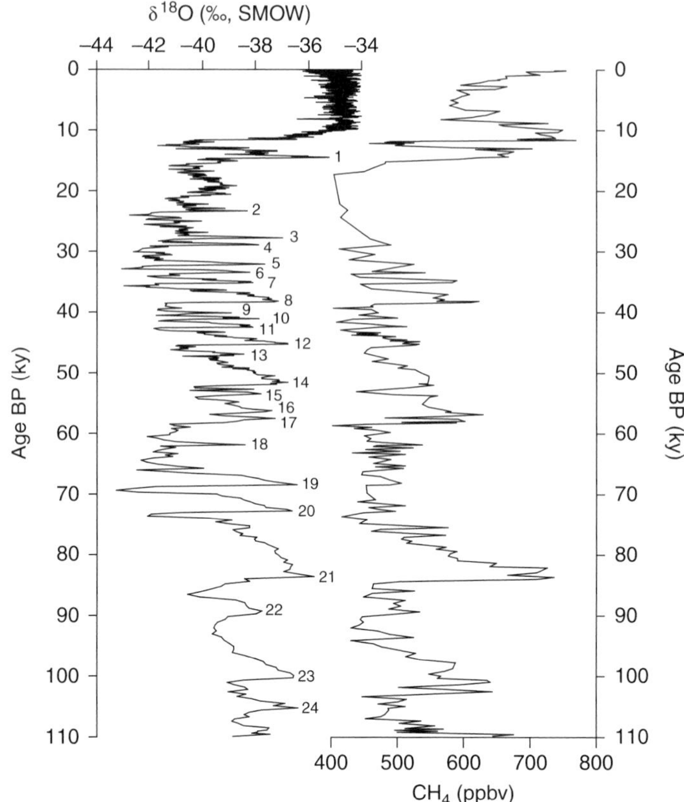

Figure 7.19. The $\delta^{18}O$-H_2O of the GISP2 ice core from Greenland as a function of age before the present (BP). Numbered intervals correspond to interstadial excursions called Dansgaard–Oeschger (D/O) events. The data show that most D/O events were associated with greater CH_4 concentration. Redrawn from Kennett *et al.* (2003).

CO_2 and CH_4 of bubbles over the last glacial–interglacial transition in a core from Dome C near the center of the East Antarctic ice sheet (75° 06′ S; 123° 24′ E) (Fig. 7.20). This core is at a location near the Antarctic Vostok core, which was discussed earlier (Fig. 7.16). In order to put the δD-ice and gas concentrations on the same time scale the differences between ice and gas ages have to be estimated by using a model for ice accumulation, which is believed to be reliable to within approximately ± 400 y. The extreme changes in CH_4, which began at about 14.5 ky BP and lasted about 2 ky (the same interval as that marked D/O #1 in the GISP2 ice core in Fig. 7.19) are synchronous in time (within present dating resolution) with the rapid warming at the end of the last glacial period referred to as the Bølling transition, which led into the Bølling–Alerod (B/A) climate optimum that is well known from terrestrial climate studies. After the B/A the $\delta^{18}O$ and CH_4 tracers in the Greenland ice (Fig. 7.19) indicate conditions plunged back into a brief period (*c.*1 ky) of ice-age conditions. This event is sychronous in time with the CH_4 minimum indicated in the Antarctic core (Fig. 7.20) and is known as the Younger Dryas (Y/D) from terrestrial climate research.

The mean oxidation time of CH_4 in the atmosphere is about 10 y and the atmospheric mixing time between the northern and southern hemispheres is about 1 y; thus the methane changes observed in

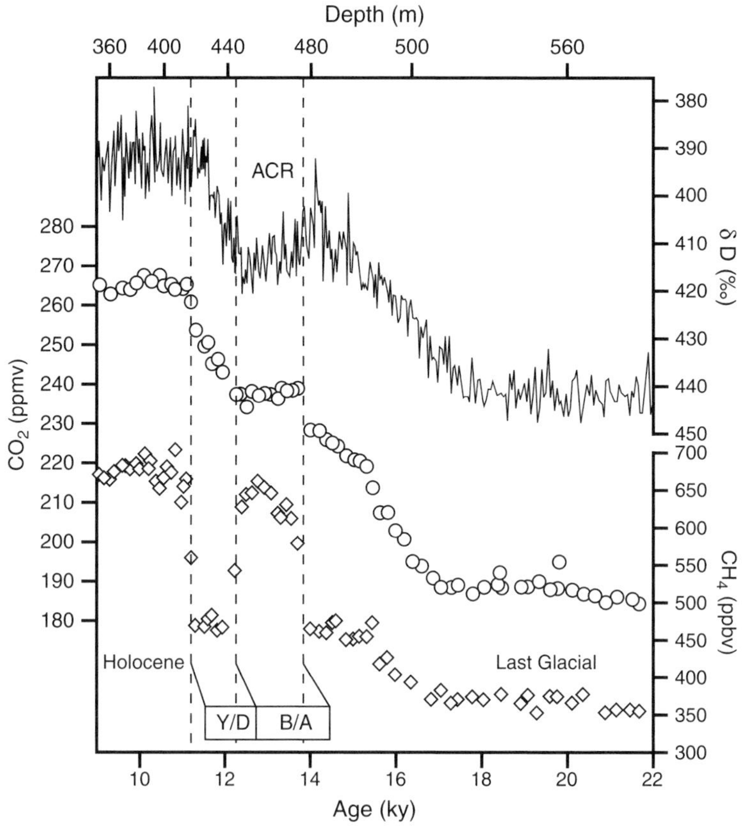

Figure 7.20. Temperature proxy (δD) and gas concentrations in bubbles from the Dome C ice core in Antarctica (75°06' S, 123°24' E) during the last glacial–interglacial transition. The Younger Dryas (Y/D) and Bϕlling/Alerod (B/A) events recorded in the Greenland ice cores are also present in Antarctic ice and are indicated according to the Greenland GRIP ice core time scale. CH_4 is a global atmospheric signal; similar CH_4 changes are observed at both poles. However, changes in temperature and CO_2 during the B/A and Y/D periods indicated here are in the opposite sense from those observed in Greenland. The ice age – gas age difference between the δD and gas records is based on an ice accumulation model. Redrawn from Monnin *et al.* (2001).

the Antarctic Dome C ice core (Fig. 7.20) are likely global and probably reflect the size of tropical wetlands, which are presently the main CH_4 source. The δD (temperature) changes observed in this core do not follow the same direction as those measured in the Greenland ice cores. As Greenland warmed during the B/A, the δD temperature proxy in Antarctica indicates a slight cooling. When Greenland plunged back into the icy Y/D, conditions in Antarctica warmed. These millennial-scale events appear out of phase, perhaps indicating that one hemisphere is forcing climate change and the other is following.

One should not expect changes in CH_4 and CO_2 in the atmosphere to resemble each other because CH_4 with its 10 y atmospheric life time can respond quickly to changes in the source strength. CO_2, on the other hand, is chemically stable in the atmosphere, but has an approximately 10 y life time with respect to atmospheric-oceanic exchange. Atmospheric CO_2 is a slave to processes that control dissolved inorganic carbon in the ocean. It is interesting that CO_2 changes shown in Fig. 7.20 coincide with Antarctic temperature changes, but not the global CH_4 concentrations, which follow the temperature changes measured in Greenland ice cores. This result has led some climate scientists to suspect that temperature changes in the Antarctic are forcing changes in the ocean that control the atmospheric CO_2 variations.

The mechanism causing the rapid changes observed in the Y/D and B/A events at the end of the last glacial period has been linked to changes in the strength of North Atlantic Deep Water formation (see, for example, Broecker, 2002). Following the rapid warming at the end of the last ice age (*c*.14.5 ky BP) and during the B/A, the continental ice sheet on North America retreated to a location just north of today's Great Lakes, where Glacial Lake Agassiz developed behind the ice sheet. After about 2 ky the warm interval was interrupted by a catastrophic release of *c*.10 000 km^3 of water from the lake into the North Atlantic. Modeling results have shown that this amount of freshwater input would make the surface waters of the North Atlantic too light to sink, inhibiting the formation of NADW.

Today NADW is formed by cooling of water that enters the North Atlantic via the Gulf Stream. This process transfers massive amounts of heat from the water to the atmosphere, warming the Northern latitudes of North America and Europe. If deep water formation slowed owing to the inundation of fresh surface waters, the region would likely cool rapidly. This connection between glacial periods and NADW formation is consistent with previous longer time scale evidence from Cd:Ca ratios, δ^{13}C-DIC and Δ^{14}C that suggests there may have been less NADW formation during the last glacial period. The origin of the repeated cycles during the time period between 20 ky and 60 ky BP may have been analogous to the most recent event, but there is little observational evidence remaining on land because remnants of previous Lake Agassiz-like dam breaks would have been either covered or destroyed by subsequent glaciations.

The rapidity and extent of temperature change associated with the glacial–interglacial transition has been put under the microscope by counting annual layers of ice and measuring the change in δ^{18}O in great detail. The chronology is solid because it is based on actually counting layers, but there is some uncertainty about the calibration of the δ^{18}O thermometer in these environments that leads to questions about the temperature record. For this reason nitrogen and argon isotopes have been used as a corroborative tool. Porous snow between the glacier surface and the depth of bubble close-off, the firn layer (Fig. 7.17) is a region where molecular diffusion dominates the transport of gases because air circulation by wind and convection is minimal. The isotope ratios of N_2 and Ar are useful indicators of rapid temperature change in this environment because they become unmixed in a diffusive medium that has a temperature gradient (Severinghaus and Brook, 1999). The heavier isotopes are slightly concentrated at the colder end of the diffusive air column, with the extent of the fractionation being dependent on the temperature gradient. Rapidly warming air sets up a thermal gradient in the firn. Because gases diffuse about ten times faster than heat, the isotope effects of the thermal gradient are registered in the bubbles as they close off long before the temperature equilibrates between the atmosphere and the base of the firn layer. This causes a heavy spike in the gas isotope ratio that is proportional to the magnitude of the rapid warming. Unfortunately, the isotope ratio of the diffusing gases is also affected by gravitational settling, which also enriches the isotopes at the base of the firn layer. By measuring the isotope ratio on two different gases it is possible to separate these effects, because the gravitational fractionation is proportional to the mass difference between the gases.

Temperature changes during the abrupt warming events have been estimated from measurements of both $\delta^{15}N$ and $\delta^{40}Ar$ in the Greenland ice core. The tracer, $\delta^{15}N_{excess}$, is calculated by subtracting $4 \times \delta^{40}Ar$ from $\delta^{15}N$ because the isotope mass differences are 4 and 1 mass units (40 – 36 and 15 – 14), respectively. A plot of $\delta^{15}N_{excess}$-atm, and the $\delta^{18}O$-ice versus depth for a roughly 300 y period between 14.5 and 14.8 ky BP (Fig. 7.21) illustrates the extreme rapidity of climate change. The oxygen isotope data in Fig. 7.21 indicate a change of 3.4‰ over a period of 60 y or less between the Oldest Dryas (the period at the very end of the last glacial period) and the Bølling. The gas isotope results are presented along with theoretical curves generated for the expected isotope change due to abrupt temperature changes of 8, 10 and 12 °C. These data indicate a temperature change of 10 °C that occurred over a period of less than 40 y! *If* this is a global phenomenon and the rule of thumb holds which states that temperature changes at the poles are 2–3 times those at the Equator, this observation represents a change of several degrees in temperate regions and a global climate change that was greater than the one we are presently experiencing. In the twentieth century (since *c.*1900) global temperatures have increased by *c.*0.7 °C.

An equally remarkable result of millennial-scale climate change research is that D/O events also occur in locations of the Northern Hemisphere that are far removed from Greenland. Thus, whatever triggered these events brought about a climate effect with a global reach. This observation was pointed out in dramatic fashion by a

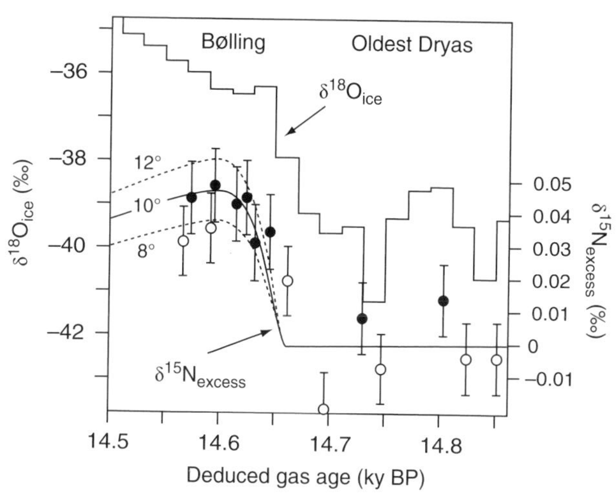

Figure 7.21. $\delta^{15}N_{excess}$ ($\delta^{15}N$ – $\delta^{40}Ar/4$) of gas bubbles (open and closed circles represent measurements from different laboratories) and $\delta^{18}O$ of ice (straight line with steps) from the GISP2 ice core in Greenland across the warming transition between the last glacial period and the Bølling Transition at 14 665 y BP. $\delta^{15}N_{excess}$ is a tracer of the magnitude of the abrupt temperature change at the beginning of the Bølling Transition. Curves are model results that predict the isotope effect for different sudden temperature changes. Redrawn from Severinghaus and Brook (1999).

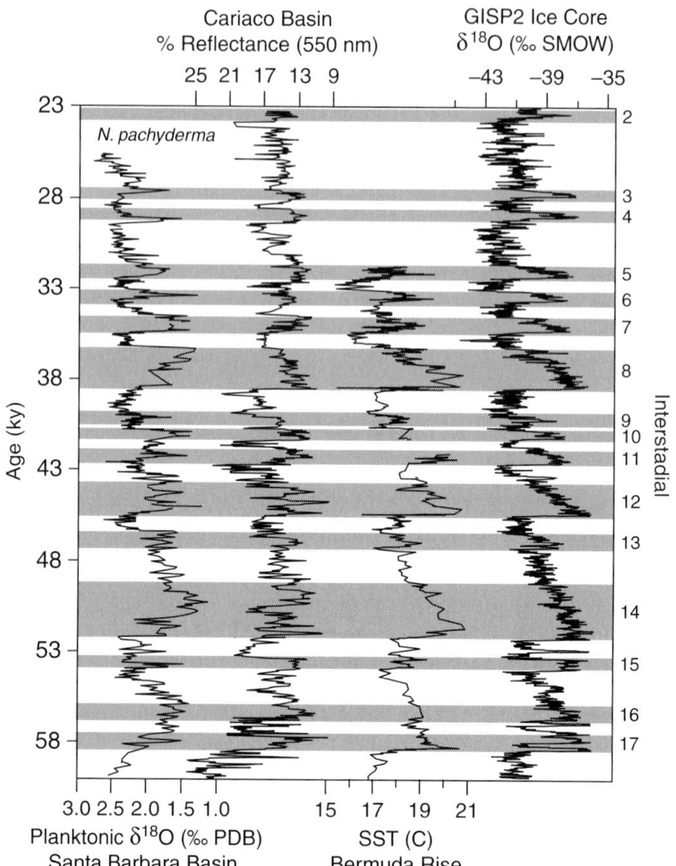

Figure 7.22. Comparison of high-resolution records showing the Dansgaard–Oeschger cycles (numbered here on the right side) from different regions of the world. The records are from left to right: $\delta^{18}O$ of planktonic Foraminifera in the Santa Barbara Basin; color reflectance of sediments from the Cariaco Basin, Venezuela; alkenone-derived sea-surface temperature estimates from a core on the Bermuda Rise; and the GISP2 ice core $\delta^{18}O$-H_2O record from Greenland. Reproduced from Kennett *et al.* (2003).

high-resolution study of intermittently anoxic marine sediments drilled from the Santa Barbara Basin (Kennett *et al.*, 2003). Varved intervals in the sediments and changes in planktonic and benthic Foraminifera $\delta^{18}O$ all follow the pattern of change revealed in the GISP2 ice core (Fig. 7.22). Apparently the mechanism that caused the $\delta^{18}O$ excursions in the Greenland ice core also affected the oxygenation and the temperature of the surface waters of the Santa Barbara Basin.

Contemporaneous D/O signals are shown in Fig. 7.22 for sediment cores from the Santa Barbara Basin (SBB), the Cariaco Basin (CB) in Venezuela, the Bermuda Rise (BR), and the GISP2 ice core. Planktonic foraminiferan $\delta^{18}O$ indicates surface water temperature change in SBB and BR; the color reflectance change of the CB sediments is controlled by the oxidation state of the overlying waters. There are many locations where Y/D events have been well documented (Broecker, 2004) and many others with the full suite of D/O events.

Additional evidence for the factors that force millennial-scale climate change comes from the study of layers of sand-sized material that are transported to the sea by ice. During the last glacial period there are six layers of ice-rafted debris (IRD) in the North Atlantic from the Labrador Sea across to France (Fig. 7.23). The layers are tens

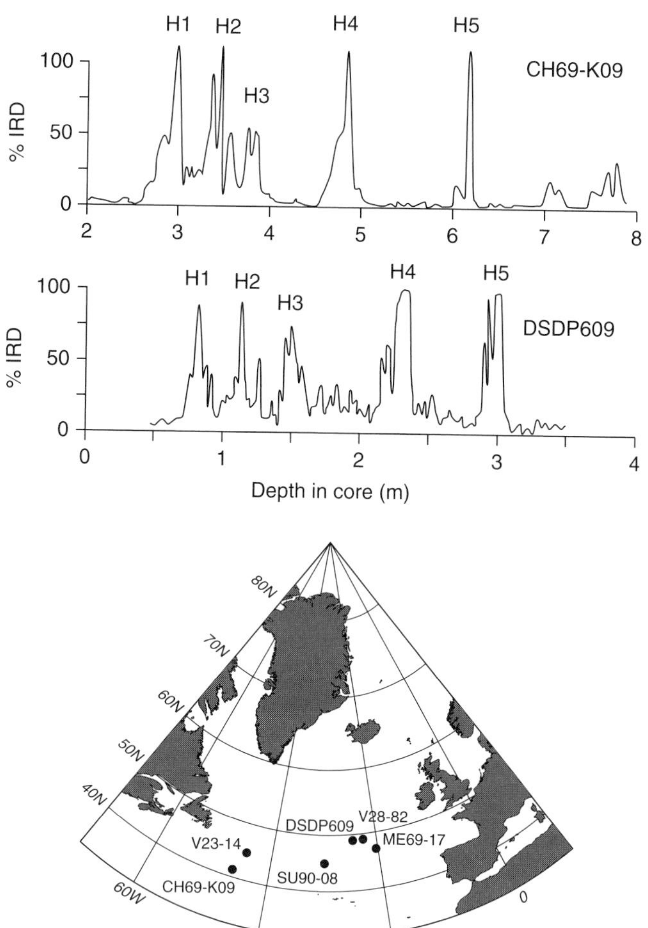

Figure 7.23. Records of the percent ice-rafted detritus (% IRD) in cores from the North Atlantic. Heinrich events 1–5 are labeled on the graphs, although longer cores indicate that there are more. The map shows a suite of cores in which these events are observed; only two representative curves of the data are shown here. Modified from Hemming (2004).

of centimeters thick on the western side and a few centimeters thick on the eastern side. This material is too large to have reached the deep ocean by wind transport or suspension in seawater: the only possible source is melting icebergs. These IRD intervals have come to be known as Heinrich Layers after their discoverer, Hartmut Heinrich. Very detailed sedimentological studies of this material indicate that IRD has the mineralogical and Pb isotope composition of the Canadian Shield in northeastern Canada. The size and abundance of the material suggests that it must be derived from armadas of icebergs that periodically swept across the Atlantic, melting as they traveled east. There is growing evidence that cold events of similar age have been found in other regions as distant as Brazil's savanna region and the South Atlantic (Sachs and Anderson, 2005).

The first five Heinrich events have been dated at calendar ages of 12, 16.5, 23, 29 and 40 ky BP. They coincide with cold intervals in between some of the D/O cycles. It is somewhat counterintuitive that armadas of icebergs would appear during cold intervals; however, the reason is that icebergs form when the ice sheets expand beyond land and are no longer grounded. The amount of fresh water contained in the number of icebergs required to deliver this much ice-rafted material to the sea floor is believed to be at least as great as that released to the North Atlantic during the purging of Lake Agassiz. If this is true, these events should also have modified the strength of the North Atlantic Deep Water flux to the deep sea. To date many details of the relationships between D/O and Heinrich events, such as why there are Heinrich events between some D/O excursions and not others, and their global manifestations, are not well understood.

The ultimate goal of paleoceanographic research is to develop a detailed enough history using a variety of tracers in strategic locations of the ocean to know how the marine system forced or responded to atmospheric chemistry and climate changes. Many plausible explanations and relatively little quantitative data existed before the rapid growth of modern paleoceanography in the 1980s. With the advent of ice core research, new paleoceanographic tracers, and collection of high-sedimentation-rate cores, the constraints have become more severe, and it has now an even greater challenge. The stakes have also become greater because of the necessity to be able to predict the response of the ocean and atmosphere to anthropogenically induced climate change. Verifying model predictions of the relations among climate and the dynamic properties of the ocean and atmosphere requires constraints on how these systems interacted in the past.

A first-order example is a mechanistic explanation for the observed past changes in atmospheric f_{CO_2}. Most climate scientists agree that a likely explanation involves changes in ocean circulation and biological processes. Some theories rely on the biological pump and the $CaCO_3$ dissolution–precipitation response; others invoke changes in ocean circulation and the $CaCO_3$ response. In Chapter 11 we will review a few of the rudiments of the theories of climate change, but suffice it to say here, that at the time of writing there is no clear winner for the dominant mechanism. Thus we have not achieved the goal of paleoclimate research, but enough new and fascinating information has been derived from tracers of ocean chemistry and climate that one expects that a consensus of how the ocean forces and/or responds to climate change cannot be far away.

References

Adkins, J.F. and D.P. Schrag (2001) Pore fluid constraints on deep ocean temperature and salinity during the last glacial maximum. *Geophys. Res. Lett.* **28**, 771–4.

Alley, R.B. (2000) *The Two Mile Time Machine*. Princeton, NJ: Princeton University Press.

Arrhenius, G. (1952) Sediment cores from the East Pacific. *Swedish Deep Sea Expedition (1947–1948) Reports*, vol. 5, pp. 1–227.

Bard, E., B. Hamelin, R. G. Fairbanks and A. Zindler (1990) Calibration of the 14 C timescale over the past 30,000 years using mass spectrometric U-Th ages from Barbados corals. *Nature* **345**, 405–10.

Berner, R. A. (1980) *Early Diagenesis: A Theoretical Approach*. Princeton, NJ: Princeton University Press.

Berner, W., B. Stauffer and H. Oeschger (1978) Past atmospheric composition and climate, gas parameters measured on ice cores. *Nature* **276**, 53–5.

Boudreau, B. P. (1997) *Diagenetic Models and their Implementation*. Berlin: Springer-Verlag.

Boyle, E. A. (1986) Paired carbon isotope and cadmium data from benthic foraminifera: implications for changes in oceanic circulation, and atmospheric carbon dioxide. *Geochim. Cosmochim. Acta* **50**, 265–76.

Boyle, E. A. (1988) Cadmium: chemical tracer in deepwater paleoceanography. *Paleoceanography* **3**, 471–89.

Boyle, E. A. (1992) Cadmium and $\delta^{13}C$ paleochemical ocean distributions during the stage 2 glacial maximum. *A. Rev. Earth Sci.* **20**, 245–87.

Boyle, E. A. and L. Keigwin (1987) North Atlantic thermohaline circulation during the past 20,000 years linked to high-latitude surface temperature. *Nature* **330**, 35–40.

Broecker, W. S. (2002) *The Glacial World according to Wally*. Palisades, NY: Eldigio Press.

Broecker, W. S. (2004) *The Role of the Ocean in Climate Yesterday, Today and Tomorrow*. Palisades, NY: Eldigio Press.

Burdige, D. J. (2006) *Geochemistry of Marine Sediments*. Princeton, NJ: Princeton University Press.

Chappell, J. and N. J. Shackleton (1986) Oxygen isotopes and sea level. *Nature* **324**, 137–40.

Crowley, T. J. (1983) The geologic record of climate change. *Rev. Geophys. Space Phys.* **21**, 828–77.

Curry, W. B., J. C. Duplessy and L. D. Labeyrie (1988) Changes in the distribution of $\delta^{13}C$ of deep water CO_2 between the last glaciation and the Holocene. *Paleoceanography* **3**, 317–41.

Duplessy, J. C., N. J. Shackleton, R. K. Matthews *et al.* (1984) ^{13}C record of benthic foraminifera in the last interglacial ocean: implications for the carbon cycle and global deep water circulation. *Quat. Res.* **21**, 222–43.

EPICA (2004) Eight glacial cycles from an Antarctic ice core. *Nature* **429**, 623–8.

Grootes, P. M., M. Stuiver, J. W. C. White, S. Johnsen and J. Jouzel (1993) Comparison of oxygen isotope records from the GISP@ and GRIP Greenland cores. *Nature* **366**, 552–4.

Hemming, S. R. (2004) Heinrich Events: massive late Pleistocene detritus layers of the North Atlantic and their global climate imprint. *Rev. Geophys.* **42**, 1–43.

Hester, K. and E. A. Boyle (1982) Water chemistry control of the Cd content of benthic foraminifera, *Nature*, **298**, 260–1.

Hughen, K., S. Lehman, J. Southon *et al.* (2004) ^{14}C activity and global carbon cycle changes over the past 50,000 years. *Science* **303**, 202–7.

Keigwin, L. D. (2004) Radiocarbon and stable isotope constraints on Last Glacial maximum and Younger Dryas ventilation in the western North Atlantic. *Paleoceanography* **19**, PA2012, doi: 10.1029/2004PA001029.

Kennett, J. P., K. G. Cannariato, I. L. Hendy and R. J. Behl (2003) *Methane Hydrates in Quaternary Climate Change: The Clathrate Gun Hypothesis*. Washington, DC: American Geophysical Union.

Ku, T. L., J. L. Bischoff and A. Boersma (1972) Age studies of mid-Atlantic ridge sediments near 42° N and 20° N. *Deep-Sea Res.* **19**, 233–47.

Lea, D. W. and E. A. Boyle (1990) Foraminiferal reconstruction of barium distributions in water masses of the glacial oceans. *Paleoceanography* **5**, 719–42.

Lea, D. W., D. K. Pak and H. J. Spero (1999) Climate impact of late Quaternary Equatorial Pacific sea surface temperature variations. *Science* **289**, 1719–24.

Mackensen, A. H., W. Hubberten, T. Bickert, G. Fisher and D. K. Futterer (1993) $\delta^{13}C$ in benthic foraminiferal tests of *Fontbotia wuellerstorfi* (Schwager) relative to $\delta^{13}C$ of dissolved inorganic carbon in the Southern Ocean deep water: implications for glacial ocean circulation models. *Paleoceanography* **8**, 587–610.

Marchitto, T. M., J. Lynch-Steiglitz and S. Hemming (2005) Deep Pacific $CaCO_3$ compensation and glacial-interglacial atmospheric CO_2. *Earth Planet Sci. Lett.* **231**, 317–36.

Martin, P. A., D. W. Lea, Y. Rosenthal *et al.* (2002) Quaternary deep sea temperature histories derived from benthic foraminiferal Mg/Ca. *Earth Planet. Sci. Lett.* **198**, 193–209.

Monnin, E. *et al.* (2001) Atmospheric CO_2 concentrations over the last glacial termination. *Science* **291**, 112–14.

Peng, T.-H. and W. S. Broecker (1984) The impacts of bioturbation on the age difference between benthic and planktonic foraminifera in deep sea sediments. *Nucl. Inst. Meth. Phys. Res.* **B5**, 346–52.

Petit, J. R. *et al.* (1999) Climate and atmospheric history of the past 420,000 years from the Vostok ice core, Antarctica. *Nature* **399**, 429–36.

Robinson, L. F., J. F. Adkins, L. D. Keigwin *et al.* (2005) Radiocarbon variability in the western North Atlantic during the last deglaciation. *Science* **310**, 1469–73.

Ruddiman, W. F. (2001) *Earth's Climate: Past and Future*. New York, NY: W. H. Freeman and Co.

Ruhlemann, C., S. Mulitza, P. J. Muller, G. Wefer and R. Zahn (1999) Warming of the tropical Atlantic Ocean and slowdown of thermohaline circulation during the last deglaciation. *Nature* **402**, 511–14.

Sachs, J. P. and R. F. Anderson (2005) Increased productivity in the subantarctic ocean during Heinrich events. *Nature* **434**, 1118–21.

Severinghaus, J. P. and E. J. Brook (1999) Abrupt climate change at the end of the last glacial period inferred from trapped air in polar ice. *Science* **286**, 930–4.

Shackleton, N. J. and N. D. Opdyke (1973) Oxygen isotope and paleomagnetic stratigraphy of equatorial Pacific core V28-238: oxygen isotope temperatures and ice volumes on a 105 year scale. *J. Quat. Res.* **3**, 39–55.

Shackleton, N. J., J. Imbrie and M. A. Hall (1983) Oxygen and carbon isotope record of East Pacific core V19-30: implications for the formation of deep water in the late Pleistocene North Atlantic. *Earth Planet. Sci. Lett.* **65**, 233–344.

Sowers, T. *et al.* (1993) A 135,000-year Vostok-Specmam common temporal framework. *Paleoceanography* **8**, 737–66.

Spero, H. J., J. Bijima, D. W. Lea and B. E. Bemia (1997) Effect of seawater carbonate ion concentration on foraminiferal carbon and oxygen isotopes. *Nature* **390**, 497–500.

Toggweiler, J. R. (1999) Variation of atmospheric CO_2 by ventilation of the ocean's deepest water. *Paleoceanography* **14**, 571–88.

II

Advanced topics in marine geochemistry

8

Marine organic geochemistry

Co-author: Kenia Whitehead

Institute for Systems Biology, Seattle, WA

The marine organic geochemistry component of chemical oceanography involves the study of ocean systems via analysis of non-living organic substances at the sub-micron size scale (e.g. molecules, atoms and nuclei). As in the field of chemistry, this is a separate branch of marine chemistry because of the complexity of compounds that are studied and the distinct nature of the analytical methods used. Organic compounds are important to a wide variety of processes that control marine and environmental chemistry. For example, approximately half of net photosynthetic production in the sea

Table 8.1. *Number of structural isomers for alkanes of increasing carbon number*

Formula	Number of isomers	Formula	Number of isomers
C_6H_{14}	5	$C_{10}H_{22}$	75
C_7H_{16}	9	$C_{15}H_{32}$	4347
C_8H_{18}	18	$C_{20}H_{42}$	366 319
C_9H_{20}	35	$C_{30}H_{62}$	4 111 846 763

is now recognized to pass via dissolved ($<0.5\,\mu m$) organic matter (DOM) into microbial heterotrophs, which are then consumed by heterotrophic Protozoa and microzooplankton. DOM also attenuates ultraviolet radiation in the surface ocean and is a catalyst for a variety of photochemical reactions. Organic molecules form complexes with a wide variety of trace elements in seawater, thereby affecting their chemical behavior and biological activity. In particular, organic substances strongly complex iron, which is limiting for photosynthesis in large areas of the ocean. Particulate ($>0.5\,\mu m$) organic matter is the main carrier of C, N, and P from the surface to the interior ocean, as well as one of the major forms for the preservation of these elements in marine sediments. Burial of organic matter in marine sediments has provided both the formation of fossil fuels and the continued presence of atmospheric O_2 for at least the past 600 million years. Finally, organic substances provide a richly detailed molecular record of past events in ocean environments.

The greatest strengths and challenges of marine organic geochemistry are traceable to the myriad of forms that organic molecules exhibit in the environment. The potential structural complexity of natural organic mixtures can be appreciated by considering the variety of molecules that can be made simply by linking carbon and hydrogen atoms together. For example, Table 8.1 lists the increasing number of simple hydrocarbons (ignoring stereoisomers) that can be made by linking the same number of carbon atoms together with single bonds, but in different arrangements (isomers). Examples of isomers for a 6-carbon alkane are shown in the first three structures in Fig. 8.1. The fact that multiple bonds, stereochemical isomers and additional elements such as N, O, S, and P often occur in organic molecules greatly adds to the potential structural diversity. There are literally millions of ways to make molecules with molecular masses in excess of several hundred atomic mass units (amu). This vast structural diversity allows the molecular blueprint for the form and function of an entire organism to be concentrated within microscopic strands of DNA and forms the basis for all life.

At another level of diversity within the structural framework are the multiple isotopes of many of the elements (except P) that make up organic compounds (^{1}H, ^{2}H, ^{3}H, ^{12}C, ^{13}C, ^{14}C, ^{14}N, ^{15}N, ^{16}O, ^{17}O, ^{18}O, and ^{32}S, ^{33}S, ^{34}S, and ^{36}S) (see Chapter 5). These isotopes occur in

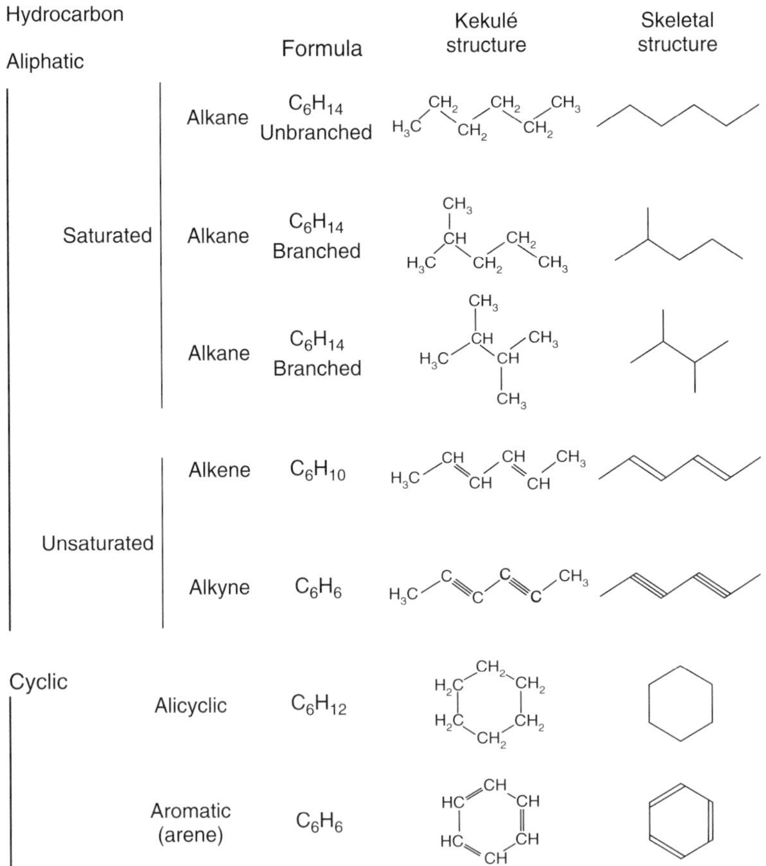

Figure 8.1. Hydrocarbons are classified by whether their carbon structure includes rings, multiple bonds, or branches. Classification of hydrocarbon structural families is shown here with the chemical formula. Structures are shown as Kekulé (also called line-bond) and skeletal structures.

characteristic abundances, reflecting their chemistry, biological source, age and/or processing history.

The vast structural diversity of organic matter in marine and terrestrial environments also poses a challenge to geochemists as they strive to identify and quantify individual components. A single comprehensive molecular analysis is essentially impossible because of the diverse types of compound, each requiring different analytical procedures. In addition, most organic matter occurs in macromolecules (e.g. proteins and polysaccharides) that must be broken down into smaller structural units (e.g. amino acids and sugars) before they can be chromatographically separated and quantified by various modes of detection. However, the procedures used to break down macromolecules are often imperfect, in part because released structures can decompose to other forms before the precursor polymer is completely dismantled. A simplifying factor in identifying and using organic molecules as tracers is that enzymes make biochemicals in a relatively limited variety of structural units (e.g. 20 protein amino acids and *c.*10 common sugars) relative to the multitude of possibilities. The usage of a limited variety of building blocks restricts the overall molecular diversity observed in nature. Major exceptions to this limited structural diversity are organic molecules that are formed cosmogenically or rearranged from biologically produced precursors by abiotic processes such pyrolysis or geothermal heating. The analytical challenge to tease individual components from complex organic matrices is a daunting task, but is made easier by the limitations imposed on organic structures by biochemical processes.

8.1 The nature of organic matter

In this section we present a brief review of the structures, characteristics and terminology used to describe the compounds in organic matter. The fundamental organic structure is the carbon-linked backbone of individual molecules. The simplest of organic substances called hydrocarbons contain only carbon chains with hydrogen appendages. Hydrogen and carbon are linked by strong covalent bonds into stable molecules with low solubilities in water and minimal tendencies to exchange atoms with the surroundings. Hydrocarbons are classified into structural families (Fig. 8.1), the terminology of which is useful to know as it often carries over to other compound classes. In essence, hydrocarbons are classified based upon whether the carbon chain includes rings, double bonds, or branches. Straight-chain hydrocarbons are called aliphatic; rings are termed cyclic. The latter are subdivided into aromatic and alicyclic, depending on whether or not they contain multiple carbon double bonds. Aliphatic compounds are further characterized as either saturated or unsaturated depending on the presence or absence of double bonds. Saturated hydrocarbons have only C–C and C–H single bonds and thus contain the maximum possible number of hydrogen atoms per carbon (i.e. the carbons are saturated with hydrogen). They have the general formula of C_nH_{2n+2} where n is any integer. Unsaturated hydrocarbons contain less hydrogen for the same number of carbons than their saturated counterparts owing to the introduction of multiple bonds (e.g. in alkenes and alkynes) or the formation of ring structures.

Many other classes of organic compounds are formed from hydrocarbon backbones by the addition of characteristic groups of atoms called functional groups. Table 8.2 is a list of common functional groups and their physical and chemical properties. More polar, and larger, molecules tend to have higher boiling and melting points. In general, the oxygen and nitrogen atoms that form functional groups impart polarity, and thus increase water solubility. The acidity, or proton-donating capacity, of a functional group is expressed as the negative log of its equilibrium dissociation constant (pK_a), so that the group will be largely dissociated (lack its proton) at aqueous pH values above its pK_a whereas at pH values less than its pK_a it will be protonated (see Chapter 4). For example, at seawater pH values of 7–8, carboxyl ($pK_a \approx 5$) and amine ($pK_a \approx 10$) functional groups exist as anions ($R - COO^-$) and cations ($R - NH_3^+$), respectively. The hydrogen (and oxygen) atoms within alcohols, phenols, carboxylic acids,

Table 8.2. *Functional groups and their corresponding properties*
Symbols and abbreviations: pK_a, log of the acidic dissociation constant; Exchange H?, exchangeable H at typical environment conditions; bold type, actual functional group.

Group	Structure	pK_a	Hydrolysis Products	Exchange H ?
Alcohol	$-\overset{\vert}{\underset{\vert}{C}}-\mathbf{O}-\mathbf{H}$	12	None	Yes
Phenol	$C_6H_5-\mathbf{O}-\mathbf{H}$	10	None	Yes
Ether	$-\overset{\vert}{\underset{\vert}{C}}-\mathbf{O}-\overset{\vert}{\underset{\vert}{C}}-$		None	
Aldehyde	$-\overset{\vert}{\underset{\vert}{C}}-\overset{\mathbf{O}}{\overset{\Vert}{\mathbf{C}}}-\mathbf{H}$		None	No
Ketone	$-\overset{\vert}{\underset{\vert}{C}}-\overset{\mathbf{O}}{\overset{\Vert}{\mathbf{C}}}-\overset{\vert}{\underset{\vert}{C}}-$		None	
Carboxyl	$-\overset{\vert}{\underset{\vert}{C}}-\overset{\mathbf{O}}{\overset{\Vert}{\mathbf{C}}}-\mathbf{O}-\mathbf{H}$	5	None	Yes
Ester	$-\overset{\vert}{\underset{\vert}{C}}-\overset{\mathbf{O}}{\overset{\Vert}{\mathbf{C}}}-\mathbf{O}-\overset{\vert}{\underset{\vert}{C}}-$		Carboxyl + Alcohol	
Amine	$-\overset{\vert}{\underset{\vert}{C}}-\mathbf{N}\langle$	10	None	Yes
Amide	$-\overset{\vert}{\underset{\vert}{C}}-\overset{\mathbf{O}}{\overset{\Vert}{\mathbf{C}}}-\mathbf{N}\langle$		Carboxyl + Amine	Yes

and amines are all exchangeable, and thus cannot be expected to retain their stable isotope compositions following synthesis. In contrast, the carbon and nitrogen constituents of organic compounds are almost never exchanged under environmental conditions.

In addition to variations in functional group distributions (i.e. atomic compositions), organic molecules also exhibit differences in the way in which atoms of the same number and type are arranged structurally and in space. These are called isomers. An example of structural isomers with different patterns of atomic linkages are the two molecules dimethyl ether (CH_3OCH_3) and ethanol (CH_3CH_2OH), which both have the formula C_2H_6O, but with the oxygen in a central versus a terminal position, respectively. Molecules that have the same linkage sequence with the same atoms can still vary by the orientation of their atom groupings about a bond or in space. Thus, "conformational isomers" vary by rotation about a single bond, "geometric isomers" vary by position on opposite sides of a double bond and "optical" (or stereo) isomers vary by order of substitution in space on a carbon that is bonded to four different groups (chiral carbon).

8.2 Methods of characterizing organic matter

Methods of characterizing organic matter in the environment are widely variable and continually expanding. Even the very basic determination of total organic matter content is not straightforward in environmental samples. It is not possible to quantify organic matter by weighing the sample, because organic matter cannot be easily separated from the greater amounts of inorganic matter associated with it in seawater and sediments. As a result, organic matter is usually quantified based on its total organic carbon (TOC) content, which can be readily measured by combustion of organic carbon to carbon dioxide after removal of any inorganic carbon present by acid treatment. This indirect method works reasonably well because all organic matter contains carbon as a major constituent (e.g. carbohydrates are *c.*45% carbon by mass and lipids are *c.*85% carbon by mass). Thus, organic matter concentrations and budgets are typically given in terms of organic carbon equivalents, with the understanding that total organic mass is on average approximately twice the mass of organic carbon.

The two general approaches used to further characterize organic matter in natural mixtures involve either bulk or molecular-level techniques. Bulk characteristics (e.g. isotope or elemental compositions) are broadly representative of the major component of organic matter and provide an overview to the system being studied (Table 8.3). A molecular approach entails chromatographically separating and quantifying a specific compound class and can provide detailed information about a process (Table 8.4). However, this specific compound class often represents only a small fraction of the whole organic mixture. Among the few compromises between these usual options of "learning a little about a lot, or a lot about a little" are spectral characterizations such as nuclear magnetic resonance (NMR) spectroscopy, or broad-based degradative analyses such as pyrolysis

Table 8.3. *Representative bulk methods of organic characterization*

MALDI, matrix-assisted laser desorption ionization; MS, mass spectrometry; CE, capillary electrophoresis.

Analytical method	Measured characteristics	Total parameters	Typical preparation	Required sample
Elemental	C, H, O, N, S	5	[a]Combustion	<1 mg
Stable isotope	^{13}C, ^{2}H, ^{16}O, ^{15}N, ^{36}S	5	[a]Combustion	<1 mg
Radioisotope	$\Delta^{3}H$ and ^{14}C,	2	[a]Combustion	<1 mg
Infrared spectra	Functional groups	~20	[b]Demineralization	<1 mg
NMR spectra	C, H, N and P types	~30	[b]Demineralization	1–10 mg
Pyrolysis-MS	Degradation products	>100	[b]Demineralization	<10 g
MALDI-MS	Intact molecules	>100	[b]Demineralization	<1 mg
CE-MS	Intact molecules	>100	[b]Demineralization	<1 mg

[a] Combustion is followed by reduction in the analysis of N and H, whereas O_2 is usually generated by pyrolysis.
[b] Demineralization includes separation from mineral phase by organic extraction or mineral dissolution and is typically necessary for sediments but not pure organic materials.

Table 8.4. *Molecular-level methods for different types of organic substance*

Method	Class	Types	[a]Preparation	[b]Chrom.	[c]Derivative	[d]Detector
Hydrocarbons	Lipid	>100	NPSE	GC	None	FI
Fatty acids	Lipid	>100	Basic Hy	GC	ME/TMS	FI
Fatty alcohols	Lipid	c.30	Basic Hy	GC	ME/TMS	FI
Sterols	Lipid	>100	Basic Hy	GC	ME/TMS	FI
Alkenones	Lipid	c.10	NPSE	GC	None	FI
Chlorophylls	Pigment	c.20	NPSE	LC	None	Flu
Carotenoids	Pigment	c.50	NPSE	LC	None	UV
Amino acids	Amine	c.20	Acid Hy	LC/GC	OPA	Flu/FI
Nucleic acids	Nucleotide	4	Isolation	LC	None/OPA	UV
Neutral sugars	Carbohydrate	20	Acid Hy	IC/GC	None/TMS	PA/FI
Acidic sugars	Carbohydrate	c.10	Acid Hy	IC/GC	None/TMS	PA/FI
Lignin phenols	Phenol	c.30	CuO-NaOH	GC/LC	None/TMS	FI
Tannins	Phenol	c.20	Acid Hy	GC	PHL/TMS	FI
Cutin acids	Polyester	c.20	MeOH-	GC	ME/TMS	FI
Pyrolysis-GC/MS	General	>100	Pyrolysis	GC	None	MS
TMAH Chemolysis	General	>100	TMAH-Heat	GC	ME	FI
CuO/NaOH	General	>100	CuO-NaOH	GC	TMS	FI

[a] Preparation indicates the usual method by which material is treated prior to chromatographic analysis: NPSE, non-polar solvent extraction; Hy, hydrolysis.
[b] Chrom. indicates the chromatographic method of choice: GC, gas chromatography; LC, liquid chromatography; IC, ion chromatography.
[c] Derivative includes: ME, methyl ester/ether; TMS, trimethylsilyl; OPA, *O*-phthaldialdehyde; PHL, phloroglucinol.
[d] Detector indicates the chromatographic detector of choice: FI, flame ionization; UV, UV adsorption; Flu, fluorescence emission; PA, pulsed amperometric; MS, mass spectrometric.

mass spectrometry (py-MS). In the last decade it has become possible to measure bulk properties (elemental compositions, spectral patterns, and isotopic compositions) on individual compounds as they elute from a gas or liquid chromatograph. These compound specific analytical methods have added a new dimension to our understanding of organic geochemistry by allowing broadly representative characterizations to be applied to molecule types of known biological or geographic origin. The following two sections are a brief description of some of the bulk and molecular methods that are commonly used to characterize organic matter in environmental samples. Example applications of many of these methods to seawater DOM are presented at the end of the chapter.

8.2.1 Bulk characterizations

The following discussion of methods by which bulk samples can be characterized is subdivided into elemental, isotopic, and spectroscopic methods. Many of these same methods can be applied to individual compound types when coupled with chromatographic techniques.

Elemental composition

The major elements occurring in pure organic substances, C, H, N, O, and S, can be routinely analyzed by a combination of methods involving combustion and pyrolysis (heating in the absence of oxygen). C, H, N, and S are measured simultaneously by high temperature ($>1000\,^{\circ}C$) combustion to CO_2, H_2O, N_2, and SO_2 gases, which are separated and quantified (Hedges and Stern, 1984; Verardo *et al.*, 1990). Typically, CHN (or CHNS) analyzers require less than a milligram of organic matter and minimal sample preparation (Cowie and Hedges, 1991). Organic oxygen can be analyzed directly in a second step by pyrolysis to carbon monoxide, which is then measured directly or after oxidization to CO_2. Alternatively, oxygen can be estimated as the difference between total sample mass and the sum of C, H, N and S. In practice these methods are compromised when dealing with samples that contain minerals and salts. The net result is that most elemental characterizations of bulk sediments are restricted to organic carbon and total nitrogen, the latter of which is often subject to appreciable error.

In spite of these potential complications, elemental analysis can meaningfully constrain the structural characteristics and biochemical compositions of individual organic materials and their mixtures. By examining the atomic quotients of different elements (i.e. H/C, O/C, N/C) distinctions between major biochemical groups can be made (Table 8.5). For example, lipids, polysaccharides, and proteins are characteristically rich in hydrogen ($H/C > 1.5$), whereas lignin and tannin are hydrogen-poor ($H/C > 1.1$). Oxygen occurs in relatively high amounts in polysaccharides ($O/C > 0.8$) and to a lesser extent in tannin, lignin, and protein ($O/C = 0.3$–0.5). Protein, RNA and DNA,

Table 8.5. *Elemental compositions of common biopolymers and plant materials*

Material	Formula	Example	H/C	O/C	N/C	S/C	[a]O_2/C
[b]Polysaccharide	$C_{100}H_{167}O_{83}$	Cellulose	1.67	0.83	0	0	1.00
[b]Protein	$C_{100}H_{158}O_{32}N_{26}S_{0.9}$	Net plankton	1.58	0.32	0.26	0.01	1.57
[b]Lipid	$C_{100}H_{189}O_{11}$	Oleic acid	1.89	0.11	0	0	1.42
[b]RNA	$C_{100}H_{100}O_{50}N_{10}$	Net plankton	0.92	0.32	0.400	0	1.57
[b]Chlorophyll	$C_{100}H_{140}O_{9.5}N_{7.5}$	Chlorophyll *a*	1.40	0.10	0.08	0	1.40
[b]Lignin	$C_{100}H_{108}O_{38}$	Gymnosperm	1.08	0.38	0	0	1.08
[b]Tannin	$C_{100}H_{66}O_{42}$	Angiosperm	0.66	0.42	0	0	0.96
[b]Marine plankton	$C_{100}H_{167}O_{35}N_{16}S_{0.4}$	Net plankton	1.67	0.35	0.16	0.004	1.45
Bacteria	$C_{100}H_{167}O_{35}N_{16}S_{0.4}$	Gram-negative	1.67	0.35	0.16	0.004	1.45
Wood	$C_{100}H_{100}O_{50}$	Gymnosperm	1.00	0.50	0	0	1.00
Tree leaf	$C_{100}H_{100}O_{50}N_{10}$	Angiosperm	1.00	0.50	0.10	0	1.13
Grass	$C_{100}H_{100}O_{50}N_{10}$	Tropical	1.00	0.50	0.10	0	1.13

[a] O_2 requirement for total respiration to CO_2, H_2O and HNO_3 was calculated by using the equation: $C_\alpha H_\beta O_\gamma N_\delta S_\sigma + \omega O_2 = \alpha CO_2 + \beta H_2O + \delta HNO_3 + \sigma SO_3$, where $\omega = 1.00\alpha + 0.25\beta + 1.25\,\delta + 1.5\sigma - 0.5\gamma$ (Hedges *et al.*, 2002).

[b] Calculated from representative structures (rather than directly measured).

and chlorophyll all contain appreciable nitrogen, which also occurs in amino-sugar polymers such as chitin.

These compositional patterns of the different biochemical types can be visualized when atomic H/C is plotted against atomic O/C (called a van Krevelen plot) and N/C vs. O/C (Fig. 8.2). Besides differentiating the relative contribution of different biochemical classes, van Krevelen plots can be used to examine changes in organic matter due to differing contributions of terrestrial to marine organic matter, or due to microbial degradation (respiration) of organic matter (Reuter and Perdue, 1984). Vascular (higher) plants, which are confined almost exclusively to land (with the exception of sea grasses), have adapted to the terrestrial environment by incorporating nitrogen-free polymers for structural support (e.g. lignin and cellulose) and protection (e.g. tannin, cutin, and suberin). Because higher plants strongly predominate on land, terrigenous organic matter is often characterized by low N/C values compared with aquatic organic matter, which is relatively nitrogen-rich owing to the contributions of plankton and bacteria (Table 8.5; Fig. 8.2). For microbial utilization of organic matter, oxygen must be added as the organic carbon is respired back to CO_2. Compounds that are rich in hydrogen and low in oxygen (e.g. in lipid, cutin, suberin and protein) have a high oxygen demand ($O_2/C > 1.4$) for complete respiration, whereas hydrogen-poor biochemicals with high oxygen content (e.g. tannin, lignin and polysaccharide) require little respirative oxygen ($O_2/C < 1.1$).

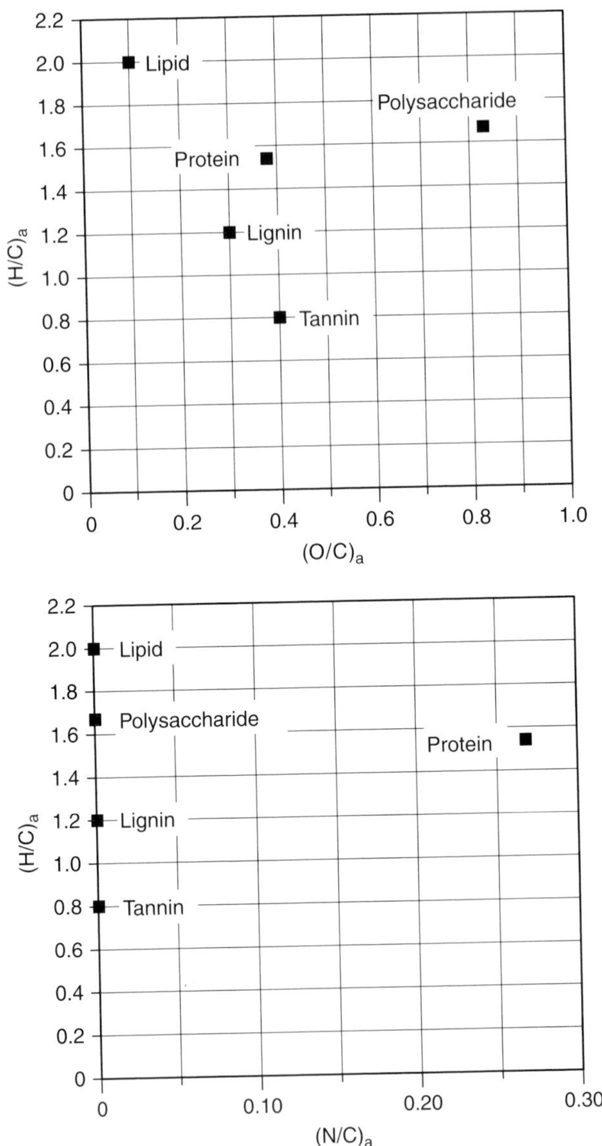

Figure 8.2. Van Krevelen plots of atomic H/C vs. O/C and H/C vs. N/C for major biochemical groups such as lipids and protein. By examining the atomic ratio of different elements, distinctions between major biochemical groups can be made. Coordinates for each biochemical indicate mean values; however, more variability exists than is shown.

Isotopic composition

Stable carbon isotope compositions of organic substances are measured via isotope ratio mass spectrometry. Samples are combusted to convert organic matter into gases (e.g. CO_2, N_2) that are then ionized and admitted into the isotope ratio-mass spectrometer (IR-MS) where the ratios of the isotopes are quantified (see Chapter 5 for further information). Owing to the low abundances of some of the isotopes, the precision and accuracy of the measurements is important (Merritt and Hayes, 1994). One of the most common and informative isotopes is ^{13}C because it is not readily altered by equilibration or degradation processes. Hence, $\delta^{13}C$ values are most commonly employed as indicators of ultimate plant or geographic sources. In

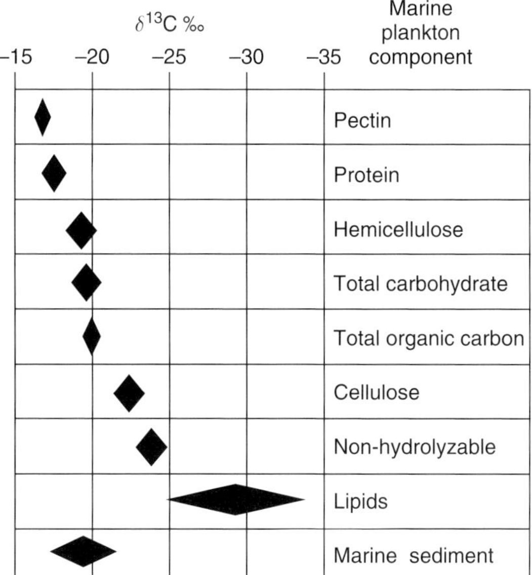

Figure 8.3. Stable carbon isotope compositions of different biochemical components found in marine plankton (from Degens, 1969). Biochemically based heterogeneity can be observed in the general trend of δ^{13}C becoming more depleted from proteins to carbohydrates to lipids, with total organic carbon representing an average of these contributions.

such applications, dissimilar carbon isotope distributions among various organic sources and biochemical compound classes can be both a complement and complication. For example, the average 6‰–7‰ more negative δ^{13}C value of common land plants compared with marine plankton (Rau, 1978; Forsberg *et al.*, 1993; Smith and Epstein, 1971) has been widely used to estimate the relative amounts of terrestrial versus marine organic matter in coastal marine sediments. The accuracy of this approach, however, depends on certainty of the isotope ratios of the two end members. For example, because of a strong input (*c.*40%) of ^{13}C-rich carbon from C4 plants growing in the US Great Plains, fine particulate organic material discharged by the Mississippi River into the Gulf of Mexico has a δ^{13}C near -20‰ (Onstad *et al.*, 2000). As a result, the contribution of this terrigenous organic matter to local marine sediments cannot be accurately determined based on stable carbon isotope measurements alone (Goñi *et al.*, 1997, 1998). In contrast, coarse particulate organic debris discharged by the Mississippi derives mainly from C3 plants and is isotopically recognizable within the river and in offshore marine sediments.

A second important consideration in stable carbon isotope analysis is that different biochemicals from the same organisms can have characteristically different δ^{13}C values (Hayes, 2001). Biochemically based heterogeneity has been recognized in both marine plankton and vascular land plant tissues (Degens, 1969). The general trend from proteins to carbohydrates to lipids is to become more depleted in ^{13}C (i.e. lighter), with total organic carbon representing an average of these contributions (Fig. 8.3). Thus, selective degradation of the major biochemical components of an organic matrix would lead to a diagenetic drift in its stable carbon isotope composition. Likewise,

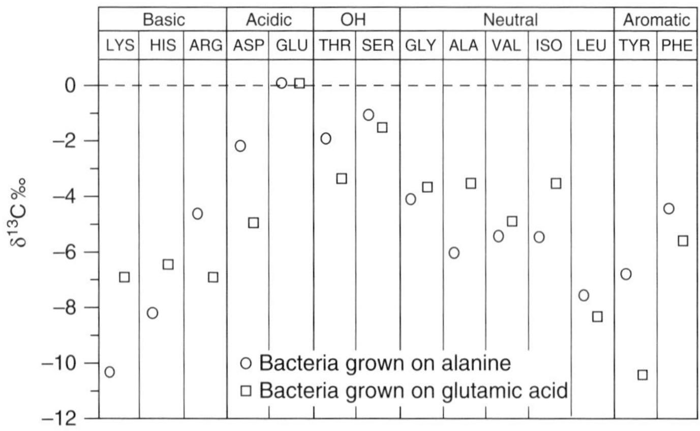

Figure 8.4. Relative stable carbon isotope compositions of individual amino acids in bacteria cultured on alanine and glutamic acid (from Macko *et al.*, 1987). Amino acids provide a good example of the wide range in $\delta^{13}C$ values that can exist within a compound class, largely owing to the many different biochemical pathways by which various amino acids are formed.

the stable carbon isotopic composition of individual organisms will change in parallel to their biochemical content, and especially with variation in the lipid to protein ratio attending reproductive cycles and dietary changes. Away from terrestrial influences, however, the $\delta^{13}C$ of organic matter in marine sediments typically reflects that of the locally predominant phytoplankton.

In some cases large contrasts in stable carbon isotope composition can be observed among individual compounds within biochemical families. The most common and pronounced example of such within-class variability is for amino acids, whose $\delta^{13}C$ values often vary over a range of more than 10‰ (Fig. 8.4). This high range of variability results largely from the many different biochemical pathways by which various amino acids are formed in living organisms. Because many of these pathways are shared by a broad diversity of organisms, the characteristic pattern of $\delta^{13}C$ offsets among individual amino acids can serve as an isotopic signature of molecules that were enzymatically formed versus abiotically produced in extraterrestrial settings. Not surprisingly, stable carbon isotope compositions also can vary systematically within particular molecules. Examples of this phenomenon are seen in the ^{13}C enrichment of the carboxyl carbon of an amino acid relative to the remaining carbons in the molecule, and the ^{13}C enrichment of alternating carbons relative to the adjacent carbon atoms in long-chain lipids, reflecting the heavy carboxyl and light methyl carbons in the acetate groups from which the lipids are formed (Hayes, 2001).

Spectroscopic methods

Spectral characterizations of organic matter involve the absorbance or emission of electromagnetic radiation from specific structural units, examples of which are the functional groups listed in Table 8.2. The methods are often grouped according to the wavelength (λ) of radiation involved, which increases from X-rays ($\lambda = 10^{-10}$–10^{-8} m) to radio waves ($\lambda = 0.1$–10^{4} m). Because the energy of electromagnetic radiation varies inversely with wavelength, energy decreases from cosmic rays ($\lambda = 10^{-12}$–10^{-14} m) to radio waves. Spectral characterizations have the advantage that they can be fast and provide unique information. They often share the disadvantages of being subject to interferences from inorganic components and being uninformative about how the detected atom groupings are linked together in large molecules. Major exceptions to the latter restriction are X-ray scattering and multi-dimensional nuclear magnetic resonance (NMR) methods. Ultraviolet, visible, infrared, Raman, and NMR spectroscopies are commonly utilized to characterize organic constituents in marine systems. Because of the increased use of NMR, we present a short review of its capabilities.

Nuclear magnetic resonance (NMR) involves exchange of extremely weak electromagnetic radiation in the radio waveband with the nucleus of atoms. NMR provides a unique perspective on electron distribution patterns about an individual atom that is a stable isotope (e.g. ^{1}H, ^{13}C). NMR is second only to X-ray crystallography in its power as a structural elucidation tool and has the distinct advantages that it can be applied to all forms of organic matter as well as to mixtures of organic compounds (Hatcher *et al.*, 1983; Preston, 1996). In contrast to mass spectral methods (discussed next), NMR does not require ionization and is applicable to any molecular size. The method is theoretically applicable to any isotope that has an odd number of either protons or neutrons in its nucleus, which rules out analysis of such common organic isotopes as ^{12}C and ^{16}O. The main disadvantages of NMR are that the instrument is expensive and that it requires milligram amounts of sample that are relatively free of paramagnetic metals such as Fe and Mn.

Much of the power of NMR spectroscopy is derived from relating measured chemical shifts, and the corresponding peak intensities, to chemical structures within organic substances. A chemical shift is the small offset of resonance (measured in parts per million, ppm) between the sample and a reference frequency. In general, nearby atoms (and functional groups) that withdraw electrons (deshield) from the target nuclide increase chemical shifts (resonance frequencies). Chemical shifts characteristic of different structural forms of ^{1}H and ^{13}C alone provide major constraints on the average structural characteristics of complex organic mixtures such as occur in biological tissues, seawater DOM and marine sediments (Benner *et al.*, 1992; Gélinas *et al.*, 2001). A typical ^{13}C spectrum (Fig. 8.5) for such a solid mixture, in this case for a marine plankton sample, exhibits much broader resonance peaks than the spike-like peaks obtained from pure organic compounds. Although such information is insufficient alone to establish the structures of large molecules in mixtures, it can constrain average structural features and the corresponding biochemical and elemental compositions. For example, the spectrum in Fig. 8.5 is sufficient to estimate that this sample contains approximately 65 weight percent (wt%) protein, 29 wt% lipid and 15 wt% carbohydrate, and has an elemental composition (in Redfield format) of $C_{106}H_{177}O_{37}N_{17}$ (Hedges *et al.*, 2002) (see Chapter 6).

Figure 8.5. A "solids" ^{13}C NMR spectrum, for a marine plankton sample (from Hedges *et al.*, 2002). Although NMR only gives a general overview, one can use this spectrum to estimate that this sample contains approximately 65 wt% protein, 29 wt% lipid and 15 wt% carbohydrate, and has an elemental composition (in Redfield format) of $C_{106}H_{177}O_{37}N_{17}$.

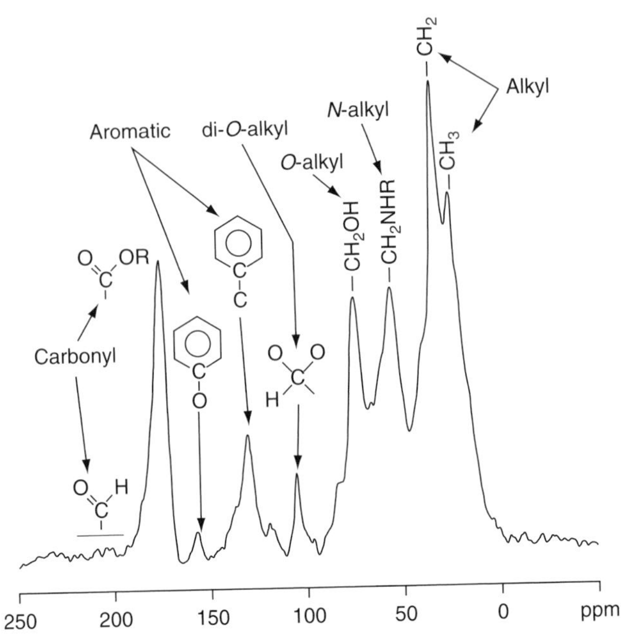

8.2.2 Molecular characterizations

For the purposes of this discussion, molecular characterization will be defined as the analysis and quantification of an essentially pure type of organic compound or compound class. Because natural organic materials usually contain many molecular components, their characterization typically involves several preparation and isolation steps prior to analysis. For biopolymers (e.g. polysaccharides and proteins), a common preparation step is to break the parent biochemical down chemically into its structural units (e.g. sugars and amino acids) that can be chromatographically separated and quantified. The analysis step may involve simple detection of individual compounds as they elute from a liquid chromatograph or other separation system, coupled with characterization methods (often spectral or mass-based) that have sufficiently high sensitivities and fast enough response times to operate in a continuous flow mode.

Chromatographic methods

The net result of chromatographic separations is that a mixture of compounds, introduced at the head of a column, is subsequently separated in space. Separation can be done in the gas (gas chromatography, GC) or liquid (liquid chromatography, LC) phase. Each technique has its advantages and problems. In both cases, separation is carried out through a column which is filled with a material designed to interact with the compounds of interest and causing them to travel at different rates so they elute (exit the column) at a characteristic time for a given set of operating conditions. There are additional variants on this general theme, as with gas chromatography, where compounds do not begin chromatographic separation until the temperature of the column head exceeds the boiling point of the individual mixture components. In essentially all chromatographic separations, the various eluents are quantified with a variety of detectors, to give a chromatographic trace in which each compound has a characteristic retention time and intensity.

Mass-based characterizations

Mass spectrometric (MS) analysis of an organic material involves the four steps: sample introduction to the spectrometer, ionization, mass analysis, and detection. Mass spectrometry is notable for its high sensitivity (to sub-nanogram amounts of material) and applicability to samples in all physical forms (liquid, solid and gas). The method can be used to (1) determine molecular masses and formulas; (2) identify compounds based on their characteristic fragmentation patterns; (3) establish structural unit sequences (as for amino acids in peptides); and (4) follow isotope incorporation and fractionation during natural processes. The first two of these applications are most broadly applied in marine organic geochemistry to characterize the structures of organic molecules; the third is a relatively new technique and is just coming into use for geochemical applications. The fourth technique, isotope ratio mass spectrometry (IR-MS), is typically used as a bulk method (discussed in the previous section).

Electron impact (EI) is the most widely used ionization method, especially for gas chromatography – mass spectrometry (GC–MS). As compounds elute from the GC column, they are introduced into the mass spectrograph, where they are bombarded by a stream of high-energy electrons. The energies of the electrons comprising the ionization beam can be adjusted from low (*c*.10 electron volts; eV) to high (*c*.80 eV) average energies, which in turn lead to "soft" versus "hard" ionization. In an ideal soft ionization, the incoming electron has just enough energy to knock a valency electron out of the target molecule, which produces only an M^+ ion. Hard ionization involves high-energy electrons that impart enough additional energy to cause the M^+ ion to subsequently shatter into multiple pieces called fragments. The contrast between mass spectra obtained from hard (EI) versus soft ionization (via chemical ionization) of a simple organic molecule is illustrated in Fig. 8.6. Molecular ions are useful because they indicate the mass of the intact molecule, but alone may not provide sufficient information for unambiguous identification. Multiple fragments on the other hand provide a characteristic pattern, but may not indicate the mass of the parent compound. Typically, choosing an intermediate stage of fragmentation, including the molecular ion and multiple high-mass fragments is ideal. Because different organic molecules exhibit differing tendencies to fragment, ionization conditions must be varied accordingly. In general, fragmentation tends to occur adjacent to functional groups and carbon branch points, but a detailed discussion of structural analysis based on mass spectral fragmentation is beyond our scope here.

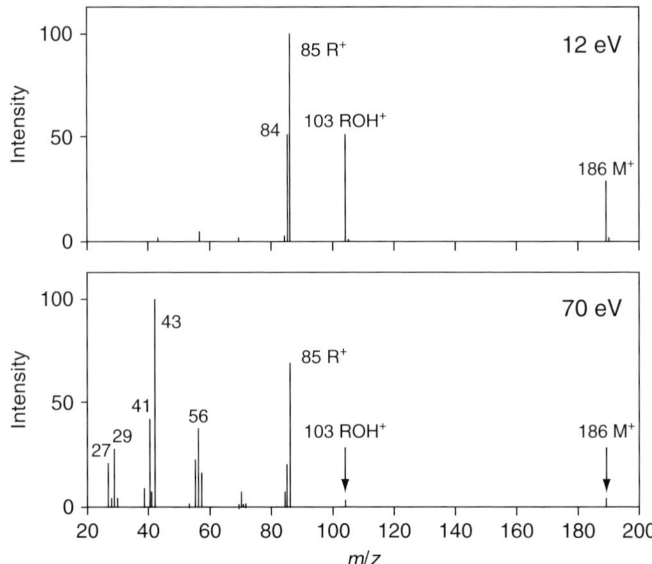

Figure 8.6. A comparison of mass spectra of dihexyl ether (molecular formula: $CH_3(CH_2)_5O(CH_2)_5CH_3$) obtained with low- (12 eV) and high- (70 eV) energy electron ionization beams. In these figures *m/z* indicates the mass to charge ratios of the measured ions, whereas the vertical axis represents their intensities relative to the "base" (most intensive) ion in that spectrum. From Lambert *et al.* (1998).

Fragmentation spectra can be considered a fingerprint for a given compound. The mass spectrum of an unknown can be compared to a library of spectra obtained under similar instrumental conditions allowing identification of the compound. If no corresponding reference spectrum is found, strong structural inferences can often be drawn from the observed fragmentation pattern and known isolation conditions. Two-stage MS analysis can also now be performed wherein the ion of interest is isolated and then fragmented. This provides an exact fingerprint for that specific compound and can be used for compound identification and structure elucidation.

Recently, powerful tools have been developed by combining MS with chromatographic techniques such as capillary electrophoresis (CE), which allows for the separation of compounds based on their polarity followed by immediate mass analysis or high-performance liquid chromatography (LC), which also separates compounds based on polarity and interaction with the column packing material. Once the compounds in a given sample are separated via CE, LC or GC techniques they are ionized at the entrance to the mass spectrometer. Recent advances have introduced electrospray ionization (ESI) and atmospheric pressure chemical ionization (APCI) as new soft ionization techniques amenable for use after separation via liquid chromatographic techniques. Another technique that is now being utilized is to imbed the sample into a crystal matrix and use a laser to impart energy to the molecules, thereby ionizing them. Once ionized, the compounds leave the matrix and enter the mass spectrometer. This technique is called matrix assisted laser desorption ionization mass spectrometry (MALDI-MS) and has the advantage that it can be applied to very large molecules and provide detailed molecular mass information. Overall, mass spectrometry is an important and highly useful tool, especially to examine biomarker compounds.

8.3 Major organic carbon compounds as biomarkers

Among the major applications of organic molecules to studies of natural systems is their use as source and process indicators. Such "biomarker" methods are based on the fact that certain types of biochemical, and their characteristic alteration products, occur only in particular types of living organism and under specific environmental conditions. Because some kinds of organism are in turn restricted in their geographic distribution (e.g. land versus aquatic) or physicochemical setting (e.g. illuminated sulfidic waters), biomarkers can have rather broad applications as environmental indicators. In general, proteins, lipids, and carbohydrates are widely distributed and abundant types of biochemicals in living organisms and their tissues, whereas other biopolymers such as lignin and tannin are unique to vascular plants and are considered indicators of terrigenous organic matter (Table 8.6). Biomarkers can also be applied as paleoindicators provided that they exist within the physical record and that their carbon skeletons are stable over long periods.

Sometimes source indicators retain fidelity through extensive alteration processes over extended time periods. The classical, and first, example of such persistence was the discovery by Alfred E. Treibs in the 1930s of tetrapyrrol ring systems derived from chlorophyll pigments in ancient rocks and petroleum (Treibs, 1934). These unmistakable "molecular fossils" proved that at least part of the organic matter in crude oil is biologically derived, putting to rest speculation that petroleum was abiotically formed. Treibs is now recognized as the father of organic geochemistry.

Biomarker compounds, and other less source specific biochemicals, often exhibit degradative changes that do not completely mask their characteristic structures. Hence, these compounds can also carry information about the extent, and sometimes the mechanism, of their alteration. The processes that produce these structural changes can be

Table 8.6. *Major biochemical compositions by % carbon for various common organisms*
See Table 8.5 for representative elemental compositions for these biochemical types.

Material	Protein	[a]Polysac.	Lipid	Pigment	Nucleic acid	Lignin	Tannin
Bacteria	55–70	3–10	5–20	2–5	20	0	0
Phytoplankton	25–50	5–50	5–20	3–20	20	0	0
Zooplankton	45–70	3–5	5–20	1–5	20	0	0
Vascular plant	2–5	37–55	<3	5–20	<1	15–40	<20
Wood	<1	40–80	<3	0	<1	20–35	<45

[a] Polysaccharides, including neutral, basic and acidic sugars.

either biological (e.g. biodegradation) or abiotic (including photochemical processes, redox reactions, and a host of spontaneous alterations during geothermal heating and incomplete combustion). The application of biomarkers as both source and degradation tracers will be presented together, because these topics are closely related and also draw on a common body of structural and physicochemical information. The following sections describe the structural characteristics and biomarker applications of the major biochemical types.

8.3.1 Amino acids

Amino acids are extremely abundant, highly polar compounds that occur predominantly in combined (protein) form within all living organisms. Because proteins account on average for 60–70 wt% of marine plankton, amino acids are the single most abundant structural units in marine biomass (Table 8.6). Individual amino acids share the same fundamental structural sequence of a carboxyl group, amine, hydrogen, and R group (of varying structure) substituted onto a single carbon (Fig. 8.7). The pK_a values of the carboxyl (COOH) and amino (CNH_2) groups of free amino acids are such that both groups occur in ionized form (COO^- and CNH_3^+) over natural pH ranges. Although there are thousands of possible ways in which these structural units can combine to form molecules, only 20 amino acids occur commonly in proteins. Over 100 additional non-protein amino acids occur in nature and have various roles in living organisms. Neutral protein amino acids, of which there are 15, have one carboxylate anion (−1) and one ammonium cation (+1) per molecule, and are called zwitterions because they have a net charge of zero. In contrast, the two acidic protein amino acids (with surpluses of carboxyl versus amine groups) and three basic counterparts (with surpluses of amine versus carboxyl groups) have negative and positive net charges, respectively. Neutral amino acids also can be further subcategorized into polar (7 total) and non-polar (8 total) groups (Fig. 8.7). The generally high polarity of all amino acids renders them extremely water soluble and non-volatile. Amino acids can be analyzed after strong acid hydrolysis and derivatization by either GC or high-pressure liquid chromatography (HPLC), although HPLC analysis with detection of fluorescent derivatives is most commonly used (Lindroth and Mopper, 1979; Cowie and Hedges, 1992a).

Even though the amino acid compositions of individual proteins can be quite different, most living organisms are compositionally quite similar (Cowie and Hedges, 1992b). The few geochemically useful exceptions to this pattern include unusually high mole percentages of aspartic acid in calcium carbonate shells, and high compositions of serine in opal shells (Constanz and Weiner, 1988; Kröger *et al.*, 1999). Apparently, the high abundances of carboxyl and hydroxyl R groups of these amino acids help them serve as templates for active precipitation of structurally similar minerals. The other major exception is the high relative abundance of the D-stereoisomers of alanine, aspartic acid, glutamic acid and serine in bacteria. These

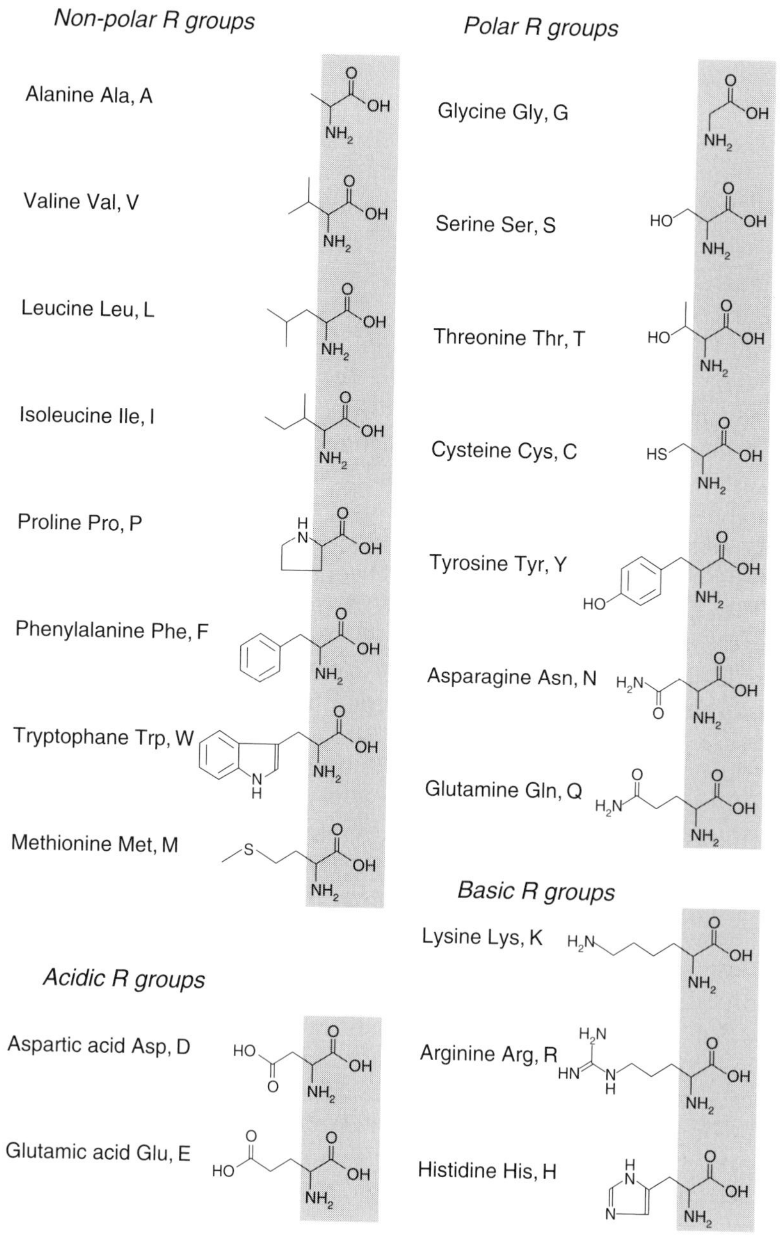

Non-polar R groups

Alanine Ala, A

Valine Val, V

Leucine Leu, L

Isoleucine Ile, I

Proline Pro, P

Phenylalanine Phe, F

Tryptophane Trp, W

Methionine Met, M

Polar R groups

Glycine Gly, G

Serine Ser, S

Threonine Thr, T

Cysteine Cys, C

Tyrosine Tyr, Y

Asparagine Asn, N

Glutamine Gln, Q

Acidic R groups

Aspartic acid Asp, D

Glutamic acid Glu, E

Basic R groups

Lysine Lys, K

Arginine Arg, R

Histidine His, H

Figure 8.7. Structures of the individual protein amino acids. Shaded area indicates the common structure present in all amino acids while the R group is given to the left. Amino acids are also classified by the functionality of their R group into non-polar, polar, acidic and basic categories.

amino acids occur in characteristically high amounts in peptidoglycan, a structurally complex biomacromolecule that makes up a portion of bacterial cell walls (Ghuysen and Shockman, 1973). D-amino acids occur (along with meso-diaminopimelic acid) in peptide bridges that crosslink polysaccharide chains of alternating *N*-acetylglucosamine and *N*-acetylmuramic acid structural units. Relatively high abundances of these specific D-amino acids are thus taken as being indicative of bacterial remains, as opposed to thermal racemization (McCarthy *et al.*, 1998; Amon *et al.*, 2001).

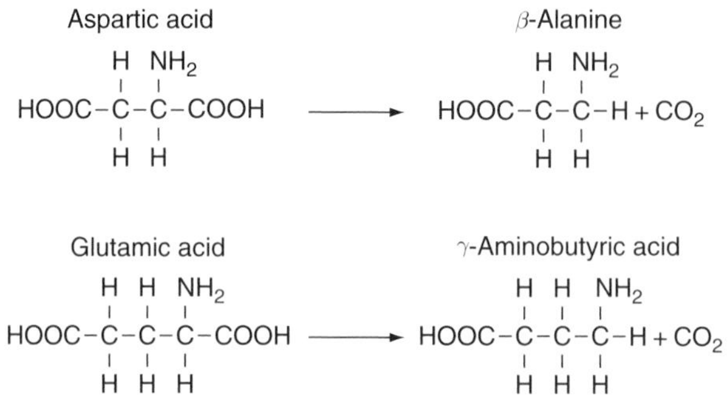

Figure 8.8. The formation of β-alanine and γ-aminobutyric acid from decarboxylation of aspartic and glutamic acid, respectively, is a widely observed trend in the degradation of amino acids. The exact mechanism of formation for these two degradation products is not known in detail.

```
Aspartic acid                            β-Alanine
   H  NH2                                   H  NH2
   |  |                                     |  |
HOOC–C–C–COOH          ——→              HOOC–C–C–H + CO2
   |  |                                     |  |
   H  H                                     H  H

Glutamic acid                            γ-Aminobutyric acid
   H  H  NH2                                H  H  NH2
   |  |  |                                  |  |  |
HOOC–C–C–C–COOH        ——→              HOOC–C–C–C–H + CO2
   |  |  |                                  |  |  |
   H  H  H                                  H  H  H
```

A major advantage of the nearly uniform amino acid compositions of living organisms is that degradation-derived changes usually are readily evident against this uniform background. One widely observed trend in amino acid composition is a steady increase in the two non-protein amino acids, β-alanine and γ-aminobutyric acid (Fig. 8.8), from mole percentages of 1 or less in living organisms to values in excess of 30 in highly degraded organic mixtures from open ocean red clays (Lee and Cronin, 1982; Wakeham *et al.*, 1997). The mechanism of formation of these two degradation indicators is not known in detail, but apparently involves loss of the carboxyl group adjacent to the amino carbon of the corresponding acidic amino acids, aspartic and glutamic acid (Fig. 8.8). Other common amino acid trends in unmineralized organic matter include increases in the mole percentages of alanine and glycine as degradation proceeds. The overall trends in amino acid composition that generally follow progressive degradation of organic matter have been consolidated in the Dauwe Index, a statistical parameter derived from principal component analysis that ranges from a value near +1 for fresh biomass, to −2 for heavily degraded organic matter such as occurs in pelagic marine sediments (Dauwe and Middelburg, 1998). The percentages of total organic carbon and nitrogen that occur in chromatographically measurable amino acids also are a useful "freshness" indicator for marine organic matter, which is characteristically protein-rich. However, this parameter does not work for terrestrially derived organic matter owing to dilution by structural biopolymers such as cellulose, hemicellulose, lignin, tannin and cutin.

8.3.2 Carbohydrates

As the name implies, carbohydrates are characterized by a nominal $(CH_2O)_n$ composition. Next to proteins, carbohydrates combined into polysaccharides are generally the most abundant (Table 8.6) and widely distributed biopolymers in living organisms. (However, there are some plankton in which the lipid concentration exceeds that of carbohydrates (Wakeham *et al.*, 1997).) The fundamental monomeric units of most carbohydrates are five- (pentose) and six- (hexose) carbon sugars, of which there are roughly ten common forms (Fig. 8.9). These structural units usually occur as either aldoses or ketoses, in which either the first (aldose) or second (ketose) carbon

Common neutral sugars

Pentoses			Hexoses				
Aldoses						Deoxy Sugars	
CHO	CHO	CHO	CHO	CHO	CHO	CHO	CHO
HCOH	HOCH	HCOH	HCOH	HOCH	HCOH	HCOH	HOCH
HCOH	HCOH	HOCH	HOCH	HOCH	HOCH	HCOH	HCOH
HCOH	HCOH	HCOH	HCOH	HCOH	HOCH	HOCH	HCOH
CH_2OH	CH_2OH	CH_2OH	HCOH	HCOH	HCOH	HOCH	HOCH
			CH_2OH	CH_2OH	CH_2OH	CH_3	CH_3
D-Ribose	D-Arabinose	D-Xylose	D-Glucose	D-Mannose	D-Galactose	L-Rhamnose	L-Fucose

Other types of sugar

Ketose	Uronic acid	Aldonic acid	Aldaric acid	Lactone	Amino sugar
CH_2OH	CHO	COOH	COOH	CHO	CHO
C=O	HCOH	HCOH	HCOH	HC	HC - NH_2
HOCH	HOCH	HOCH	HOCH	HOCH	HOCH
HCOH	HCOH	HCOH	HCOH	HCOH O	HCOH
HCOH	HCOH	HCOH	HCOH	HCOH	HCOH
CH_2OH	COOH	CH_2OH	COOH	HC	CH_2OH

Figure 8.9. Structures of common neutral aldoses and other types of simple sugars. Five- (pentose) as well as six- (hexose) carbon sugars are common, as are deoxy sugars, where the oxygen is lost from the last carbon. Other sugars involve further modifications in the location and number of oxygen atoms within the sugar.

has a carbonyl (C = O) substitution. Less common than these neutral sugars are acidic sugars containing one or two carboxyl groups (COOH) on the terminal carbons, and basic sugars with one or more amine groups (NH_2) attached. Individual sugars are usually linked together in short (oligosaccharide) to long (polysaccharide) chains, in which they occur almost exclusively in the ring form. The linkage between carbohydrate monomers is formed via a dehydration reaction and can be reversed by chemical or enzymatic hydrolysis to yield individual sugars. Because most carbohydrates are biosynthesized in the form of polysaccharides, their analysis necessitates acid hydrolysis (often with H_2SO_4) to release the carbohydrate monomers, which are then quantified by various types of liquid or gas chromatography (Cowie and Hedges, 1984a,b; Skoog and Benner, 1997). Owing to their high polarity and potential for hydrogen bonding, simple sugars and saccharides are non-volatile and highly soluble in water.

Owing to the wide distribution of individual sugars in living organisms, seldom is an individual compound unambiguously characteristic of a given biological or geographic source. However, there are a few exceptions to this, one of which is muramic acid, a component of the peptidoglycan of bacterial cell walls. But for most other purposes, source analysis is based on characteristic abundance patterns among the major ten or so sugars that make up living organisms (Fig. 8.9). For example, high percentages of glucose are typical of vascular land plants. In particular, woods contain approximately 40 wt% of cellulose, a polysaccharide made up only of glucose. Because trees contain by far most of the biomass on Earth, cellulose is the most abundant polysaccharide, and glucose the most abundant biochemical, in existence. Almost as abundant is hemicellulose, a type of more water-soluble polysaccharide that is rich in mannose among flowering vascular plants (angiosperms) and in xylose among non-flowering vascular plants (gymnosperms). Owing largely to the combined contributions of cellulose and hemicelluloses, fresh vascular plant tissues exhibit characteristically high yields of total neutral sugars that are often in excess of 25 wt% of total tissue mass (Hedges *et al.*, 1985). In addition to giving much lower total yields of neutral sugars, marine organisms tend to yield relatively high amounts of ribose, which is the only sugar component of ribonucleic acid (RNA). The few types of bacteria that have been studied yield high levels of the two deoxy-sugars, rhamnose and fucose. High percentage yields of these two six-carbon sugars have been applied as indicators of more highly degraded marine and terrestrial organic matter, as both have low overall yields of glucose and total sugars (Hedges *et al.*, 1988; Amon *et al.*, 2001). Because carbohydrates are relatively labile and prone to selective degradation, they have found limited use as source indicators in degraded organic matter and sediments.

8.3.3 Lipids

Lipids encompass a large group of compounds that are categorized into two main structural families. Straight-chain lipids are referred to as being normal, which is indicated by the prefix n. The subscript on the carbon symbol, C, indicates the number of carbon atoms in the chain. Thus a straight-chain hydrocarbon containing 17 carbon atoms is represented as n-C_{17} (Fig. 8.1). Such linear compounds are biosynthesized from acetyl CoA subunits ($CoA - CO_2CH_3$) that are linked together (with loss of internal oxygen) to form unbranched hydrocarbon chains with different terminal functionalities (e.g. alcohol, ketone and acid groups). Sometimes, especially in bacteria, the two-carbon chain elongation process begins with a larger structural unit that has a methyl branch at the second (iso-) or third (anteiso-)carbon from the unsubstituted end, thereby producing an iso- or anteiso-carbon skeleton.

The second major structural family is produced by a biosynthetic pathway that involves joining together various numbers of a methyl-branched, doubly unsaturated, five-carbon structural unit called an isoprene. Several isoprene units are joined to form a variety of isoprenoid molecules, such as squalene, which contain a total number of carbons that is a multiple of five (in this case 30). The "ene" ending indicates that the hydrocarbon is an alkene, containing one or more carbon double bonds (Fig. 8.1). Such isoprenoid molecules are often intermediates in the formation of cyclic products containing one to five or more rings and various types of functional groups. Isoprenoid compounds are classified based on the number (n) of isoprene units they contain, with commonly occurring examples being monoterpenoid ($n = 2$), diterpenoid ($n = 4$), and triterpenoid ($n = 6$) substances. These are cyclic molecules in which fused rings are in a flat conformation (all trans-form) such as sterols, which are all trans-triterpenoid compounds with three methyl groups removed from their tetracyclic ring system. Sites of ring formation, as well as most of the internal methyl-branched carbons, in isoprenoid compounds are chiral carbons. Hence, these types of molecules typically carry a wealth of embedded stereochemical information.

Hydrocarbons

The hydrocarbons, which occur in ancient rocks and petroleum and thus have extensive economic and paleoceanographic importance, are the most studied class of lipids. Their broad distribution in ancient samples results in part because, upon heating in reducing environments, organic substances often lose polar functional groups such as OH, NH_2, and COOH in the form of stable small molecules such as water, amines, and carbon dioxide. Common co-products are aliphatic hydrocarbons, aromatic hydrocarbons and (in extreme cases) graphite. Mild loss of functional groups causes minimal changes in carbon backbones, such that characteristic structural features of parent alcohols, amines, and acids are retained by their hydrocarbon products. More advanced structural alterations include the formation of aromatic ring systems and sometimes the loss of alkyl (hydrocarbon) side chains. Thus source-specific structural patterns may be expressed across multiple compound classes as original biochemicals are altered to various forms as a result of biological degradation or thermal alteration. As shown in Fig. 8.10, biochemicals such as sterols can undergo

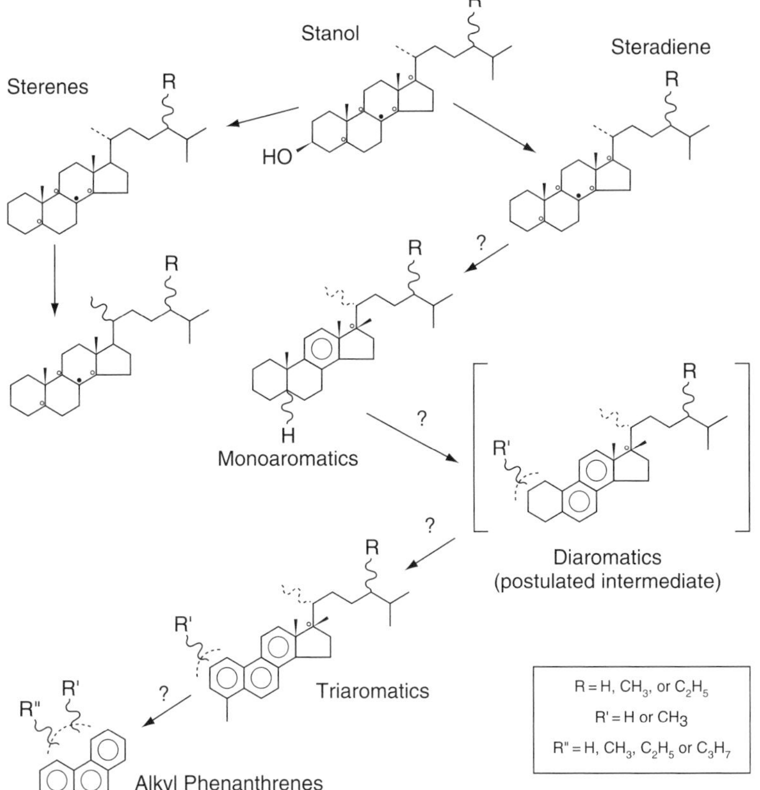

Figure 8.10. Stages in the biogeochemical alteration of sterols to aliphatic and aromatic hydrocarbon products (from Mackenzie *et al.*, 1982). Sterols provide a good example of how some compounds can undergo substantial biogeochemical alteration and yet retain structural characteristics that are clearly traceable to their hydrocarbon source.

Table 8.7. *Selected hydrocarbon biomarkers for different biological sources*

Organism	Major hydrocarbons	Comments
Diatoms and dinoflagellates	n-$C_{21:6}$ (HEH[a])	plus highly branched, C_{20}, C_{25} and C_{30} alkenes in some diatoms
Red, green and yellow algae	n-C_{15} and n-C_{17}	some of these algal types also contain unsaturated n-C_{17} hydrocarbons
Cyanobacteria	n-C_{15} and n-C_{17}	also sometimes contain n-$C_{19:1}$ and n-$C_{19:2}$, and 7- and 8-methylheptane
Bacteria	n-alkanes	typically have smooth distributions of *n*-alkanes over the range of C_{15}–C_{30+}
Zooplankton	pristane in copepods, HEH in zooplankton-eating diatoms	pristane is derived from the phytol side chain of chlorophyll pigments
Vascular land plants (especially leaf cuticles)	n-alkanes in the range of n-C_{25} to n-C_{35+}	characterized by a high odd-carbon preference index (CPI) of 5–10

[a] HEH, all-cis-3,6,9,12,15,18-heneicosahexaene.

substantial biogeochemical alteration and yet retain structural characteristics that are clearly traceable to this source.

Hydrocarbons occur in numerous subclasses (Fig. 8.1) characterized by the presence of carbon rings, double bonds, and branching (near-terminal and isoprenoid), all of which can carry some source information. Table 8.7 is a list of some of the major types of living organism and selected hydrocarbons they contain that, when abundant, may be characteristic of the biological source. Hydrocarbons in living organisms contain an odd number of carbon atoms because they are formed by removal of the last (functionalized) carbon from even carbon numbered intermediates that were biosynthesized from acetate subunits. In general, polyunsaturated hydrocarbons are of planktonic origin. Marine algae tend to biosynthesize relatively short-chain ($<C_{24}$) hydrocarbons, whereas vascular land plants make longer-chain analogs. The hydrocarbons produced by vascular plants occur primarily in association with leaf waxes (see below, fatty acids). Such so-called plant wax *n*-alkanes exhibit a strong predominance of odd versus even carbon numbered molecules. This characteristic distribution can be quantified by a carbon preference index (CPI), which for any range of carbon numbers (*n* to *m*) in a series is defined as:

$$CPI_{m \to n} = \frac{2\sum_{i=0}^{i=(n-m)/2} C_{m+2i}}{\sum_{i=0}^{i=(n-m)/2} C_{m-1+2i} + \sum_{i=0}^{i=(n-m)/2} C_{m+1+2i}}, \tag{8.1}$$

$$n > m \text{ and } \mathrm{mod}(m+n)/2 = 0.$$

The $CPI_{25\text{-}35}$ for plant wax *n*-alkanes is in the range of 5–10, meaning on average that each odd-numbered hydrocarbon is 5–10 times more abundant than the even-numbered molecules with one more or one fewer CH_2 group in the structure. Plant wax *n*-alkanes are remarkably resistant to biodegradation, in part because they are embedded in small waxy particles from leaf surfaces. Many of the hydrocarbons occurring in animals are ultimately obtained via their diets from plants. For example, the most abundant hydrocarbon in calanoid copepods is pristane, a saturated C_{19} isoprenoid derived from the phytol side chain of chlorophyll *a* in their phytoplankton diet. Zooplankton feeding on a diet of diatoms can similarly accumulate an n-C_{21} hydrocarbon with six double bonds (n-$C_{21:6}$; the number after the colon indicates the number of double bonds).

In contrast to these relatively simple biochemical patterns, hydrocarbon mixtures in fossil materials such as petroleum and ancient rocks are characteristically complex in composition. This complexity results in part from mixed biological and biochemical sources to sedimentary materials. Because heating causes loss of functional groups, fossil hydrocarbons also are often produced from polar compounds that give rise to structural characteristics never seen in biosynthetic counterparts. At least equally important, however, are extensive structural rearrangements of both biosynthetic and thermally formed hydrocarbons, which include chain cleavage, alkyl group migration, double bond formation, and destruction, plus extensive inversions of stereochemistry. One manifestation of such thermal scrambling is production of extended homologous series of hydrocarbon structural types such as the *n*-alkane series with CPI values near one (e.g. near-equal abundances of adjacent odd and even carbon numbered molecules).

Aromatic hydrocarbons, characterized by the presence of fused benzene-like rings (Fig. 8.1), are referred to as being polynuclear if composed of more than two rings. Polynuclear aromatic hydrocarbons (PAHs) are geochemically interesting because they have essentially no direct biosynthetic sources and yet are widely distributed in sediments of essentially all geological ages. Some aromatic hydrocarbons owe their existence to formation during relatively low-temperature biodegradation and geothermal heating from a variety of cyclic precursors such as sterols (see, for example, Fig. 8.10), resins and pigments. Such PAHs usually occur in restricted suites characterized by the predominance of a limited number of ring types with extensive ring substitution by alkyl groups. In contrast, incomplete combustion (pyrolysis) of a wide variety of organic fuels (e.g. vegetation, petroleum and coals) produces an extended suite of PAHs that includes essentially all imaginable combinations of five- and six-ring structures. As opposed to low-temperature counterparts, combustion-derived PAHs exhibit a strong predominance of rings lacking alkyl substitution.

Fatty acids, fats and waxes

By definition, a fatty acid is a carboxylic acid that contains twelve or more carbons terminating in a carboxyl group. Because carboxyl groups disassociate at pH above 4–5, free fatty acids occur as negative carboxylate anions ($R-COO^-$) in seawater and marine sediments. Although fatty acid salts are more water-soluble than hydrocarbons,

Figure 8.11. Chemical structures of representative triacylglycerols and waxes, which are formed by fatty acids linked via ester bonds to alcohols (to form triacylglycerols or waxes) or to phosphate groups (to form phospholipids). Triacylglycerols consist of one to three fatty acids linked to the three-carbon alcohol glycerol.

Triacylglycerol

Wax

Phospholipid

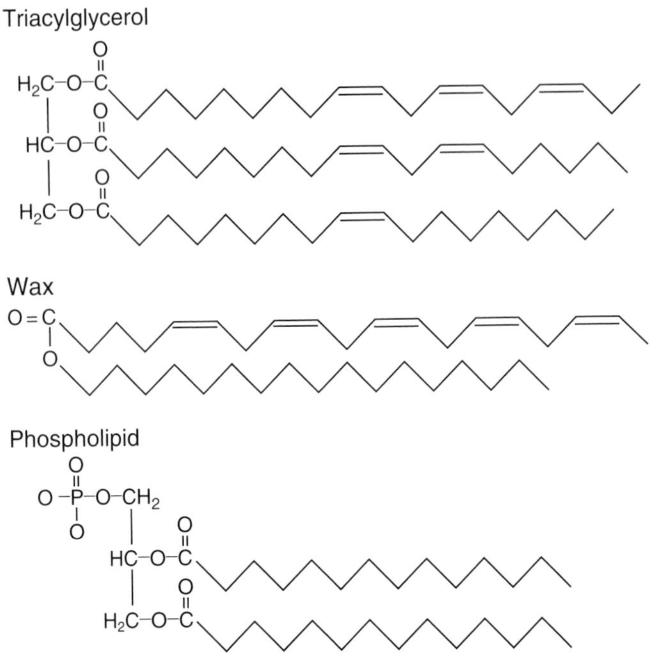

their solubilities still are only of the order of tens of milligrams per liter (ppm). Linear fatty acids are made from the same biosynthetic pathways as hydrocarbons, however, without the final decarboxylation step. Thus, plant fatty acids contain predominantly even numbers of carbon atoms, but with the same pattern of longer chain length that is typical of vascular land plants. Likewise, high degrees of polyunsaturation are typical of plants, with n-$C_{18:2}$ and n-$C_{18:3}$ being common among both marine and vascular plants, with increasing unsaturation (e.g. n-$C_{18:4}$, n-$C_{20:4}$, n-$C_{20:5}$, n-$C_{22:5}$ and n-$C_{22:6}$) among various phytoplankton.

As opposed to hydrocarbons, which are not associated with other compounds, fatty acids in nature are usually linked by ester bonds (R–COO–R′) to alcohols to form triacylglycerols or waxes, to sugars to form glycolipids and to phosphate groups to form phospholipids. Triacylglycerols consist of one to three fatty acids linked to the three-carbon alcohol glycerol (Fig. 8.11) and are referred to as fats if solid, and oils if liquid, at room temperature. Waxes are esters of fatty acids with fatty alcohols, which likewise contain more than twelve carbons per molecule (Fig. 8.11). Waxes are formed by most land plants and some marine animals, such as copepods, which use them for energy storage. Because polar lipids (phospholipids) are degraded extremely rapidly outside of cells and have many source-specific structural units, they are often used to quantify and identify living biomass in natural samples.

Lipids can also be combined with saccharides to form lipopolysaccharides (lipids covalently linked to small saccharides). Although little work has been done on the geochemistry of lipopolysaccharides, they can be used as specific source indicators that contain a broad spectrum of ester-bound components such as normal and branched fatty acids and also hydroxy-fatty acids. The presence of hydroxy-fatty acids has even been used as an indicator for the presence of Gram-negative bacterial cell wall material in dissolved organic matter (Wakeham *et al.*, 2003).

Sterols

Sterols are a diverse class of demethylated triterpenoid alcohols that occur in a diverse array of living organisms (Volkman, 2005). These polar molecules perform multiple roles in living organisms, including serving as hormones and as stabilizers of lipid bilayers. Sterols are primarily biosynthesized by plants, from which most animals derive their counterparts by conversion to simpler, cholesterol-dominated assemblages. Bacteria do not biosynthesize or accumulate sterols. Because sterols are often biosynthesized in taxonomically related patterns, they have broad potential as plant biomarkers. The ability of sterols to retain characteristic structural features through relatively severe alteration conditions, including prolonged geothermal heating, has already been stressed (Fig. 8.10). Sterol sources, however, are often so diverse as to make source assignments challenging. Some most commonly applied tracer applications include ergosterol for fungi, dinosterol for dinoflagellates, and β-sitosterol for vascular land plants (Volkman, 2005).

Alkenones

Long-chain alkenones containing unbranched sequences of 37 carbons with 2–4 unconjugated trans (E) double bonds (Fig. 8.12) have sparked great geochemical interest over the past two decades. This attention results from the observation that these structurally unusual lipids are biosynthesized by only a few marine haptophyte algae (e.g. *Emiliania huxleyi* and *Gephyrocapsa oceanica*) in patterns that reflect ambient water temperature. The general pattern is for unsaturated alkenones to be produced in greater relative abundance at lower water temperatures because the introduction of the double bonds increases the fluidity of the membrane (Prahl and Wakeham, 1987). Remarkably, despite substantial degradation there appears to be little selectivity during degradation of alkenones containing two, three, or four double bonds, so that the temperature signal persists relatively

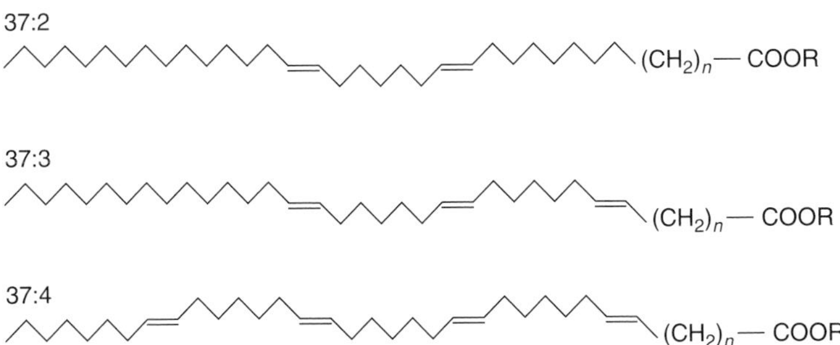

Figure 8.12. The three most commonly occurring C_{37} alkenones. The designation 37:2, 37:3 and 37:4 indicates the number of carbons (37) and the number of double bonds (2, 3, and 4).

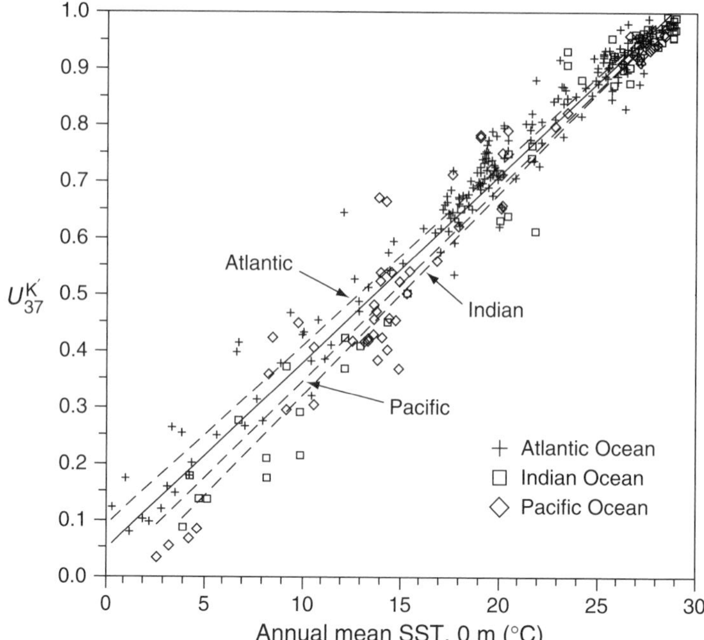

Figure 8.13. The relation between alkenone unsaturation ($U_{37}^{K'} = \frac{37:2}{37:2+37:3}$) and annual sea surface temperature (SST) for samples from various locations in the World Ocean (from Müller *et al.*, 1998). In general, unsaturated alkenones are produced in greater relative abundance at lower water temperatures to maintain cell membrane fluidity.

unchanged through marine water columns and into sedimentary records (Brassell, 1993). These favorable relationships have led to formulation of two different forms of alkenone unsaturation indices:

$$U_{37}^{K} = \frac{37:2 - 37:4}{37:2 + 37:3 + 37:4} \tag{8.2}$$

$$U_{37}^{K'} = \frac{37:2}{37:2 + 37:3} \tag{8.3}$$

that can be related to water temperature. As illustrated in Fig. 8.13, the direct correspondence of C-37 alkenone unsaturation to annual mean sea surface temperature (SST) is remarkably consistent over the entire temperature range of 0–30° C (Müller *et al.*, 1998). This relation is the basis for the alkenone paleotemperature scale, which is now being used by paleoceanographers as a complement to the more conventional stable oxygen isotope method described in Chapters 5 and 7. Two major advantages of the alkenone-based versus the $\delta^{18}O$ paleotemperature method are that the molecular method (a) is independent of the effects of "light" water storage in continental glaciers, and (b) can be used for carbonate-poor continental margin and pelagic clay sediments. In addition, being relatively stable molecules of unambiguous marine origin, alkenones can serve as excellent molecular targets for compound-specific stable isotope analysis.

8.3.4 Pigments

Pigments include a wide variety of organic compounds with the common characteristic that they selectively absorb light in the visible range of the electromagnetic spectrum (400–700 nm) and therefore exhibit color. All these substances have highly conjugated systems of alternating carbon (and often nitrogen) double bonds that readily absorb relatively long-wavelength (low-energy) light. This unique property has been a major reason for studying pigments, as well as a useful tool in their detection and characterization. In fact, the technique of chromatography was named because it was capable of separating differently colored pigments from natural samples. Although pigments include a wide range of organic compound classes, the present discussion will be confined to the organic geochemistry of chlorophylls. These pigments are involved in photosynthesis, and thus are of broad functional and chemical interest. Chlorophylls are also useful source and reaction indicators, because the ultimate source is restricted to biosynthesis by plants and some bacteria. Pigments are sparingly soluble in water, non-volatile and thermally sensitive.

The chlorophylls characteristically contain four pyrrole groups (labeled A–D) linked into a tetrapyrrol ring with various modifications (Fig. 8.14). The most common variants are phorbin-type compounds with one saturated carbon double bond in the D ring and an isocyclic ring bridging the 6 γ-carbons. Chlorophylls *a* and *b* vary only by substitution (CH_3 versus CHO) at the third carbon on ring B. Chlorophylls *c*1 and *c*2 vary by having fully aromatic D rings, a methyl ester instead of a phytol ester on the propionic side chain of this ring, and substitution of either an ethyl or an ethylene at carbon 4 on the B ring. Bacterial chlorophylls *a* and *b* have only one carbon double bond in both the B and D rings, and a 15-carbon (farnesyl) ester in place of the 20-carbon phytol ester commonly found in other chlorophylls. They also vary from each other depending on whether the ethyl group at carbon 4 is saturated or unsaturated (Fig. 8.14). Structural variations involving carbon double bonds conjugated with the tetrapyrrol ring typically cause shifts in the light absorbance patterns of the chlorophyll that allow different organisms to selectively utilize different wavelength bands of light. All of these chlorophylls coordinate with a magnesium ion located in the middle of the tetrapyrrol ring.

Although chlorophyll *a* is ubiquitous, chlorophyll *b* occurs primarily in green algae and vascular plants and chlorophyll *c* is characteristically found in diatoms and dinoflagellates. The bacteriochlorophylls occur in purple, green, and brown sulfur photosynthetic bacteria. Because such distributions are taxonomically meaningful and can be rapidly measured by HPLC analysis with combined UV detection (Llewellyn and Mantoura, 1996), molecular-level pigment analyses are now often carried out in complement to microscopic algae counts.

Chlorophyll pigments undergo alteration reactions in the water and surface sediments that record specific processes or conditions. Among the best known and most useful of these alterations is the conversion of chlorophyll pigments to corresponding phaeophorbides

Figure 8.14. Structures of different chlorophyll pigments found in marine phytoplankton, algae and bacteria (bacteriochlorophylls). The chlorophylls characteristically contain four pyrrole groups (labeled A–D) linked into a tetrapyrrol ring. Most are phorbin-type compounds with one saturated carbon double bond in the D ring and an isocyclic ring bridging the 6 γ-carbons. Modifications to ring B delineate the various chlorophyll compounds and are shown (R groups). Another difference between the various chlorophyll compounds is the presence of a 20-carbon phytol ester versus a 15-carbon (farnesyl) as in the bacteriochlorophylls.

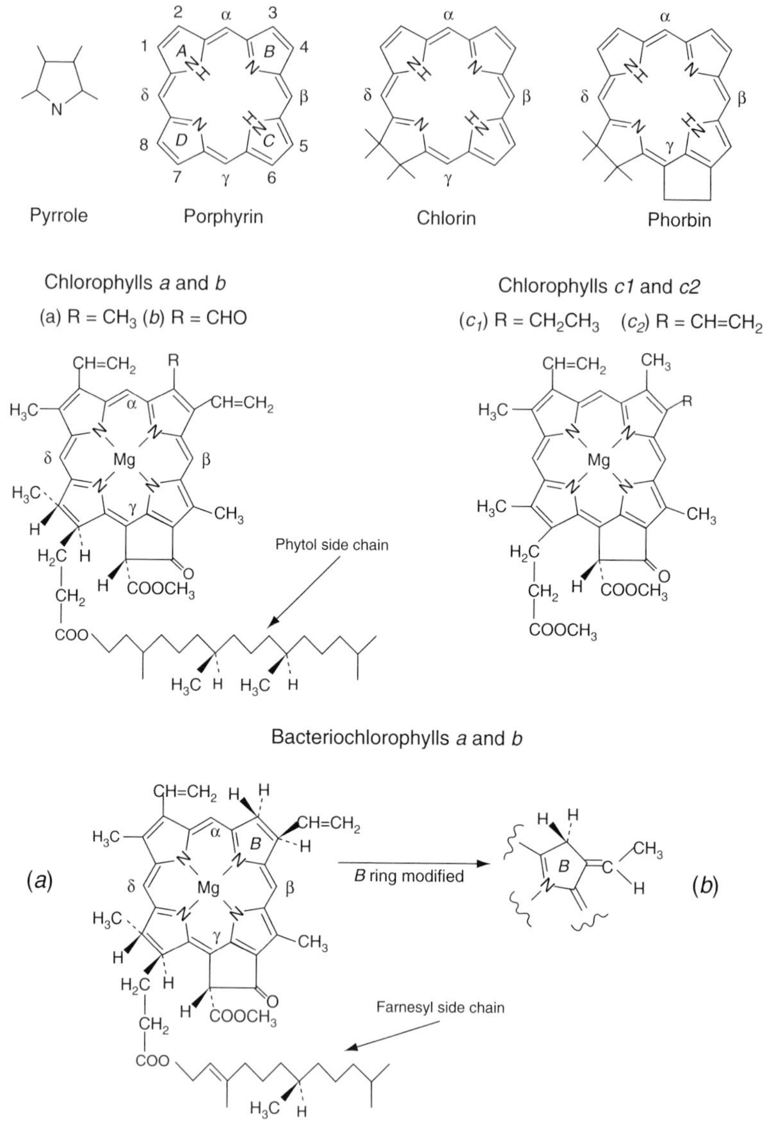

within zooplankton guts (Fig. 8.15). In the course of digestion by herbivores, dietary chlorophylls lose Mg^{2+} (demetallation) and phytol, the latter by enzymatic ester cleavage (Shuman and Lorenzen, 1975; Ziegler *et al.*, 1988). During algal senescence, either conversion can occur alone, to generate the corresponding phaeophytin (Mg^{2+} loss) or chlorophyllide (phytol loss) products (Sun *et al.*, 1994) (Fig. 8.15).

8.3.5 Lignin and tannins

One of the most clear-cut patterns of biochemical distribution between land and marine organisms is the high concentration of phenolic materials in vascular land plants and the absence of these

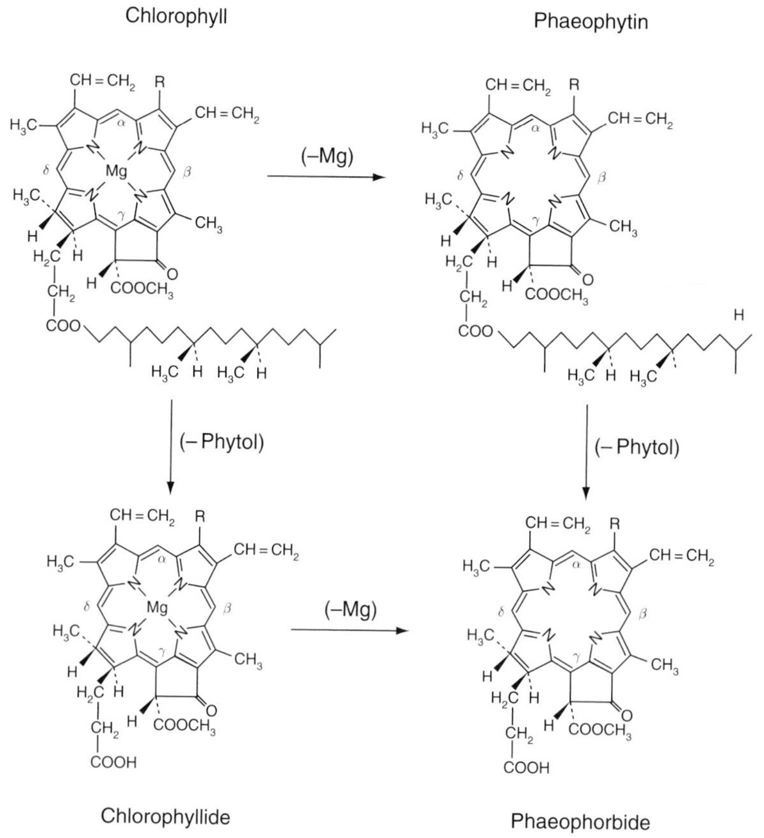

Figure 8.15. Alteration pathway of chlorophyll pigments to corresponding phaeophorbides. Loss of Mg from chlorophyll results in phaeophytin while the loss of the phytol ester from either chlorophyll or phaeophytin results in chlorophyllide and phaeophorbide, respectively.

substances in marine organisms (Table 8.6). Phenols contain a benzene aromatic ring substituted by one or more hydroxyl groups that are weakly acidic and dissociate at pH values near ten. The two most abundant types of phenol biopolymer are lignins and tannins, both of which occur only in vascular plants and thus have potential for indicating the presence of land-derived organic matter in marine systems. Tannins can make up as much as 20% of leaf and bark tissue (Benner *et al.*, 1990; Kelsey and Harmon, 1989). There are three types of tannin: condensed and hydrolyzable tannins are found in vascular plants and a third, less common type, phlorotannins, are found in brown algae. The structure of condensed tannin is shown in Fig. 8.16.

Lignin, by far the most widely studied phenol type, is an extended polymer composed of structural units with a phenolic ring and a three-carbon side chain (Fig. 8.16). Lignin occurs in cell walls of vascular plants, where it adds rigidity and protection against biodegradation. Woods often contain 20–30 wt% lignin, making it one of the most abundant biopolymers on land. The benzene ring of lignin structural units may be substituted by 0, 1 or 2 methoxyl $(-OHC_3)$ groups in positions flanking the phenolic hydroxyl group. The extent of substitution on the aromatic ring is related to the taxonomy of vascular plants. Gymnosperms (e.g. coniferous trees and ferns) primarily have one methoxyl whereas angiosperms (e.g. grasses and

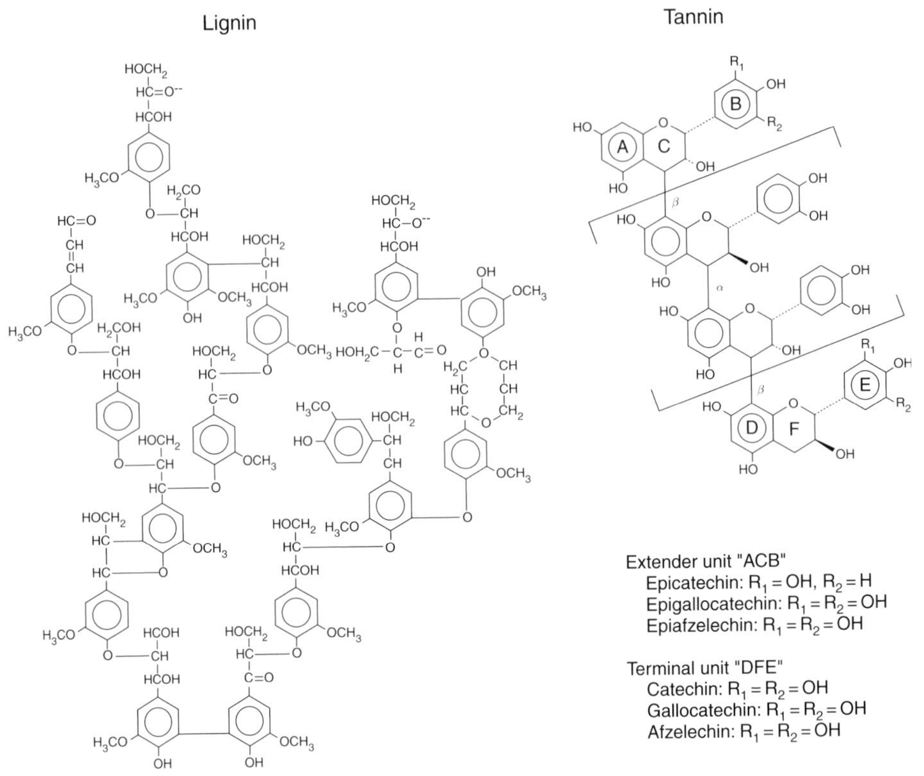

Lignin

Tannin

Extender unit "ACB"
Epicatechin: $R_1 = OH$, $R_2 = H$
Epigallocatechin: $R_1 = R_2 = OH$
Epiafzelechin: $R_1 = R_2 = OH$

Terminal unit "DFE"
Catechin: $R_1 = R_2 = OH$
Gallocatechin: $R_1 = R_2 = OH$
Afzelechin: $R_1 = R_2 = OH$

Figure 8.16. Schematic chemical structures of lignin and tannin polyphenols. The benzene ring of lignin structural units can be substituted by 0, 1 or 2 methoxyl ($-OHC_3$) groups, in positions flanking the phenolic hydroxyl group. The extent of substitution is related to the taxonomy of vascular plants. The structure of condensed tannin is shown here along with its building blocks of three-ring flavonols (see figure for terminology of extender and terminal units). Several variations of these stereochemically active compounds occur in condensed tannin.

broadleaf trees) uniquely carry two methoxyl groups. These structural units are then linked by a variety of ether and carbon–carbon linkages (Fig. 8.16) whose diversity and overall chemical stability impart relatively high resistance to biodegradation (Hedges *et al.*, 1985). Lignin is insoluble in water and one of the few biopolymers that can resist hydrolysis by concentrated sulfuric acid.

Most molecular-level methods for using lignin as a geochemical indicator involve chemical degradation of the polymer to simple (one-ring) phenols that still carry the previously mentioned patterns in methoxylation ($-OCH_3$) substitution. The most commonly applied method of degradation involves high-temperature oxidation with cupric oxide (CuO) in basic solution (Hedges and Ertel, 1982), although chemolysis with tetramethylammonium hydroxide (TMAH) also yields parallel information. The CuO method releases the eight simple phenols illustrated in Fig. 8.17 as major reaction products. Six of these reaction products comprise monomethoxylated (vanillyl) and dimethoxylated (syringyl) phenols, which occur primarily as aldehydes, along with smaller amounts of the corresponding ketones and acids. In addition, two cinnamyl phenols are produced that retain the three-carbon side chain of the original structural units and terminate in a carboxyl group. These two compounds, *p*-coumaric and ferulic acid, are produced only from non-woody tissues of vascular plants, and thus complement the taxonomic constraints derived from methoxylation patterns with information about plant tissue type.

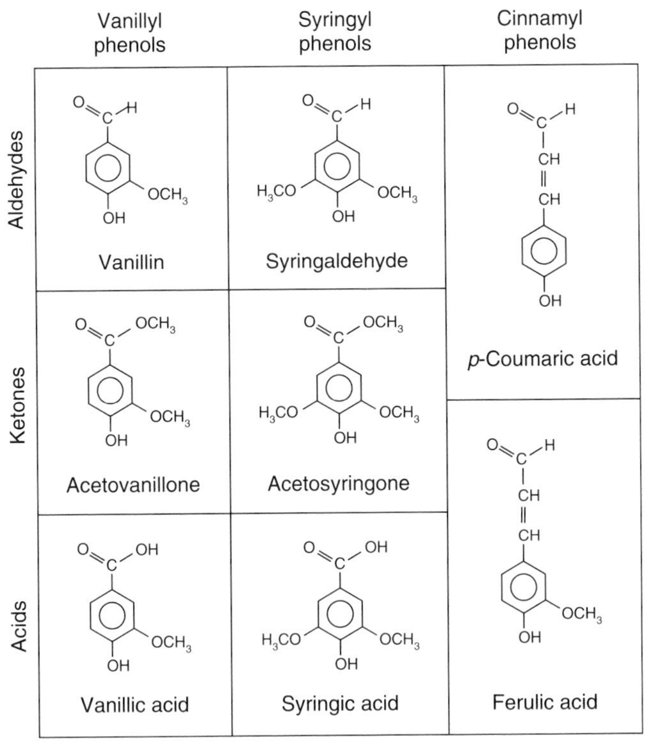

	Vanillyl phenols	Syringyl phenols	Cinnamyl phenols
Aldehydes	Vanillin	Syringaldehyde	*p*-Coumaric acid
Ketones	Acetovanillone	Acetosyringone	
Acids	Vanillic acid	Syringic acid	Ferulic acid

Figure 8.17. High-temperature cupric oxide (CuO) oxidation of vascular plant lignins yields the eight simple phenols illustrated here as major reaction products. These reaction products comprise monomethoxylated (vanillyl) and dimethoxylated (syringyl) phenols, which occur primarily as aldehydes, along with smaller amounts of the corresponding ketones and acids. Two cinnamyl phenols (*p*-coumaric and ferulic acid) are also produced that retain the three-carbon side chain of the original structural units and terminate in a carboxyl.

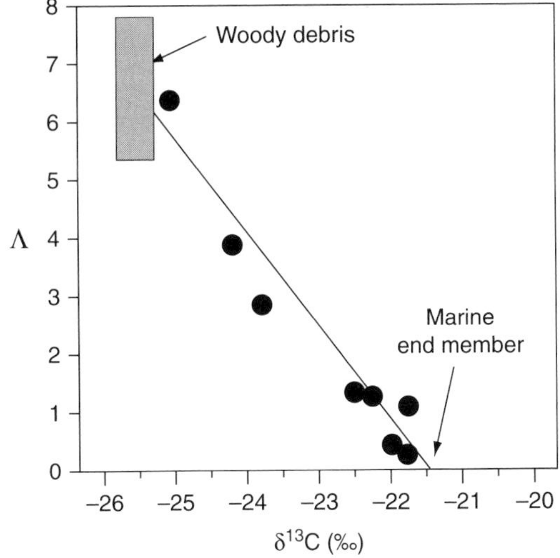

Figure 8.18. Plot of total organic carbon-normalized yield, Λ (mg of the CuO reaction products per 100 mg OC) versus $\delta^{13}C$ for surface sediments from the Washington State continental shelf and slope (from Hedges and Mann, 1979). See text and Fig. 8.17 for description of oxidation products.

Owing to its unambiguous source, lignin is used to trace the input of terrestrial organic matter into marine systems. The parameter lambda (Λ) represents the total yield of lignin phenols per 100 mg of organic carbon, and can be used to estimate the relative amounts of marine- and terrestrial-derived organic matter in marine sediments from a particular coastal zone. Under favorable circumstances, correlation plots of *Λ* versus $\delta^{13}C$ (Fig. 8.18) can be used to test for simple mixing of compositionally uniform marine and land-derived organic matter sources, and to estimate the lignin composition of the terrestrial component and the $\delta^{13}C$ of the marine component (Hedges and Mann, 1979). Among the phenols in Fig. 8.17, the mass ratio of vanillic acid to vanillin $(Ad/Al)_v$, also yields useful information about the average degradative state for the lignin component of the total organic matter in environmental samples. This method is based on the observation that most fresh vascular plant tissues yield $(Ad/Al)_v$ ratios in the range of 0.15–0.25. Acid/aldehyde ratios increase systematically within the lignin remnant as woody tissues are more extensively degraded by fungi. In addition, photodegradation can also increase $(Ad/Al)_v$ ratios. Information about the extent of lignin alteration can be useful in biodegradation studies, as well as to assess the extent to which source-related patterns such as syringyl to vanillyl phenol ratios can be altered by selective biodegradation of lignin structural units.

Tannins are highly ordered biopolymers that are widespread in nature, but have been studied less extensively than lignin. There are two major types of tannin, condensed and hydrolyzable, which can account for up to 20 wt% of leaf, needle and bark tissues. Condensed tannin is composed of three-ring flavanols; hydrolyzable tannin is made primarily of galic acid or its derivatives, which are often esterified to polyols such as glucose (Fig. 8.16). A third, less common type of tannin, phlorotannin, is found in brown algae. Tannins carry complementary source and degradation information that can be molecularly resolved by acid hydrolysis of the initial condensed ring structures to various phenol structural units that can be measured by gas chromatography (Hernes and Hedges, 2000). In particular, some distinctions between vascular plant sources that are not feasible by using lignin phenols can be made with tannins (Hernes *et al.*, 2001). Tannins appear to undergo rapid degradation in aquatic environments with attending depolymerization and immobilization of nitrogenous substances, although the mechanisms involved are yet to be worked out in detail.

8.4 Dissolved organic matter in seawater

The oceans hold one of the largest reservoirs of organic carbon. Thus far, the majority of organic compounds we have discussed are hydrophobic and are associated predominantly with marine particles that have a relatively short life time as they sink from surface waters to the sediments. Besides particles, there is a vast amount of carbon stored in ocean waters in a dissolved form. The separation of particulate and dissolved phases is most commonly accomplished by passage of seawater samples through a filter with a pore size in the range of 0.2–1.0 μm (see Fig. 6.6). With the exception of viruses and small bacteria, components in the dissolved organic matter (DOM) pool are nonliving and too small to sink in water at appreciable rates, hence the term "dissolved." Regardless of this small size and a low average concentration near 1 mg l^{-1} (1 ppm), seawater DOM contains approximately the same amount of carbon as in atmospheric CO_2 and more than two orders of magnitude more carbon than occurs in living marine biomass. In addition to its importance as a major reservoir of bioactive elements, seawater DOM also affects light transmission in the ocean, is involved in photochemical reactions, complexes metals, and carries biological and geochemical information. About 25%–35% of seawater DOM occurs in molecules with masses greater than 1000 amu that correspond to nominal molecular sizes of a nanometer or more. These larger DOM components are referred to as being high molecular mass or colloidal, where the latter term refers to objects large enough to scatter light and to have physicochemical behaviors that are strongly affected by surface properties such as charge density. Seawater colloids also are sufficiently large to be effectively separated from other DOM components by ultrafilters with pore sizes as small as a nanometer.

Seawater DOM has numerous sources and sinks and a range of potential reactions. Sources of the sub-micron components include exudation from phytoplankton, microbial degradation of bioparticles, animal wastes (excretion), viral infection of bacteria, sloppy feeding by zooplankton and other animals, and input of dissolved molecules from rivers and surface sediments. Removal mechanisms include photodegradation, sorption to sinking particles, and microbial utilization. Tritiated thymidine and leucine uptake experiments indicate that up to half of the carbon formed by photosynthesis is shunted via dissolved organic molecule intermediates into bacteria.

The distribution of DOM in the ocean is only partly consistent with the evidence for an extremely labile component. A typical vertical profile of DOM concentration in the temperate Pacific Ocean (Fig. 8.19) shows a surface maximum near 2 mg l^{-1} of total mass, which at an average carbon content near 50 wt% is equivalent to roughly 1 mg OC l^{-1} (1 mg C = 83.3 μM). This concentration drops by about half in the upper 500–1000 m of the water column and remains essentially constant near 0.5 mg l^{-1} (*c.*40 μM) DOC to the ocean floor. Although the surface maximum points toward a biological source of reactive DOM, the persistence of DOM in deep ocean waters with cycling times of the order of 500–1000 y indicates a major fraction that is stable, at least on this time scale. A bulk DOM property that pertains to this quandary is its ^{14}C content as a function of depth (Fig. 8.19). The radiocarbon content of DOC drops from a $\Delta^{14}C$ (see Chapter 5) near −200‰ in near-surface waters to between −500 and −600‰ in the deep Pacific, where DOC concentrations have dropped to a low constant value. If converted to an "age," which can be misleading for a heterogeneous organic mixture, the deep-ocean $\Delta^{14}C$ value corresponds to roughly 6000 y. Thus on average molecules forming deep ocean DOM appear to have survived numerous cycles of water through the global ocean circulation system (Chapter 1). Such a long mean residence time would allow

Table 8.8. *Selected bulk chemical compositions of DOM of river and seawater origin*

Abbreviations: C=C, percentage of unsaturated carbon (largely aromatic); LOPs, lignin oxidation products; %BOC, percentage of total dissolved organic carbon that occurs in chromatographically measurable biochemicals.

DOM source	(C/N)	Δ^{14}C, ‰	δ^{13}C, ‰	δ^{15}N, ‰	C=C	LOPs, $ng\,l^{-1}$	%BOC
Amazon River	34	+265	−29.2	+4.0	31	2200	25
Surface Pacific	16.5	−263	−21.7	+7.9	7	5–15	7
Deep Pacific	18.6	−546	−21.7	+8.1	19	7–10	3

Sources: Hedges *et al.* (1997) (and references therein); Benner *et al.* (1992); Opsahl and Benner (1997).

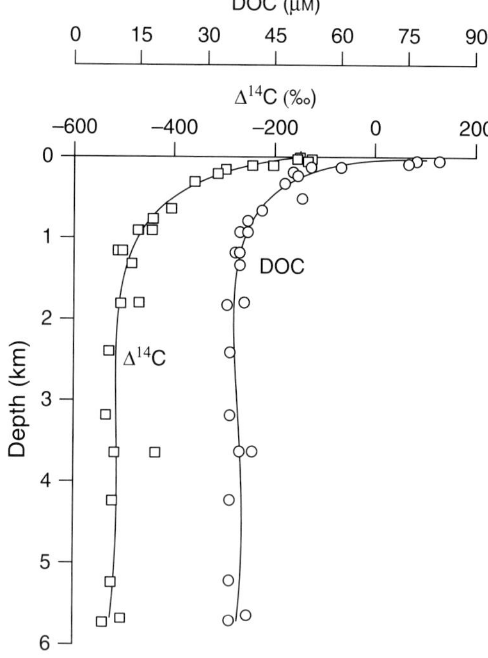

Figure 8.19. Vertical profiles of concentration (circles) and Δ^{14}C (squares) versus water column depth for dissolved organic carbon in the temperate Pacific Ocean (after Druffel *et al.*, 1989).

for the pool of oceanic DOC to be supported solely at steady state by the global discharge of DOC by rivers ($c.0.25 \times 10^{15}\ gC\,y^{-1}$).

The bulk chemical composition of seawater DOM (Table 8.8), however, is not consistent with a predominant riverine origin. In particular, seawater DOM is depleted in ^{14}C and enriched in ^{13}C and ^{15}N compared with most DOM discharged by rivers (with the Amazon being taken here as an example). Seawater DOM is also depleted in aromatic carbon (as measured by ^{13}C NMR) and lignin phenol structural units (as determined by CuO oxidation), possibly as a consequence both of the low phenol content of marine plankton, and of selective alteration of aromatic carbon by photodegradation. Thus all evidence to date, including elevated total concentrations in surface marine waters (Fig. 8.19), indicates that seawater DOM is largely

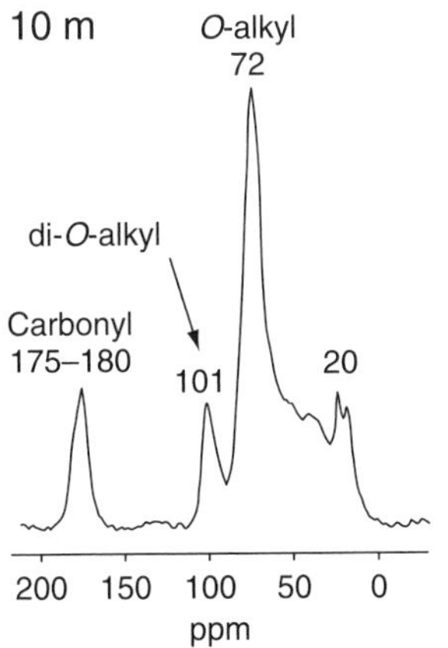

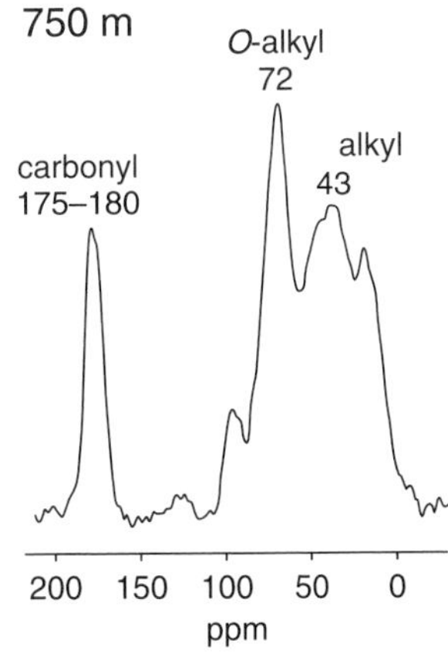

Figure 8.20. Solid-state ^{13}C NMR spectra of the high molecular weight (HMW) fraction (>1000 Da) of DOM isolated from surface (10 m) and interior (750 m) ocean water from the Gulf of Mexico (from Dr R. Benner, unpublished). Observed shifts indicate large concentrations of carbohydrates (70–110 ppm) that decrease with depth, whereas very little aromatic carbon is observed. Most of the uncharacterized fraction in HMW DOM consists of alkyl carbon (0–45 ppm) at a 2:1 atomic basis with carboxyl carbon.

autochthonous (locally derived), rather than being allochthonous (exotic) material of an ultimate terrestrial origin.

Many different types of measurements indicate that seawater DOM is compositionally complex and is thus far poorly characterized at the molecular level (Benner, 2002). For example, the percentage of total HMW DOC that can be accounted for as chromatographically measured amino acids plus neutral and basic sugars is only 15% in surface waters and 6% in the deep ocean. When compared with corresponding values near 80% for fresh marine plankton and of 30% and 20% for sinking and sedimentary marine particles, the extremely low biochemical content of HMW seawater DOM points toward a highly altered chemical structure. Direct molecular analyses of the same biochemicals in total seawater DOM, which has lower molecular mass, indicate concentrations that are no more than half of the HMW DOC values. These results, and steeper decreases of HMW DOC into the interior ocean, together indicate that lower molecular mass DOM is more refractory. Although the extremely low fraction of recognizable biochemicals suggests the presence of structurally complex "humic-like" material of abiotic origin, ^{15}N NMR analyses indicate that essentially all the nitrogen in HMW DOM is in amide structures that are almost certainly of biochemical origin. Corresponding ^{13}C NMR spectra (Fig. 8.20) indicate large concentrations of carbohydrate (chemical shift range of 70–110 ppm) that decrease with depth and thus are relatively labile over this interval. These spectra indicate very little aromatic carbon, as is generally associated with humic substances. Other than carbohydrate-like material, most of the uncharacterized fraction in HMW DOM consists of alkyl carbon (chemical shift range of 0–45 ppm), which covaries on a 2:1 atomic basis with carboxyl carbon that appears to occur in carboxylic acids rather than esters or amides. This carboxyl-rich aliphatic fraction, which increases in relative abundance with depth, has no known biochemical source or formation mechanism.

The molecular composition of the high molecular weight (HMW) fraction of seawater DOM that can be isolated by ultrafiltration with

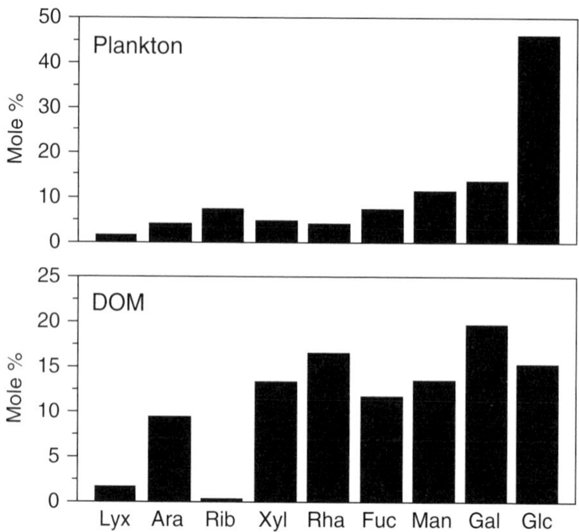

Figure 8.21. Neutral aldose compositions of hydrolysate mixtures from marine plankton and HMW DOM from surface seawater. Both samples are characterized by similar relative sugar compositions with the major exceptions of lyxose, ribose and galactose relative to glucose. Abbreviations: Lyx, lyxose; Ara, arabinose; Rib, ribose; Xyl, xylose; Rha, rhamnose; Fuc, fucose; Man, mannose; Gal, galactose; Glc, glucose. From McCarthy *et al.* (1998).

1000 amu cut-off filters also indicates a biologically atypical material. Neutral sugars, the largest measurable biochemical component of HMW seawater DOM, have an aldose composition (Fig. 8.21) unlike that of plankton, or any other type of particulate marine matter. The hydrolysate mixtures are characterized by similar amounts of most common sugars (except lyxose and ribose), and a predominance of galactose versus glucose. The amino acid composition of seawater DOM is similar to that of living organisms, although with somewhat elevated concentrations of glycine, alanine and other amino acids that are indicative of advanced biodegradation. Among the individual amino acids, alanine, serine and aspartic and glutamic acid are enriched ($>$10 mol%) in D- versus L-enantiomers, as is typical of bacterial cell walls. However, muraminic acid is present in much lower relative concentrations than would be expected for intact peptidoglycan, meaning that molecularly intact peptidoglycan no longer exists in DOM. Likewise, molecular-level measurements of individual amino acids account for only 10% of the total nitrogen in the HMW DOM fraction.

Although there is likely a relation between the high level of molecularly uncharacterized organic matter in seawater DOM and its low ^{14}C content and bulk reactivity, the mechanisms and reaction histories that lead to these intriguing characteristics are at present far from clear. One of the greatest challenges in modern oceanography is to bring the powerful analytical methods described in this chapter together in a concerted effort to define the sources, reactions and fates of seawater DOM.

Overall, recent studies involving a variety of measurements and experimental or field strategies have provided consistent evidence that both the dissolved and particulate pools of organic carbon are altered and remineralized in the ocean. However, fully understanding and identifying the processes involved and the conditions under which they occur has proven to be extremely difficult and will continue to be an ongoing challenge for marine organic geochemists. Future work will involve the development of improved bulk measurements and novel applications for the detection of molecular tracers as well as overall advances in analytical techniques and instrumentation. The key, as in most geochemical studies, will be to identify the processes controlling DOM and POM distributions so that well-chosen measurements can accurately describe these complex processes and provide information on the reaction history as well as their future states.

References

Amon, R. M. W., H. P. Fitzner and R. Benner (2001) Linkages among the bioreactivity, chemical composition and diagenetic state of marine dissolved organic matter. *Limnol. Oceanogr.* **46**, 287–97.

Benner, R. (2002) Chemical composition and reactivity. In *Biogeochemistry of Marine Dissolved Organic Matter* (ed. D. A. Hansell and C. A. Carlson), pp. 59–90. New York, NY: Elsevier Science.

Benner, R., P. G. Hatcher and J. I. Hedges (1990) Early diagenesis of mangrove leaves in a tropical estuary: bulk chemical characterization using solid-state 13 C NMR and elemental analysis. *Geochim. Cosmochim. Acta* **54**, 2003–13.

Benner, R., J. D. Pakulski, M. McCarthy, J. I. Hedges and P. G. Hatcher (1992) Bulk chemical characteristics of dissolved organic matter in the ocean. *Science* **255**, 1561–4.

Brassell, S. C. (1993) Application of biomarkers for delineating marine paleoclimatic fluctuations during the Pleistocene. In *Organic Geochemistry. Principles and Applications* (ed. M. H. Engel and S. A. Macko), pp. 699–738. New York, NY: Plenum Press.

Constanz, B. and S. Weiner (1988) Acidic macromolecules associated with the mineral phase of Scleractinian coral skeletons. *J. Exp. Zool.* **248**, 253–8.

Cowie, G. L. and J. I. Hedges (1984a) Determination of neutral sugars in plankton, sediments and wood by capillary gas chromatography of equilibrated isomeric mixtures. *Analyt. Chem.* **56**, 497–504.

Cowie, G. L. and J. I. Hedges (1984b) Carbohydrate sources in a coastal marine environment. *Geochim. Cosmochim. Acta* **48**, 2075–87.

Cowie, G. L. and J. I. Hedges (1991) Organic carbon and nitrogen geochemistry of Black Sea surface sediments from stations spanning the oxic:anoxic boundary. In *Black Sea Oceanography* (ed. E. Izdar and J. W. Murray), pp. 343–9. Boston, MA: Kluwer.

Cowie, G. L. and J. I. Hedges (1992a) Improved amino acid quantification in environmental samples: charge-matched recovery standards and reduced analysis time. *Mar. Chem.* **37**, 223–38.

Cowie, G. L. and J. I. Hedges (1992b) Sources and reactivities of amino acids in a coastal marine environment. *Limnol. Oceanogr.* **37**, 703–24.

Dauwe, B. and J. J. Middelburg (1998) Amino acids and hexosamines as indicators of organic matter degradation state in North Sea sediments. *Limnol. Oceanogr.* **43**, 782–98.

Degens, E. E. (1969) The biochemistry of stable carbon isotopes. In *Organic Geochemistry* (ed. G. Eglinton and M. T. J. Murphy), pp. 304–28. New York, NY: Springer-Verlag.

Druffel, E. R., P. M. Williams and Y. Suzuki (1989) Concentrations and radiocarbon signatures of dissolved organic matter in the Pacific Ocean. *Geophys. Res. Lett.* **16**, 991–4.

Forsberg, B. R., C. A. Araujo-Lima, R. M. Martinelli *et al.* (1993) Autotrophic carbon sources for fish of the central Amazon. *Ecology* **74**, 643–52.

Gélinas, Y., J. A. Baldock and J. I. Hedges (2001) Demineralization of marine and freshwater sediments for CP/MAS 13 C NMR analysis. *Org. Geochem.* **32**, 677–93.

Ghuysen, J. M. and G. D. Shockman (1973) In *Bacterial Membranes and Cell Walls* (ed. L. Leive), vol. 1, pp. 37–130. New York, NY: Dekker.

Goñi, M. A., K. C. Ruttenberg and T. I. Eglinton (1997) Sources and contribution of terrigenous organic carbon to surface sediments in the Gulf of Mexico. *Nature* **389**, 275–8.

Goñi, M. A., K. C. Ruttenberg and T. I. Eglinton (1998) A reassessment of the sources and importance of land-derived organic matter in surface sediments from the Gulf of Mexico. *Geochim. Cosmochim. Acta* **62**, 3055–75.

Hatcher, P. G., E. C. Spiker, N. M. Szeverenyi and G. E. Macial (1983) Selective preservation and origin of petroleum-forming aquatic kerogen. *Nature* **305**, 498–501.

Hayes, J. M. (2001) Fractionation of the isotopes of carbon and hydrogen in biosynthetic processes. Valley, J. W. and Cole, D. R. (organizers) for the Mineralogical Society of America. Presented at the National Meeting of the Geological Society of America, Boston, USA.

Hedges, J. I. and J. R. Ertel (1982) Characterization of lignin by gas capillary chromatography of cupric oxide oxidation products. *Analyt. Chem.* **54**, 174–8.

Hedges, J. I. and D. C. Mann (1979) The lignin geochemistry of marine sediments from the southern Washington coast. *Geochim. Cosmochim. Acta* **43**, 1809–18.

Hedges, J. I. and J. H. Stern (1984) Carbon and nitrogen determination of carbonate containing solids. *Limnol. Oceanogr.* **29**, 657–63.

Hedges, J. I., J. A. Baldock, Y. Gelinas *et al.* (2002) The biochemical and elemental compositions of marine plankton: a NMR perspective. *Mar. Chem.* **78**, 47–63.

Hedges, J. I., W. A. Clark and G. L. Cowie (1988) Fluxes and reactivities of organic matter in a coastal marine bay. *Limnol. Oceanogr.* **33**, 1137–52.

Hedges, J. I., G. L. Cowie, J. R. Ertel, R. J. Barbour and P. G. Hatcher (1985) Degradation of carbohydrates and lignins in buried woods. *Geochim. Cosmochim. Acta* **49**, 701–11.

Hedges, J. I., R. G. Keil and R. Benner (1997) What happens to terrestrial organic matter in the ocean? *Org. Geochem.* **27**, 195–212.

Hernes, P. J., R. Benner, G. L. Cowie *et al.* (2001) Tannin diagenesis in mangrove leaves from a tropical estuary: a novel molecular approach. *Geochim. Cosmochim. Acta* **65**, 3109–22.

Hernes, P. J. and J. I. Hedges (2000) Determination of condensed tannin monomers in plant tissues, soils and sediments by capillary gas chromatography of acid depolymerization extracts. *Analyt. Chem.* **72**, 5115–24.

Kelsey, R. G. and M. E. Harmon (1989) Distribution and variation of extractable total phenols and tannins in the logs of four conifers after 1 year on the ground. *Can. J. For. Res.* **19**, 1030–6.

Kröger, N., R. Deutzmann and M. Sumper (1999) Polycationic peptides from diatom biosilica that direct silica nanosphere formation. *Science* **286**, 1129–32.

Lambert, J. B., H. F. Shurvell, D. A. Lightner and R. G. Cooks (1998) *Organic Structural Spectroscopy*, part IV, *Mass Spectrometry*. Englewood Cliffs, NJ: Prentice-Hall.

Lee, C. and C. Cronin (1982) The vertical flux of particulate organic nitrogen in the sea: decomposition of amino acids in the Peru upwelling area and the equatorial Atlantic. *J. Mar. Res.* **40**, 227–51.

Lindroth, P. and K. Mopper (1979) HPLC determination of sub-picomole amounts of amino acids by precolumn fluorescence derivatization with O-phthaldialdehyde. *Analyt. Chem.* **51**, 1667–74.

Llewellyn, C. A. and R. F. C. Mantoura (1996) Pigment biomarkers and particulate carbon in the upper water column compared to the ocean interior of the northeast Atlantic. *Deep-Sea Res.* **43**, 1165–84.

Mackenzie, A. S., S. C. Brassell, G. Eglinton and J. R. Maxwell (1982) Chemical fossils: the geological fate of steroids. *Science* **217**, 491–504.

Macko, S. A., M. L. F. Estep, P. E. Hare and T. C. Hoering (1987) Isotopic fractionation of nitrogen and carbon in the synthesis of amino acids by microorganisms. *Isotope Geosci.* **65**, 79–92.

McCarthy, M. D., J. I. Hedges and R. Benner (1998) Major bacterial contribution to marine dissolved organic nitrogen. *Science* **281**, 231–4.

Merritt, D. A. and J. M. Hayes (1994) Factors controlling precision and accuracy in isotope-ratio-monitoring mass spectrometry. *Analyt Chem.* **66**, 2336–47.

Müller, P. J., G. Kirst, G. Ruhland, I. von Storch and A. Rosell-Mele (1998) Calibration of the alkenone paleotemperature index UK'37 based on core-tops from the eastern South Atlantic and the global ocean (60° N – 60° S). *Geochim. Cosmochim. Acta* **62**, 1757–72.

Onstad, G. D., D. E. Canfield, P. D. Quay and J. I. Hedges (2000) Sources of particulate organic matter in rivers from the continental USA: lignin phenol and stable carbon isotope compositions. *Geochim. Cosmochim. Acta* **64**, 3539–46.

Opsahl, S. and R. Benner (1997) Distribution and cycling of terrigenous dissolved organic matter in the ocean. *Nature* **386**, 480–2.

Prahl, F. G. and S. G. Wakeham (1987) Calibration of unsaturation patterns in long-chain ketone compositions for palaeotemperature assessment. *Nature* **330**, 367–9.

Preston, C. M. (1996) Applications of NMR to soil organic matter analysis: history and prospects. *Soil Sci.* **161**, 144–66.

Rau, G. (1978) Carbon-13 depletion in a subalpine lake: carbon flow implications. *Science* **201**, 901–2.

Reuter, J. H. and E. M. Perdue (1984) A chemical structural model of early diagenesis of sedimentary humus/proto-kerogens. *Mitt. Geol.-Paläont. Inst., Univ. Hamburg* **56**, 249–62.

Shuman, F. R. and C. J. Lorenzen (1975) Quantitative degradation of chlorophyll by a marine herbivore. *Limnol. Oceanogr.* **20**, 580–6.

Skoog, A. and R. Benner (1997) Aldoses in various size fractions of marine organic matter: implications for carbon cycling. *Limnol. Oceanogr.* **42**, 1803–13.

Smith, B. N. and S. Epstein (1971) Two categories of $^{13}C/^{12}C$ ratios for higher plants. *Pl. Physiol.* **47**, 380–4.

Sun, M. Y., R. C. Aller and C. Lee (1994) Spatial and temporal distributions of sedimentary chloropigments as indicators of benthic processes in Long Island Sound. *J. Mar. Res.* **49**, 57–80.

Treibs, A. (1934) Chlorophyll und Häminderivate in bituminösen Gesteinen, Erdölen, Erdwäschen und Asphalten. *Ann. Chem.* **510**, 42–62.

Verardo, D.J., P.N. Froelich and A. McIntyre (1990) Determination of organic carbon and nitrogen in marine sediments using the Carlo Erba NA-1500 Analyzer. *Deep-Sea Res.* **37**, 157–65.

Volkman, J.K. (2005) Sterols and other triterpenoids: source specificity and evolution of biosynthetic pathways. *Org. Geochem.* **36**, 139–59.

Wakeham, S.G., C. Lee, J.I. Hedges, P.J. Hernes and M.L. Peterson (1997) Molecular indicators of diagenetic status in marine organic matter. *Geochim. Cosmochim. Acta* **61**, 5363–9.

Wakeham, S.G., T.K. Pease and R. Benner (2003) Hydroxy fatty acids in marine dissolved organic matter as indicators of bacterial membrane material. *Org. Geochem.* **34**, 857–68.

Ziegler, R., A. Blaheta, N. Guha and B. Schonegge (1988) Enzymatic formation of pheophorbide and pyropheophorbide during chlorophyll degradation in a mutant of *Chlorella fusca* Shihira et Kraus. *Pl. Physiol.* **132**, 327–32.

9

Molecular diffusion and reaction rates

Chemical kinetics is the study of the rate and mechanism of reaction. At ocean boundaries chemical fluxes are often determined by the interplay of molecular diffusion and reaction rates. Both of these topics are important to the chemical perspective of oceanography because they provide the necessary mechanistic and mathematical background for the study of chemical fluxes.

Labile chemical reactions near chemical equilibrium are a special case of kinetics, in which concentrations do not change with time, because the rates of the reactions producing and destroying a substance are equal. The study of chemical equilibrium is much more advanced than kinetics in its application to marine chemistry, because of the availability of thermodynamic state functions (e.g. free energies) for the calculation of equilibrium constants (Chapter 3). The added degree of complexity in the study of kinetics involves understanding the pathway and rate of approach to equilibrium. For example, it is much simpler to calculate the potential energy change of a boulder that once was part of a mountain peak and is now at rest in a river bed than it is to determine the pathway and speed of its fall.

Chemical thermodynamic models and increasingly sophisticated analytical measurements reveal the marine environment as a dynamic system that is rarely truly at equilibrium. Photosynthesis captures the energy from the sun and produces highly reduced organic molecules, forcing the Earth away from the state of equilibrium. This activity provides the environment with a constant source of energy for spontaneous reactions that tend to return the system to a state of lower free energy and provide the driving force for chemical reactions. Although the subject of reaction rate kinetics is highly formalized in the discipline of chemistry, the utility of theoretical constructions is reduced when dealing with reactions in the environment because of their complicated nature and reaction rate catalysis.

Even though it is difficult to predict reaction rates in marine systems, the concepts of molecular diffusion and mechanisms of reaction underpin much of geochemical research at the air–water and sea floor–ocean boundaries. A basic knowledge of molecular diffusion and chemical kinetics is essential for understanding the processes that control these fluxes. This chapter explores the topics of molecular diffusion, reaction rate mechanisms and reaction rate catalysis. Catalysis is presented in a separate section because nearly all chemical reactions in nature with characteristic life times of more than a few minutes are catalyzed.

9.1 Molecular diffusion

9.1.1 The diffusion equation

The average velocity of atoms and molecules in a gas can be described by probability theory because they occur in large numbers and undergo changes in speed and direction on time scales that are short relative to the period of observation. This is true not only for gases but also for solids and liquids, although the velocity distributions in these phases are different from that in the ideal gas. The mean linear velocity of ions in water is on the order of tens or hundreds of meters per second, yet the average net movement, for instance, of a single ion in totally quiescent water is only about 0.05 mm s^{-1}. The reason for the difference between velocity and net movement is again related to the random nature of thermal kinetic energy. Although molecules have high average velocities, they change direction very frequently and randomly. As the result of collisions with other molecules, particles lose part of their memory of the original movement, and this occurs until they are "randomized" again. The mean free path is the distance between the several collisions needed to lose track of the earlier movement. The average distance over which an ion in water moves in any given direction before encountering another atom is very short, on the order of 10^{-12} m.

A simple model demonstrating the essential ideas of the theory of random walk is illustrated in Fig. 9.1 (Csanady, 1973). Let us assume that a particle is located at $x = 0$ at time $t = 0$. Its movements, which

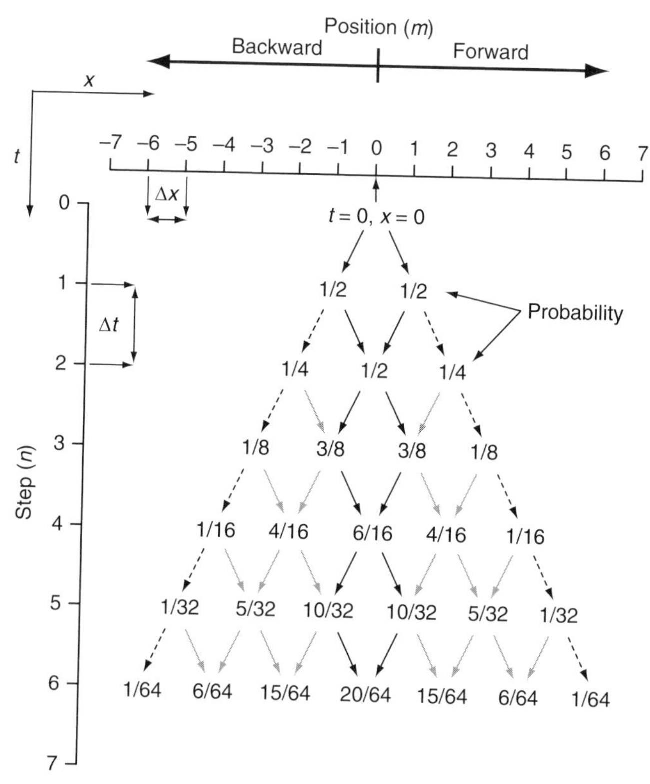

Figure 9.1. Schematic description of the random walk of a particle along a straight line. The pyramid shows the probabilities of a molecule reaching position m ($x = -6, -5, \ldots, 0, \ldots, 5, 6$) along the x axis after n equal steps of distance $\Delta x = 1$. At each step the probability of moving forward or backward is equal. Fractions represent the probability of a particle reaching that location starting from $x = 0$, $t = 0$. The dashed lines are the least probable paths and the solid lines represent the most probable paths (see text).

can only occur along a straight line, shall consist of steps of size Δx at equal time intervals Δt. For every step there is equal probability for either a forward or backward movement (Fig. 9.1). In the light of the prior discussion, we can say that every single collision completely randomizes the particle again; there is no memory effect. If the experiment is repeated over and over again (each time starting with one single particle at $x = 0$) and the location of the particle after n steps is registered, the probability of finding the particle at $x = m \times \Delta x$ (where $m = -n \cdots +n$) is proportional to the number of different pathways leading to this particular position. In Fig. 9.1 there are many pathways to positions in the interior (the solid line, high probability), but few along the outside (dashed line, low probability). This example illustrates the basic tenet of probability: that the likelihood of an occurrence is proportional to the number of ways in which it can happen. (The chance of having a boating accident is greater if both the motor and rudder are old and might fail than it would be if only one of these parts were in bad repair.)

The continuum of probabilities, p, of achieving any repeated event, m, in which two possible outcomes are equally possible describes a normal density function and can be expressed by the equation

$$\mathrm{p}_n(m) = \sqrt{\frac{2}{\pi \times n}} \exp\left(-\frac{m^2}{2n}\right). \tag{9.1}$$

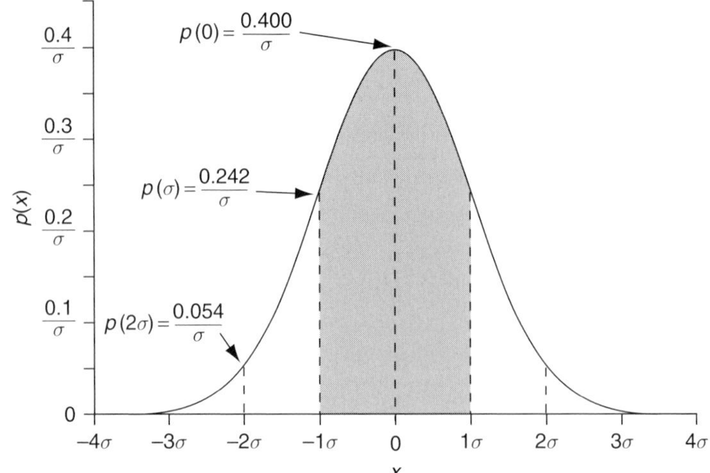

Figure 9.2. A normal distribution (Eq. (9.2)) in which $\mu = 0$. The x axis is scaled by the standard deviation, σ, and the y axis is scaled by σ^{-1}. The shaded area is the integral under the curve for plus or minus one standard deviation, $\pm\sigma$, and represents 68.2% of the total area under the curve. The area under the curve bounded by $\pm 2\sigma$ is 95.4% of the total area.

In the random walk example described above, n is the number of steps and m the position on the x axis. The general equation for the normal distribution is (Bevington and Robinson, 2003):

$$p(x) = \frac{1}{\sigma\sqrt{2\pi}} \exp\left(-\frac{(x-\mu)^2}{2\sigma^2}\right), \tag{9.2}$$

where μ is the value of the origin ($\mu = 0$ in the random walk experiment) (Fig. 9.1), and σ is the standard deviation. A plot of p(x) scaled by σ^{-1} versus x scaled by σ (Fig. 9.2) yields a bell-shaped curve in which 68.2% of the area under the central portion of the curve is bounded by plus or minus one standard deviation, $\pm\sigma$. By inspection of Eqs. (9.1) and (9.2), the standard deviation of the random walk experiment is $\sqrt{n}$. After 100 steps, 68.2% of the random walk particles will be within ± 10 steps of the origin. After 10^4 steps the standard deviation is ± 100, meaning that there is a 68.2% probability that the particle will be found within ± 100 steps of the origin. The particle spreads along the x axis with time, but the distance from the origin traveled per step becomes smaller with time (10% for 100 steps and 1% for 10^4 steps).

Equation (9.1) can be transformed from the m and n integer values to a continuous distribution by introducing the relation between the integer values and the continuous space, x, and time, t, coordinates. The position, m, is the distance, x, along the line divided by the step length, Δx (Fig. 9.1)

$$m = x/\Delta x. \tag{9.3}$$

The number of steps, n, is equal to the product of the step frequency, ν (time^{-1}), and time, t ($n = \nu \times t$) and if we define the average velocity, u (distance/time), as the product of the step frequency and the step distance, ($u = \nu \times \Delta x$), then

$$n = \frac{u}{\Delta x} \times t. \tag{9.4}$$

Before we can make the substitution replacing Eq. (9.1) by a continuous distribution in x and t, we must realize that the two variables are related via the expression

$$p_n(m) = 2\Delta x \times p(x, t) \tag{9.5}$$

because from the position m the two possible new positions, $m + \Delta x$ and $m - \Delta x$ (Fig. 9.1) after any single time step, Δt, are a distance $2\Delta x$ apart. Substituting for m and n in Eq. (9.1) and combining with substituting Eq. (9.1) into Eq. (9.5) gives

$$p(x, t) = \sqrt{\frac{1}{2\pi \times \Delta x \times u \times t}} \exp\left(-\frac{x^2}{2\Delta x \times u \times t}\right) \tag{9.6}$$

which is a normal distribution (Eq. (9.2)) with a standard deviation of

$$\sigma = \sqrt{\Delta x \times u \times t}. \tag{9.7}$$

This result describes the slow mixing velocities of dissolved molecules since the value for the standard deviation, σ, is a measure of the spatial spreading of a given tracer in water. The patch grows in proportion to $\gamma \times \sqrt{t}$ where the spreading factor, γ, is given by, $\sqrt{\Delta x \times u}$. In spite of the large value of u, the patch grows slowly because the molecules are continuously changing direction. Because Δx, a measure of the mean free path length or average step size of the molecules between successive collisions, is extremely small (10^{-12} m), γ is small. Note also that the units of $\Delta x \times u$ are distance squared per time, equivalent to those of a diffusion coefficient.

We can now show the relation between the random walk analogy (Fig. 9.1) and the diffusion coefficient. If we released a cluster of "random walk" particles with total mass Q at $x = 0$ and $t = 0$, they would describe a normal distribution about $x = 0$ for $t > 0$. The number of particles, M, occupying a distance ΔL along the line is given by

$$M = Q \times p(x, t) \times \Delta L. \tag{9.8}$$

The mean concentration, C, of the particles over the distance ΔL is thus

$$C = \frac{M}{\Delta L} = Q \times p(x, t) = \frac{Q}{\sqrt{2\pi \times \Delta x \times u \times t}} \exp\left(\frac{-x^2}{2\Delta x \times u \times t}\right). \tag{9.9}$$

Exactly the same problem can be approached by using the equation for molecular diffusion in one spatial dimension, x,

$$\frac{\partial C}{\partial t} = D \times \frac{\partial^2 C}{\partial x^2}, \tag{9.10}$$

where D is the molecular diffusion coefficient. If the total quantity of diffusing particles, Q, is released at time $t = 0$ and at $x = 0$, then the concentration, C, at x and t is (see Crank, 1975):

$$C(x, t) = \frac{Q}{2}\left(\frac{1}{\sqrt{\pi \times D \times t}}\right) \exp\left(\frac{-x^2}{4D \times t}\right). \tag{9.11}$$

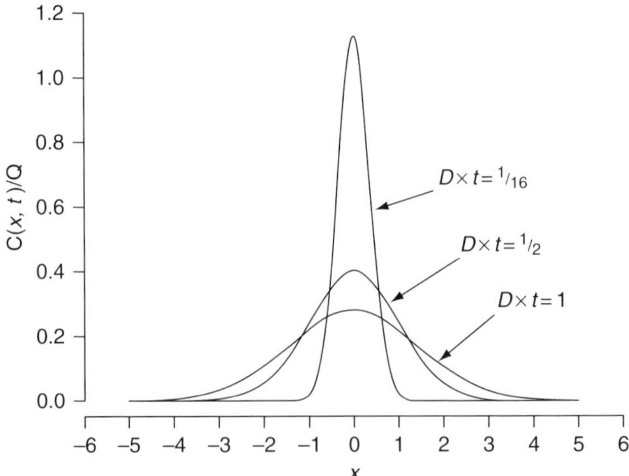

Figure 9.3. The solution to the molecular diffusion equation, Eq. (9.1). The ratio of the concentration, C, at time x and t to the total quantity of diffusing particles, Q, released at time $t = 0$ along a plane at $x = 0$. Numbers on the curves are values of $D \times t$.

The concentration distribution for this equation is symmetrical about $x = 0$ and becomes more spread out and flatter with time (Fig. 9.3). Comparing the result from the diffusion equation with Eq. (9.9), which is in exactly the same form, we see that the diffusion coefficient is

$$D = \frac{\Delta x \times u}{2} = \frac{\Delta x^2}{2 \times \Delta t}. \tag{9.12}$$

This is the classic equation relating distance and time for species in a totally quiescent medium where transport is controlled by molecular diffusion. The simple formula, sometimes referred to as Einstein's equation, is very useful for order of magnitude calculations determining the distance a substance will travel by molecular diffusion. For example, if one wanted to know if a chemical species with rate constant k would react significantly during transport across a molecular diffusion boundary layer of thickness Δx one would compare the time calculated from the above equation with the reciprocal of the reaction rate constant, k (see below).

9.1.2 Molecular diffusion coefficients in water

Molecular diffusion coefficients in water are usually determined in the laboratory by using a tracer in agar or gel to insure that the solution is totally free of turbulent motion. Values for gases and ions in pure water are presented in units of $\text{cm}^2\,\text{s}^{-1}$ in Table 9.1. The molecular diffusion coefficients and their temperature dependence for gases were calculated from the Eyring equation,

$$D = A \exp\left(\frac{E_a}{RT}\right) \tag{9.13}$$

using the pre-exponential factor, A, and activation energy, E_a, terms presented in Table 9.1. Diffusion coefficients for the rare gases are inversely proportional to the atomic mass with a difference of

Table 9.1. *Molecular diffusion coefficients, D, for gases and ions in pure water*

Diffusion coefficients, D, have units of $10^{-6}\ cm^2\ s^{-1}$ and were multiplied by 10^6 before tabulating. For example, D_{H^+} (25 °c) = $93.1 \times 10^{-6}\ cm^2\ s^{-1}$.

Gases[a]					Ions[b]					
Species	$A^c \times 10^5$ ($cm^2\ s^{-1}$)	E_a (kJ mol^{-1})	$D\times10^6$ ($cm^2\ s^{-1}$) $A\exp(-E_a/RT)$		Species	$D\times10^6$ ($cm^2\ s^{-1}$) $RT\lambda/F^2Z$		Species	$D\times10^6$ ($cm^2\ s^{-1}$) $RT\lambda/F^2Z$	
			$T=0\ °C$	$T=25\ °C$		$T=0\ °C$	$T=25\ °C$		$T=0\ °C$	$T=25\ °C$
N_2	7900	19.62	14.0	28.8	H^+	56.1	93.1			
O_2	4200	18.37	12.9	25.4	Li^+	4.7	10.3	OH^-	25.6	52.7
CO_2	5019	19.51	9.3	19.2	Na^+	6.3	13.3	F^-	—	14.7
CH_4	3047	18.36	9.4	18.5	K^+	9.9	19.6	Cl^-	10.1	20.3
He	818	11.70	47	72.9	NH_4^+	9.8	19.8	Br^-	10.5	20.1
Ne	1608	14.84	23.0	40.4	Mg^{2+}	3.6	7.0	HS^-	9.8	17.3
Ar	10600	20.63	12.0	25.8	Ca^{2+}	3.7	7.9	NO_3^-	9.8	19.0
Kr	6393	20.20	8.8	18.5	Sr^{2+}	3.7	7.9	HCO_3^-	—	11.8
Xe	9007	21.61	6.6	14.7	Mn^{2+}	3.0	6.9	SO_4^{2-}	5.0	10.7
					Fe^{2+}	3.4	7.2	CO_3^{2-}	4.4	9.6
					Al^{3+}	2.4	5.6			

[a] The diffusion coefficients for gases were calculated from the Eyring equation, and A and E_a were determined from experimental results of Wise and Houghton (1966) for N_2, O_2, and Ar, and from Jahne and Dietrich (1987) for the rest. $R = 0.008\,314\,510$ (kJ mol^{-1} K^{-1}).

[b] Ion diffusion coefficients were determined from the Nernst equation (Eq. (9.14)) and data for the equivalent conductance, λ ($\Omega\ m^2\ mol^{-1}$) (see text and Li and Gregory, 1974; *Handbook of Chemistry and Physics*, 1970). $R = 8.314\ 510$ (J mol^{-1} K^{-1}), T is temperature (K), $F = 964\,86.309$ (coulombs mol^{-1}), and Z is the absolute value of the charge on the ion.

[c] Values of A are multiplied by 10^5 for tabulation here. For example, $A_{N_2} = 0.079\,00\ cm^2\ s^{-1}$.

slightly more than a factor of five between He and Xe. With respect to temperature, diffusion coefficients for inert gases at 25 °C are about twice those at 0 °C.

The diffusion coefficient for an ion, j, can be determined by using the Nernst equation,

$$D_j = \frac{RT \times \lambda_j}{Z_j \times F^2} \tag{9.14}$$

and the tabulated values of equivalent conductances, λ ($\Omega\,\mathrm{m}^{-2}\,\mathrm{mol}^{-1}$), and the Faraday and gas constants, F and R, respectively. T is the absolute temperature and Z is the absolute value of the charge. Ions with low charge to radius ratios (low ionic potential) such as Br^-, NH_4^+, and K^+ diffuse faster than small, highly charged ions such as Mg^{2+} and Al^{3+}. This is because a higher charge creates a thicker hydration layer of water molecules that must diffuse along with the ion (Li and Gregory, 1974).

The temperature dependence of ionic diffusion is about the same as that for gases, with diffusion rates increasing by roughly a factor of two between 0 and 25 °C. The temperature dependence has primarily to do with the change in the viscosity of water, which doubles as the temperature is lowered from 25 to 0 °C. The effect of salinity on molecular diffusion is also related to the change in viscosity. Diffusion coefficients of gases and ions are believed to be between 5% and 10% slower, respectively, in seawater because the viscosity increases owing to the presence of ions (Jahne and Dietrich, 1987; Li and Gregory, 1974).

It is often suggested that the diffusion coefficient of an ion in water depends on its own mobility as well as that of the ions that diffuse along with it, because charge balance must be maintained in the solution. Indeed, experiments in which a high-mobility cation and a low-mobility anion diffuse into a dilute solution indicate that the faster-diffusing ion is retarded whereas the slower-diffusing ion is accelerated (Vinograd and McBain, 1941; Ben-Yaakov, 1972). Although the effect has important consequences in freshwater systems, it is less critical in seawater because seawater is a strong electrolyte. Essentially, the small gradient in charge potential created by diffusion of trace concentrations of cations and anions of different mobility in seawater is neutralized by other ions that are in great abundance (primarily Na^+ and Cl^-).

9.2 Reaction rates

9.2.1 Reaction mechanisms and reaction order

Chemical reaction between dissolved species normally requires the interaction of molecules or atoms with each other, with a solid, or with a surface not involved in the reaction (i.e. the walls of the container). Experimentally determined reaction rates are thus written as a proportionality between the time rate of change of the reactant species and their concentrations. The proportionality term is the reaction rate constant and the exponent to which the reactant concentration(s) are raised depends upon the stoichiometry of the rate-limiting (elementary) reaction. There is often no clear relation between the overall stoichiometric reaction equation and the experimentally observed reaction rate equation. The reason for this is that the overall reaction mechanism is often a combination of elementary reactions, of which only one is rate-limiting. The difference between overall and elementary reactions is one of the main themes of chemical reaction rates in water.

Consider the elementary chemical reaction between atoms X and Y to produce products P and Q,

$$xX + yY \underset{k_-}{\overset{k_+}{\rightleftharpoons}} pP + qQ. \tag{9.15}$$

The rate of change of the concentration (indicated by square brackets, [ ]) of each reactant and product is related by the proportionality represented in the lower-case letters, which symbolize stoichiometric coefficients:

$$\frac{1}{x}\frac{d[X]}{dt} = k_+[X]^x[Y]^y - k_-[P]^p[Q]^q. \tag{9.16}$$

Each of the rate terms, k_+ and k_-, in Eq. (9.16) is related to concentrations in a way that one would predict if the probability of reaction were dependent on the collision of randomly moving particles: the rate is proportional to the product of the number of entities involved in the reaction. All other factors that determine the reaction rate (energy barriers, temperature dependence, the effect of other species in solution, catalysis, etc.) are represented in the rate constant, k, which has units necessary to balance the left- and right-hand sides of the rate expression. Because ion interaction effects that are accounted for by activity coefficients in chemical equilibrium calculations (Chapter 3) are all incorporated into the rate constant, concentrations and not activities are used on the right-hand side of the reaction rate equation.

Equation (9.15) describes a reversible reaction, whereby the reaction can proceed to the right as well as to the left. Not all kinetic reactions are reversible. For example, radioactive decay, many oxidation reactions, and organic matter degradation proceed, for all practical purposes, in only one direction until the reaction is inhibited or the reactant is effectively exhausted. Reversible reactions, such as CO_2 hydration, other acid–base relationships and some precipitation–dissolution reactions will attain, at some point, a steady state in which both the forward and reverse reactions occur at the same rate and the concentrations of both reactants and products no longer change. This is the state of chemical equilibrium at which the product of the reaction products raised to the exponent of their stoichiometric coefficients divided by a similar arrangement for the reactants is equal to the apparent equilibrium constant, K' (see Chapter 3):

$$K' = \frac{[\mathrm{P}]^p[\mathrm{Q}]^q}{[\mathrm{X}]^x[\mathrm{Y}]^y}.$$

An important link between kinetics and thermodynamics is the relation between the rate constants and the apparent equilibrium constant:

$$K' = \frac{[\mathrm{P}]^p[\mathrm{Q}]^q}{[\mathrm{X}]^x[\mathrm{Y}]^y} = \frac{k_+}{k_-}. \tag{9.17}$$

This relation is called the "principle of detailed balancing" and holds true only when the concentrations are at or near equilibrium. A practical application of this equation is that, with values for the rate constant in one direction and the equilibrium constant of an elementary reaction, one can calculate the other rate constant for a reversible reaction near equilibrium.

The order of a reaction is the sum of the concentration coefficients for the reacting species. The order of the reaction in Eq. (9.15) is therefore $(x+y)$ in the forward direction and $(p+q)$ in the reverse direction. Reaction order is also used to describe the coefficient of the individual species in the reaction. The order of reaction (9.15) with respect to [X] is x, with respect to [P] is p, and similarly for the other species. Thus, the rate at which a zero-order reaction proceeds is a constant and independent of the concentration of the reacting substance. The degradation of organic matter is said to be zero order with respect to O_2 or SO_4^{2-}, because the rate of the process is independent of the O_2 or SO_4^{2-} concentrations above critical, low level concentrations. A first-order reaction is one that proceeds spontaneously or by an encounter with an unreactive surface such as the walls of a container. Radioactive decay is a first-order reaction; the rate of decay is determined by energetic factors within the nucleus of the atom, and the only factor influencing the reaction is the concentration of the element undergoing decay. Second-order reactions are caused by interaction between two atoms or molecules. This type of reaction is very common in solutions because it is often the most probable type of encounter. Elementary reactions rarely exceed zero-, first-, and second-order, especially in solutions. There are higher-order elementary reactions; however, they are unusual because of the low probability for simultaneous interaction among three or more species. Envision the game of pool, in which there are continuous frequent collisions between two balls, or a ball and the rail, but almost never a simultaneous collision of three balls.

Reactions in a single phase are said to be homogeneous. Examples of experimentally observed homogeneous reaction rates in aqueous solution for carbon dioxide hydration, and ferrous iron oxidation in water are presented in Tables 9.2 and 9.3, respectively. The overall reaction is presented first, followed by the reaction rate equation and then some suggestions for a reaction mechanism that shows the connection between these two. The reaction mechanism is sometimes a hypothesis if it is not based on experimental evidence.

Table 9.2. *CO_2 hydration: the relations among (a) reaction stoichiometry; (b) the observed reaction rate equation; and (c) reaction mechanisms*

K and *k* are equilibrium and rate constants, respectively. Combination of equations for both mechanisms (c_1) and (c_2) results in the overall reaction shown in (a). Mechanism (c_2) represents a base catalysis of the hydration reaction.

(a)	$CO_2 + H_2O \underset{k_{CO_{2r}}}{\overset{k_{CO_2}}{\rightleftharpoons}} HCO_3^- + H^+$	
(b)	$\frac{dCO_2}{dt} = -(k_{CO_2} + k_{OH^-}[OH^-]) \times CO_2 + (k_{CO_2,r}[H^+] + k_{HCO_3}) \times [HCO_3^-]$	
$(c_1)^a$	$CO_2 + H_2O \underset{k_{13}}{\overset{k_{31}}{\rightleftharpoons}} HCO_3^- + H^+$; $CO_2 + H_2O \underset{k_{23}}{\overset{k_{32}}{\rightleftharpoons}} H_2CO_3$; $H_2CO_3 \underset{k_{21}}{\overset{k_{12}}{\rightleftharpoons}} HCO_3^- + H^+$	$K_{H_2CO_3} = \frac{[H^+][HCO_3^-]}{[H_2CO_3]} = \frac{k_{21}}{k_{12}}$ $k_{CO_2} = k_{31} + k_{32}$ $k_{CO_{2,r}} = k_{13} + \frac{k_{23}}{K_{H_2CO_3}}$
(c_2)	$CO_2 + OH^- \underset{k_{HCO_3^-}}{\overset{k_{OH^-}}{\rightleftharpoons}} HCO_3^-$	$K_W = [H^+][OH^-]$ $k_{OH^-}[OH^-] = k_{OH^-}\frac{K_W}{[H^+]}$

[a] $k_{12} \approx 5\times10^{10}$ ($M^{-1}s^{-1}$) and $K_{H_2CO_3} \approx 1.7\times10^{-4}$ M (Kern, 1960), thus $k_{21} \approx 10^7\ s^{-1}$. The reaction is very fast and at pH ≈7, $[H_2CO_3]/[HCO_3^-] \approx 10^{-3}$. For these reasons the reaction pathways for CO_2 to HCO_3^- in (c_1) are indistinguishable and the rate constants are combined into k_{CO_2} and $k_{CO_2,r}$.

The reversible reaction of carbon dioxide with water and OH^- ions has the overall reaction equation (Table 9.2a)

$$CO_2 + H_2O \underset{k_{CO_{2r}}}{\overset{k_{CO_2}}{\rightleftharpoons}} HCO_3^- + H^+. \tag{9.18}$$

The reaction rate equation (Table 9.2b) is more complicated than would be guessed from this overall reaction because there are several mechanisms of CO_2 reaction with water to form HCO_3^-. The reaction rate dependency on the concentrations of the different species is determined by measuring the effect on the rate as a function of varying one concentration while all others are held constant. For example, CO_2 was varied while the pH remained constant to determine the effect of CO_2 on the rate in Table 9.2. The reaction of CO_2

with H_2O to form HCO_3^- takes place either directly or through the intermediate H_2CO_3. Because carbonic acid, H_2CO_3, and bicarbonate, HCO_3^-, are at equilibrium (via $K_{H_2CO_3}$), and $[H_2CO_3]$ is about 1000 times smaller than $[HCO_3^-]$, at pH = 7, these two mechanisms are not experimentally distinguishable. The pathways are thus combined into the mechanism described by the rate constants k_{CO_2} and $k_{CO_2,r}$, which are a combination of the pathways indicated by k_{31}, k_{13} and k_{23}, k_{32}, respectively. The other rate mechanism that is obvious from equation (b) in Table 9.2 is illustrated in (c_2). The hydroxylation of CO_2 dominates at pH > 10. The equilibrium expression for equation (c_2) can be transformed to that for the overall reaction (a) by substituting for OH^- by using the dissociation constant for water.

A second example of the difference between overall reactions and reaction mechanisms is the irreversible oxidation of ferrous iron by oxygen. The overall reaction is (Table 9.3a):

$$Fe^{2+} + \frac{1}{2}H_2O + \frac{1}{4}O_2 \xrightarrow{k} Fe^{3+} + OH^-. \quad (9.19)$$

The experimentally determined reaction rate equation is fourth-order and dependent on $[Fe^{2+}]$, $[O_2]$, and $[OH^-]^2$ (Table 9.3b). The rate dependence on $[Fe^{2+}]$ and $[O_2]$ was determined by measuring the rate of disappearance of Fe(II) (Roman numerals are used to indicate the oxidation state without specifying the actual species in solution) at various pHs while O_2 remained constant (and vice versa for the O_2 dependency) (Stumm and Lee, 1961). We know from previous discussion that the fourth-order reaction mechanism in Table 9.3b would be nearly impossible. Millero *et al.* (1987b) suggested a rate-limiting mechanism that is a second-order reaction between

Table 9.3. *Ferrous iron oxidation by oxygen above pH = 5*

The relations are shown among: (a) stoichiometry; (b) the observed reaction rate equation; and (c) a possible mechanism suggested by Millero *et al.* (1987b).

(a) $Fe^{2+} + \frac{1}{2}H_2O + \frac{1}{4}O_2 \xrightarrow{k} Fe^{3+} + OH^-$

(b) $\frac{d[Fe^{2+}]}{dt} = k[Fe^{2+}][O_2][OH^-]^2$

(c) equilibrium:

$$Fe^{2+} + 2OH^- \rightleftharpoons Fe(OH)_2 \qquad K_{FeOH_2} = \frac{[Fe(OH)_2]}{[Fe^{2+}][OH^-]^2}$$

rate-limiting reaction:

$$Fe(OH)_2 + O_2 \rightleftharpoons Fe^{3+} + O_2^- + 2OH^-$$

$$\frac{d[Fe(OH)_2]}{dt} = k_{Fe(OH)_2}[Fe(OH)_2][O_2], \quad \text{where}$$

$$k = k_{Fe(OH)_2}K_{Fe(OH)_2}$$

$Fe(OH)_2$ and oxygen that results in the oxidation of Fe(II) to Fe(III) along with the reduction of O_2 to superoxide, O_2^-. The superoxide then reacts with other species in solution to return to a stable molecule. The combination of this reaction with the equilibrium between OH^- and Fe^{2+} results in the experimentally observed fourth-order rate equation (Table 9.3). At pHs below about 5 the reaction rates were much slower and attributed (Millero *et al.*, 1987b) to slower reactivity of the solution species, Fe^{2+}, $FeOH^+$ and $Fe(OH)_2$. Thus, the equation in Table 9.3b holds only above pH *c*.5.

These examples illustrate that reaction order is determined by rate experiments and cannot be inferred from knowledge of the stoichiometry of the overall reaction. Furthermore, the order of the rate equation, and the order with respect to reactants or products, provide information about reaction mechanism, but usually not about the order of the rate-limiting reaction because of accompanying equilibria with short-lived reaction intermediaries.

9.2.2 Mineral–water reaction

Heterogeneous reactions involve interaction between dissolved and solid phases. There are two principal rate-limiting, mechanistic steps in this type of reaction. The first is transport of ions or molecules to and away from the solid surface, and the second is reaction at the surface of the solid. (Adsorption of reactants to the solid surface could also be included as a mechanism, but this process is relatively fast.) The kinetics of dissolution and precipitation of minerals in water follow the same rules regarding mechanisms and order as homogeneous reactions. The general equation for the change in concentration of species X due to the reaction of mineral M is first-order with respect to the portion of the solid that is in contact with water. It is proportional to the mineral surface area, *S*, per unit of water volume, *V*, in contact with the mineral. Concentration of species X in the mineral (mol $m^{-3}_{solution}$) is indicated in the following equations with the subscript *m*, and the solution concentration is in square brackets.

$$-\frac{dX_m}{dt} = \left(\frac{d[X]}{dt}\right)_{dissolution} + \left(\frac{d[X]}{dt}\right)_{precipitation} \tag{9.20}$$

$$\left(\frac{d[X]}{dt}\right)_{dissolution} = k_{d,x} \times \left(\frac{S}{V}\right) \tag{9.21}$$

$$\left(\frac{d[X]}{dt}\right)_{precipitation} = k_{p,x} \times \left(\frac{S}{V}\right). \tag{9.22}$$

The rate constants for dissolution, $k_{d,x}$, and precipitation, $k_{p,x}$ (mol m^{-2} s^{-1}) refer only to the species X. If the mineral dissolves congruently, that is, if the ratio of the constituents released to solution is the same as in the solid, then the solid stoichiometric coefficient times the "total" mineral dissolution and precipitation rate constants can be used.

Mineral–water reactions that are controlled by transport in the dissolved phase are characterized by a rate equation that describes the flux of reactants through a laminar layer of thickness δ adjacent to the mineral. Transport in this very thin layer is by molecular diffusion; therefore Eq. (9.20) becomes

$$\frac{d[X]}{dt} = \frac{D}{\delta} \times \left(\frac{S}{V}\right) \times ([X]_s - [X]) \tag{9.23}$$

where D is the molecular diffusion coefficient of dissolved species X and the subscript s indicates the dissolved concentration at thermodynamic equilibrium. Since diffusion is rate-limiting, reaction at the surface is assumed to be fast enough to maintain the solute concentrations immediately adjacent to the mineral surface at saturation. Thus, the concentration difference, $[X]_s - [X]$, represents the gradient in concentration across the diffusion layer. It is not possible to independently determine δ, making the above formulation of the rate constant, by itself, insufficient to distinguish the mechanism of the reaction. A diffusion-controlled reaction, however, should be dependent upon the rate of stirring of the solution, since the thickness of the molecular diffusion layer, δ, is determined by the turbulence surrounding solid particles. Indeed, the most common experimental method for determining the role of diffusion in controlling water–rock reactions is to measure the rate as a function of stirring with all other conditions constant. We shall see that the reaction order and the reaction rate temperature dependence also provide clues to the rate-limiting mechanism.

Most experimental studies conclude that the rate of dissolution and precipitation of minerals is stirring-independent in solutions that are near the chemical conditions of natural waters (Berner, 1978). Dissolution studies with solids that are prepared by etching with acid to remove fine grains resulting from grinding commonly reveal a first-order rate with respect to the solid surface to solution volume ratio (Eq. (9.21)) and a zero-order rate with respect to the concentration of the products of dissolution for conditions far from equilibrium. The changes in concentrations of silicate and aluminum as a function of time during Na-feldspar and aluminum oxide dissolution, respectively (see Fig. 9.4) are linear, indicating that there is no dependency on the solution concentrations (i.e. zero-order). If these reactions were dependent on the solution concentrations, the rates would become slower as the dissolved concentrations increased. The actual mechanism of the dissolution reactions involves, in many cases, the formation of crystallographically controlled etch pits in surface regions of active dissolution (Berner, 1978). The process of dissolution is thus one of localized reaction in regions of surface defects rather than a uniform detachment of the reaction products.

Although dissolution reactions are frequently zero-order with respect to solution concentrations of the species that make up the solid, they are commonly observed to be pH-dependent, so that the observed rate equations take the form

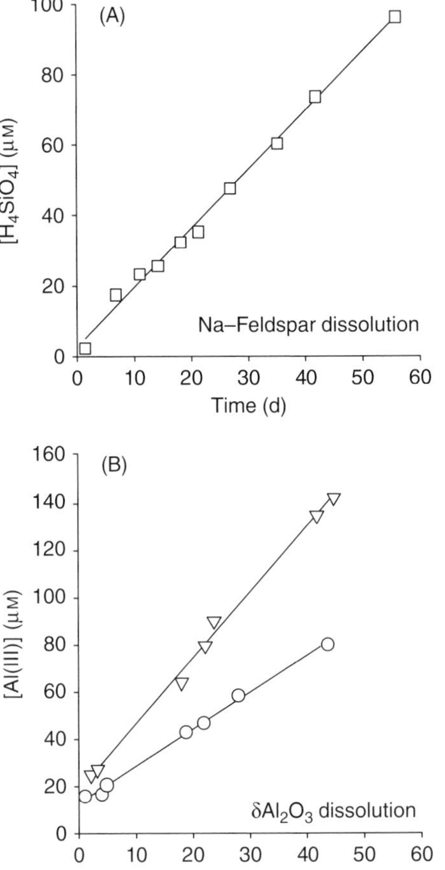

Figure 9.4. Demonstration of the zero-order dissolution rate of feldspar and aluminum oxide with respect to the solute concentrations of H_4SiO_4 and Al (III). (A) The increase in the concentration of silicic acid in solution during the dissolution of Amelia Courthouse Albite (sodium feldspar) (Holdren and Berner, 1970). (B) The increase in dissolved aluminum as a function of time in the presence of δAl_2O_3 (2.2 g l^{-1} Al_2O_3 with a surface area of 113 $m^2 g^{-1}$) (Furrer and Stumm, 1986).

$$\left(\frac{d[X]}{dt}\right)_{\text{dissolution}} = k_{d,x} \times \left(\frac{S}{V}\right) \times [H]^n \quad (9.24)$$

(Lasaga, 1984). The order, n, with respect to the hydrogen ion concentration is not necessarily an integer and is usually in the range $0 < n < 1$. Some of the important mineral–water reactions in aquatic systems are listed in Table 9.4 along with the experimentally observed reaction order with respect to hydrogen ion activity, where available. Even in this limited number of examples there is no general functionality with respect to pH. The fractional order has been attributed to the catalysis of the reaction rate by adsorption of hydrogen ions at the solid interface (Furrer and Stumm, 1986). If this is the case, the order with respect to hydrogen ion is determined by the relation between the bulk pH, the H^+ concentration at the surface, and the mechanism of the surface reaction. We return to the discussion of proton catalysis of mineral–water dissolution reactions in the section on heterogeneous catalysis.

Rates of precipitation are more difficult to study than dissolution because they involve the steps of nucleation and crystal growth.

Table 9.4. *Mineral–water dissolution reactions, their activation energies, E_a, and the order of the reaction with respect to hydrogen ion activity, $a^n_{H^+}$*

Reaction	E_a (kJ mol^{-1})	Order	Reference
$CaCO_3(s) + H^+ \rightarrow Ca^{2+} + HCO_3^-$ Calcite	34, 41		Sjoberg (1976): pH = 8.0–10.1. Plummer *et al.* (1978): pH > 5.5. (E_a is for only one of the proposed dissolution mechanisms.)
$CaMg(CO_3)_2(s) + 2H^+ \rightarrow Ca^{2+} + Mg^{2+} + 2HCO_3^-$ Dolomite	77–160	$a^{0.5}_{H^+}$	Busenberg and Plummer (1982): pH = 0–7. (The reaction order and E_a apply to $T < 45\,°C$.)
$FeMgSiO_4(s) + 4H^+ \rightarrow Fe^{2+} + Mg^{2+} + H_4SiO_4$ Olivine	38	$a^{1.0}_{H^+}$	Grandstaff (1981): pH = 3–5.
$CaAl_2Si_2O_8(s) + 3H_2O \rightarrow Ca^{2+} + 2OH^- + Al_2Si_2O_5(OH)_4(s)$ Ca feldspar (anorthite) … Kaolinite	35	$a^{0.54}_{H^+}$	Fleer (1982): pH = 2–5.6.
$KAlSi_3O_8(s) + H^+ + \frac{9}{2}H_2O \rightarrow K^+ + 2H_4SiO_4 + \frac{1}{2}Al_2Si_2O_5(OH)_4(s)$ Feldspar (orthoclase) … Kaolinite	38	$a^{1.0}_{H^+}$	Helgeson *et al.* (1984): pH = 3–8.
$\delta Al_2O_3(s) + 6H^+ \rightarrow 2Al^{3+} + 3H_2O$ Al oxide	57	$a^{0.4}_{H^+}$	Furrer and Stumm (1986).
$SiO_2(s) + 2H_2O \rightarrow H_4SiO_4$ Silicate oxide, quartz	61–75	$a^{0.0}_{H^+}$	Rimstidt and Barns (1980). (E_a is for both quartz and amorphous silica at pH < 7.0.)

Modified from Schnoor and Stumm (1986).

High degrees of supersaturation commonly precede the initiation of precipitation because of the energy barriers to the initial reaction. Precipitation is sometimes inhibited in natural waters by dissolved species that are foreign to the crystal but which occupy surface sites favorable for crystal growth. For example, it has been demonstrated that high concentrations of Mg^{2+} in seawater inhibit the precipitation of $CaCO_3$ in surface waters of the ocean. This is the reason for observed high degrees of $CaCO_3$ supersaturation. In general, the growth rate of a solid from solution after the initial nucleation step has been shown to be a high-order function of the dissolved species that form the solid (Lasaga, 1984).

9.2.3 Characteristic life time

The characteristic life time of a reaction is a measure of the time required after initiation for it to reach completion. This period is frequently related to the rate constant for the reaction in a very clear and specific way. Solutions to some of the common zero-, first- and second-order rate equations are presented in Table 9.5. Examples of zero- and first-order reactions are discussed in this section; application of the second-order equations to general catalytic processes will be presented in the section on catalysis. The last column of Table 9.5 lists the relations between τ, the characteristic life time of the reactant with respect to the chemical reaction, and the rate constant for the reaction. The meaning of the characteristic life time depends upon the order and reversibility of the reaction.

As shown in the previous section, the dissolution rate of minerals is frequently zero-order with respect to the concentration of the element(s) entering solution (Fig. 9.3). The change in concentration of Al as a function of time in the dissolution of aluminosilicate is equal to the rate of Al dissolution, J, which has units of concentration per unit time

$$\frac{\mathrm{d[Al]}}{\mathrm{d}t} = J. \tag{9.25}$$

The solution to this equation describes a straight line beginning at the initial ($t = 0$) condition, $[\mathrm{Al}]_0$ (Table 9.5, Fig. 9.4). If this equation were to describe the Al concentration throughout the reaction, then the characteristic life time of Al would be equal to the final Al concentration (that achieved when the mineral reaction stops for some reason, $[\mathrm{Al}]_f$) divided by the dissolution rate, $[\mathrm{Al}]_f/J$ (Table 9.5). In practice, this relation holds only at the beginning of the dissolution reaction. If the reaction continues and the solid is in excess, the reaction order is likely to change as saturation equilibrium between the concentration in solution and the solid is approached and precipitation and dissolution rates become comparable.

The characteristic life time, τ, of reactants with rate equations that have exponential solutions is less clear than in the zero-order case because the concentration asymptotically approaches the final value (equilibrium or zero concentration). In this situation, one

Table 9.5. *Analytical solutions to zero-, first-, and second-order kinetic reaction equations*

The stoichiometric equations represent elementary reactions where k is the reaction rate constant and τ the reaction mean life. See the text for explanation of these values. The zero subscripts indicate initial ($t = 0$) concentrations and the subscript "f" indicates final concentrations. Second-order equation (b) is the solution to the catalytic reaction.

Equation	Solution	τ
Zero order:		
$\frac{dX}{dt} = J$	$X(t) = X_0 + Jt$	X_f/J
First order:		
Irreversible		
(a) $X \xrightarrow{k} P$		
$-\frac{dX}{dt} = kX$	$X(t) = X_0 \exp(-kt)$	$1/k$
Reversible		
(b) $X \underset{k_{-1}}{\overset{k_1}{\rightleftharpoons}} Y$		
$-\frac{dX}{dt} = k_1 X - k_1 Y$ where: $X_0 - X = Y - Y_0$	$X(t) = X_0 \exp[-(k_1 + k_{-1})t] - \frac{k_{-1}}{(k_1 + k_{-1})}(X_0 + Y_0)(1 - \exp[-(k_1 + k_{-1})t])$	$1/(k_1 + k_{-1})$
Second order:		
(a) $X + X \xrightarrow{k} P$		
$-\frac{dX}{dt} = kX^2$	$\frac{1}{X} = \frac{1}{X_0} + kt$	
(b) $X + Y \xrightarrow{k} P$		
$-\frac{dX}{dt} = kXY$ where: $X_0 - X = Y_0 - Y$	$\frac{1}{(Y_0 - X_0)} \ln \frac{X_0 Y}{Y_0 X} = kt$	

must choose the point in the reaction progress that is to be defined by the characteristic time, τ. The common (but arbitrary) choice, and the one illustrated in Table 9.5, is the time required for the decaying exponential term to reach 1/e of its initial value. This "e-folding" relation between the concentration at time t and the initial or final concentration is dependent on reaction order and reversibility.

An example of a nearly pure first-order irreversible reaction is the decay of radioactive isotopes. These reactions depend on instabilities within the nucleus rather than the electronic configuration of the atom and are thus totally uninfluenced by the surroundings. The reaction rate is dependent on the concentration of the isotope and the probability of spontaneous decay. An important example in aquatic systems is the decay of carbon-14 (Fig. 9.5), which is formed in the upper atmosphere by interaction between cosmic ray particles and nitrogen atoms. ^{14}C is then mixed into the lower atmosphere and incorporated into organic and mineral material containing carbon while it decays by the first-order process

$$\frac{d[^{14}C]}{dt} = -\lambda \times [^{14}C], \tag{9.26}$$

which has an exponential solution, given in Table 9.5. The rate constant, k, in radioactive decay reactions is referred to as a decay constant and given the symbol λ. When half of the isotope has disappeared, $t_{1/2}$, the natural logarithm of the remaining fraction is 0.693, which leads to a simple relation between the decay constant and the radioactive half life (see also Chapter 5):

$$\ln\left(\frac{[^{14}C]}{[^{14}C]_0}\right) = \ln\tfrac{1}{2} = -0.693 = -\lambda \times t_{1/2} \text{ or } t_{1/2} = \frac{0.693}{\lambda}. \tag{9.27}$$

The characteristic life time, $\tau = 1/k = 1/\lambda$, of the decay reaction is referred to as the mean life and is equal to the time required for the concentration to reach $1/e$ or *c.*0.37 of its initial value.

Whereas radioactive decay is never a reversible reaction, many first-order chemical reactions are reversible. In this case the characteristic life time is determined by the sum of the forward and reverse reaction rate constants (Table 9.5). The reason for this may be understood by a simple thought experiment. Consider two reactions that have the same rate constant driving them to the right, but one is irreversible and one is reversible (e.g. k in first-order equation (a) of Table 9.5 and k_1 in first-order reversible equation (b) of the same table). The characteristic time to steady state will be shorter for the reversible reaction because the difference between the initial and final concentrations of the reactant has to be less if the reaction goes both ways. In the irreversible case all reactant will be consumed; in the irreversible case the system will come to an equilibrium in which the reactant will be of some greater value. The difference in the characteristic life time between the two examples is determined by the magnitude of the reverse reaction rate constant, k_-. If k_- were zero the characteristic life times for the reversible and irreversible reactions would be the same. If $k_- = k_+$ then the characteristic time for the reversible reaction is half that of the irreversible rate.

Solutions for two types of irreversible second-order reaction are presented in Table 9.5. The first (second-order reaction (a)) is a

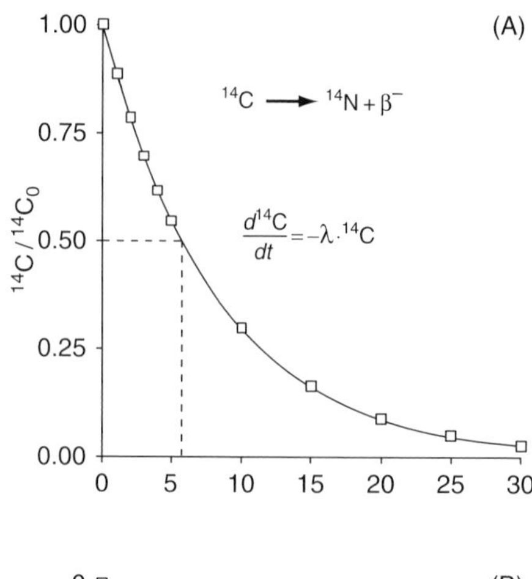

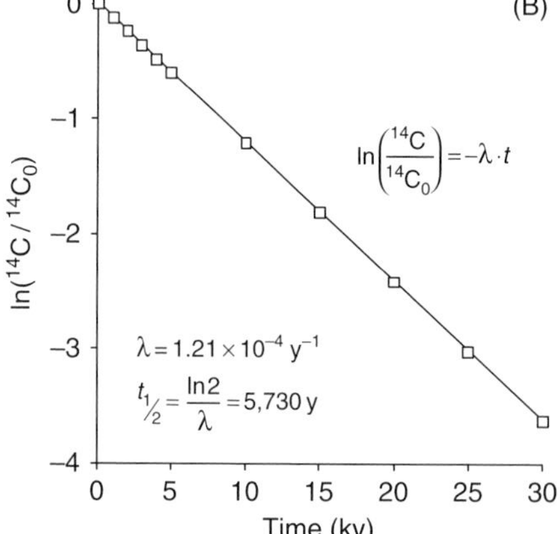

Figure 9.5. The solution to the first-order reaction rate equation describing the radioactive decay of carbon-14.

reaction that involves a single reactant, which combines with itself to form products, and the second (second-order reaction (b)) is the equation describing general catalysis (see later). There are no unambiguous definitions for the characteristic life time for these reactions. For cases in which one of the reactants is in great excess, the value of τ approaches the reciprocal of the rate constant multiplied by the initial concentration of the reactant in excess. This approximation is used in Table 9.6 to define the characteristic life time of some environmentally common homogeneous second-order reactions. Note that the life times in Table 9.6 vary from less than a microsecond for fast reactions that are diffusion-controlled to greater than ten thousand seconds, or about eleven orders of magnitude!

9.2.4 The temperature dependence of reaction rates

Reaction rates in water increase by roughly a factor of two for every 10 K temperature increase. Possible reasons for this are that molecules move faster at higher temperature and there are more collisions, or that the probability of encounters with sufficient energy for reaction increase with temperature. To determine

Table 9.6. *First- and second-order rate constants, k, for some important reactions in marine chemistry*

Unless otherwise indicated, the constants are presumed to be elementary reactions at $T = 25\,^\circ\mathrm{C}$ and zero ionic strength. Mean life times for the second-order reactions, τ, are given for the case in which one of the reactants has an initial concentration (assumed to be 10^{-6} M) that is in great excess of the other.

Reaction	k 1st-order (s^{-1})	k 2nd-order ($M^{-1}\,s^{-1}$)	τ (s)	Reference
$e^-_{aq} + OH \longrightarrow OH^-$		3×10^{10}	3×10^{-5}	Gardiner (1972)
$H^+ + HS^- \longrightarrow H_2S$		7.5×10^{10}	1×10^{-5}	Pankow and Morgan (1981); Eigen and Kustin (1962)
$H_2S \longrightarrow H^+ + HS^-$	1×10^4		10^{-4}	Pankow and Morgan (1981); Eigen and Kustin (1962)
$FeOH^{2+} + H^+ \longrightarrow Fe^{3+} + H_2O$		7×10^{10}	1.4×10^{-5}	Pankow and Morgan (1981); Eigen and Kustin (1962)
$H^+ + HCO_3^- \longrightarrow H_2CO_3$		5×10^{10}	2×10^{-5}	Eigen *et al.* (1961)
$CO_2 + H_2O \longrightarrow H_2CO_3$			27	Johnson (1982)
$H_2CO_3 \longrightarrow CO_2 + H_2O$	11–25		0.04–0.1	Pocker and Bjorkquist (1977)
$H^+ + HCO_3^- \longrightarrow CO_2 + H_2O$		$4–8 \times 10^4$	12–25	Johnson (1982)
$CO_2 + OH^- \longrightarrow HCO_3^-$		$7–11 \times 10^3$	90–140	Johnson (1982)
$HCO_3^- \longrightarrow CO_2 + OH^-$	2×10^{-4}		5×10^3	Johnson (1982); Pocker and Bjorkquist (1977)
$HS^- + O_2 \longrightarrow$ prod.		$3–6 \times 10^{-3}$	$2–3 \times 10^3$	Chen and Morris (1972); pH = 7.9. This is not a true second-order reaction: $d[HS^-]/dt = k \times [HS^-]^{1.4}[O_2]^{0.6}$.
$Fe^{2+} + O_2 \longrightarrow$ prod.	pH = 7 pH = 8	2–3 $2–13 \times 10^2$	$8–50 \times 10^4$ 80–500	Sung and Morgan (1980). The reaction rate equation is not second-order: $d[Fe^{+2}]/dt = k \times [Fe^{2+}][O_2][OH^-]^2$.

which of these is the limiting factor one can compare the formula for reaction rate dependency with those for the temperature dependence of molecular velocities in a gas and the probability that the molecules will have a velocity greater than some value, u.

Arrhenius proposed a widely used and fundamental equation for the temperature dependence of reaction rates in the year 1889, which was based entirely on experimental measurements:

$$k(T) = A_{\mathrm{Ar}} \exp\left(\frac{-E_{\mathrm{a}}}{RT}\right). \tag{9.28}$$

E_{a} is called the activation energy with dimensions of energy per amount of substance; A_{Ar} is referred to as the pre-exponential (or frequency) factor and has the same dimensions as the rate constant; R is the gas constant; and T is the absolute temperature.

The mean velocity of a molecule in an ideal gas is proportional to the square root of temperature, T:

$$\bar{u} = \sqrt{\frac{8k \times T}{\pi \times m}}, \tag{9.29}$$

where k is Boltzmann's constant and m is the mass of the molecule. A similar temperature dependency would hold for ions in water. On the other hand, the probability of the molecule having a velocity u is determined by the Maxwell–Boltzmann probability distribution of velocities of a gas

$$p(u) = A \times u^2 \times \exp\left(\frac{m \times u^2}{2k \times T}\right), \tag{9.30}$$

where A is a constant that is different from the pre-exponential term A_{Ar} in the Arrhenius equation (Eq. (9.28)). The similarity of the form of the Arrhenius and Maxwell–Boltzmann equations indicates that the probability of high-energy encounters is the most important factor in determining the temperature dependence of bond breaking and formation.

The method of experimentally determining the activation energy of a reaction is to plot the observed rate against temperature. Using Eq. (9.28) above, one derives a linear relation between $\ln(k)$ and $1/T$, with the intercept equal to the pre-exponential factor, A_{Ar}. This relation is demonstrated for the rate constants of CO_2 hydration, Fe^{2+} and HS^- oxidation, and opaline silica dissolution in Fig. 9.6. Plots of this type are used by kineticists to derive information about the mechanism of reaction because the activation energy refers to the energy barrier for the rate-limiting elementary reaction. For example, the activation energy for diffusion in aqueous solutions is about 20 kJ mol^{-1} (*c.* 5 kcal mol^{-1}). If the temperature dependence of a reaction is in this range, it may be assumed that the rate of the elementary reaction is diffusion-controlled. A change in the linear relation between $\ln(k)$ and $1/T$ could

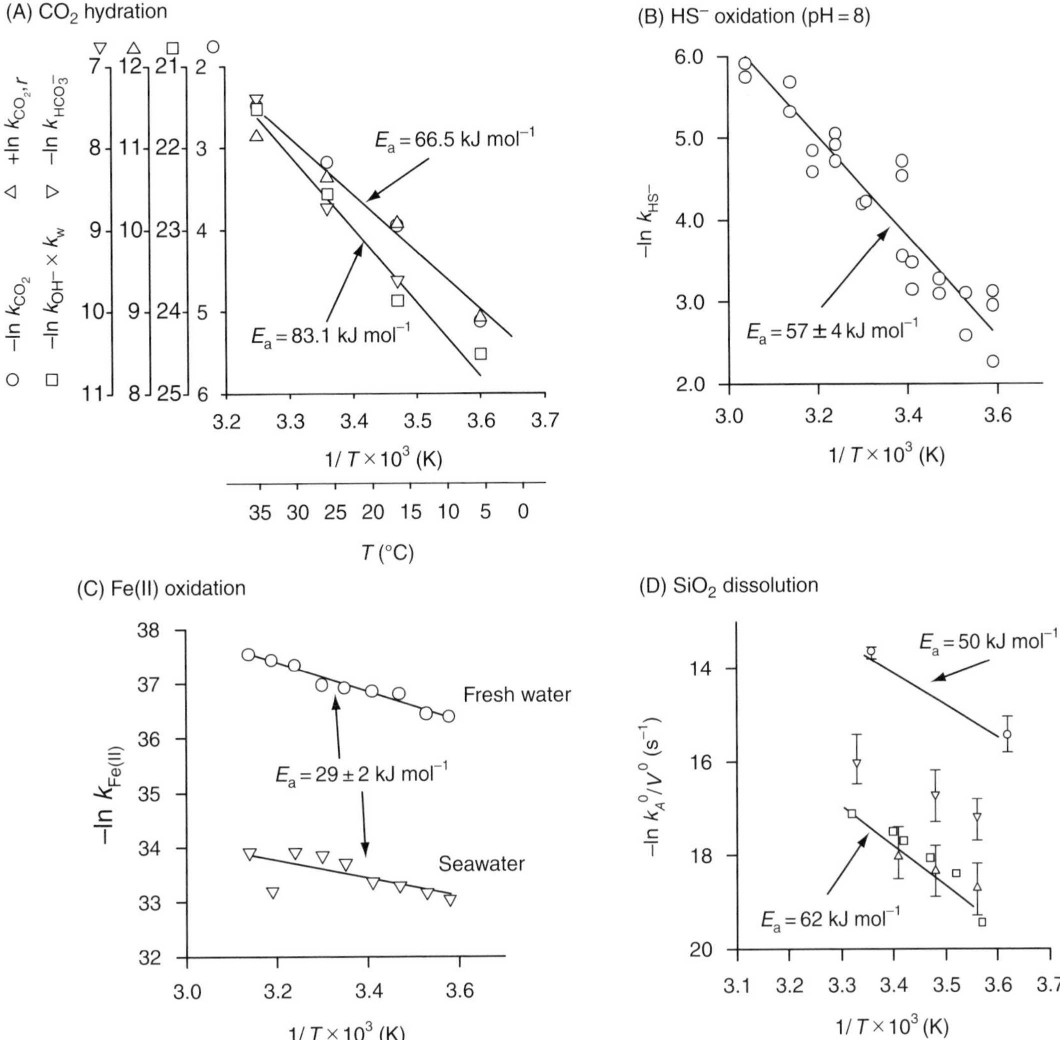

Figure 9.6. Temperature dependence of environmentally important reaction rate constants. (A) CO_2 reaction with H_2O or OH^- in seawater. The data are plotted so that the *slope* of the line is the Arrhenius activation energy, E_a, divided by the gas constant, R (8.315 J mol^{-1} deg^{-1}) (Johnson, 1982). (B) HS^- oxidation (Millero *et al.*, 1987a). (C) Fe(II) oxidation in seawater (Millero *et al.*, 1987b). (D) The dissolution rate constants for opal as a function of temperature. The values plotted here are normalized to a constant area: volume ratio, $A^0/V^0 = 10\ cm^{-1}$. The high dissolution rate constants (circles) are for acid-cleaned siliceous sediments (Hurd, 1972); the squares are data for peroxide and acid-washed plankton (Lawson *et al.*, 1978); and the other results (triangles) are for two untreated single species of diatom (Kamatani, 1982).

indicate a change in the mechanism controlling the reaction rate. The results shown in Fig. 9.6 indicate that the CO_2 hydration reactions, the two oxidation reactions, and the opal dissolution reaction all follow a single mechanism over the temperature range shown in the figure, and in each case activation energies are significantly greater than 20 kJ mol^{-1}, indicating that the reaction rates are limited by bond breaking rather than by molecular diffusion.

Activation energies for some important water–rock reactions in nature were presented along with pH dependency in Table 9.4. The values are mostly in the range 35–80 kJ mol^{-1}, indicating that these reactions are not diffusion-controlled for the pH range in which they were studied. These values, however, are smaller than typical energies required for bond breaking in crystals, which are in the range 80–400 kJ mol^{-1} (Lasaga, 1984). The likely explanation for the observed activation energy is that the dissolution reactions are catalyzed by adsorption of solutes, such as hydrogen ions and organic molecules, on the surface.

9.3 Reaction rate catalysis

Most marine chemical reaction rates with mean lives greater than tens to hundreds of seconds have been observed to be catalyzed. This observation implies that laboratory experiments in pure solutions are apt to overestimate the mean lives of reacting species. For this reason the study of chemical kinetics in aquatic systems must include a thorough appreciation of reaction rate catalysis. The definition of a catalyst has evolved from a rather narrow view of a substance that changes the speed of reaction without undergoing chemical change itself, to a broader interpretation in which the catalyst is any substance that alters the mechanism of the chemical reaction, causing the activation energy to be lowered without changing the free energy change for equilibrium between reactants and products (Moore and Pearson, 1981).

The process of catalysis involves a chemical reaction of a substrate with the catalyst, forming a complex, which then reacts to form products. The mechanism by which this occurs is different for homogeneous, heterogeneous and enzymatic catalysis. A probable mechanism for homogeneous catalysis in solution involves binding of the catalyst (for example, hydrogen ions, hydroxyl ions and some metals) to the substrate, thus lowering the electron density at the site subject to bond breaking and requiring less energy for reaction (Basolo and Pearson, 1968). One of the reasons that protons, hydroxyl ions, and trace metals are effective homogeneous catalysts is their high charge density. Other factors such as electron structure also come into play, particularly for transition metals (Moore and Pearson, 1981). Heterogeneous catalysis (e.g. reactions facilitated by solid surfaces) involves a series of steps in which the substrate must diffuse to the surface, adsorb, react, and then diffuse away. Any of these steps can be rate-limiting, but chemical reaction at the surface usually proves to be the slowest step. A third type of catalytic action is enzymatic catalysis. Enzymes are large proteins produced by living systems, sometimes for the specific purpose of catalyzing chemical reactions that are necessary for life. The mechanism often involves reaction at a site in the catalytic molecule that is particularly suited to the geometry of the substrate, causing enzymatic catalysis to be highly specific and very effective. The actual mechanisms of lowering the energy required for reaction are understood in only a few cases.

Observed first-order rate constants for uncatalyzed and catalyzed reactions of CO_2 hydration, hydrogen sulfide oxidation, and Mn^{2+} oxidation in water are presented in Table 9.7. The catalytic turnover number (CTN) that accompanies the rate constants is defined as the initial rate of reaction (M, s^{-1}) divided by the concentration of catalyst. The CTN is a measure of the power of the catalyst, but its effectiveness in seawater also depends on the catalyst concentration. Since there are few generalities regarding the magnitude of enhancement of reaction rates by catalysis in nature, the subject is highly empirical. For example, there is no way to predict the vast range of variation of the observed rate constants in Table 9.7. The very high turnover numbers (up to nearly 10^8) by some transition metals and enzymes indicate the extreme potential of reaction rate enhancement by trace quantities of catalyst.

9.3.1 The mechanism of reaction rate catalysis

Rate constants for catalytic reactions, k_{cat}, are normally reported separately from those for the uncatalyzed reaction, k_0, if both are known:

$$k_{obs} = k_0 + k_{cat}. \tag{9.31}$$

Although the actual physical mechanism varies for homogeneous, surface, and enzyme catalysis, the processes all involve reaction of the substrate, X, with the catalyst, C, which is usually reversible, and then further reaction of the catalyst–substrate complex, X ∗ C, perhaps with another solute, W, to products, P. Mathematically, the catalytic equations can often be represented in a general way as

$$X + C \underset{k_{-1}}{\overset{k_1}{\rightleftharpoons}} X * C \tag{9.32}$$

and

$$X * C + W \xrightarrow{k_2} P + C, \tag{9.33}$$

in which the rate-limiting step is usually the second of these reactions. With these elementary reactions as a guide, the rate equation for the change in the substrate–catalyst complex as a function of time is

$$\frac{d[X * C]}{dt} = -k_1[X][C] + k_{-1}[X * C] - k_2[X * C][W]. \tag{9.34}$$

A common feature of the initial catalyzed reaction rate, before significant amounts of C are consumed, is the rate dependency on substrate concentration. The rate is first-order with respect to [X] when [X] is relatively low and independent of [X] when [X] is high (Fig. 9.7). The reason for this rate dependence is that the

Table 9.7. *Homogeneous, surface and enzyme catalysis of carbon dioxide hydration, hydrogen sulfide oxidation and Mn^{+2} oxidation*

The observed reaction rate constants, k_{obs}, have been normalized to 1st-order constants when necessary. The catalytic turnover number, CTN, is the number of moles of substrate reacted per minute divided by the moles of enzyme present.

Reaction	Catalyst (concentration) (mol l^{-1})	k_{obs} (s^{-1})	CTN[a] (min^{-1})	References
CO_2 hydrolysis	None	2.6×10^{-2}	1.8×10^{1}	Pocker and Bjorkquist (1977); conditions were $T = 25\,°C$,
	Cu^{2+} (5×10^{-3})	4×10^{-2}	2.4×10^{1}	pH = 6–7 except for $[OH^-]$ catalysis (pH = 8 and 10),
	Zn^{2+} (5×10^{-3})	6×10^{-2}	1.2×10^{4}	$[CO_2]_0 = 3.4 \times 10^{-2}$.
	OH^- (10^{-6})	6×10^{-3}	1.2×10^{4}	
	OH^- (10^{-4})	6×10^{-1}	1.0×10^{7}	
	$Co(NH_3)_5OH_2^{3+}$ (5×10^{-8})	2.5×10^{-1}	8.0×10^{7}	Khalifah (1971); Coleman (1973)
	Carbonic anhydrase	5×10^{5}		
HS^- oxidation	None	$3 - 6 \times 10^{-6}$		Chen and Morris (1972)
	Cu TSP (5×10^{-6})	1.5×10^{-4}	1.8	Hoffman and Lim (1979); conditions were $T = 25\,°C$, pH ≈ 8,
	Ni TSP (5×10^{-6})	1.9×10^{-3}	1.7×10^{1}	$[O_2] = 10^{-3}$ and $HS^- = 10^{-3}$.
	Co TSP (5×10^{-10})	1.5×10^{-3}	1.3×10^{5}	
Mn^{2+} oxidation	None[b]	$0.2 - 1.3 \times 10^{-4}$	1×10^{-2}	Constant is from Morgan (1964);
	Silica (1.5×10^{-5})	5.5×10^{-5}	5×10^{-2}	[c]Davies (1985); conditions were $T = 2\,°C$, $[O_2] = 1.26$ mM,
	Goethite (1.5×10^{-5})	2.3×10^{-4}	5×10^{-2}	pH = 8.5–9.1, $[Mn^{2+}]_0 = 50 \times 10^{-6}$.
	Lepidocrocite (2.0×10^{-5})	3.0×10^{-4}	1×10^{-2}	
	Bacteria (5.0×10^{-8})	1.0×10^{-6}		[d]Tebo and Emerson (1986); conditions were $T = 9\,°C$,
				pH ≈ 7.4, $[O_2] = 15$ μM.

[a] CTN = $[S]_0\, k_{obs}/[C]$ (min^{-1}) where $[S]_o$ is the initial substrate concentration and [C] is the catalyst concentration.

[b] Diem and Stumm (1984) found no auto-oxidation over 7y.

[c] The catalyst concentrations represent the concentrations of surface sites $[\equiv(SO)_2\,Mn]$ measured by titration on the various solids.

[d] Values are from tracer addition experiments in water from the O_2–H_2S interface in Saanich Inlet. The oxidation rate was c.2 nM h^{-1} at a substrate concentration of about 0.5 μM and the bacterial catalytic site concentration, [C] = 50 nM, was estimated from zero-oxygen Mn^{2+} uptake experiments.

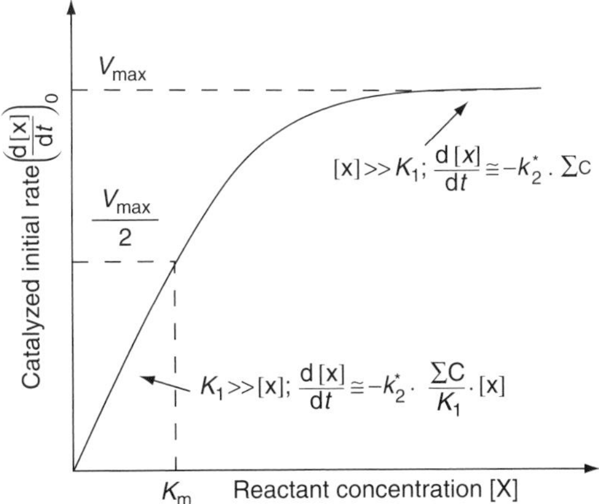

Figure 9.7. The initial reaction rate of a catalyzed reaction versus the concentration of the substrate [X] (Eq. (9.39), where $K_1 = k_{-1}/k_1$). The catalytic reaction could be homogeneous, heterogeneous or enzyme catalysis so long as it follows the simple catalytic mechanism. The substrate concentration, [X], at a rate of half the maximum reaction rate, $V_{max}/2$, defines K_m in Michaelis–Menten enzyme kinetics.

total catalyst concentration is very often much less than the substrate concentration. Because the reaction proceeds through the complex $X*C$, the initial rate will be limited by the substrate concentration if [X] is lower than [C] and will be independent of the substrate concentration when $[X*C]$ attains its maximum value (i.e. the catalyst is saturated). The case in which the reaction identified by k_2 is rate-limiting is easily demonstrated mathematically, assuming that the formation of the substrate–catalyst complex is at steady state

$$\frac{[X*C]}{[X]\,[C]} = \frac{k_1}{k_{-1}} \tag{9.35}$$

and that the concentration of the reactant, W, is constant (i.e. $k_2^* = k_2[W] = \text{constant}$). The total catalyst concentration, $\sum C$, is defined as the sum of free catalyst and complex

$$\sum C = [C] + [X*C]. \tag{9.36}$$

Combining the above two equations yields a relation between the substrate–catalyst complex, $[X*C]$, and the substrate, [X]:

$$[X*C] = \sum C \frac{k_1[X]}{k_1[X] + k_{-1}}. \tag{9.37}$$

Because we assumed that the catalyst–complex formation was fast, the overall initial rate,

$$\left(\frac{d[X*C]}{dt}\right)_0$$

of the substrate reaction is determined by the rate-limiting reaction (9.33), i.e. $X*C$ and X are consumed at equal rates

$$\left(\frac{d[X]}{dt}\right)_0 = \left(\frac{d[X*C]}{dt}\right)_0 = -k_2^*[X*C], \tag{9.38}$$

which, when combined with Eq. (9.39), yields

$$\left(\frac{d[X]}{dt}\right)_0 = -k_2^* \sum C \left(\frac{[X]}{[X] + k_{-1}/k_1}\right). \tag{9.39}$$

The shape of the curve of the initial reaction rate versus substrate concentration is illustrated schematically in Fig. 9.7. If the substrate concentration is sufficiently lower than the ratio of the rate constants, (k_{-1}/k_1), then the reaction rate is first-order with respect to [X]; if it is sufficiently greater, the reaction rate is independent of [X] and approaches the constant value $k_2^* \sum C$. The equation can usually be simplified by considering the relative magnitudes of the terms k_{-1} and $k_2 \times [W]$.

Although it is often unnecessary, for interpreting initial time-dependent experimental results, to solve Eq. (9.34) in its complete form, the full time-dependent solution for the case in which the reaction of the substrate–catalyst concentration (Eq. (9.33)) is rate-limiting has been used to describe surface catalysis of the oxidation of Fe^{2+} and Mn^{2+} (Sung and Morgan, 1980). In this case the substrate and catalyst are stoichiometrically related to their initial values, $[X_0]$ and $[C_0]$

$$[X_0] - [X] = [C_0] - [C], \tag{9.40}$$

and the solution is presented as (b) in the second-order reactions in Table 9.5.

9.3.2 Homogeneous catalysis

The most important types of homogeneous catalysis in water are performed by acids, bases and trace metals. A wide variety of mechanisms have been outlined for acid/base catalysis and are presented in kinetics texts (e.g. Moore and Pearson, 1981; Laidler, 1965). A number of bases have been observed to catalyze the hydration of carbon dioxide (Moore and Pearson, 1981; Dennard and Williams, 1966). Examples are listed in Table 9.7 for OH^- and the base $Co(NH_3)_5OH_2^{3+}$. The most dramatic effect is the catalysis of HS-oxidation by cobalt-4,4′,4″,4″-tetrasulfophthalocyanine ($Co\text{-}TSP^{2-}$). At concentrations of 0.1 nM $Co\text{-}TSP^{2-}$ the reaction rate was catalyzed from a mean life of roughly 50 h to about 5 min. The investigators attributed the reason for historically inconsistent experimentally determined reaction rates for the H_2S–O_2 system by different researchers partly to contamination by metals. Clearly, catalysis by metal concentrations that are present in less than nanomolar concentrations is likely to be effective in aquatic systems. We shall see that similar arguments apply to catalysis by surfaces and enzymes.

9.3.3 Heterogeneous catalysis

One of the main differences between laboratory experimental systems and natural waters is the presence in nature of a large variety of particles. It was originally believed that the mechanism for the catalysis of reactions by solid surfaces was that the surfaces served as localized regions of high concentration. Because reaction orders with respect to the substrate are usually one or greater, surface concentration would cause an enhancement of the reaction rate. It is now known that this is not the mechanism of heterogeneous catalysis, primarily because the same substrate with a variety of catalytic surfaces can result in different rates and products (Laidler, 1965). The general reason for the catalytic effect of surfaces is the same as that for homogeneous catalysis in that the surface provides an alternative mechanism with a lower activation energy for the reaction.

The surface adsorption model described in Chapter 3 (Section 3.4.4) has been used to describe observed rates of surface catalysis of the oxidation of Mn^{2+} by silica and two iron oxide solids (Davies and Morgan, 1989). The observed oxidation rate constants are compared with that for the maximum suggested rate of the homogeneous reaction in Table 9.7. Iron oxides were the most effective of the catalytic surfaces; at a concentration of 20 μM, lepidocrocite enhanced the reaction rate by nearly a factor of 10. The observed rate law for the surface reaction was

$$\frac{d[Mn^{2+}]}{dt} = k^* \frac{\langle \equiv SOH \rangle [Mn^{2+}]}{[H^+]^2} \times a \times p_{O_2}, \tag{9.41}$$

where, as before, $\langle \equiv SOH \rangle$ is the concentration of surface species that exchange protons with manganese during adsorption. The solid concentration is denoted by a.

The rate law and experimentally determined surface characteristics can be combined to determine the mechanism of this environmentally important reaction. The equations describing the amphoteric (acid/base) properties of the solid surfaces have been previously presented in Eqs. (3.47) and (3.48). One interpretation of the second-order dependency of manganese exchange with respect to hydrogen ion is that two surface sites are required to adsorb one dissolved ion, which is known as a bidentate complex of Mn^{2+}, on the surface. Thus, two previously adsorbed hydrogen ions exchange places with Mn^{2+} in solution, which is denoted by the equilibrium constant β_2^S:

$$\begin{matrix} \equiv S\text{-}OH \\ \equiv S\text{-}OH \end{matrix} + Mn^{2+} \overset{\beta_2^S}{\rightleftarrows} \begin{matrix} \equiv S\text{-}O \\ \equiv S\text{-}O \end{matrix} \triangleright Mn + 2H^+. \tag{9.42}$$

The mechanistic sequence for adsorption of Mn^{2+} on the surface and then oxidation proposed by Davies and Morgan (1989) is illustrated in Fig. 9.8. After Mn^{2+} adsorption there is adsorption of O_2 molecules on the surface (Fig. 9.8b) and then an electron transfer between adsorbed O_2 and Mn^{2+} (Fig. 9.8c). The mechanism is completed by reaction of the adsorbed Mn(III)–superoxide complex (Fig. 9.8d). If the rate-limiting step is the reaction of the adsorbed Mn(III)–superoxide complex to products, then the rate for the elementary reaction is

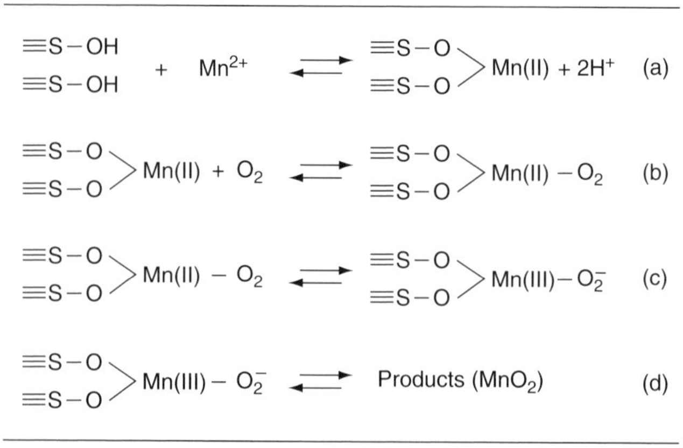

Figure 9.8. A possible mechanism for surface (≡S) catalysis of Mn(II) (Davies, 1985). The surface is partly covered by OH^- ions, which are the sites of reactions. The steps are: (a) Mn^{2+} complexation on the surface by an exchange reaction with two H^+ ions; (b) formation of a Mn^{2+}–O_2 complex on the surface; (c) transfer of electron from Mn^{2+} to O_2, forming a superoxide–Mn(III) complex; and (d) further electron transfer to create the oxidized manganese product.

$$\begin{matrix}\equiv S-OH \\ \equiv S-OH\end{matrix} + Mn^{2+} \rightleftharpoons \begin{matrix}\equiv S-O \\ \equiv S-O\end{matrix} > Mn(II) + 2H^+ \quad \text{(a)}$$

$$\begin{matrix}\equiv S-O \\ \equiv S-O\end{matrix} > Mn(II) + O_2 \rightleftharpoons \begin{matrix}\equiv S-O \\ \equiv S-O\end{matrix} > Mn(II) - O_2 \quad \text{(b)}$$

$$\begin{matrix}\equiv S-O \\ \equiv S-O\end{matrix} > Mn(II) - O_2 \rightleftharpoons \begin{matrix}\equiv S-O \\ \equiv S-O\end{matrix} > Mn(III) - O_2^- \quad \text{(c)}$$

$$\begin{matrix}\equiv S-O \\ \equiv S-O\end{matrix} > Mn(III) - O_2^- \rightleftharpoons \text{Products } (MnO_2) \quad \text{(d)}$$

$$-\frac{d[Mn^{2+}]}{dt} = \bar{k} \times \left[{}^{\equiv S-O}_{\equiv S-O} \triangleright Mn(III) - O_2^-\right]. \tag{9.43}$$

Using the equilibrium constants for reactions a, b, and c in Fig. 9.8 (denoted as, β_2^S, K_b, and K_c^*, respectively) the above equation becomes

$$-\frac{d[Mn^{2+}]}{dt} = \bar{k}K_b K_c^* \beta_2^S \left(\frac{[O_2]\langle\equiv SOH\rangle[Mn^{2+}]}{[H^+]^2}\right). \tag{9.44}$$

This equation explains the observed rate dependency in Eq. (9.41), where the rate constant in the observed reaction, k^*, is equal to the product, $\bar{k}K_b K_c^* \beta_2^S$. The surface catalysis of manganese oxidation is very effective in natural waters. The strength of the catalysis as indicated by the catalytic turnover number (CTN) in Table 9.7 is less than some of the other catalysts listed, but particle surfaces are ubiquitous.

Experimental studies have shown that not only oxidation reactions, but nearly all mineral dissolution reactions in nature, can be interpreted as a heterogeneous surface rate catalysis. As discussed earlier these reactions are usually stirring-independent and zero-order with respect to the products of dissolution (Fig. 9.4). The functionality of the dissolution reaction with respect to pH, and the fact that activation energies are significantly lower than the strength of metal oxide bonds (Table 9.4), suggests that the surface reactions controlling dissolution are catalyzed.

The view of this process forwarded by Werner Stumm and coworkers is that the surface of the oxide continuously exchanges protons and other potential catalysts with the solution phase. The surfaces possess the same amphoteric behavior as that described in the last section for solids catalyzing manganese oxidation. Exchange of the catalyst (either protons or organic molecules) is a fast reaction relative to the rate-limiting detachment of the metal from the oxide surface. In this view, the reason for catalysis of the dissolution reaction is that the reacting ion attaches to the surface, drawing away some of the electron charge from the metal–oxide bond, weakening it and facilitating detachment of the metal ions.

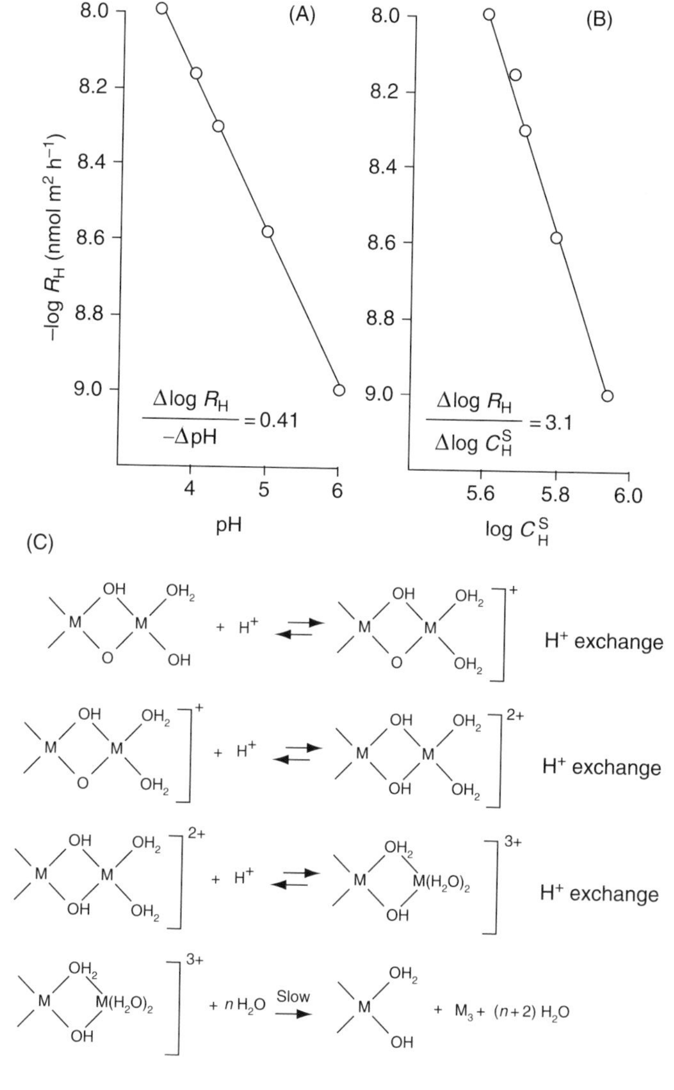

Figure 9.9. Proton catalysis of the dissolution rate of aluminum oxide. The logarithm of the dissolution rate, R_H, is plotted as a function of pH (A) and the logarithm of the surface concentration of hydrogen ion, C_H^s (B). The non-integral rate with respect to pH (A) but nearly integral rate with respect to log C_H^s is common in mineral dissolution reactions. A schematic representation of the four steps proposed to explain the data in (A) and (B) is presented in (C). The first three steps are surface proton exchange reactions and are followed by the rate-determining detachment of the metal from the surface. Adapted from Furrer and Stumm (1986).

Studies of the dissolution rates of aluminum oxide and beryllium oxide (Furrer and Stumm, 1986) showed that the rates of these reactions are facilitated by increased concentrations of protons and various aliphatic and aromatic ligands at the oxide surface. The key link for determining the mechanism of reaction catalysis is establishing the relation between the concentration of these species in solution and the concentration at the solid surface. This relation is determined by potentiometric titration of the solution–solid mixture, just as described earlier for the surface catalysis of Mn^{2+} oxidation. The relations between the dissolution rate of aluminum oxide and the hydrogen concentration both in solution and on the solid surface are presented along with the proposed mechanism in Fig. 9.9. The solution dependence of the dissolution rate on pH (Fig. 9.9A) has a fractional order of 0.4, which is similar to that of other oxide dissolution reactions (Table 9.4). But when the same relation is plotted as a function of the concentration of protons at the solid surface (Fig. 9.9B), the reaction order is about 3.

The reaction rate equation for the dissolution of aluminum oxide can be written as a function of the concentration of hydrogen ions at the surface, $[H^+]_s$, following the result in Fig. 9.9:

$$\frac{d[Al^{3+}]}{dt} = k_{cat,\,H}[H^+]_s^3, \tag{9.45}$$

where the rate constant $k_{cat,\,H}$ refers to the catalysis of the reaction by protons. The proposed mechanism is a sequence of rapid proton exchange reactions followed by the rate-limiting detachment of the metal ion, M (Fig. 9.9C). The rate expression for the last reaction is

$$\frac{d[M]}{dt} = k_{cat}\left[\begin{matrix}OH_2\\OH\end{matrix}\; M(H_2O)_2\right] = k_{cat}K_1K_2K_3\left[\begin{matrix}OH\\OH\end{matrix}\; M \;\begin{matrix}OH_2\\OH\end{matrix}\right][H^+]^3, \tag{9.46}$$

where the surface complex represented in the detachment reaction of Fig. 9.9 is the concentration in brackets in the middle on the right side and the equilibrium constants (K_1, K_2, K_3) represent the hydrogen ion exchange reactions.

9.3.4 Enzyme catalysis

Enzymes are the most effective catalysts known and control the rates of many reactions in living systems. (The enzyme carbonic anhydrase, which facilitates the transfer of carbon from HCO_3^- in blood to CO_2 in our lungs, is so efficient that each molecule of the enzyme can hydrate 10^6 molecules of CO_2 per second!) Enzymes consist of high molecular mass (10^5 to $>10^6$ amu) protein molecules that are generally between 30 and 1000 Å in diameter. Since they are so large, their behavior as catalysts lies somewhere between the homogeneous and heterogeneous systems discussed so far. Enzymes are characterized by extreme specificity with regard to the catalyzed reactions and can be regulated by inhibitors that are similar enough to the substrate to interfere with the catalytic process. Although the molecular mass and size of enzymes indicates that they comprise very long chains of amino acids, the active catalytic sites are only a very small portion of the total molecule (Fig. 9.10).

Studies by X-ray crystallography have shown that the reason for the extreme specificity of the catalytic activity has to do with the

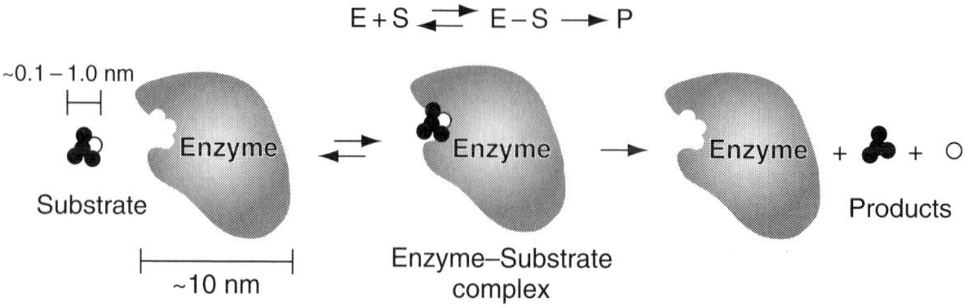

Figure 9.10. A schematic diagram of the reaction of an enzyme with a substrate, indicating their relative sizes (typically about 10 nm for a medium-sized enzyme and 0.1–1.0 nm for the substrate) and the "lock and key" fit.

geometry of the active site of the enzyme. This characteristic is often referred to by using the "lock and key" analogy. Although the study of enzyme catalysis has in recent years made great progress in understanding the mechanisms of some reactions, very basic questions such as the reason the molecules are so large with only a few active sites, and elucidation of the actual processes of reaction enhancement once the molecule is bound to the enzyme, are still, for most systems, unanswered. This is not surprising considering the enormous task of finding and studying the active site on these very large molecules.

The kinetic mechanism of enzyme reaction in its most simple form is exactly analogous to the two-step process of rapid exchange with an intermediate and then reaction to products described in the general catalysis equations. The formulation was introduced by Michaelis in 1913 to present evidence for the substrate–enzyme complex. There are a great number of modifications of the Michaelis–Menten equations describing the various mechanisms that have been discovered for different enzymes. Some of the most significant are the kinetics for multi-intermediates, for competitive and non-competitive inhibition (Lehninger, 1975), and the allosteric enzyme kinetics of Monod *et al.* (1963). Our brief discussion of the mechanism will deal only with the simple Michaelis–Menten kinetics for a one-substrate reaction.

Reaction rates by enzyme catalysis are described by Eqs. (9.32) and (9.33) with some minor modifications of the terminology

$$X + C \underset{k_{-1}}{\overset{k_1}{\rightleftarrows}} X * C \tag{9.47}$$

$$X * E \xrightarrow{k_2} E + \text{product}. \tag{9.48}$$

We have substituted the symbol E (enzyme), for the catalyst, C, and it is explicitly shown that the enzyme is regenerated by the rate-limiting reaction (9.48), which could also be written with the reactant W if necessary. The initial reaction rate after the formation of the enzyme complex, $X * E$, is analogous to Eq. (9.39) where the ratio of the reaction rates, (k_{-1}/k_1), is equal to the Michaelis–Menten constant, K_m:

$$-\left(\frac{d[X * E]}{dt}\right)_0 = k_2 \times \sum E \times \left(\frac{[X]}{[X] + K_m}\right). \tag{9.49}$$

The maximal rate occurs when the enzyme sites are virtually all occupied (saturated) by substrate; i.e. if $K_m << [X]$ then $[X * E] \sim \sum E$ and the reaction rate is $k_2 \times \sum E$, which is termed V_{max} in enzyme catalysis and is analogous to the maximum rate of homogeneous and surface catalysis when the dissolved or surface-bound catalysts are saturated. The rate of reaction at the substrate concentration equal to K_m is $V_{max}/2$. If $K_m >> [X]$, the rate becomes linear in the substrate concentration, [X]

$$-\left(\frac{d[X * E]}{dt}\right)_0 \cong \frac{V_{max}}{K_m} \times [X]. \tag{9.50}$$

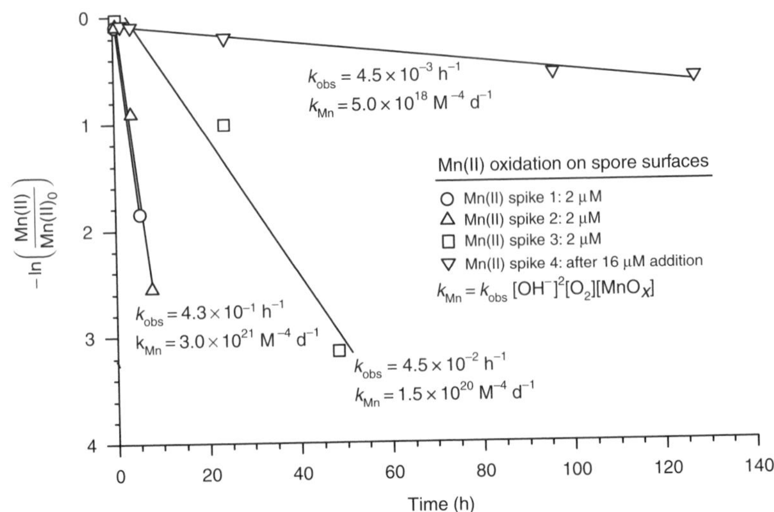

Figure 9.11. The progressively slower oxidation rate of Mn^{2+} in the presence of spores that catalyze the reaction and become coated with oxidized manganese. The spore surfaces were initially bare of MnO_X and bacame progressively coated over time as concentrations of Mn(II) indicated in the figure were added. The symbols represent successive additions of dissolved Mn^{2+}. Original oxidation rates are about 10 times greater than the final rates, which approach rates similar to those observed on the surface of colloidal MnO_2. The solid Mn precipitate had an oxidation state of between 3 and 3.5. Experiments were performed in DOC-free seawater at 23° C, pH = 7.8 and $[O_2] \approx 0.23$ atm (Hastings and Emerson, 1986).

Catalysis of carbon dioxide hydration by carbonic anhydrase is the most rapid enzymatic reaction known (Table 9.7). It has been shown (Khalifah, 1971) that the K_m for the human forms of the enzyme are between 4 and 9 mM, and the catalytic turnover number, $V_{max}/\sum E$, is 8×10^7 (Table 9.7).

Among the metal oxidation and reduction reactions that occur in the environment, the most intensely studied enzyme-catalyzed reaction is the oxidation of divalent manganese. Bacteria and bacterial spores that carry out this function have been isolated from natural waters. These microbes are believed to produce proteins that enhance the rate, but the actual substances and the biochemical mechanism responsible for the reaction are not well characterized.

Laboratory Mn^{2+} oxidation rate studies in the presence of the manganese-oxidizing spores of a marine bacillus (Fig. 9.11) revealed a progressively decreasing oxidation rate with successive additions of Mn^{2+}. Oxidation state measurements of the solid that progressively coated the spores during the course of the experiment indicated that the product was oxidized manganese. Initial oxidation rates were comparable to those observed for bacterial catalysis in the environment and the final rates were nearly identical to those expected for heterogeneous catalysis of the oxidation rate on the surface of MnO_2. Field studies of the oxidation rate of Mn^{2+} utilizing water samples labeled with the radiotracer ^{54}Mn from a location of rapid Mn oxidation in an anoxic fjord indicated an oxidation rate that was roughly 10^5 greater than that which would be expected without catalysis (Tebo and Emerson, 1986).

Catalysis by enzymes requires an empirical interpretation of reaction rates because it is difficult to transfer generalizations derived from laboratory experiments to natural waters when the reaction rates are microbially catalyzed. At the present state of our knowledge, the most pressing problem is to determine which reaction rates are influenced by the process of enzyme catalysis so that it is clear when laboratory-derived mechanisms can be applied. The next task, which presently seems far away, will be to generate predictive expressions for the effect of enzyme catalysis on well-defined substrates so that generalizations for quantifying reaction rates controlled by this process in nature can be developed.

References

Basolo, F. and R. G. Pearson (1968) *Mechanisms of Inorganic Reactions*, 2nd edn. New York, NY: Wiley.

Ben-Yaakov, S. (1972) Diffusion of sea water ions. I. Diffusion of sea water into dilute solution. *Geochim. Cosmochim. Acta* **36**, 1395–406.

Berner, R. A. (1978) Rate control of mineral dissolution under earth surface conditions. *Am. J. Sci.* **278**, 1235–52.

Bevington, P. C. and D. K. Robinson (2003) *Data Reduction and Error Analysis*. Boston, MA: McGraw-Hill.

Busenberg, E. and N. L. Plummer (1982) The kinetics of dissolution of dolomite in CO_2-H_2O systems at 1.5 to 65 °C and 0 to 1 atm. pCO_2. *Am. J. Sci.* **282**, 45–78.

Chen, K. and J. C. Morris (1972) Kinetics of oxidation of aqueous sulfide by O_2. *Envir. Sci. Technol.* **6**, 529–37.

Coleman, J. E. (1973) Carbonic anhydrase. In *Inorganic Biochemistry*, vol. 1 (ed. G. L. Eichorn), pp. 489–548. Amsterdam: Elsevier.

Crank, J. (1975) *The Mathematics of Diffusion*, 2nd edn. Oxford: Clarendon Press.

Csanady, G. T. (1973) *Turbulent Diffusion in the Environment*. Dordrecht, Holland: Reidel.

Davies, S. H. R. (1985) Mn(II) oxidation in the presence of metal oxides. Ph.D. thesis, California Institute of Technology, Pasadena, CA.

Davies, S. and J. J. Morgan (1989) Manganese (II) oxidation kinetics on metal oxide surfaces. *J. Colloid Interface Sci.* **129**, 63–77.

Dennard, A. E. and J. P. Williams (1966) The catalysis of reaction between carbon dioxide and water. *J. Chem. Soc.* **A**, 812–16.

Diem, D. and W. Stumm (1984) Is dissolved Mn^{2+} being oxidized by O_2 in absence of Mn-bacteria or surface catalysis? *Geochim. Cosmochim. Acta* **48**, 1571–3.

Eigen, M. and K. Kustin (1962) The influence of steric factors in fast protolytic reactions as studied with HF, H_2S and substituted phenols. *J. Am. Chem. Soc.* **82**, 5952–3.

Eigen, M., K. Kustin and G. Maass (1961) Die Geschwindigkeit der Hydration von SO_2 in waesseriges Loesung. *Z. Physik. Chem.* (N.F.) **30**, 130–6.

Fleer, V. N. (1982) The dissolution kinetics of anorthite ($CaAl_2Si_2O_8$) and synthetic strontium feldspar ($SrAl_2Si_2O_8$) in aqueous solution at temperatures below 100 °C: with applications to the geological disposal of radioactive nuclear wastes. Ph.D. thesis, Pennsylvania State University, University Park, PA.

Furrer, G. and W. Stumm (1986) The coordination chemistry of weathering. I. Dissolution kinetics of δ-Al_2O_3 and BeO. *Geochim. Cosmochim. Acta* **50**, 1847–60.

Gardiner, W. C. (1972) *Rates and Mechanisms of Chemical Reactions*. W. A. Benjamin.

Grandstaff, D. E. (1981) The dissolution rate of forsteric olivine from Hawaiian beach sand. *Third Int. Symp. Water-Rock Interaction, Proc.*, pp. 72–4. Edmonton: Alberta Research Council.

Handbook of Chemistry and Physics (1970) Cleveland, OH: The Chemical Rubber Publishing Co.

Hastings, D. and S. Emerson (1986) Oxidation of manganese by spores of a marine Bacillus: kinetic and thermodynamic considerations. *Geochim. Cosmochim. Acta* **50**, 1819–24.

Helgeson, H. C., W. M. Murphy and P. Aagaard (1984) Thermodynamic and kinetic constraints on reaction rates among minerals and aqueous solutions. II. Rate constants, effective surface area, and the hydrolysis of feldspar. *Geochim. Cosmochim. Acta* **48**, 2405–32.

Hemmes, P. *et al.* (1971) Kinetics of hydrolysis of ferric ion in dilute solutions. *J. Phys. Chem.* **75**, 929–32.

Hoffman, M. R. and X. Lim (1979) Kinetics and mechanism of the oxidation of sulfide by oxygen. *Envir. Sci. Technol.* **13**, 1406–12.

Holdren, G. R. and R. A. Berner (1970) Mechanism of feldspar weathering. I. Experimental studies. *Geochim. Cosmochim. Acta* **43**, 1161–71.

Hurd, D. C. (1972) Factors affecting solution rate of biogenic opal in seawater. *Earth Planet. Sci. Lett.* **15**, 411–17.

Jahne, B. H. G. and W. Dietrich (1987) Measurement of the diffusion coefficients of sparingly soluble gases in water. *J. Geophys. Res.* **92**, 10 767–76.

Johnson, K. S. (1982) Carbon dioxide hydration and dehydration kinetics in seawater. *Limnol. Oceanogr.* **27**, 849–55.

Kamatani, A. (1982) Dissolution rates of silica from diatoms decomposing at various temperatures. *Mar. Biol.* **68**, 91–6.

Kern, D. M. (1960) The hydration of carbon dioxide. *J. Chem. Ed.* **37**, 14–23.

Khalifah, R. G. (1971) The carbon dioxide hydration activity of carbonic anhydrase. *J. Biol. Chem.* **246**, 2561–73.

Laidler, K. J. (1965) *Chemical Kinetics*. New York, NY: McGraw-Hill.

Lasaga, A. C. (1984) Chemical kinetics of water-rock interactions. *J. Geophys. Res.* **89**, 4009–25.

Lawson, D. D., D. C. Hurd and S. Punkrutz (1978) Silica dissolution rates of decomposing phytoplankton assemblages at various temperatures. *Am. J. Sci.* **278**, 1373–93.

Lehninger, A. L. (1975) *Biochemistry*. New York, NY: Worth.

Li, Y.-H. and S. Gregory (1974) Diffusion of ions in seawater and in deep-sea sediments. *Geochim. Cosmochim. Acta* **38**, 703–14.

Millero, F., S. Hubinger, M. Fernandez and S. Garnett (1987a) Oxidation of H_2S in seawater as a function of temperature, pH and ionic strength. *Envir. Sci. Technol.* **21**, 439–43.

Millero, F., S. Sotolongo and M. Izaguirre (1987b) The oxidation kinetics of Fe(II) in seawater. *Geochim. Cosmochim. Acta* **51**, 793–801.

Monod, J., J.-P. Changeaux and F. Jacob (1963) Allosteric proteins and cellular control systems. *J. Mol. Biol.* **6**, 306–29.

Moore, J. W. and R. C. Pearson (1981) *Kinetics and Mechanism*. New York, NY: John Wiley & Sons.

Morgan, J. J. (1964) Chemistry of aqueous manganese (II) and (IV). Ph.D. thesis, Harvard University, Cambridge, MA.

Pankow, J. F. and J. J. Morgan (1981) Kinetics for the aquatic environment. *Envir. Sci. Technol.* **15**, 1155–64.

Plummer, L. N., T. M. L. Wigley and D. L. Parkhurst (1978) The kinetics of calcite dissolution in CO_2-water systems at 5 to 60 °C and 0.0 to 1.0 atm CO_2. *Am. J. Sci.* **278**, 179–216.

Pocker, Y. and D. W. Bjorkquist (1977) Stopped-flow studies of carbon dioxide hydration and bicarbonate dehydration in H_2O and D_2O. Acid-base and metal ion catalysis. *J. Am. Chem. Soc.* **99**, 6537–43.

Rimstidt, J. D. and H. L. Barns (1980) The kinetics of silica-water reactions. *Geochim. Cosmochim. Acta* **44**, 1683–99.

Schnoor, J. L. and W. Stumm (1986) The role of chemical weathering in the neutralization of acid deposition. *Schweiz. Z. Hydrol.* **48**, 171–95.

Sjoberg, E. L. (1976) A fundamental equation for calcite dissolution kinetics. *Geochim. Cosmochim. Acta* **40**, 441–7.

Stumm, W. and G. F. Lee (1961) Oxygenation of ferrous iron. *Industr. Eng. Chem.* **53**, 143–6.

Sung, W. and J. J. Morgan (1980) Kinetics and products of ferrous ion oxygenation in aqueous systems. *Envir. Sci. Technol.* **14**, 561–8.

Tebo, B. and S. Emerson (1986) Microbiol manganese (II) oxidation in the marine environment: a quantitative study. *Biogeochemistry* **2**, 149–62.

Vinograd, J. and J. McBain (1941) Diffusion of electrolytes and ions in their mixtures. *J. Am. Chem. Soc.* **63**, 2008–15.

Wise, D. L. and H. G. Houghton (1966) The diffusion coefficients of ten slightly soluble gases in water at 10–60 °C. *Chem. Eng. Sci.* **21**, 999–1010.

10

Gases and air–water exchange

In many ways the atmosphere is the great integrator of global metabolic processes and human influences on the environment. The gaseous products of extremely heterogeneous reactions on land and in the sea are averaged into a nearly homogeneous signal in the atmosphere because of its rapid circulation. Changes in atmospheric gas concentrations play a delicate and sometimes uncertain role in the processes that dominate global climate. In order to understand the mechanisms that control the concentration of gases in the atmosphere one must understand the importance of the bottom boundary, three-fourths of which is the ocean. Gases play important roles as tracers of air–water gas exchange in the chemical perspective of oceanography. Among other things these fluxes are used to determine the rate of marine biological production and the fate of anthropogenically produced gases in the sea.

Much of atmospheric chemistry deals with the concentrations and processes that control trace gases (less than several ppm, parts per million, e.g., $moles_{gas}\ mole_{air}^{-1}$) such as ozone, O_3, carbon monoxide, CO, and many other trace compounds of oxygen, hydrogen, nitrogen, and sulfur. Most of the reactions that create and destroy these compounds in the troposphere are driven by photochemical processes. Reaction times are often very fast and in some locations natural processes are heavily impacted by fossil fuel burning and other anthropogenic effects. This subject is a main topic of atmospheric chemistry, and we refer the reader to books on this subject (e.g. Jacob, 1999) for a discussion of these processes. Our concern here is mainly with gases that make up the bulk of the atmosphere, and are at least partly controlled by the mechanisms of air–water exchange.

Gases with concentrations near and above 1 ppm in the troposphere (see Table 3.6) are N_2, O_2, H_2O, Ar, CO_2, Ne, He, CH_4, Kr, and N_2O. The thermodynamic solubilities of these gases in water and seawater are discussed in Chapter 3 and their Henry's Law coefficients are presented in Table 3.6. Of the gases in the list above, the noble gases, Ar, Ne, He, and Kr, are inert, i.e. they do not participate in chemical and biological reactions because their outer electron shells are full. Noble gases are not very soluble and their inventories are primarily in the atmosphere. For example, ten times more Xe, the most soluble non-radioactive noble gas, resides in the atmosphere than in the ocean whereas there is 400 times more He, the least soluble noble gas, in the atmosphere than in the ocean. Water vapor has the third highest concentration in the atmosphere, but the factors determining its concentration are dominated by the thermodynamics of the water cycle. The other gases in Table 3.6 are influenced by biological processes, but only O_2, CO_2, CH_4, and N_2O are useful tracers of marine and terrestrial biological processes. The atmospheric pressure of N_2 is so large that changes in concentration by the microbial processes of nitrogen fixation and denitrification are not presently observable.

Three of the four biologically active gases in Table 3.6 (CO_2, CH_4, and N_2O) are important "greenhouse gases." They absorb long-wave radiation reflected from the Earth's surface and in doing so warm the atmosphere. This is also true of chlorofluorocarbons, which have atmospheric concentrations less than parts per billion and no natural source. Water and CO_2 are the most important gases that affect radiative forcing in the atmosphere, with water trapping about 94 W m^{-2} and CO_2 50 W m^{-2} of outgoing radiation (Dickinson and Cicerone, 1986). (The total incoming radiation to the top of the atmosphere is *c.*342 W m^{-2}.) The next most important gases in the radiation balance are CH_4 and N_2O, both of which trap about 1 W m^{-2}. CO_2, CH_4, and N_2O are all increasing with time because of anthropogenic processes. The increase in radiative forcing between 1765 and 1990 due to the increases in concentrations of these gases is estimated to be 1.5, 0.5, and 0.15 W m^{-2}, respectively (IPCC, 2001). The effect of CH_4 and N_2O is proportionally larger than their concentrations because they have a larger radiative forcing per molecule and they have increased dramatically in the industrial period. Although water vapor is actually the most important greenhouse gas, its changes are most likely to be in response to climate change rather than forcing it. Thus, CO_2 is clearly the most important player in greenhouse forcing caused by anthropogenic increases.

Of the four atmospheric gases that are tracers of biological processes, three, O_2, CO_2, and N_2O, have important sources and sinks in the ocean. The ocean produces some CH_4; however, production of this gas is believed to be primarily (about 98%) in terrestrial environments. N_2O is produced as an intermediate in the microbial processes of nitrification and denitrification and is supersaturated in regions of strong upwelling like the equatorial ocean. O_2 and CO_2 are produced and consumed by photosynthesis on land and in the sea. O_2, like the noble gases, is relatively insoluble, so it resides primarily in the atmosphere. CO_2 is much more soluble and reacts with water to form HCO_3^-. The different sizes of these reservoirs and the rates of exchange between the air and sea greatly affect the response of these two gases to terrestrial and oceanic forcing. (See Chapter 11 on the marine carbon cycle for a discussion.)

Most of the discussion in this chapter will be about evaluating the magnitude and discerning the mechanism of air–water gas exchange. Two applications for which this process is important to oceanography are the invasion of anthropogenically produced CO_2 into the ocean, and evasion of biologically produced O_2 to the atmosphere. These processes are discussed in Chapters 11 and 6, respectively. Briefly, global maps of surface water–atmosphere difference in f_{CO_2} (see Fig. 11.8) can be used along with the air–sea gas exchange rate to evaluate the mechanisms controlling the invasion rate of anthropogenic CO_2. If this can be done well enough it will be possible to predict the effects of climate change on the future ocean uptake rate of anthropogenic CO_2. Probably the most promising tracer for determining the net global ocean primary production is the seasonal change of O_2 in the euphotic zone of the ocean. The main flux controlling the upper ocean mass balance of oxygen is the exchange of oxygen between the ocean and atmosphere. With estimates of biologically produced oxygen excess in the surface ocean and the air–water gas exchange rate, it should be possible to go a long way toward determining the areal distribution of the global net biological O_2 production.

Determining the gas exchange rate between the atmosphere and oceans, lakes and rivers evolved along with the study of radioisotopes in the environment, some of which are gases. The first reliable global fluxes between the air and sea were determined in the 1950s from the distribution of natural, and later bomb-produced, radiocarbon (^{14}C). Later, localized gas exchange rates in surface waters of the oceans were evaluated by using measurements of ^{222}Rn, the natural decay product of ^{226}Ra. Much effort has gone into comparing global gas exchange rates determined by atmospheric ^{14}C invasion with global extrapolation of regionally determined gas exchange rates. To do this one must understand the relation between gas exchange and some parameter that can be determined globally, such as wind speed. This relation has been investigated by using a series of purposeful tracer releases into the ocean. In these experiments, trace gases are added to the ocean and followed over time to directly determine the rate of gas evasion while the atmospheric wind speed and other ocean surface parameters are measured. Gas exchange rates and wind speed are to a first order correlated; however, these relations do not capture all the observed variability in environmental gas exchange rates. There is still plenty to do to understand all the important mechanisms involved in this process.

10.1 Air–sea gas transfer models

The net flux of gases between air and water depends on the driving force, which is the gas concentration or pressure gradient, and a transfer coefficient, G, that has units of length divided by time (e.g. m d^{-1}). The concentration of a gas at the surface skin of the water is assumed to be at equilibrium with the fugacity of the gas in air. (The fugacity is very nearly equal to the partial pressure for atmospheric gases; see Section 3.4.5). Since turbulent transport in the air and water decays toward the interface, at some point the dominant transport process will be by molecular diffusion, which is the limiting step in gas exchange. Because molecular diffusion of gases is typically four orders of magnitude greater in air than that in water, the liquid interface is the limiting factor to gas exchange for all but extremely soluble gases and water vapor. The flux of gas C across the air–water interface, F_C, is proportional to the difference in gas fugacity between the air and water and the mass transfer coefficient.

$$\begin{aligned} F_C &= G_C \times \{ [C] - [C^{Sat}] \} \\ &= G_C \times \{ [C] - K_{H,C} \times f_C^{a} \} \\ &= G_C \times K_{H,C} (f_C^{w} - f_C^{a}), \end{aligned} \qquad (10.1)$$

where [C] is the concentration of gas C (mol m^{-3}) in the liquid nearest the interface and [C^{Sat}] is the concentration at solubility equilibrium, which is equal to the Henry's Law constant, $K_{H,C}$ times the atmospheric fugacity, f^a (see Eq. (3.54) and Table 3.6). This much of the theory of gas transfer is accepted and has been used in the engineering and environmental science communities for many decades. Most of the discussion in the rest of this chapter will deal with the dependence of the transfer velocity, G, on the molecular diffusion coefficient of the gas, estimating the value of G and how G varies with environmental variables such as wind speed.

Molecular processes at interfaces dominate the exchange mechanisms of momentum, heat, and gases. Turbulence in the fluid on either side of the air–sea interface is determined by eddies of many sizes (Fig. 10.1). As one approaches the interface from the

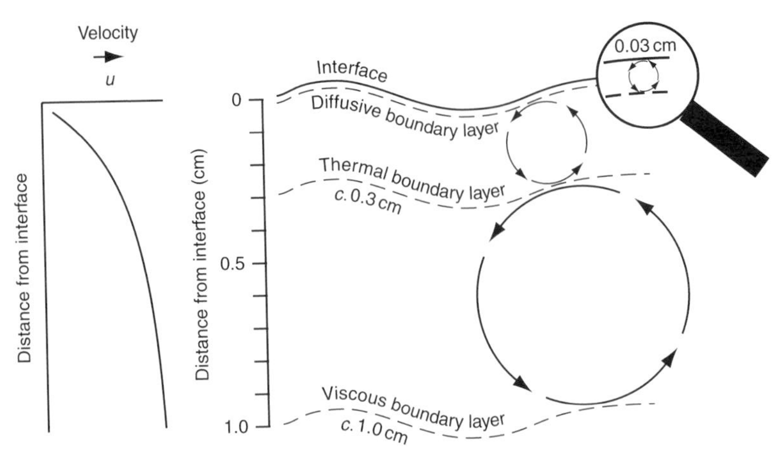

Figure 10.1. A schematic representation of the boundary layers for momentum, heat and mass near the air–water interface. The velocity of the water and the size of eddies in the water decrease as the air–water interface is approached. The larger eddies have greater velocity, which is indicated here by the length of the arrow in the eddy. Because random molecular motions of momentum, heat and mass are characterized by molecular diffusion coefficients of different magnitude (0.01 $cm^2 s^{-1}$ for momentum, 0.001 $cm^2 s^{-1}$ for heat and $10^{-5} cm^2 s^{-1}$ for mass), there are three different distances from the wall where molecular motions become as important as eddy motions for transport. The scales are called the viscous (momentum), thermal (heat) and diffusive (molecular) boundary layers near the interface.

interior, smaller and smaller eddies with progressively less velocity dominate the mixing process. The *diffusive boundary layer* is the depth at which molecular processes become a more important mechanism of transport than eddy mixing (Fig. 10.1). The transition from eddy to molecular diffusion for momentum, heat, and gases occurs at different depths because the molecular diffusion of momentum (the kinematic viscosity, *c.*$10^{-2} cm^2 s^{-1}$) is about 10 times greater than that of heat (D_{heat}, *c.*$10^{-3} cm^2 s^{-1}$) and about 10^3 times greater than molecular diffusion of gases (D_{gas}, *c.*$10^{-5} cm^2 s^{-1}$). The viscous (momentum) boundary layer is thus about three times larger than the thermal diffusion boundary layer and about thirty times that of the diffusive boundary layer for gases (Fig. 10.1). The different boundary layers have characteristic distances from the solid interface of approximately 1.0, 0.3, and 0.03 cm, respectively. (The distance, z, scales roughly with the square root of the molecular diffusion coefficient, $z = (2Dt)^{1/2}$, where t is a common time for each process.) Within the viscous boundary layer, momentum flux is controlled by the kinematic viscosity, ν, which is a measure of the resistance of flow to the influence of gravity. "Stickier" substances have higher kinematic viscosity. In this region, eddy mixing of heat and gases decreases to zero as the interface is approached. Because it is more difficult to diffuse through "stickier" substances, the resistance to heat and gas exchange at the interface is proportional to their diffusion coefficients but inversely proportional to the kinematic viscosity, ν. Whereas diffusion coefficients increase with temperature because the energy input to random molecular motions increases, the kinematic viscosity decreases with temperature along with the decline in the resistance to flow. The mass transfer coefficient of gas C is related to a gas- and fluid-normalized transfer coefficient, G^*, times the diffusion coefficient of the gas divided by the kinematic viscosity, all to the power n:

$$G_C = G^* \times \left(\frac{D_C}{\nu}\right)^n. \quad (10.2)$$

Table 10.1. *Molecular diffusion coefficients, D, and Schmidt numbers, Sc, for gases*

The molecular diffusion coefficients, *D* (in units of 10^{-5} cm^2 s^{-1}; see note[a]), were determined from the equations presented in Chapter 9, Table 9.1. The kinematic viscosity of water is from Pilson (1998). The kinematic viscosity is 3%–5% greater in seawater than in freshwater, and we assume here that this is the only factor causing a salinity dependence on Sc. Opposite trends with *T* for diffusion coefficients and kinematic viscosity create greater temperature dependence for Schmidt numbers than for the molecular diffusion coefficients.

	D ($\times 10^5$)[a] (cm^2 s^{-1})		Sc[b] ($S=0$)		Sc[c] ($S=35$)	
Gas	5 °C	20 °C	5 °C	20 °C	5 °C	20 °C
N_2	1.63	2.52	931	398	958	415
O_2	1.49	2.24	1019	448	1048	467
Ar	1.42	2.23	1070	450	1100	469
CO_2	1.09	1.68	1394	598	1433	623
Ne	2.62	3.65	580	275	596	287
He	5.19	6.73	293	149	310	155
Kr	1.03	1.61	1475	624	1516	650
Xe	0.79	1.27	1923	791	1977	823
CH_4	1.09	1.63	1394	616	1433	642

[a] The value of the diffusion coefficient, *D*, is multiplied by 10^5 before tabulating. For example, $D_{N_2} = 1.63 \times 10^{-5}$ at 5°C.
[b] For $S=0$, $\nu = 1.519 \times 10^{-2}$ cm^2 s^{-1} (5 °C) and 1.004×10^{-2} (20°C).
[c] For $S=35$, $\nu = 1.562 \times 10^{-2}$ cm^2 s^{-1} (5 °C) and 1.046×10^{-2} (20°C).

G^* is an empirical constant, with units of length per unit time, determined experimentally, and the exponent, *n*, is predicted to vary between ½ and 1. The ratio of the kinematic viscosity to diffusion coefficient is called the Schmidt number, Sc (Table 10.1):

$$\mathrm{Sc_C} = \frac{\nu}{D_\mathrm{C}}. \qquad (10.3)$$

Thus, G_C is proportional to $\mathrm{Sc_C}^{-n}$. Comparison of transfer coefficients of the same gas at different temperatures, and/or in different liquids, requires knowledge of the correct dependence of the gas exchange rate on the kinematic viscosity. However, when comparing the gas transfer coefficients of different gases in the same solution at the same temperature, the dependence on kinematic viscosity plays no role because it is a property of the fluid and cancels.

Different gas exchange models suggest a variety of dependencies of the mass transfer coefficient, G_C, on the Schmidt number. We present three different models of the mechanism of gas exchange along with evidence for their importance in nature.

10.1.1 Rigid wall model

This first model is the closest conceptually to the description of turbulent decay as the interface is approached (Fig. 10.1). It is also the most complicated mathematically, so we present only the result. Deacon (1977) developed a theory for gas transfer toward or away from a smooth plane by using experimentally derived heat fluxes at smooth surfaces. He assumed that the theory for momentum transfer near a rigid wall, and the empirical relation between the viscous boundary layer thickness and the flow velocity of liquids determined by Reichardt, are also true for heat and gases (see Hinze, 1959). By integrating the resistance to transfer across the viscous boundary layer and plotting laboratory experimental results for heat transfer coefficients versus the Schmidt number, he demonstrated a $-^2/_3$ dependence of the gas or heat transfer coefficient on Sc ($n = ^2/_3$):

$$G_{\mathrm{C,Wall}} = \frac{U^*}{12} \times \mathrm{Sc_C}^{-2/3}, \tag{10.4}$$

where U^* is the friction velocity at the air–liquid interface. (The friction velocity is the velocity extrapolated to the interface from wind speed at some height above the interface multiplied by a constant called the drag coefficient.) He further argued that the same dependency should be important for air–water transfer at moderate wind speeds.

10.1.2 Stagnant film model

A gas transfer model at the other extreme with regard to both complexity and reality is the stagnant film model. It is very simple and has a long history that began in the application to industrial processes (Whitman, 1923). This model (Fig. 10.2) assumes that the air–liquid interface is bounded by a film in the air and one in the water through which gases travel by molecular diffusion. Since molecular diffusion in the water is much slower, the water side is usually the limiting factor to gas exchange in this model. Applying Eq. (10.1) to this model along with Fick's second law of diffusion yields:

$$G_{\mathrm{C,sf}} = D_{\mathrm{C}}/\delta, \tag{10.5}$$

where δ is the thickness of the stagnant layer in the liquid phase. All properties that control the exchange of gases are incorporated into the stagnant boundary layer thickness, δ, which is thinner for faster gas exchange rates. The model is too simple to predict a dependence

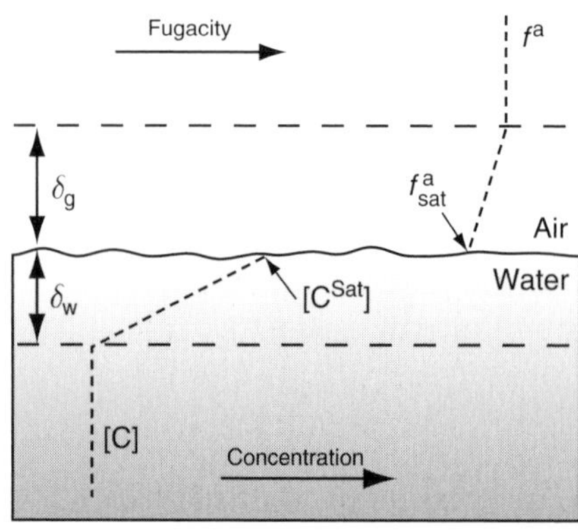

Figure 10.2. A schematic diagram of the stagnant molecular layers of thickness δ_g and δ_w near the air–water interface. Fugacity in the bulk atmosphere and concentration in the water are indicated by f^a and [C], respectively. Chemical equilibrium at the air–water interface is described by Henry's Law, where $K_H f^a_{sat} = [C^{Sat}]$.

of the gas exchange rate on wind speed, as in the Deacon model. Gas exchange in this model is dependent on the molecular diffusion coefficient to the first power and thus the Schmidt number to the −1 ($n = 1$). This model has been applied to a variety of situations in the ocean and lakes; values for δ of 20–40 μm are common for the ocean and 50–100 μm for more sheltered lake systems. Although the parameterization is easy to conceptualize, it is not realistic to expect a laminar layer at the interface between the air and water for all conditions of turbulence found in the environment.

10.1.3 Surface renewal model

The third mechanistic model of gas transfer at the air-water interface, the surface renewal model, is of intermediate complexity. There are several versions of the surface renewal model (Dankwertz, 1970) but the basic assumptions are the same. One envisions the surface of the liquid as a location that is repeatedly replaced by pristine parcels of water from below with the bulk gas concentration (Fig. 10.3). While the water parcel is at the interface it exchanges gases as though it is infinitely deep and stagnant. The equations of molecular diffusion in a semi-infinite space can thus be used to characterize the transfer process. The parameter that characterizes the rate of gas

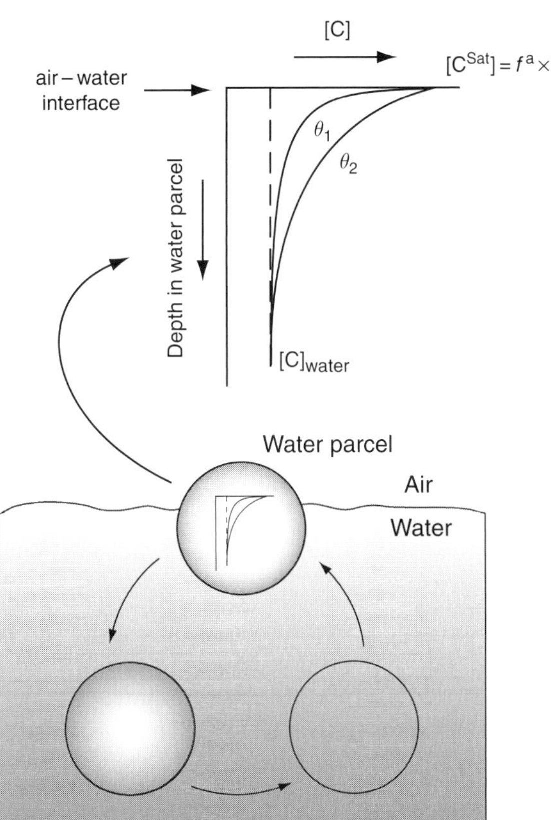

Figure 10.3. A schematic representation of the surface renewal model of air–water gas transfer. Water parcels arrive at the interface from the water interior, where they remain for some time period, θ. The fugacity and concentration at the air–water interface are at equilibrium and atmospheric gas diffuses into or out of the liquid as though it were an infinitely deep layer.

exchange is the rate of renewal of the water parcels. If gas transfer is fast the parcels are renewed frequently so the gradient between the concentration at the interface and the bulk is steep. If gas exchange is slow the parcels remain at the surface for a long time while the interfacial gradient becomes less and less steep with time.

The general time-dependent, one-dimensional diffusion equation for a gas C with concentration [C] is

$$\frac{\partial[\mathrm{C}]}{\partial t} = D_{\mathrm{C}} \times \frac{\partial^2[\mathrm{C}]}{\partial z^2}. \tag{10.6}$$

With the boundary conditions such that the surface gas concentration is equal to the saturation value

$$z = 0, \quad [\mathrm{C}] = [\mathrm{C}^{\mathrm{Sat}}] \tag{10.7}$$

and the concentration far from the interface is equal to the bulk value

$$z \gg 0, \ [\mathrm{C}] = [\mathrm{C}^0] \tag{10.8}$$

the solution is

$$[\mathrm{C}] - [\mathrm{C}^0] = \{\ [\mathrm{C}^{\mathrm{Sat}}] - [\mathrm{C}^0]\ \} \times \mathrm{erfc}\left\{\frac{z}{2 \times (D_{\mathrm{C}} \times t)^{0.5}}\right\} \tag{10.9}$$

where $\mathrm{erfc}(z) = (1 - \mathrm{erf}(z))$ and erf is the error function. We see, before finishing the derivation, that the flux into the liquid is going to depend on the square root of the diffusion coefficient ($n = ½$) because that is the form of the solution to the diffusion equation into a semi-infinite space. The flux at the air–water interface ($z = 0$) is

$$F = D \times \left.\frac{\mathrm{d}[\mathrm{C}]}{\mathrm{d}z}\right|_{z=0} = \left(\frac{D_{\mathrm{C}}}{\pi \times t}\right)^{0.5} ([\mathrm{C}^{\mathrm{Sat}}] - [\mathrm{C}^0]). \tag{10.10}$$

If each parcel is exposed at the interface for the same time, θ, the amount absorbed during this time is

$$Q = \int_0^{\theta} F\mathrm{d}t = 2 \times \theta \times \left(\frac{D_{\mathrm{C}}}{\pi \times \theta}\right)^{0.5} ([\mathrm{C}^{\mathrm{Sat}}] - [\mathrm{C}]). \tag{10.11}$$

The average rate of influx is thus

$$F_{\mathrm{avg}} = \frac{Q}{\theta} = 2 \times \left(\frac{D_{\mathrm{C}}}{\pi \times \theta}\right)^{0.5} ([\mathrm{C}^{\mathrm{Sat}}] - [\mathrm{C}]). \tag{10.12}$$

In this model the parameters controlling the gas exchange rate are characterized by the replacement time, θ, rather than a boundary layer thickness. Comparing this equation with Eq. (10.1) we see that the mass transfer coefficient is proportional to the square root of the molecular diffusion coefficient.

$$G_{\mathrm{C,sr}} = \frac{F_{\mathrm{avg}}}{[\mathrm{C}^{\mathrm{Sat}}] - [\mathrm{C}^0]} = 2 \times \left(\frac{D_{\mathrm{C}}}{\pi \times \theta}\right)^{0.5}. \tag{10.13}$$

Table 10.2. *The gas transfer dependence on the molecular diffusion coefficient, D (and hence Schmidt number, Sc) predicted by the three air–water gas transfer models described in the text*

n, the exponent in the term D^n (or Sc^{-n}).

Model (reference)	Characteristics	n
Rigid wall (Deacon, 1977)	Derived from dependence of turbulent flow away from a rigid wall	$^2/_3$
Stagnant boundary layer (Whitman, 1923)	Assumes stagnant layer at the air-water interface	1
Surface renewal (Higby, 1935)	Assumes diffusion into a semi-infinite liquid that is periodically replaced	½

Thus, the flux depends on the negative ½ power of the Schmidt number, $Sc^{-1/2}$ and $n = ½$.

The three different models give dependencies on the Schmidt number between $n = ½$ and $n = 1$ (Table 10.2). Laboratory experiments have been designed to determine which of the models most closely represents gas exchange in nature. One approach (Ledwell, 1984) was designed to measure the exchange rate of gases that have a wide difference in molecular diffusion coefficients (He, CH_4, and N_2O). This experiment demonstrated an unambiguous square root dependency ($n = ½$) of the mass transfer coefficient on the Schmidt number. In a more extensive series of experiments using a circular wind tunnel Jahne *et al.* (1987) measured the exchange rate of a variety of gases and heat as a function of wind speed, friction velocity and surface contaminants. They observed that when the water was smooth, either because the wind speed was low or there was a slick of oil on the surface, the gas transfer rate varied with the Schmidt number raised to the $-^2/_3$ power ($n = ^2/_3$). When the wind speed was normalized to friction velocity there was a common value of U^* at which the surface developed capillary waves and the Schmidt number dependency changed to $n = ½$ (Fig. 10.4). This result indicated that the mechanism of gas exchange depends on the presence of capillary waves, which develop at different wind speeds depending on the presence of surfactants. Since a surface film of oil decreased the effective friction of the wind on the water surface, this contamination delayed the onset of waves and decreased the rate of gas exchange. Because in most natural settings the air–water interface is occupied by waves, the square root dependency dominates in nature. However, this result may have importance regarding the influence on gas exchange of natural and anthropogenic slicks of organic molecules on the surface ocean (see later).

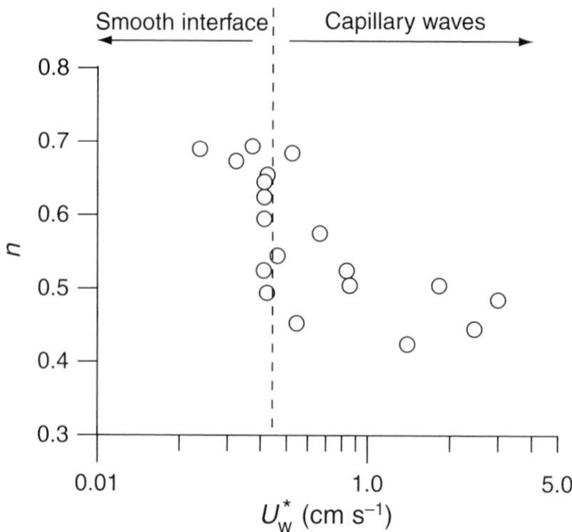

Figure 10.4. The Schmidt number dependence, n, derived from gas exchange of He and CH_4 in a wind tunnel experiment as a function of the friction velocity, U_w^* (the wind speed extrapolated to the air–water interface via the drag coefficient). The break in the exponent from 0.7 to 0.5 occurs at the transition from a smooth surface to one dominated by capillary waves. Redrawn from Jahne *et al.* (1987).

In no case has the gas exchange rate been observed to be linearly dependent on the Schmidt number. Thus, the simplest model of gas exchange, the stagnant boundary layer, is the big loser in the competition to parameterize the mechanism in nature. The most realistic representation is that of the square root dependency; this has been adopted in the interpretation of tracers of gas exchange in nature.

10.2 Measurements of gas exchange rates in nature

In order to determine the flux of gases across the air-water interface in nature one must know both the mass transfer coefficient and the concentration gradient between the surface ocean and atmospheric saturation equilibrium (Eq. (10.1)). The gas transfer coefficient has been measured in a number of wind tunnel experiments and by using both naturally occurring radioisotopes and artificially added tracers in the environment. Measurements made in the field most faithfully reproduce the actual environmental situation. There have been three main methods used to determine the mass transfer coefficient in the sea. Two of the methods involve the mass balance of the radioactive tracers ^{14}C and ^{222}Rn between the atmosphere and ocean. Since ^{14}C has a half life of 5730 y and ^{222}Rn has a half life of 3.85 d, these results average the gas exchange process over dramatically different time and space scales. Much of the effort during the past several decades of the study of air–sea gas exchange has been to rationalize apparent discrepancies between the results derived from ^{14}C and ^{222}Rn measurements. The great mediator in this debate has been determinations of the gas exchange mass transfer coefficients from direct tracer release experiments in the ocean. The length of these experiments is limited by the amount of time one can follow a tagged patch of seawater before it dissipates or the length of time a ship can remain at sea. The duration of such experiments is typically on the order of a few weeks to one month, which is not greatly different from the characteristic time scale for ^{222}Rn exchange. The advantage of the tracer release experiments is that they are Lagrangian by their very nature, so changes in concentration associated with advection do not create uncertainties in the interpretation of the gas concentration.

10.2.1 The radiocarbon method

The distributions of both natural and bomb-produced radiocarbon in the ocean have been used to determine the mean global gas exchange mass transfer coefficient. The calculation based on natural radiocarbon assumes a steady-state mass balance between the production of ^{14}C in the atmosphere by cosmic ray spallation (see Chapter 5) and radioactive decay of ^{14}C in the ocean. This simple balance works because the DIC carbon reservoir in the ocean is by far the largest carbon pool that exchanges with the atmosphere on the time scale of the ^{14}C mean life (see Table 11.1). Assuming that at steady state the decay of DI^{14}C in the ocean must be equal to the flux of ^{14}C atoms across the air–water interface, $F_{\text{atm-ocean}}$ (mol m^{-2} d^{-1}),

$$A^{\text{o}} \times F^{14}_{\text{atm-ocean}} = \lambda_{14} \times \overline{\text{DI}^{14}\text{C}} \times V^{\text{o}} \qquad (\text{mol d}^{-1}), \tag{10.14}$$

where A^{o} (m^2) and V^{o} (m^3) are the ocean area and volume, λ_{14} (d^{-1}) is the decay constant for ^{14}C and $\overline{\text{DI}^{14}\text{C}}$ is the mean value of DI^{14}C in the deep ocean (the ocean below 100 m that is isolated from the atmosphere). The flux, $F^{14}_{\text{atm-ocean}}$, is equal to the mass transfer coefficient, G^{14} (m d^{-1}), multiplied by the concentration difference (mol m^{-3}) in ^{14}C between the atmosphere and ocean:

$$F^{14}_{\text{atm-ocean}} = G^{14} \times \left(f^{\text{a}}_{^{14}\text{CO}_2} \times K_{\text{H},^{14}\text{C}} - \left[^{14}\text{CO}_2\right]^{\text{so}} \right) \quad (\text{mol m}^{-2}\text{ d}^{-1}), \tag{10.15}$$

where the superscripts ^{so} and ^{a} indicate surface ocean and atmosphere, respectively. It is advantageous to write the ^{14}C concentrations in terms of the ^{12}C values and ratios of radiocarbon to stable carbon because these are the values actually measured,

$$R = \frac{^{14}\text{C}}{^{12}\text{C}}. \tag{10.16}$$

Using this formulation, the $^{14}CO_2$ fugacity in the atmosphere becomes

$$f^{\text{a}}_{^{14}\text{CO}_2} = f^{\text{a}}_{\text{CO}_2} \times R^{\text{a}}, \tag{10.17}$$

where R^{a} is the atmospheric $^{14}CO_2/^{12}CO_2$ ratio. Expressing the surface ocean $^{14}CO_2$ concentration in these terms is somewhat complicated because the value measured is the ^{14}C content of the DIC in the surface ocean, not the $^{14}CO_2$. Furthermore, there is an equilibrium isotope fractionation between CO_2 and DIC analogous to that for the ^{13}C/^{12}C ratio (Chapter 5). Designating this fractionation factor as $\alpha^{\text{eq},14}$, the concentration of $^{14}CO_2$ can be written as

$$[^{14}CO_2]^{so} = \frac{[^{14}CO_2]^{so}}{[DI^{14}C]^{so}} \times [DI^{14}C]^{so} = \frac{[CO_2]^{so}}{[DIC]^{so}} \times \alpha^{eq,14} \times [DIC]^{so} \times R^{so} \quad (\text{mol m}^{-3}), \tag{10.18}$$

where R^{so} is the $^{14}C/^{12}C$ ratio in surface water DIC. Canceling DIC on the right side of Eq. (10.18) and changing the CO_2 concentration in surface waters from concentration to fugacity results in

$$[^{14}CO_2]^{so} = f^{w}_{CO_2} \times K_H \times \alpha^{eq,14} \times R^{so}. \tag{10.19}$$

The pre-industrial atmosphere and mean surface ocean f_{CO_2} values were within a few percent of being the same globally, so we will assume that $f^{a}_{CO_2} = f^{w}_{CO_2}$. Now, substituting Eqs. (10.17) and (10.19) into Eq. (10.15) and combining it with Eq. (10.14) gives an expression for the global mass transfer coefficient for $^{14}CO_2$ in terms of known quantities:

$$G^{14} = \lambda_{14} \times \frac{[\overline{DIC}]}{[CO_2]^{so}} \times \frac{V^o}{A^o} \times \left\{ \frac{\left(\frac{R^{do}}{R^a}\right)}{\left(\frac{K_{H,14}}{K_{H,12}}\right) - \alpha^{eq,14} \times \frac{R^{so}}{R^a}} \right\} \quad (\text{m d}^{-1}). \tag{10.20}$$

Since the diffusion coefficients of $^{14}CO_2$ and CO_2 are about the same ($^{14}CO_2$ diffuses 0.2% slower than $^{12}CO_2$), this is also the mass transfer coefficient for CO_2.

The decay constant for ^{14}C is 1/8267 y^{-1} and the mean depth of the ocean (V^o/A^o) is 3800 m. The average ocean concentration of DIC is 2.2 mol m^{-3} and the CO_2 concentration at equilibrium with a pre-industrial atmospheric f_{CO_2} of 280 ppm at the mean surface ocean temperature and salinity of 17.6 °C and S = 34.8 is 0.0097 mol m^{-3} (from Appendix 4.1).

The pre-industrial atmospheric $^{14}C/^{12}C$ values, R^a, have been determined from radiocarbon measurements of tree rings, and the value of R^{do} is based on hundreds of deep ocean $DI^{14}C$ measurements, which are still uncontaminated by the ^{14}C from nuclear weapons testing in the 1950s and 1960s. From these measurements, the $^{14}C/^{12}C$ ratio between the mean deep ocean (do) DIC and that in the atmosphere, R^{do}/R^a, is 0.83.

The surface ocean (so) ratio R^{so}/R^a in Eq. (10.20) is the most difficult to estimate, because there are few surface ocean $DI^{14}C$ samples from the ocean before nuclear weapons testing. The surface ocean pre-industrial radiocarbon value, R^{so}, has been determined from the average of 41 surface values measured in the Atlantic Ocean before nuclear weapons testing and data from a banded coral that grew in the Florida Straits and recorded the surface ocean ^{14}C activity before 1900. Using this average value for R^{so} and the value determined from tree rings for R^a, the ratio R^{so}/R^a is about 0.954 (Broecker *et al.*, 1986).

The other terms in the right-hand side of Eq. (10.20) are the equilibrium fractionation factors for $^{14}CO_2$ gas between the atmosphere and ocean, $K_{H,14}/K_{H,12}$, and the equilibrium fractionation between CO_2 and DIC ($\alpha^{eq,14}$). These values are determined from measurements of the fractionation factors for $^{13}C/^{12}C$ and assuming that those for $^{14}C/^{12}C$ are twice as large because the mass difference is twice as great. Approximating DIC as HCO_3^-, the $^{13}C/^{12}C$ fractionation factors are 0.9989 for $K_{H,13}/K_{H,12}$ and 0.991 for $(CO_2/H^{13}CO_3^-)/(^{13}CO_2/HCO_3^-)$. This results in a value of 0.9978 for $K_{H,14}/K_{H,12}$ and 0.982 for $\alpha^{eq,14}$.

Combining these values in Eq. (10.20) gives a CO_2 gas exchange mass transfer coefficient G^{14} of 3.9 m d^{-1} (16.2 cm h^{-1}). In order to compare exchange coefficients determined from different tracers at different temperatures the values are normalized to a Schmidt number of 600 (very close to that for CO_2 at 20 °C in fresh water). Therefore, using the relations in Eqs. (10.2) and (10.3) and data from Table 10.1:

$$G_{CO_2,\,20\,C,\,S=0}/G_{CO_2,\,17.6\,C,\,S=35} = (761/603)^{0.5} = 1.12; \qquad (10.21)$$

$$G_{CO_2,\,20\,C,\,S=0} = 3.9\ \text{m d}^{-1} \times 1.12 = 4.36\ \text{m d}^{-1} = G_{600}. \qquad (10.22)$$

Broecker and Peng (1982) assessed the errors in this calculation and suggest that the difference in ^{14}C between the ocean and atmosphere, R^{so}/R^{a}, is the least well known with an uncertainty of about ± 10‰. This creates an uncertainty in the gas transfer coefficient of ± 25% or about ± 1.2 mm d^{-1}.

An independent check on the steady-state ^{14}C-determined mass transfer coefficient has been made by adding up the inventory of bomb-produced ^{14}C in the ocean and calculating the flux across the air–water interface necessary to account for this value. The resulting mass transfer coefficient, when normalized to a Schmidt number of 600, is 3.5 ± 1.2 m d^{-1}, not significantly different from that determined above (Sweeney *et al.*, 2007).

10.2.2 The ^{222}Rn method

The other naturally occurring tracer of gas exchange in the environment is the relatively short-lived radioactive gas ^{222}Rn. Seawater has a rather high and readily measurable activity of ^{226}Ra (5–15 dpm kg^{-1}), the immediate parent of ^{222}Rn (Fig. 5.18).

$$^{226}\text{Ra}\ \left(t_{1/2} = 1620\,\text{y}\right) \rightarrow {}^{222}\text{Rn}\ \left(t_{1/2} = 3.85\ \text{d}\right). \qquad (10.23)$$

As we discussed in Chapter 5 the activities of the two isotopes would be in secular equilibrium in a closed system, which is the case for ^{222}Rn and ^{226}Ra within the ocean's interior away from the air–water interface. At the surface, however, ^{222}Rn escapes to the atmosphere because its activity in the marine atmosphere far from land is very low. Exchange to the atmosphere creates a deficit in the activity of ^{222}Rn relative to the activity of ^{226}Ra in the surface ocean, which is proportional to the rate of gas exchange across the air–water interface.

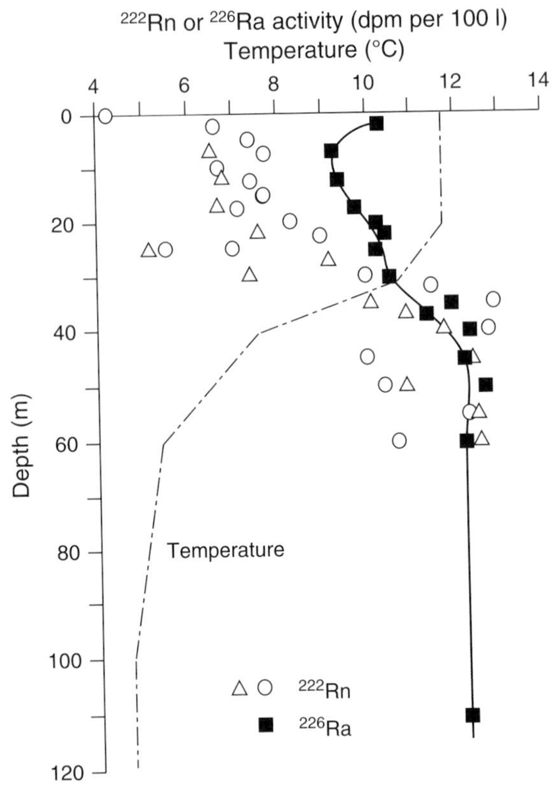

Figure 10.5. Profiles of ^{222}Rn, ^{226}Ra, and temperature in the surface waters of the subarctic Pacific Ocean. Open symbols are ^{222}Rn data from profiles taken on different days; filled squares represent ^{226}Ra data. From Emerson *et al.* (1991).

There have been hundreds of radon profiles measured in the surface ocean. Examples are presented in Fig. 10.5. These data are used to determine the gas exchange mass transfer coefficient by calculating a mass balance for ^{222}Rn. The change in concentration (mol m^{-3}) of ^{222}Rn in surface waters with time integrated to the depth, z^*, at which secular equilibrium is achieved, usually 200 m or less, is equal to the sum of three terms: production by the integrated ^{226}Ra concentration, and losses due to radioactive decay (mol m^{-2} d^{-1}) and flux to the atmosphere:

$$\int_{z=0}^{z=z^*} \frac{d[^{222}\mathrm{Rn}]}{dt} dz = \int_{z=0}^{z=z*} \lambda_{\mathrm{Ra}}[\mathrm{Ra}] dz - \int_{z=0}^{z=z*} \lambda_{\mathrm{Rn}}[\mathrm{Rn}] dz - F_{\text{air-water}} \ (\text{mol m}^{-2}\ \text{d}^{-1}). \tag{10.24}$$

The air–sea flux using Eq. (10.1) is equal to the mass transfer coefficient times the concentration difference, which can be simplified since the atmospheric partial pressure of radon, p^{222}Rn, is negligible several tens of kilometers away from land:

$$F_{\text{air-water}} = G_{\mathrm{Rn}} \times \left\{ \left[^{222}\mathrm{Rn}^{\mathrm{so}}\right] - p^{222}\mathrm{Rn} \times \mathrm{K}_{\mathrm{H,Rn}} \right\} = G_{\mathrm{Rn}} \times \left[^{222}\mathrm{Rn}^{\mathrm{so}}\right] \quad (\text{mol m}^{-2}\ \text{d}^{-1}). \tag{10.25}$$

Combining the above two equations, multiplying by λ_{Rn} to change all concentrations to activity concentrations (dpm m^{-3}; see Chapter 5) gives:

$$G_{Rn}\, A_{222}^{so} = \lambda_{Rn} \int_{z=0}^{z=z^*} (A_{226} - A_{222})\mathrm{d}z - \int_{z=0}^{z=z^*} \frac{\mathrm{d}A_{222}}{\mathrm{d}t}\mathrm{d}z \quad (\text{dpm m}^{-2}\ \text{d}^{-1}). \tag{10.26}$$

In this equation A_{222} and A_{226} are shorthand notation for the activities of ^{222}Rn and ^{226}Ra, respectively, and A_{222}^{so} is the radon activity in the well-mixed surface waters of the ocean. If activity profiles of radon and radium are determined, the only remaining unknown in Eq. (10.26) is the mass transfer coefficient, G_{Rn}.

Because of the rather short half life of ^{222}Rn, there can be significant changes with time in its activity, compared with the gas exchange time scale, and therefore the non-steady-state term (the last one in Eq. (10.26)) must be evaluated to determine accurate values of G_{Rn}. This has been done both by measuring a series of radon profiles at a single location and by averaging many individual profiles in different ocean basins using steady-state assumptions.

A summary of the values of the mass transfer coefficient for radon determined from time-series experiments and basin-scale averaging as a function of the measured or averaged wind speed during the experiments is presented in Fig. 10.6. The data are plotted along with the global average value determined from natural and bomb ^{14}C measurements (Eq. (10.20)). There is clearly an increase in the mass transfer coefficient with increasing wind speed. Also, the values from the ^{14}C mass balances appear to be somewhat higher than those determined by the ^{222}Rn method, although, not surprisingly, there is a lot of scatter in the latter values. The scatter in the shorter-term measurements almost certainly represents real environmental variability in the gas exchange rate that is due to factors other than the gas exchange rate.

The lower line labeled (L&M) in Fig. 10.6 is a compilation of results from gas exchange rates determined in lakes and wind tunnel experiments (Liss and Merlivat, 1986). It consists of three linear regions of increasing gas exchange rate : wind speed ratio. The first regime represents gas exchange when the surface of the water is smooth ($U_{10} = 0$–3.6 m s^{-1}) and is from the equation of Deacon (1977; our Eq. (10.4)). The next regime is for gas exchange from rough surfaces ($U_{10} = 3.6$–13 m s^{-1}) and was derived from gas exchange rates determined by adding a gas tracer to a small lake and determining the gas loss and wind speed as a function of time. The third linear regime ($U_{10} > 13$ m s^{-1}) is based on wind tunnel measurements in which breaking waves and production of bubbles were observed.

The Liss–Merivat gas exchange versus wind speed relation has been challenged as more data become available from field measurements. A widely accepted alternative at the writing of this book

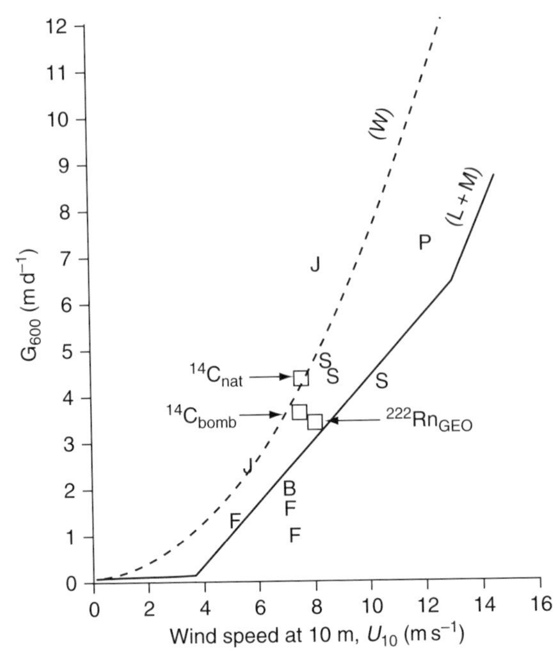

Figure 10.6. Gas transfer rates normalized to a Schmidt number of 600, G_{600}, from global ^{14}C and localized ^{222}Rn measurements in the ocean as a function of wind speed measured at 10 m above the air–water interface, U_{10}. $^{14}C_{nat}$ and $^{14}C_{bomb}$ are the gas exchange rates determined from natural and bomb-produced ^{14}C discussed in the text. The box labeled Rn_{GEO} represents the average result from over 100 ^{222}Rn measurements in the oceans during the GEOSECS program (Peng *et al.*, 1979). P and S are the results of repeated ^{222}Rn profiles measured in the subarctic Pacific (Peng *et al.*, 1974 and Emerson *et al.*, 1991, respectively). B is the result of repeated profiles in the subtropical Atlantic (Broecker and Peng, 1971). F and J are the results of "continuous" ^{222}Rn measurements in the North Atlantic (Kromer and Roether, 1983). The solid lines are the proposed gas exchange–wind speed relations of Liss and Merlivat (1986) (L+M) (for $U < 3.6$ m s^{-1}, G(cm h^{-1}) $= 0.17 \times (Sc/600)^{-2/3}$; for $3.6 < U_{10} < 13$, G(cm h^{-1}) $= 2.8 \times (U - 3.6) \times (Sc/600)^{-1/2}$; for $U > 13$, G(cm h^{-1}) $= 5.9 \times (U - 8.4) \times (Sc/600)^{-1/2}$) and Wanninkhof (1992) (W) (for short-term average winds, G(cm h^{-1}) $= 0.31 \times U^2 \times (Sc/600)^{-1/2}$).

is the curve presented by Wanninkhof (1992) in which the gas transfer–wind speed relation is assumed to depend on the square of the wind speed, as observed in some wind tunnel experiments. This quadratic relation, indicated by the line labeled (W) in Fig. 10.6, is forced to pass through the gas transfer velocities determined by using natural ^{14}C data derived from whole-ocean and the Red Sea ^{14}C mass balances. The difference between the two parameterizations of the dependence of the gas exchange rate on wind speed increases dramatically at high wind speeds. They differ by about 30% at the global mean wind speed of 7 m s^{-1} and a factor of two at 11 m s^{-1}. Because of this rather large discrepancy, alternative methods believed to eliminate some of the inadequacies of the radon and ^{14}C methods have been devised.

10.2.3 Tracer release experiments

Notice that in the compilation of ^{222}Rn and ^{14}C-determined gas exchange rates (Fig. 10.6) only one measurement has been made at wind speeds above $U_{10} = 10$ m s^{-1}. Since this regime is the most important for gas exchange and parameterizations in Fig. 10.6 differ by the most in this range, tracer release experiments have been devised to further refine the gas transfer–wind speed relations. In these experiments two tracers, one of which must be a gas, are released into the surface waters of the ocean (Watson *et al.*, 1991). After injection of the tracers, the ship follows the center of the patch, while the progressive decreases in tracer concentrations are measured. The decreases are caused by dilution from mixing and the gas flux to the atmosphere. Since there are two tracers and two unknowns, the gas transfer velocity can be determined. Following a tracer patch and measuring the concentration decrease with time has the advantage over the radon method that the gas exchange experiment follows the same water parcel rather than measuring concentrations at a stationary location, which is likely to sample water masses with different gas exchange histories.

Because the tracers used in these experiments must be detectable at very low levels, have no natural background, and be inert to the aquatic environment, the gases ^3He and sulfurhexafluoride (SF_6) have been most often used. Few other tracers have been found that meet these criteria. An important caveat of the dual tracer experiment is that the exact relation of the gas exchange rates of the two tracers must be known in order to couple the equations describing transport and gas exchange. The square root dependency (Eq. (10.2) where $n = ½$) has been assumed, but there are data and theories that suggest that this formulation breaks down in the presence of bubbles. Laboratory determinations of the effect of bubbles on the Schmidt-number dependence indicate that a value of n in the range of -0.44 to -0.50 should be used for dual tracer experiments (Asher and Wanninkhof, 1998). One field experiment in which a non-volatile tracer (bacterial spores) was released in addition to the two gases suggested an optimum value for n of -0.51 (Nightingale *et al.*, 2000). Both of these results indicate that the departure from square root dependency causes less error than that associated with keeping track of the tracer patch and is thus not a significant problem for the ^3He and SF_6 experiments. However, when extrapolating the dual tracer results to CO_2 gas exchange, errors of up to 18% could result if a square root dependency on Sc is assumed.

Results of all the dual tracer experiments to date (Fig. 10.7) suggest a dependence on wind speed that falls between the two relations presented in Fig. 10.6. Using this compromise curve and global wind speed data to determine the mean ocean transfer velocity, one derives a global G_{600} value of 4.1 m d^{-1} (Nightingale *et al.*, 2000), which is certainly within the error of the values determined by both natural and bomb ^{14}C uptake ($G_{600}^{14} = 4.4 \pm 1.2$ m d^{-1} and 3.5 ± 1.2 mol^{-1}, respectively).

10.3 Gas saturation in the oceans

One of the most promising tracers of net biological productivity in the ocean is the flux of oxygen from surface waters. The method of estimating oxygen mass balance in the surface ocean was described in Section 6.3.3. A major component of the O_2 mass balance described there is the gas exchange rate between the ocean and atmosphere. A complicating factor in determining the biologically produced flux of oxygen to the atmosphere is that only about half of the observed O_2 supersaturation in the ocean's surface waters (Fig. 10.8) is caused by biological processes. The other mechanisms

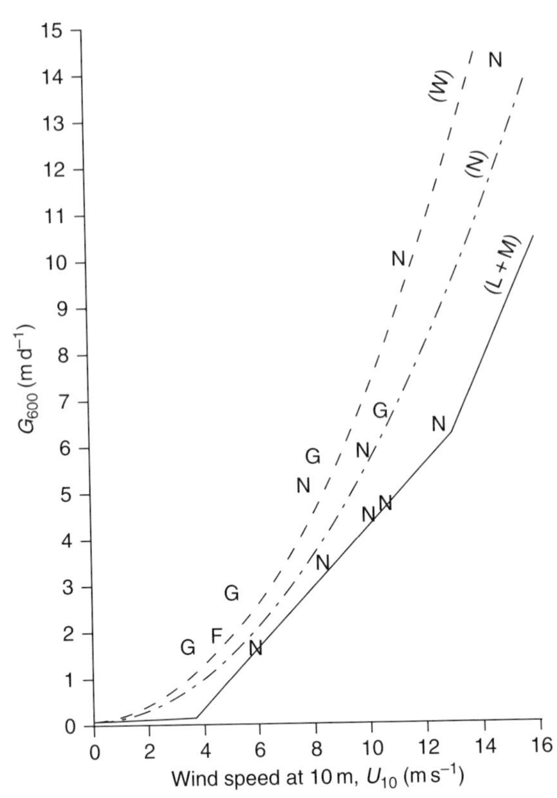

Figure 10.7. Gas transfer velocity, G_{600}, as a function of wind speed, U_{10}, for all dual tracer experiments. N is from the North Sea, G is from the Georges Bank and F from the Florida Shelf. Lines are the same as those described in the caption to Fig. 10.6, with the addition of the one in the middle, which is the best fit through the dual tracer data $G_{600} = \{0.22 \times U^2 + 0.33\,U\}\,(Sc/600)^{-1/2}$, where U is in m s^{-1} and G is in cm h^{-1}). Redrawn from Nightingale *et al.* (2000).

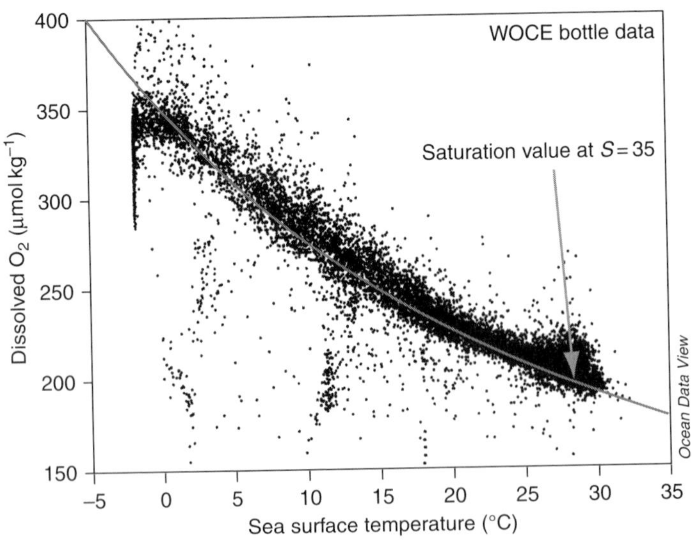

Figure 10.8. Measured oxygen concentrations in ocean surface waters as a function of temperature from the WOCE data base. The line marks the saturation value with atmospheric oxygen at a salinity of 35.

creating supersaturation are warming of the surface ocean in summer in mid- and high-latitude regions and the injection of atmospheric gases into surface waters from bubbles formed by breaking waves. To determine the biologically produced O_2 supersaturation and its flux to the atmosphere, one must determine the portion of the oxygen supersaturation that is caused by the thermal and bubble mechanisms.

A dramatic example of the importance of physical processes in producing gas supersaturation is the several percent supersaturation of inert gases N_2, Ar, and Ne measured in surface waters. Data of this type from the subtropical Pacific are presented in Fig. 6.12 and the oxygen mass balance model is described in Chapter 6. Since most gases are less soluble in warmer water, the degree of supersaturation increases during summer warming because the transfer of the excess gas to the atmosphere is not rapid enough to maintain saturation equilibrium. The warming process, however, cannot be the entire story for two reasons. First, Ne is supersaturated (Fig. 6.12) and it is a gas with a Henry's Law coefficient that is nearly temperature-independent (see Fig. 3.11). Secondly, both N_2 and Ar remain supersaturated even in the fall and winter when the temperature trend is cooling rather than warming (Fig. 6.12). Both of these observations require an alternative mechanism to create supersaturation, which is the entrainment of air bubbles caused by breaking waves. Departures from saturation equilibrium caused by temperature changes can be readily determined because temperature changes are measured and the temperature dependence of gas solubility is precisely known. The bubble mechanism, however, is much less well understood.

10.3.1 Bubble processes

Most readers who have seen the surface of the ocean during a moderate wind have noticed that breaking waves introduce lots of small air bubbles into the water (Fig. 10.9). The hydrostatic pressure only 1 m below the ocean surface is already equal to about 10% of the entire atmospheric pressure, so the pressure inside a bubble is *c.*110% of that at the ocean surface. Thus, there is a strong tendency for bubbles entrained in the downwelling limb of a breaking wave to lose some or all of their gas to the surrounding fluid by diffusion across the bubble surface. This causes the surface waters to be supersaturated with respect to saturation equilibrium. We define the degree of supersaturation, Δ (%), as

$$\Delta = \frac{[C] - [C^{Sat}]}{[C^{Sat}]} \times 100, \qquad (10.27)$$

where a positive value indicates supersaturation and a negative value undersaturation.

To account for the process of bubble-induced gas exchange we modify the gas transfer equation to include both the transfer at the air–water interface, F_{awi} (Eq. (10.1)), and the flux caused by bubbles, F_B. The total flux, F_T, is now

$$F_T = F_{awi} + F_B \qquad (\text{mol m}^{-2}\ \text{d}^{-1}). \qquad (10.28)$$

Figure 10.9. The intensity of acoustic backscatter as a function of depth in the ocean at Stn. P in the subarctic Pacific at a wind speed of 12 m s^{-1}. Backscatter intensity is an indication of the depth of penetration of bubbles caused by breaking waves. (Data courtesy of Sven Vegel of the Institute of Ocean Sciences, Sidney, BC.) (See Plate 7.)

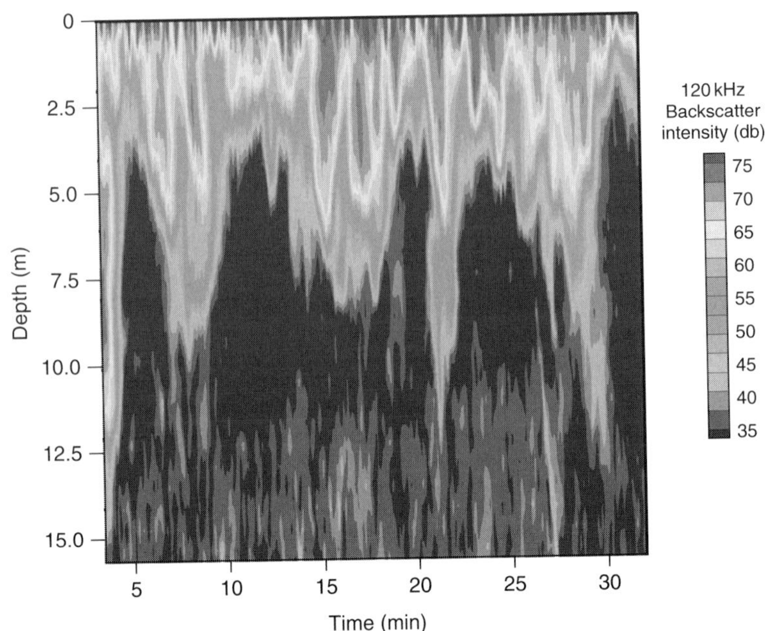

The flux created by bubbles has been mathematically described in many ways, but all present theories are strongly dependent on assumptions regarding the nature of the bubble surface, the initial size spectra of the bubbles, and the distributions of bubbles with depth. A model that has been used to predict the effect of bubbles on gas saturation (Keeling, 1993, as modified from Fuchs *et al.*, 1987) assumes that the full spectrum of bubble process can be described by a combination of two bubble transfer processes (Fig. 10.10). The first is the mechanism by which small bubbles, $< 50\,\mu\text{m}$ in diameter, completely collapse and inject their contents into the water. This mechanism has been called "air injection" or "total trapping" by bubbles. In this case flux of gas from the bubble depends only on the total volume of air transferred by these bubbles, which is described by an empirical transfer velocity, V_{inj} ($\text{mol m}^{-2}\text{ d}^{-1}\text{ atm}^{-1}$) and the mole fraction, X^{a}, of the gas in the air

$$F = V_{\text{inj}} \times X^{\text{a}}_{\text{C}}. \qquad (10.29)$$

The second mechanism, called "exchange" or "partial trapping" describes the process of bubble transfer caused by larger bubbles, 50–500 μm in diameter, that do not collapse but exchange gases across the bubble-water interface and then rejoin the atmosphere. In this case the flux depends on a different empirical constant, V_{ex} ($\text{mol m}^{-2}\text{ d}^{-1}\text{ atm}^{-1}$), the atmospheric gas mole fraction, X^{a}, and the degree of overpressure of the gas in the bubble, ΨP, caused by hydrostatic pressure and surface tension compared with the atmospheric pressure, P. The entire bubble flux can now be written as

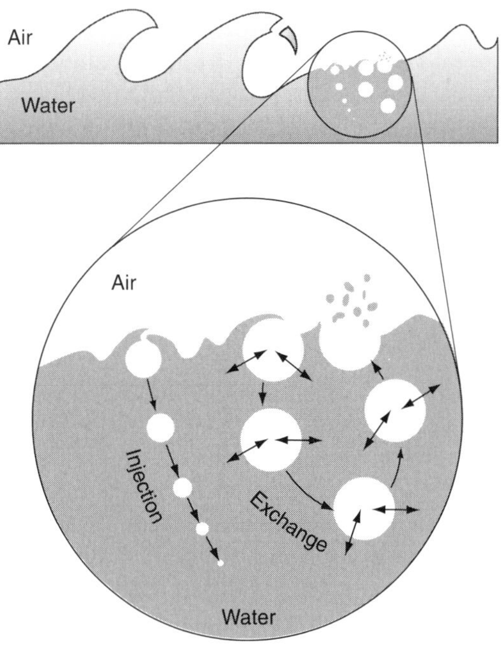

Figure 10.10. A schematic diagram illustrating two mechanisms of bubble-induced gas exchange. Air injection (total trapping) indicated by the empirical constant V_{inj} is indicated by smaller bubbles that collapse totally as they are subducted into the water by a wave. Exchange (partial trapping) indicated by the empirical constant V_{ex} is for larger bubbles that are submerged by waves and exchange their contents partly with the surrounding water before they resurface.

$$F_B = V_{inj} \times X_C^a + V_{ex} \times \left(\frac{\Psi P}{P} - \Delta \times 100\right) \times X_C^a, \qquad (10.30)$$

where Δ is the degree of supersaturation (in percent) in the bulk water. Theories of the process of gas transfer across the bubble interface (e.g. Levich, 1962; Merlivat and Memery, 1983) indicate that the exchange coefficient V_{ex} is dependent on the molecular diffusion coefficient, D, and the Henry's Law coefficient, K_H, just as it is for gas transfer at the ocean surface. Thus we can write

$$F_B = (V_{inj} \times X_C^a) + \left(V_{ex}^* \times D_C^n \times K_{H,C}^m \times \left(\frac{\Psi P}{P} - \Delta\right) \times X_C^a\right). \qquad (10.31)$$

Now the constant V_{ex}^* takes on very strange units in order to give F_B the normal units of flux (mol m^{-2} d^{-1}). The various models of the gas exchange process from a plume of bubbles suggest different exponents m and n for the dependence on the diffusion and solubility properties of different gases. The largest possible values of m and n for bubbles that are assumed to have clean surfaces are $n = 0.5$ and $m = 1$ (Merlivat and Memery, 1983), which is the same as the case for surface gas transfer (Eq. (10.13)). Partial equilibration of the bubble gases with the water reduces the exponents. The model of Keeling (1993) suggests that the bubble exchange process should scale with the diffusion coefficient to the exponent $n = 0.35$ and Henry's Law coefficient to the exponent $m = 0.7$.

In practice the relation between V_{inj} and V^*_{ex}, the exponents n and m, and the gas overpressure, ΨP, derived from models depend on many assumptions. To simplify the matter without doing undue damage to the description of the mechanisms involved, we assume that the bubble overpressure does not have a strong influence on the exponents m and n, so it can be included in the empirical constant,

$$V'_{\text{ex}} = V^*_{\text{ex}} \times \left(\frac{\Psi P}{P} - \Delta \times 100\right). \tag{10.32}$$

Now Eq. (10.24) can be written

$$F_{\text{B}} = \left(V_{\text{inj}} + V'_{\text{ex}} \times D^n \times K^m_{\text{H}}\right) \times X^{\text{a}}_{\text{C}}. \tag{10.33}$$

The total flux of gas through bubbles is represented by a combination of two categories of process: one in which small bubbles dissolve entirely, represented by the transfer coefficient, V_{inj}, and the other in which larger bubbles partly dissolve, represented by the transfer coefficient, V'_{ex}. The exponents on D_{C} and $K_{\text{H,C}}$ depend on the model of the transport process. One end member of gas flux through bubbles is air injection or total trapping only, i.e. $V'_{\text{ex}} = 0$, where the ratio of the gases injected is equal to their ratios in the atmosphere. The other end member is the exchange process only, i.e. $V_{\text{inj}} = 0$, where $n = 0.5$ and $m = 1.0$. The main importance of the exchange (or partial trapping) mechanism is that the bubble flux depends not only on the atmospheric pressures but also on both the Henry's Law coefficient and molecular diffusivity.

One of the most important results of including the bubble mechanism in the description of gas exchange is that it explains why gases are supersaturated when there is no net gas flux at the air–water interface. This situation is illustrated by considering a body of water in contact with the atmosphere at constant temperature where there is a sustained, constant breaking of waves that creates a transfer of gases by bubbles (Fig. 10.11). At steady state there will be no net flux of inert gases across the air–water interface. The flux to the water by bubbles will be exactly balanced by the flux out via diffusive transfer across the surface ($F_{\text{T}} = 0$, $F_{\text{awi}} = F_{\text{B}}$) and from Eqs. (10.1) and (10.33)

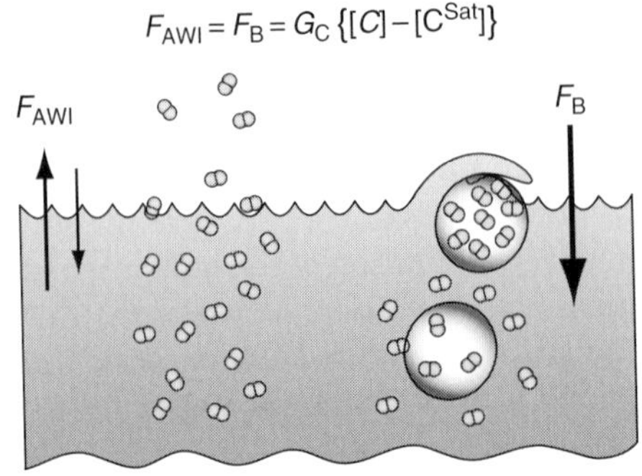

Figure 10.11. A schematic diagram of steady-state gas supersaturation caused by bubble processes when there is no net flux at the air–water interface. The small symbols represent gas molecules in the air and dissolved in the water. The greater concentration of these symbols in the water relative to air on the left side of the diagram indicates that the dissolved gas is supersaturated in the water. The bubble-induced flux, F_{B}, into surface waters, illustrated on the right side of the diagram, is balanced by a diffusive flux across the air–water interface, F_{AWI}, indicated on the left side.

$$G_C\{[C]-[C^{Sat}]\} = \left(V_{inj} + V'_{ex} \times D^n_C \times K^m_{H,C}\right) \times X^a_C. \tag{10.34}$$

If the right side is non-zero there must be a gradient in concentration, $[C]-[C^{Sat}] > 0$, in order for there to be no net flux across the air–water interface, i.e. the water must be supersaturated.

The degree of supersaturation necessary to maintain a diffusive flux equal to the bubble-induced flux is determined by dividing both sides of the above equation by $[C^{Sat}]$, substituting the Henry's Law relation, assuming $f^a_c = X^a_c$ and using the definition of G_C in Eq. (10.2) with $n = 0.5$ and Δ in Eq. (10.27):

$$\begin{aligned}\Delta/100 &= \frac{1}{G_C} \times \left(V_{inj} + V'_{ex} \times D^n \times K^m_H\right) \times \frac{X^a_C}{C^{Sat}} \\ &= \frac{\nu^{0.5}}{G^*} \times \left(V_{inj} \times D^{-0.5} \times K_H^{-1} + V'_{ex} \times D^{n-0.5} \times K_H^{m-1}\right).\end{aligned} \tag{10.35}$$

If the bubble flux is zero, $V_{inj} = V'_{ex} = 0$, then the degree of supersaturation is zero. If the bubble flux is not zero, the degree of supersaturation necessary to maintain steady state increases as V_{inj} and/or V'_{ex} increase, as the Henry's Law constant or diffusion coefficient decreases or as the mass transfer coefficient, G^*, decreases. The diffusion coefficient dependency stems from the fact that molecules that diffuse more slowly (higher molecular mass) must have a larger air–sea gradient ($[C]-[C^{Sat}]$) to maintain the same flux.

The limiting cases for the effect of bubbles on the degree of saturation occur when one or the other of the two bubble mechanisms dominates. When $V'_{ex} = 0$, air injection (complete trapping) is the most important mechanism controlling bubble fluxes.

$$\Delta/100 = \frac{V_{inj} \times \nu^{0.5}}{G^*} \times D^{-0.5} \times K_H^{-1}. \tag{10.36}$$

In this situation the degree of supersaturation is highly dependent on the physicochemical properties of the gases (diffusion coefficient and solubility). Less soluble gases are more supersaturated than soluble ones at steady state because the water has lower relative concentrations of these gases to start with. For this reason, a small amount of air forced into the water by bubble processes has a much larger effect on the relative concentrations of Ne or N_2 than it does on CO_2.

In the other extreme, when the exchange mechanism of bubble processes dominates ($V_{inj} = 0$), and the bubble mechanism has the diffusion coefficient and solubility dependence prescribed for initial gas transfer across a clean bubble ($m = 1$ and $n = 0.5$), the gas supersaturation has no dependence on the physicochemical properties of the gases!

$$\Delta/100 = \frac{V_{ex} \times \nu^{0.5}}{G^*}. \tag{10.37}$$

Here the degree of saturation depends only on the ratio of the coefficients for the bubble flux and surface gas transfer. In this case

Table 10.3. *The relative degree of supersaturation of N_2, Ar, and Ne at 20 °C created by bubble processes at steady state when there is no net gas flux across the air–water interface*

See Fig. 10.11. The equations describing these cases are given in the text.

	m	n	Gas ratios		$(D_1/D_2)^{n-0.5}$		$(K_{H,1}/K_{H,2})^{m-1}$		
Air injection only	0	0	Δ_{N_2}/Δ_{Ar}	=	0.94	×	2.22	=	2.1
			Δ_{Ne}/Δ_{Ar}	=	0.79	×	3.22	=	2.6
Exchange only (Keeling, 1993)	0.7	0.35	Δ_{N_2}/Δ_{Ar}	=	0.99	×	1.28	=	1.3
			Δ_{Ne}/Δ_{Ar}	=	0.86	×	1.41	=	1.2
Exchange only (maximum)	1.0	0.5	Δ_{N_2}/Δ_{Ar}	=	1.0	×	1.0	=	1.0
			Δ_{Ne}/Δ_{Ar}	=	1.0	×	1.0	=	1.0

there are no differences among the degrees of supersaturation for different gases. The reason for this is that the diffusion and Henry's Law dependence of surface exchange and the bubble exchange mechanism are exactly the same ($D^{0.5} \times K_H$). Because using exponents of $m=1$ and $n=0.5$ represents an extreme case for the exchange (partial trapping) mechanism, it is probably more likely that this process alone would yield a relation more like that derived by Keeling (1993):

$$\Delta = \frac{V_{ex} \times \nu^{0.5}}{G^*} \times D^{n-0.5} \times K_H^{m-1}, \tag{10.38}$$

where $n = 0.35, m = 0.7$.

The expected effects of these three different cases on the relative degrees of supersaturation of the N_2/Ar and the Ne/Ar ratios are listed in Table 10.3. The largest differences among the supersaturations of the different gases are created by the mechanism of air injection ($V'_{ex}=0$), where N_2 and Ne achieve more than twice the supersaturation of Ar. The differences among the three gases become more subtle as the exchange mechanism becomes more important. Notice also that solubility differences dominate the control of supersaturations because the diffusion coefficients of these gases are about the same. This will be true for all the gases in Table 10.1 except He, which has a much higher diffusion coefficient.

10.3.2 Inert gas saturation in the ocean

Measurements of inert gases in the subtropical surface ocean, and models that include both the effects of temperature and bubble processes, have shown that the air injection and exchange mechanisms of bubbles are about equally important for explaining surface ocean gas concentrations (Hamme and Emerson, 2006). The relative importance of these mechanisms is less clear in high latitudes in winter where the deep waters of the ocean are formed. It is, however, a good bet that the mechanism of air injection, V_{inj}, is, if anything, more important there because of the high winds associated with high-latitude winter conditions. These conditions are much more

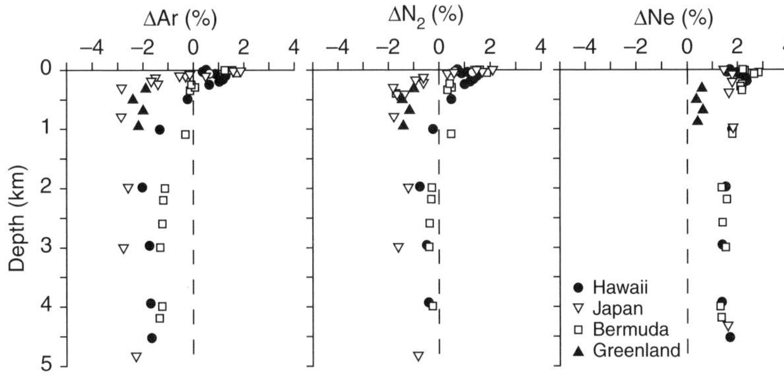

Figure 10.12. Depth profiles of the supersaturation (%) of argon, nitrogen, and neon in the world's ocean. Symbols are listed in the key. From Hamme and Emerson (2002) except for the Greenland Sea, which is unpublished.

likely to produce large waves and bubble plumes that penetrate to great depth.

Only recently have gases been measured accurately enough to determine the differences between their concentrations in the ocean (see, for example, Emerson *et al.*, 1998). The degrees of supersaturation of N_2, Ar, and Ne (Fig. 10.12) reveal very different behavior in waters below the thermocline. Deep-water gas saturations are set at the ocean's surface in the North Atlantic and under the ice shelves of Antarctica (see Chapter 1) during deep-water formation. The process in the North Atlantic occurs in the Norwegian, Greenland and Labrador Seas during winter when surface waters cool and subsequently sink. The gas saturations are most influenced by the processes of cooling and bubble-mediated gas exchange. Wintertime cooling increases the equilibrium saturation state of gases in the water, creating dissolved gas undersaturation in surface waters and invasion from the atmosphere. Because gas exchange with the atmosphere is not fast enough to re-equilibrate the dissolved gases with the atmosphere before the water sinks, this process tends to create undersaturated deep water. Bubble processes, in the meantime, act to increase the degree of saturation. Ne is the most insoluble of the three gases discussed here and the temperature dependence of its Henry's Law coefficient is about three times less than those of N_2 and Ar (Fig. 3.11). Thus, bubble processes dominate the mechanisms influencing the degree of saturation of Ne, and it is nearly uniformly 1.5% supersaturated. N_2 is about half as soluble as Ar, but they share very similar solubility temperature dependencies. Like Ne, N_2 feels the bubble process more strongly than Ar, but unlike Ne, N_2 also responds to the cooling effect. Ar is affected most by cooling and as a result is more undersaturated than the others.

One can make an educated guess of the importance of cooling and bubble processes in creating the degrees of supersaturation in Fig. 10.12. If we assume that air injection dominates the bubble mechanisms because of the deep mixing, then solubilities should have a large influence in determining the gas supersaturations (Eq. (10.36)). The supersaturation effect for Ne should be about twice that for N_2 and about 4 times that for Ar at 2 °C. Assuming there is no temperature effect for Ne and its 1.5% supersaturation is solely the result of air injection by bubbles results in a bubble effect that is 0.8% for N_2 and about 0.4% for Ar. The temperature effect for Ar is thus about −2.1% and that for N_2 is about −1.2%. These values are approximate, because there is a small temperature effect for Ne and there may be some role for the bubble exchange mechanism in addition to air injection. We have also not considered the possibility that the atmospheric pressure in the region of deep water subduction is probably below one atmosphere. Since all calculations are done assuming one atmosphere pressure, this effect would be equal for all the gases, but would change the calculated importance of the effects of bubbles and cooling.

Marine chemists are presently exploring the use of an extended suite of inert gas tracers to separate the effects of bubbles, gas exchange, temperature change and atmospheric pressure on the degree of gas supersaturation. When more high quality data become available for more gases the importance of the different mechanisms will be clearer.

10.4 Surface films and chemical reactions

We end this chapter on gases and gas exchange with a brief discussion of two processes that are known to have an effect on air–water gas exchange rates in nature but are presently not well enough characterized in the ocean to estimate their global importance. The effects of both surface films and chemical reactions on the gas exchange process have been studied for many years, but given the available data and models, neither process appears to be of first-order importance in the ocean; however, they may be of second-order importance. Surface films affect surface roughness and the formation of waves, which have been shown to be important to the exchange rates of all gases. Chemical reactions affect only those gases that react with water, which include CO_2 and some other gases with much lower concentrations (e.g. SO_2).

10.4.1 The effect of surfactants on air–sea transfer

Non-polar, organic molecules in seawater originate from natural populations of plankton and also from spillage of petroleum into the ocean from oil tankers. These molecules are hydrophobic and thus tend to collect at the air–water interface. There are two main possibilities for how these substances might affect the rate of gas exchange. First, there is an additional film on the surface through which gases must pass; and second, surfactants dampen the formation of waves. Because organic surfactants are highly permeable to gases, it is the second effect that is most important to gas exchange in the ocean.

Laboratory studies have clearly established that the onset of waves occurs at higher wind speeds in the presence of organic compounds and that this reduces the gas exchange rate at a given wind

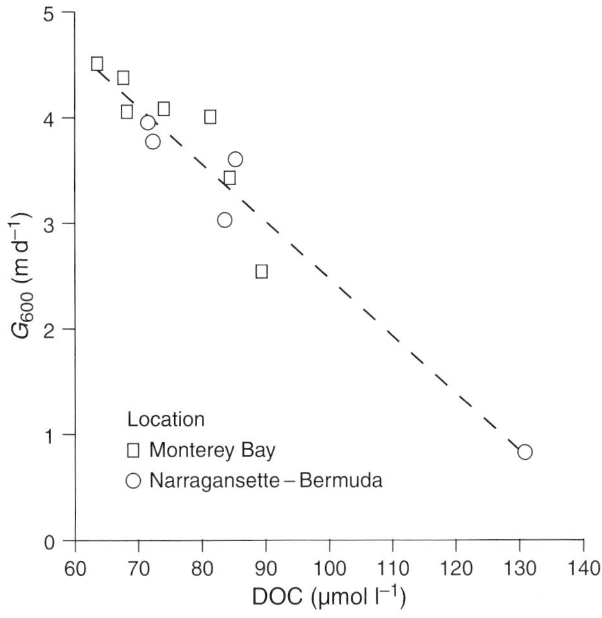

Figure 10.13. The gas exchange mass transfer coefficient in different seawater samples from the continental margin regions of North America as a function of the dissolved organic carbon (DOC) concentration of the water. Exchange rates were determined in laboratory experiments in which gas exchange conditions were identical for different water samples from the locations indicated. The measured gas exchange rates of oxygen are inversely proportional to the DOC concentration of the water, implying that the organic content affects the interface properties that control gas transfer. Redrawn from Frew (1997).

speed (see Section 10.1). The question then becomes, how important is this effect in nature? The effect of *in situ* levels of surfactants on gas exchange has been studied in the ocean by determining the gas transfer velocity in seawater samples with different concentrations of dissolved organic carbon (DOC) (Frew, 1997). In this study the gas exchange mass transfer coefficient was measured in water samples taken from the surface ocean in a transect away from the New England coast, where there is a strong gradient in DOC concentration in surface waters. Higher productivity rates near the continent produce a higher concentration of dissolved organic matter in the euphotic zone. The gas exchange transfer velocities measured in this experiment varied inversely with the concentration of DOC (Fig. 10.13), and the magnitude of the effect was very dramatic, nearly a factor of two!

These experiments open the possibility that some of the scatter in the comparison of natural gas exchange measurements with wind speed may be due to the effect of surfactants. If a way to remotely characterize wave slopes is devised, then it may give insight into second-order effects on the transfer velocity caused by surfactants. Another implication is that pollution of the seas by petroleum may have an inhibitory effect on the global gas exchange rate.

10.4.2 The effect of chemical reactions on the gas exchange rate of CO_2

Calculation of CO_2 gas exchange rates in nature from air–water gradients in f_{CO_2} implicitly assumes that CO_2 does not react during the transfer across the interface. If reaction did occur it would steepen the CO_2 gradient very near the interface (within tens of microns)

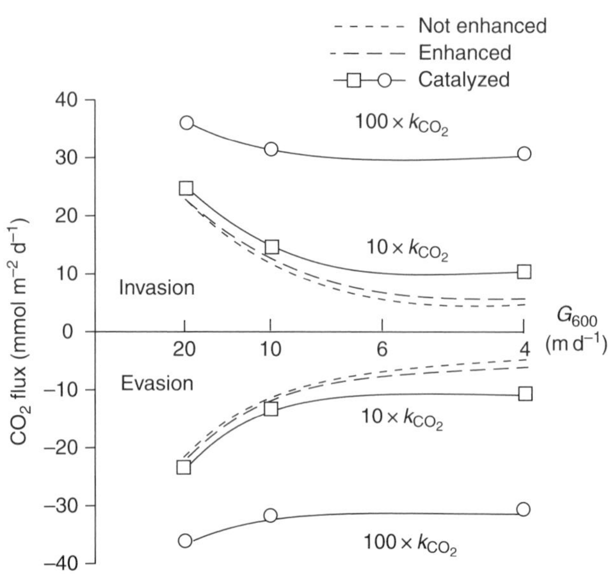

Figure 10.14. The calculated effect of chemical enhancement of CO_2 gas exchange in the ocean as a function of the gas exchange mass transfer coefficient, G_{600} (m d^{-1}). The air–water f_{CO_2} gradient is 40% over- or undersaturation for evasion and invasion, respectively. Lines are drawn for different forward rate constants of CO_2 hydration. Short dashed lines indicate CO_2 net flux for no reaction of CO_2 with water. Longer dashed lines were calculated by using laboratory-determined values for the hydration rate of CO_2 (Table 4.6). CO_2 gas exchange rates are chemically enhanced by roughly 1, 3 and 15% for mass transfer coefficients of 20, 10 and 4 m d^{-1}, respectively. The solid lines indicate CO_2 exchange fluxes calculated for CO_2 hydration rates increased by 10 times ($10 \times k_{CO_2}$) and 100 times ($100 \times k_{CO_2}$). In order for the chemical reaction to have a significant effect on CO_2 gas transfer in the range of mass transfer coefficients that dominate gas transfer in the ocean, $G_{600} > 5$ m d^{-1}, the rate constant would have to be catalyzed by at least a factor of 10. Redrawn from Emerson (1995).

and create a gradient in HCO_3^- into the solution in this region. The scale of this gradient is much too small to measure, so the importance of this process must be estimated from models and laboratory experiments. The CO_2 reaction with water will be unimportant to the gas exchange rate if the residence time of CO_2 with respect to transport across the air-water interface is shorter than the reaction time. This process has been shown to be important for determining CO_2 gas exchange rates in lakes, where the physical gas transfer rates are slower than in the ocean and pH can become very high in the summer (Emerson, 1995). These same calculations, however, suggest that the chemical reaction of CO_2 during gas exchange in the ocean is probably unimportant. The reason for this can be demonstrated by comparing the reaction rates for CO_2 at seawater pH and the residence time of CO_2 at the interface with respect to diffusive transfer. The reaction time of CO_2 with water at pH = 8 is about 0.037 s^{-1} (Fig. 4.3) for a reaction-rate residence time of about 25 s. The gas exchange mass transfer coefficients for CO_2 in the ocean are 5–7 m d^{-1} ($6–8 \times 10^{-3}$ cm s^{-1}) at a Schmidt number of 600 (Fig. 10.6). Since the characteristic thickness of the liquid boundary layer at the marine air-water interface in the stagnant film model is 20–40 μm (Broecker and Peng, 1982), the exchange residence time is less than one second (30 μm / 7×10^{-3} cm s^{-1} = 0.4 s).

This simple calculation suggests that there should be little CO_2 reaction during gas transfer for conditions in the ocean. The residence time of a gas in the diffusive layer is much shorter with respect to diffusion than with respect to reaction. To determine whether this is generally true, CO_2 gas exchange rates including the effects of CO_2 reaction have been calculated by using a reaction diffusion model for the chemistry of seawater as a function of the gas exchange mass transfer coefficient (Emerson, 1995). Results (Fig. 10.14) indicate that fluxes of CO_2 across the air–water interface with and without reaction are only a few percent different for mass transfer coefficients of 5–10 m d^{-1} and 20%–30% different for a mass transfer coefficient of c.1 m d^{-1}. Reaction rates are only significant to gas transfer at the very slowest mass transfer coefficients observed in the ocean. Since the total integrated rate of exchange is highly weighted toward the higher range of mass transfer coefficients (Fig. 10.14), the maximum error in assuming no reaction during CO_2 transfer should be less than 10%.

The kinetic rate constants for CO_2 hydration determined in the laboratory in sterile seawater (Table 4.6) are known sufficiently well that this value should create little uncertainty in the above calculation. However, in natural waters the reaction rates may be enzymatically catalyzed. Carbon dioxide hydration catalysis by carbonic anhydrase (CA) is the most powerful enzyme reaction known (see the discussion in Section 9.3). The catalytic turnover number (the number of moles of substrate reacted, divided by the number of moles of enzyme present) is 8×10^7 min^{-1} for CA (Table 9.7), and marine diatoms are known to produce carbonic anhydrase (Morel *et al.*, 1994). The calculations presented in Fig. 10.14 indicate that increasing the CO_2 hydration rate constant by 10-fold should increase the gas exchange rate of CO_2 in the ocean by 10%–50%.

One can make a first-order assessment of whether a 10-fold enhancement in the reaction rate is feasible by knowing how much carbonic anhydrase it would require in solution. Fortunately, scientists have studied the catalytic effect of CA in water and found that the hydration rate of CO_2 is increased by 30 times in water for five nanomolar concentrations of CA at pH = 8.35 (Silverman and Tu, 1976). Their result is a lower limit for our purposes as the rate of catalysis increases with decreasing pH. Thus, only nanomolar levels of CA are necessary to create sufficient CO_2 catalysis to make a difference. Presently it is unknown whether these levels of CA exist near the air–sea interface anywhere in the ocean. Until one is able to determine low levels of this enzyme in seawater, the potential importance of catalyzed CO_2 reaction kinetics on local air–sea CO_2 exchange will remain uncertain. The relatively good agreement between gas exchange rates determined by natural and bomb-produced ^{14}C (which would both be enhanced by CO_2 reaction during gas exchange) with those determined from the gas exchange–wind speed relation by using inert gas tracers argues against a significant global chemical effect.

References

Asher, W. L. and R. Wanninkhof (1998) The effect of bubble-mediated gas transfer on purposeful dual-gaseous tracer experiments. *J. Geophys. Res.* **103**, 10 555–60.

Broecker, W. S. and T.-H. Peng (1971) The vertical distribution of radon in the Bomex area. *Earth Planet. Sci. Lett.* **11**, 99–108.

Broecker, W. S. and T.-H. Peng (1982) *Tracers in the Sea*. Lamont–Doherty Geological Observatory.

Broecker, W. S., J. R. Ledwell, T. Takahashi *et al.* (1986) Isotopic versus micrometeorologic ocean CO_2 fluxes: a serious conflict. *J. Geophys. Res.* **91**, 10 517–27.

Dankwertz, P. V. (1970) *Gas-Liquid Reactions*. New York, NY: McGraw-Hill.

Deacon, E. L. (1977) Gas transfer to and across the air-water interface. *Tellus* **29**, 363–74.

Dickinson, R. E. and R. J. Cicerone (1986) Future global warming from atmospheric trace gases. *Nature* **319**, 109–15.

Emerson, S. (1995) Enhanced transport of carbon dioxide during gas exchange. In *Proceedings of the 3rd International Symposium on Air–Water Gas Transfer, Heidelberg, Germany* (ed. J. A. Monahan), pp. 23–35. AEON Verlag.

Emerson, S., P. D. Quay, C. Stump, D. O. Wilbur and M. Knox (1991) O_2, Ar, N_2 and ^{222}Rn in surface waters of the subarctic Pacific Ocean: net biological O_2 production. *Global Biogeochem. Cycles* **5**, 49–70.

Emerson, S., C. Stump, D. O. Wilbur and P. D. Quay (1998) Accurate measurements of O_2, N_2, and Ar gases in water and the solubility of N_2. *Mar. Chem.* **64**, 337–47.

Frew, N. M. (1997) The role of organic films in air-sea gas exchange. In *The Sea Surface and Global Change* (ed. P. S. Liss and R. A. Duce), pp. 121–72. Cambridge: Cambridge University Press.

Fuchs, G., W. Roether and P. Schlosser (1987) Excess 3He in the ocean surface layer. *J. Geophys. Res.* **92**, 6559–68.

Hamme, R. C. and S. R. Emerson (2002) Mechanism controlling the global oceanic distribution of the inert gases argon, nitrogen and neon. *Geophys. Res. Lett.* **29** (23), 2120, doi: 10.1029/2002GL015273.

Hamme, R. C. and S. R. Emerson (2006) Constraining bubble dynamics and mixing with dissolved gases: implications for productivity measurements by oxygen mass balance. *J. Mar. Res.* **64** (1), 73–95.

Higby, R. (1935) The rate of adsorption of a pure gas into a still liquid during short periods of exposure. *Trans. Am. Inst. Chem. Eng.* **35**, 365–89.

Hinze, J. O. (1959) *Turbulence: an Introduction to its Mechanism and Theory*. New York, NY: McGraw-Hill.

IPCC (2001) *Climate Change 2001: The Scientific Basis*. Contribution of working group I to the Third Assessment Report of the Intergovernmental Panel on Climate Change (ed. J. T. Houghton, Y. Ding, D. J. Griggs *et al.*). Cambridge: Cambridge University Press.

Jacob, D. J. (1999) *Introduction to Atmospheric Chemistry*. Princeton, NJ: Princeton University Press.

Jahne, B., G. Heinz and W. Dietrich (1987) Measurement of the diffusion coefficients of sparingly soluble gases in water. *J. Geophys. Res.* **92**, 10 767–76.

Keeling, R. F. (1993) On the role of bubbles in air-sea gas exchange and supersaturation in the ocean. *J. Mar. Res.* **51**, 237–71.

Kromer, B. and W. Roether (1983) Field measurements of air-sea gas exchange by the radon deficit method during JASIN, 1978 and FGGE, 1979. *Meteor. Forsch.-Ergebn.* **24**, 55–75.

Ledwell, J. R. (1984) The variation of gas transfer coefficient with molecular diffusivity. In *Gas Transfer at Water Surfaces* (ed. W. Brutsart and G. H. Jirka), pp. 293–302. Dordrecht: D. Reidel.

Levich, V.G. (1962) *Physicochemical Hydrodynamics.* Englewood Cliffs, NJ: Prentice-Hall.

Liss, P.S. and L. Merlivat (1986) Air-sea gas exchange rates: introduction and synthesis. In *The Role of Air-Sea Gas Exchange in Geochemical Cycling* (ed. P. Buat-Menard), pp. 113–27. Hingham, MA: Reidel.

Merlivat, L. and L. Memery (1983) Gas exchange across an air-water interface: experimental results and modeling of the bubble contribution to transfer. *J. Geophys. Res.* **88**, 707–24.

Morel, F.M.M., J.R. Reinfelder, S.B. Roberts *et al.* (1994) Zinc and carbon co-limitation of marine phytoplankton. *Nature* **369**, 740–2.

Nightingale, P.D., G. Malin, C.S. Law *et al.* (2000) In situ evaluation of air-sea gas exchange parameterizations using novel conservative and volatile tracers. *Global Biogeochem. Cycles* **14**, 373–87.

Peng, T.-H., W.S. Broecker, G.G. Matjoei, Y.H. Li and A.E. Bainbridge (1979) Radon evasion rates in the Atlantic and Pacific Oceans as determined during the GEOSECS Program. *J. Geophys. Res.* **84**, 2471–86.

Peng, T.-H., T. Takahashi and W.S. Broecker (1974) Surface radon measurements in the north Pacific Ocean Station Papa. *J. Geophys. Res.* **79**, 1772–80.

Pilson, M.E.Q. (1998) *An Introduction to the Chemistry of the Sea.* Upper Saddle River, NJ: Prentice-Hall.

Silverman, D.N. and C.K. Tu (1976) Buffer dependence of carbonic anhydrase catalyzed oxygen-18 exchange at equilibrium. *J. Am. Chem. Soc.* **97**, 2263–9.

Sweeney, C., E. Gloor, A.R. Jacobson *et al.* (2007) Constraining global air-sea gas exchange for CO_2 with recent bomb ^{14}C measurements. *Global Biogeochem. Cycles* (in press).

Wanninkhof, R. (1992) Relationship between wind speed and gas exchange over the ocean. *J. Geophys. Res.* **97**, 7373–82.

Watson, A.J., R.C. Upstill-Goddard and P.S. Liss (1991) Air-sea gas exchange in rough and stormy seas measured by a dual tracer technique. *Nature* **349**, 145–7.

Whitman, W.G. (1923) The two-film theory for gas adsorption. *Chem. Metall. Engng.* **29**, 146–8.

11

The global carbon cycle: interactions between the atmosphere and ocean

Cycling of carbon among the ocean, atmosphere and land is a fundamental component of the chemical perspective of oceanography because carbon dioxide is the most important greenhouse gas in the atmosphere (except for H_2O, which behaves in a feedback rather than forcing capacity). Since there is about 50 times as much inorganic carbon dissolved in the sea as there is CO_2 in the atmosphere, ocean carbonate chemistry has a great impact on f_{CO_2} in the atmosphere. On time scales of hundreds to a thousand years the main marine processes that influence f_{CO_2} in the atmosphere are the thermodynamic temperature dependence of CO_2 solubility in seawater (the *solubility pump*), and the interplay between the rate of ocean circulation and the rate of biological carbon removal from the euphotic zone to the deeper reservoirs of the ocean (the *biological pump*). On longer time scales, of the order of one to tens of thousands of years, the preservation and dissolution of calcium carbonate along with the rate of weathering and the transport of bicarbonate to the sea come more into play.

The Earth is presently in the early stages of a grand acid–base titration of seawater by CO_2. Anthropogenic CO_2 is being added to the atmosphere at a rate fast enough to have resulted in an approximately 30% increase in the f_{CO_2} of the atmosphere since pre-industrial time. Only about 40% of the CO_2 added to the atmosphere has remained there: the rest has gone into the land and ocean carbon reservoirs. It is extremely important that we understand the processes controlling the uptake of CO_2 because future prediction of global climate will depend on knowing how CO_2 partitions between the atmosphere, land, and ocean.

The long-term response of the ocean's carbonate system to past climate perturbations, on time scales ranging from ten thousand to hundreds of millions of years, involves the interaction of the seawater carbonate system with the carbonate solids deposited in marine sediments. For example, there are very large changes in the calcium carbonate content of deep-sea sediment cores that span past glacial–interglacial times from all the ocean basins that tell us something about past excursions in the chemistry of the sea and atmosphere. Interpreting these records requires an understanding of the present-day relation between carbonate chemistry and $CaCO_3$ preservation in marine sediments.

In this chapter we explain the most important processes controlling the interaction of carbon between the atmosphere and ocean. Processes affecting $CaCO_3$ burial are reviewed in the following chapter on sediment diagenesis. We begin this chapter with an overview of the global carbon cycle, then describe the role of steady-state processes involving the ocean's solubility and biological pump, and end with a discussion of the fate of CO_2 from anthropogenic perturbation.

11.1 The global carbon cycle

The global carbon cycle involves reactions within and exchange among the major global reservoirs: atmosphere, ocean and land (Fig. 11.1). The important reactions are formation and destruction of organic matter and calcium carbonate via photosynthesis/respiration and precipitation/dissolution, respectively. Exchange among the reservoirs is primarily via CO_2 gas exchange, flow of dissolved inorganic carbon ($DIC = HCO_3^- + CO_3^{2-} + CO_2$) in rivers, and the burial of the mineral $CaCO_3$.

The amount of carbon in each of the main global reservoirs and the exchange fluxes (Table 11.1, Fig. 11.1) provides a qualitative impression of how much the reservoirs depend on one another. Among the atmosphere, land and ocean the carbon reservoir size of the atmosphere is by far the smallest (600 Pg before anthropogenic changes, where $1\ \text{Pg} = 10^{15}$ g); DIC of the ocean the largest (38 000 Pg) and the exchangeable reservoirs in land plants and soils are somewhere in between (*c.*2000 Pg). Since atmosphere–land and atmosphere–ocean CO_2 exchange rates are about the same (on the order of $100\ \text{Pg y}^{-1}$), it is pretty clear that the pressure of CO_2 in the atmosphere is a slave to processes that occur in the larger reservoirs.

There are two categories of flux depicted in Fig. 11.1: "long-term" fluxes, indicated by dashed arrows, and "short-term" fluxes, indicated by solid arrows. The "long-term" fluxes represent fluxes

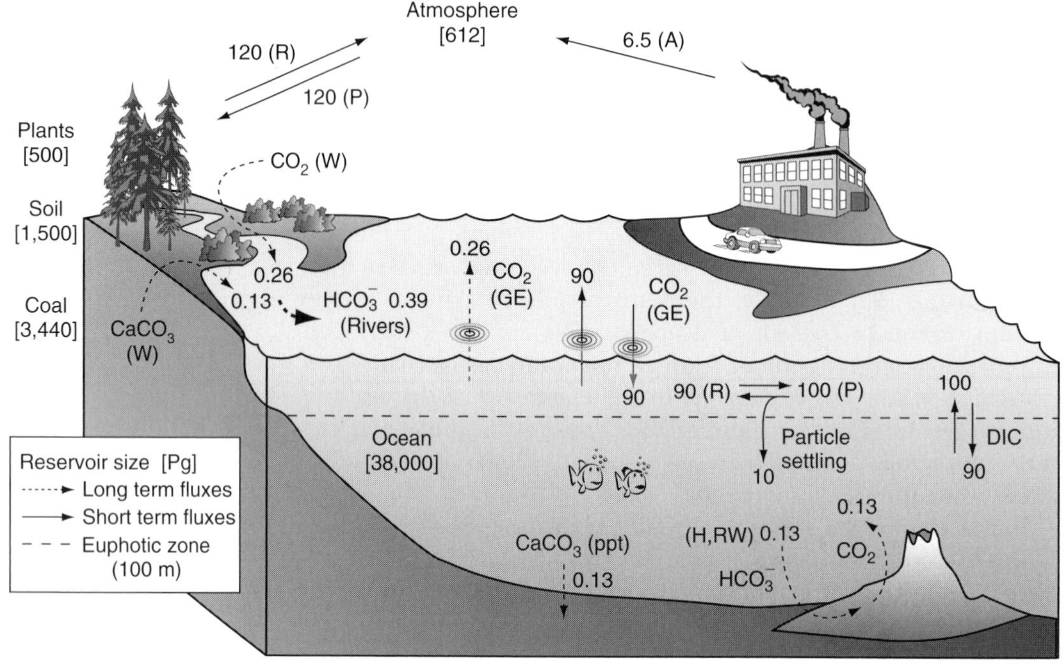

Figure 11.1. The global carbon cycle. Values in brackets are preanthropogenic reservoir sizes in Pg (10^{15} g); values on the arrows are fluxes in Pg y^{-1}. Dashed lines represent the long-term carbon cycle determined by weathering. Values are normalized to the flux of DIC from rivers (see Chapter 2). Solid arrows are the shorter-term carbon fluxes associated with photosynthesis and respiration. The wiggly vertical line indicates particulate C and DOC transport from the ocean euphotic zone to deep water. Symbols: W, weathering of carbonates ($CaCO_3 + CO_2 + H_2O \rightarrow 2HCO_3^- + Ca^{2+}$) and silicates (silicate $+ CO_2 + H_2O \rightarrow$ clay $+ HCO_3^- +$ cations); GE, gas exchange; P, gross photosynthesis ($CO_2 + H_2O \rightarrow CH_2O$ (OM) $+ O_2$); R, respiration (CH_2O (OM) $+ O_2 \rightarrow CO_2 + H_2O$); pptn, calcite precipitation (the reverse of carbonate weathering); H, hydrothermal processes; RW, reverse weathering (the reverse of silicate weathering).

associated with processes of weathering; the "short-term" fluxes are those that are driven primarily by photosynthesis and respiration. Each of these mechanisms controls the f_{CO_2} in the atmosphere on different time scales. Changes in weathering reactions due to sea-floor spreading and tectonics such as mountain building are believed to have been the major factors for regulating the CO_2 content of the atmosphere through the early history of the Earth on 10–100 million year time scales (see, for example, Berner, 1990). More recent fluctuations of f_{CO_2} on glacial–interglacial time scales (10–100 000 years) observed in trapped atmospheric bubbles in ice cores were probably influenced by the shorter-term fluxes involving ocean circulation and biological processes in the ocean and on land (see Chapter 7).

The present-day weathering fluxes depicted in Fig. 11.1 are normalized to the flux of DIC (primarily HCO_3^-) that enters the ocean via the world's rivers (*c.*0.4 Pg y^{-1}) (Table 2.3). Since we know that the ultimate origin of weathering-produced carbon is from the atmosphere and $CaCO_3$ rocks on an approximately 2:1 basis (see Fig. 2.2), all other fluxes can be scaled accordingly. The flux of CO_2 out of the atmosphere (0.26 Pg y^{-1}), that provides the acid for dissolution of rocks in soils via weathering and flows to the ocean as HCO_3^-, must be matched by an equal return flux to the atmosphere. This return flux derives partly from precipitation of $CaCO_3$ in the ocean (reaction (iii) in Table 2.2, also given in the caption to Fig. 11.1), which is supported by the Ca^{2+} and HCO_3^- flow from rivers. The other contribution to the return of CO_2 to the atmosphere is via a reaction to form silicate rocks that is the equivalent of reverse weathering (the reverse of reaction (vi) in Table 2.2, also reproduced in the caption to Fig. 11.1) and occurs either in ocean sediments or at hydrothermal areas. Even

Table 11.1. *Carbon reservoirs (excluding terrestrial rocks other than coal) and fluxes*

Reservoirs (Pg)		
Atmosphere: CO_2 (288 ppm in 1850)		612
(369 ppm in 2000)		784
Oceans:	Biota	1–2
	DOC	700
	Org. C in sediments (1 m)	1000
	DIC	38 000
Terrestrial:	Biota	600
	Soil humus (1 m)	1500
	Fossil fuels (identified reserves): gas	44
	oil	90
	coal, oil sand and shale	3440
Fluxes ($Pg\ y^{-1}$)		
Atmosphere–ocean exchange		90
Gross primary production: ocean		100
land		120
Net primary production: ocean		45
land		60
Net C export from the surface ocean		8–15
Sedimentation of org. C in the ocean		0.2
Anthropogenic changes (Pg or $Pg\ y^{-1}$)		
Cumulative changes (Pg) (1800–1994)		
Fossil fuels burnt and cement produced		244
Atmospheric increase		165
Storage in the ocean		118
Inferred terrestrial change		−39
Partitioning of anthropogenic fluxes (1990s) ($Pg\ y^{-1}$)		
Fossil fuel and cement production		6.3 ± 0.4
Atmosphere accumulation		3.2 ± 0.1
Uptake by terrestrial biosphere		-1.4 ± 0.7
Ocean uptake		-1.7 ± 0.5

Source: The data are from the compilations of Pilson (1998), IPCC (2001), and Sabine *et al*. (2004).

though these fluxes are much smaller than those involved in the "short-term" CO_2 exchange, they are known at least as well because they are constrained by the chemical concentrations and flow of rivers.

The larger "short-term" fluxes are mainly controlled by the processes of photosynthesis and respiration. In the case of the atmosphere–land exchange CO_2 is taken up directly from the atmosphere by photosynthesis in leaves and released from the plants and soils to the atmosphere via respiration. The uptake and release are balanced (except for the 0.26 Pg of CO_2 that ultimately goes to weathering rocks in the soil) as long as the plant and soil reservoirs are not changing. The exchange of CO_2 between the atmosphere and ocean represents the one-way gas exchange flux and is controlled by the f_{CO_2}

and the gas exchange mass transfer coefficient (see Chapter 10). In pre-industrial times the mean f_{CO_2} of the atmosphere and ocean were about the same, so the one-way fluxes across the air–water interface were also about the same (*c.*90 Pg y^{-1}). It is fortuitous that the atmosphere–ocean exchange rate of CO_2 is about equal to marine euphotic zone gross photosynthesis and respiration rates (*c.*100 Pg y^{-1}).

Categorization of carbon fluxes in the ocean (Fig. 11.1) is more detailed than on land because rates of gas exchange and biological processes can be separated. The difference between organic carbon production by photosynthesis and respiration in the marine euphotic zone is about 10% of the rate of photosynthesis and equal to the transport of organic carbon to the ocean interior by sinking particulate material and mixing of dissolved organic carbon (DOC). This is the marine "biological carbon pump." A similar process occurs on land between the living biomass and humic material, but it is more heterogeneous and difficult to characterize. Although the marine biological carbon pump is only about one tenth of the photosynthesis in the ocean, it represents a relatively large flux (*c.*10 Pg y^{-1}) away from contact with the atmosphere into the long-term deep-ocean DIC reservoir. The analogous reservoir on land is soil carbon, which is about 30 times smaller and has a shorter residence time. Thus, changes in the magnitude of the ocean's biological pump may have had a large influence on the f_{CO_2} of the atmosphere in the past and could also be important in controlling future f_{CO_2}.

Fluxes and reservoirs in Fig. 11.1 are those believed to be representative of pre-industrial values. Comparison of these values with the rate of input of CO_2 to the atmosphere via fossil fuel burning and cement production (*c.*6.2 Pg y^{-1}) places the anthropogenic input into perspective with natural processes. The anthropogenic perturbation is about 20 times smaller than the one-way exchange of CO_2 between the atmosphere and ocean and the rates of photosynthesis on land or in the ocean, but about 20 times larger than the fluxes of CO_2 from global weathering and about half the present biological carbon pump to the deep ocean. This perturbation is by all measures very large compared with natural fluxes. The amount of fossil fuel that has been burnt since 1800 (*c.*244 Pg) (Table 11.1), is greater than the known reserves of gas and oil but less than *c.*10% of the known coal reserves. Coal reserves, albeit very large, are still only about one tenth the size of the ocean's DIC reservoir. In the following sections we demonstrate in a simple quantitative way the influence of steady-state processes on the f_{CO_2} of the atmosphere and review what is known about the fate of carbon anthropogenically introduced into the atmosphere.

11.2 The biological and solubility pumps of the ocean

The carbon pump in the ocean refers to processes that create a vertical gradient in the dissolved inorganic carbon (DIC) concentration. These

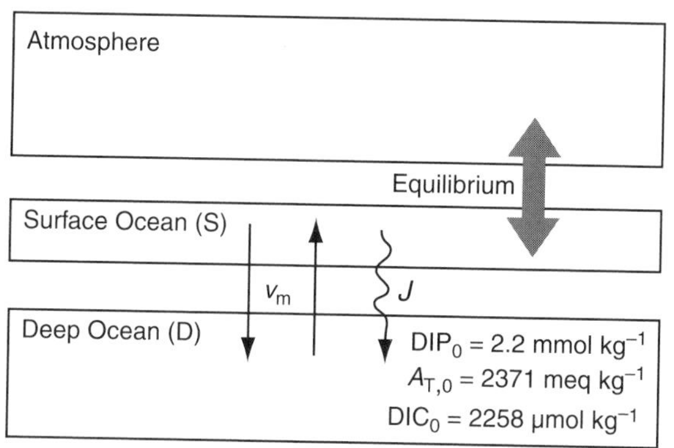

Figure 11.2. Sketch of the three-box model of the atmosphere, surface and deep ocean. Equations indicate the circulation dynamics (V_m in y^{-1}, is the mixing rate between the surface and deep ocean); stoichiometry of the particulate transport (J in mol y^{-1}); and chemical equilibria of the carbonate system.

Dynamics: $V_D \times \dfrac{d[C_D]}{dt} = 0 = V_m \times ([C_S] - [C_D]) + J$

Stoichiometry: $\Delta P : \Delta N : \Delta DIC : \Delta A_T : \Delta Ca$

$1 : 16 : 136 : 44 : 30$

Equilibrium:

$$DIC = [HCO_3^-] + [CO_3^{2-}] + [CO_2]$$

$$A_{C\&B} = [HCO_3^-] + 2 \times [CO_3^{2-}] + [B(OH)_4^-]$$

$$B_T = B(OH)_3 + B(OH_4^-)$$

$$K_{H,CO_2} = \frac{[CO_2]}{f^a_{CO_2}} \qquad K_2' = \frac{[CO_3^{2-}][H^+]}{[HCO_3^-]}$$

$$K_1' = \frac{[HCO_3^-][H^+]}{[CO_2]} \qquad K_B' = \frac{[B(OH)_4^-][H^+]}{[B(OH)_3]}$$

processes move DIC from the upper *c.*100 m of the ocean to the ocean's interior and thus decrease the amount of carbon in the surface ocean and atmosphere. This carbon transfer has been categorized into the "solubility" and "biological" carbon pumps (Volk and Hoffert, 1985). The strength of the solubility pump is measured by the gradient in temperature between the surface and deep ocean. Dense colder water can hold more DIC at thermodynamic equilibrium with atmospheric f_{CO_2} so this process increases the DIC content of deeper waters. The strength of the biological pump is determined by the vertical gradient of nutrients created by the transport of organic matter in the form of particles and dissolved organic matter (DOM) from the euphotic zone to deeper waters. The stoichiometric relation between nutrient and DIC uptake and regeneration during biological processes then determines the biologically produced vertical gradient of DIC in the ocean.

A very simple demonstration of the importance of ocean temperature, marine biology and ocean dynamics to atmospheric f_{CO_2} is achieved by using a three-box model (Fig. 11.2) of the ocean and atmosphere that is similar to the two-layer model introduced in Chapter 6. This construct can be used to show the sensitivity of atmospheric f_{CO_2} to ocean surface temperature, biological processes and overturning circulation. In this calculation we concern ourselves only with steady-state natural concentrations that are uninfluenced by human activities. Although this model involves gross approximations, it is a jewel in demonstrating the mechanisms that are necessary for the implementation of these concepts in more complicated models with more boxes (see, for example, Toggweiler, 1999) and global circulation models (see, for example, Archer *et al.*, 2000).

In this simple three-box model one assumes the entire atmosphere, surface and deep ocean each have homogeneous concentrations. The well-mixed approximation for the atmosphere is not bad, as measurements of f_{CO_2} indicate that differences from place to place are only a few percent. The north–south hemispheric gradient is only 1%–2%, and this is primarily caused by greater fossil fuel usage in the north. The homogeneous surface and deep ocean assumptions, however, are extreme approximations because of the vast differences in global temperature and biological productivity. We use averages, which undoubtedly lead to results that are only crudely accurate.

It is assumed in this calculation that the atmosphere and surface ocean are in chemical equilibrium with respect to carbon. That is, the f_{CO_2} of the atmosphere equals the f_{CO_2} in the surface ocean. Although this is not the case in many areas, ocean surveys indicate that atmospheric f_{CO_2} is on average about 8 ppm higher than that of the surface ocean, or only 2% greater in the atmosphere. Again, most of the atmosphere excess is caused by anthropogenic fossil fuel burning: the pre-industrial mean surface ocean and atmosphere f_{CO_2} values were probably nearly equal.

Since we assume the atmosphere and ocean are in chemical equilibrium, and the input from rivers and burial in the sediments are small compared to the other fluxes, the entire dynamics of the model is reduced to the rate of surface–deep mixing and the sinking of particles. (For simplicity, DOC transport is not considered in this simple model.) One can see that for a steady state to be achieved the flux of carbon to the surface ocean must equal the sinking flux of particles. The mean residence time for deep water is that determined by natural ^{14}C measurements (see Chapter 6): 500–1000 y.

The chemical currency of the ocean model consists of three dissolved constituents: DIC and alkalinity for carbon, and phosphate for nutrients. DIC and alkalinity are used to represent the carbon system because they are total carbon and charge quantities that are independent of temperature and pressure (i.e. they do not vary in concentration because of temperature differences between surface and deep waters as do their constituents, HCO_3^-, CO_3^{2-} and CO_2). There is also a thermodynamic temperature dependence between these chemical parameters and f_{CO_2}, and they change in clearly defined ways relative to nutrients during biological transformations. Dissolved inorganic phosphate (we use the symbol DIP here, but it is often represented as simply PO_4^{3-}) represents nutrients in the model. We could also have chosen nitrate (NO_3^-); however, we avoid the complications of nitrogen fixation and denitrification by choosing phosphate, and biological uptake and release of both nutrients are related by a constant ratio (Fig. 6.2). Micronutrients like iron are not considered in this very simplified model. The deep concentrations of alkalinity, DIC and DIP are the mean values for the world's ocean: 2371 eq kg^{-1}, 2258 mol kg^{-1} and 2.2 mol kg^{-1}, respectively.

Model geometry and parameters are compiled in Fig. 11.2. The strategy is to determine the concentrations of alkalinity and DIC in the surface ocean layer from the equations that describe the model dynamics and the deep ocean values, and then to calculate the f_{CO_2} in thermodynamic equilibrium with these values at the mean surface water temperature and salinity. Carbonate plus borate alkalinity, $A_{C\&B}$, will be employed in our carbonate equilibrium calculations. (See Chapter 6 for a discussion of the carbonate equilibrium calculations.)

The change in concentration of a dissolved constituent, [C], in the deep reservoir is:

$$V_D \frac{d[C]}{dt} = 0 = v_m([C]_S - [C]_D) + J_C \qquad (11.1)$$

where V_D is the deep reservoir volume, v_m is the mixing rate ($m^3\,y^{-1}$), $[C]_S$ and $[C]_D$ are the surface and deep concentrations, respectively (mol m^{-3}) and J_C is the biological flux to the deep reservoir in units of mol y^{-1}. Because this is a steady-state calculation we assume the left side of Eq. (11.1) is equal to zero. To a good first approximation, the only concentration variable is $[C]_S$ because the volume of the deep ocean is so much greater than the surface that the deep concentration cannot change appreciably from the total average values. One can write three equations of this type for DIP, DIC and $A_{C\&B}$. These equations are related through the biological flux terms, J_C, and the stoichiometric ratios, r, described later in the Section 11.2.2.

11.2.1 The solubility pump

The results of calculations to determine the sensitivity of atmospheric f_{CO_2} to changes of the solubility and biological pumps are presented in Table 11.2. An accurate model estimate of change requires a much more elaborate resolution of surface temperature, nutrient concentration and circulation than is possible with this crude model, but the basic principles and trends can be demonstrated. The *standard case* is determined by using a weighted mean surface temperature of 20 °C. This is a crude estimate, as equatorial mean values are 27 °C, subtropical values are on annual average 23 °C and high-latitude annual mean values are <10 °C. The surface-water phosphate value also represents a weighted mean between near zero values in subtropical surface waters and higher values in the Equator and high latitudes. The $[DIP]_S$ value of 0.5 mol kg^{-1} is chosen to generate an f_{CO_2} value that is near that measured in the atmosphere by using the calculation procedures described in the next section.

Changes in f_{CO_2} brought about by solubility are caused by the temperature dependence of the equilibrium constants. The model

Table 11.2. *The effect of the solubility and biological pumps on the fugacity of CO_2 in the atmosphere, f_{CO_2}, determined by the simple two-layer ocean model depicted in Fig. 11.2*

The first row is the standard case and the rows under this indicate changes due to temperature, carbon flux, circulation rate and the organic carbon to $CaCO_3$ ratio of the particle flux, OC : $CaCO_3$.

Case	Temp °C	$[DIP]_S$ µmol kg^{-1}	τ_{mix} y	$R_{OC:CA}$	DIC_S µmol kg^{-1}	$A_{T,S}$ µeq kg^{-1}	f_{CO_2} atm
Standard	20	0.5	1000	3.5	2027	2296	375
Temp. effect	15						304
	25						460
Biol. pump							
Carbon flux	20	2.2			2258	2371	1184
		0.0			1959	2274	293
Circulation		0.85	500		2074	2312	446
		0.0	1500		1959	2274	291
OC:$CaCO_3$ (P:OC = 106)		0.5	1000	10:1	2059	2361	337
				1.5:1	1957	2157	485

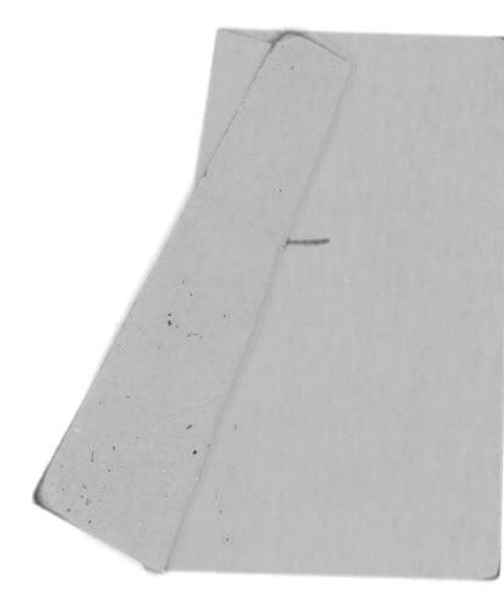

results indicate that atmospheric f_{CO_2} in equilibrium with seawater decreases when the temperature of the water is colder. Although the temperature response of CO_2 depends on all three carbonate equilibrium constants, K_H, K'_1 and K'_2, and the borate equilibrium constant, K_B, (Chapter 4) the combined temperature dependence of CO_2 solubility trends in the same direction as that for unreactive gases: colder water can accommodate more CO_2. Atmospheric f_{CO_2} changes by about 15 µatm per degree change in surface water temperature. This result suggests that global warming has a positive feedback: higher temperatures will expel more CO_2 from the ocean. Decreases in global temperature during the last ice age would tend to draw down atmospheric f_{CO_2} as more would have been dissolved in the ocean. This is in the direction observed for f_{CO_2} changes during glacial periods (Chapter 7), but the change in our simple model is about twice that of more realistic global models. It has been shown in many different more realistic models that a reduction in global ocean surface temperatures of 3–5 degrees (coupled with an increase in salinity due to the change in global ice volume) can account for a decrease in atmospheric f_{CO_2} of between 20 and 30 ppm out of a measured decrease of about 80 ppm (Sigman and Boyle, 2000).

Probably the most important dynamic of the solubility pump that the simple model in Fig. 11.2 misses is a large horizontal temperature gradient in the surface ocean. Water upwelled at the Equator is heated and degasses to the atmosphere as it flows north and south. At high latitudes the same water is cooled to the point where it reabsorbs some of the CO_2 it lost to the atmosphere before it down-wells to the ocean interior. Thus, any change in the f_{CO_2} of the atmosphere resulting from global temperature changes and the consequent effect on the solubility pump would also depend on the latitudinal distribution of the temperature change. In spite of this weakness the sense of the relative changes to expect in the solubility pump are clear from the simple model.

11.2.2 The biological pump

Modern studies of the biological pump began with investigations of the mechanism of surface to deep carbon transfer used in box models (e.g. Broecker, 1971) together with the conceptual link between the experimental investigations of particle fluxes and biological production (Eppley and Peterson, 1979). These studies approached the problem from the points of view of both biological and chemical oceanography and have become in recent years prime examples of interdisciplinary oceanographic investigations.

In order to determine the sensitivity of atmospheric f_{CO_2} to the flux of biological particles, J, in Fig. 11.2, one must establish the chemical content of the sinking material. More precisely, one needs to know how the degradation and dissolution of the particulate matter changes the water chemistry of the deep ocean. In the discussions of life in the ocean (Chapter 6) we established that the Redfield ratios for changes in P, N, and C during organic matter production and degradation are close to their traditional values:

$$\Delta P : \Delta N : \Delta C_{organic} = 1 : 16 : 106. \tag{11.2}$$

For every mole of DIP added to the water by organic matter degradation, 106 moles of C (as DIC) are added. The oxidation of NH_3 to NO_3^- generates a proton and thus decreases alkalinity by one equivalent for each of the 16 moles of NH_3 oxidized per mol of PO_4 added during organic matter degradation. This creates a stoichiometry for change in deep-water DIC and alkalinity with respect to phosphorus by organic matter degradation of:

$$\Delta P : \Delta N : \Delta DIC : \Delta A_{C\&B} = 1 : 16 : 106 : -16. \tag{11.3}$$

Calcium carbonate is the only other constituent of biologically produced matter other than organic carbon and nitrogen that creates a substantial change in the DIC and alkalinity of seawater. Observed alkalinity : DIC ratio changes in the ocean below the euphotic zone require degradation of organic carbon and dissolution of $CaCO_3$ in a ratio of about 10 to explain the surface to deep Atlantic changes and between 1 and 2 to explain the trends following the deep ocean circulation (between 2–4 km depth) (Fig. 4.6). For the purposes of a two-layer ocean one needs a composite value for the ocean below 100 m, which skews the average OC : $CaCO_3$ ratio in sinking particulate material toward those below the thermocline, and the ratio for this model becomes 3.5 : 1. Thus, one mole of $CaCO_3$ is produced or dissolves for every 3.5 moles of organic matter produced or degraded. The DIC change attributed to $CaCO_3$ for one mole of phosphorus change is 106/3.5 = 30. Since production or dissolution of $CaCO_3$ creates a 2 : 1 change in Alk : DIC, the composite stoichiometric change for the modern ocean in our model becomes:

$$\Delta P : \Delta N : \Delta DIC : \Delta A_{C\&B} = 1 : 16 : 136 : 44. \quad (11.4)$$

To determine the surface ocean concentration of DIC and $A_{C\&B}$ as a function of surface–deep gradient of DIP, one can write three steady-state versions of Eq. (11.1) for DIP, DIC and $A_{C\&B}$. These equations are related through the biological flux terms J_C, and the stoichiometric ratios, r, given in equations like that in (11.4):

$$J_{DIC} = J_P \, r_{DIC:P} \, ; \, J_A = J_P \, r_{A:P}. \quad (11.5)$$

By combining the phosphorus equation with those for alkalinity and DIC, the ratio v_m/J_C can be eliminated to demonstrate the steady-state surface–deep water differences for DIP, DIC, and $A_{C\&B}$:

$$DIC_S - DIC_D = \left([DIP]_S - [DIP]_D \right) r_{DIC:P} \quad (11.6)$$

$$A_{C\&B,\,S} - A_{C\&B,\,D} = \left([DIP]_S - [DIP]_D \right) r_{A_T:P}. \quad (11.7)$$

Assuming that the mean deep-water concentrations are constants because this reservoir is so large compared with the surface layer, one can determine the surface-water values of DIC and $A_{C\&B}$ as a function of the surface–deep differences in DIP. When this is done the carbonate equilibrium equations (Chapter 4, Fig. 11.2) can be used to determine f_{CO_2}.

The response of our simple ocean and atmosphere model to changes in the biological pump (Table 11.2) is manifested in three ways. The best indicator of the effect of the biological pump is the concentration difference in phosphate between the surface and deep ocean. Because of Eq. (11.1), any change in the ratio of the particle export to surface–deep mixing, v_m/J_C, must be accompanied by a change in the surface–deep differences. In response to an increase in particle flux or a decrease in circulation the gradient in limiting nutrient concentration between the surface and deep ocean must increase. The biological pump (particle flux) change scenarios in Table 11.2 represent the most extreme cases in surface ocean nutrient concentrations in which the DIP concentration in surface waters is either equal to the deep value or zero. The former case is the "Strangelove Ocean" in which there is no biological pump at all. This is the chemical equilibrium ocean. Something like this may have happened at the time of the last mass extinctions in geologic history. Sixty million years ago at the Cretaceous–Tertiary boundary it is believed that a meteor collided with the Earth and 80% of existing species became extinct. Without the biological carbon pump, the model predicts that the atmosphere would have an f_{CO_2} of about 1000 atm. The actual value may not be accurate because of the simple model architecture, but the trend toward increasing f_{CO_2} is clear. In the other extreme, when the nutrients in surface waters are entirely depleted, the maximum effect of the biological pump is realized. In our three-box model this causes an 80 ppm decrease from the standard case. Given this large effect, one can readily understand why marine scientists and paleoclimatologists have focused on the change in efficiency of high-latitude nutrient utilization in surface waters as a means for explaining past changes in atmospheric f_{CO_2} (see Chapter 7).

If the particle export remains unchanged but the mixing rate is altered, the DIP gradient between surface and deep waters changes. This is demonstrated in the biological pump (circulation) section of Table 11.2 by varying the residence time of the water in the deep ocean between 500 and 1500 y. If the residence time becomes smaller (i.e. circulation faster), the gradient from surface to deep concentrations becomes smaller and atmospheric f_{CO_2} increases, creeping in the direction of the chemical equilibrium ocean. The opposite is true if circulation decreases. In this case atmospheric f_{CO_2} drops until all the nutrients are consumed in the surface waters and then the surface–deep gradient can increase no more. When surface nutrients are totally depleted, any decrease in mixing must be matched exactly by a decrease in J_C to maintain a constant v_m/J_C ratio.

As long as the nutrient concentration in surface waters is greater than zero the biological term, J_C, and the mixing term, v_m, are independent: changes in either term can cause a change in the atmospheric f_{CO_2}. This case is analogous to today's ocean at high latitudes, where productivity and the biological pump are limited by factors other than the flux of nutrients from the deep ocean to surface waters. Surface nutrient concentrations are high in these regions because the probability of their removal by biological production is of the same magnitude as exchange with deeper waters by mixing. As long as our simple two-layer model has finite nutrient concentration in the surface ocean, its response is most analogous to today's high-latitude ocean. On the other hand, when nutrient concentrations in the surface ocean are totally depleted so that the biological pump is limited by the flux of nutrients from deep waters to the surface ocean, there is no atmospheric response to circulation changes. The reason for this insensitivity is that when surface nutrients are totally depleted, the surface ocean–deep ocean nutrient gradient is constant, so a change in the nutrient flux to the surface must be exactly compensated by changes in the particle export rate (see Eq. (11.1)). This might be similar to the situation in the subtropical oceans today, where nutrient concentrations in surface waters are very low.

The dependence of f_{CO_2} on the chemical character of the biological debris that is produced in the surface ocean is demonstrated in the model by the last sensitivity analysis in Table 11.2 (biological pump – OC : $CaCO_3$). If the organic carbon to $CaCO_3$ ratio of the particulate material, $r_{OC:Ca}$, becomes greater than in the standard case (poorer in $CaCO_3$) while maintaining the same phosphate : organic matter ratio, there will be less alkalinity removed from surface waters per mole of DIC removed. Since the $A_{C\&B}$ – DIC ratio in surface waters is about equal to $[CO_3^{2-}]$ (see Chapter 4), this difference and the carbonate ion concentration will become greater under this scenario. Higher CO_3^{2-} concentrations mean more basic waters and lower f_{CO_2} (Fig. 4.2). Changing the particulate carbon flux ratio $r_{OC:Ca}$ from 3.5 to 10 in the model decreases the atmospheric f_{CO_2} by about 40 μatm.

If the particulate matter carbonate : carbon ratio changed in the opposite sense to near unity (i.e. a much greater flux of $CaCO_3$ relative to organic matter), this would decrease the alkalinity of surface waters relative to DIC because the formation and removal of $CaCO_3$ decreases $A_{C\&B}$ and DIC in a ratio of 2 : 1. A greater decrease in $A_{C\&B}$ relative to DIC would make the $A_{C\&B}$ – DIC difference smaller, decreasing the surface water CO_3^{2-} concentration and increasing f_{CO_2}. In the model a decrease in the $r_{OC:Ca}$ ratio from 3.5 to 1.5 causes an increase in f_{CO_2} by 110 μatm.

Formation and removal of organic matter and $CaCO_3$ have opposite effects on the f_{CO_2} of the atmosphere. An organic carbon-rich flux from the euphotic zone decreases atmospheric CO_2, whereas a $CaCO_3$-rich flux increases it. Thus, the ratio of coccolithophorids ($CaCO_3$-secreting autotrophic plankton) to other non-$CaCO_3$-secreting organisms that grow in the ocean's surface waters has important consequences for the f_{CO_2} in the atmosphere. If surface waters become more hostile to $CaCO_3$-secreting organisms because of increases in anthropogenic CO_2, as some controlled experiments on the effects of lower marine surface-water pH suggest, and the organic carbon portion of the biological pump remains the same, the ocean will respond with a negative feedback, tending to lower the f_{CO_2} of the atmosphere by slowing down the carbonate-related portion of the biological pump. Paleoceanographers search deep-sea sediment cores for evidence of changes in the rates of burial of diatoms (opal shells) and coccolithophorids through geologic time because of their potential ramifications for past levels of atmospheric CO_2. However, relating the preservation of carbonate and opal shells to their rain rates from the euphotic zone has so far been achieved only qualitatively because a large portion of the tests of both these phytoplankton taxa dissolve before they are buried in marine sediments.

Results from the simple ocean–atmosphere model in Fig. 11.2 and Table 11.2 demonstrate the fascinating ramifications of changing sea-surface temperatures and the interplay between biological fluxes and ocean circulation for atmospheric f_{CO_2}. Present coupled atmosphere–ocean models suggest that a doubling of atmospheric f_{CO_2} would result in a 2–4 °C mean global temperature increase. Ocean circulation and sea-surface ecology are also sensitive to temperature and f_{CO_2} changes, which will provide feedbacks that oceanographers are striving to understand quantitatively.

11.3 The fate of anthropogenic CO_2 in the ocean

Great advances in our understanding of the fate of fossil fuel CO_2 began with two seminal studies: the onset of monitoring the f_{CO_2} of the atmosphere by Dave Keeling, starting in 1958 (see, for example, Keeling, 1960) and studies of the CO_2 buffering capacity of the ocean (Revelle and Suess, 1957). These investigations demonstrated that only a portion of the CO_2 introduced to the atmosphere was accumulating there, and that the ocean's surface layer had limited potential to adsorb the rest. Subsequent descriptions of the dynamics of the process of CO_2 adsorption by the ocean by using a series of box and multi-layer models were forerunners for introducing chemistry into present-day global circulation models (GCMs).

Even though the magnitude of the flux controlled by the biological pump (10–15 Pg y^{-1}) is similar to the anthropogenic CO_2 flux to the atmosphere (*c.* 6 Pg y^{-1}), it is by no means clear that the latter has any effect on the former. Since the limiting factors for biological organic carbon export from the surface ocean are nutrients and light, there is not a direct effect of anthropogenic CO_2 contamination on the rate of biological processes. However, warming of the Earth caused by greenhouse gas increase may change the circulation of the ocean and atmosphere, and this could in turn affect the rate of nutrient delivery to the ocean's euphotic zone. The uncertainties in the processes involved are presently so great that we are not sure of the magnitude or even the direction of this potential effect. We shall assume here that the processes controlling anthropogenic CO_2 uptake in the ocean are physical and thermodynamic, and that there is no effect on the biological pump.

The atmospheric burden of anthropogenic CO_2 is very well known because it has been measured since 1958; and recent ice core studies indicate how the atmospheric CO_2 evolved between the pre-industrial period and 1958 (Fig. 11.3). Because industries and governments keep records of the amount of fossil fuel recovered, this number is also known. The amount of atmospheric CO_2 increase over the period 1860–1989 is estimated to be only 40% of total emissions (Table 11.1). Where has the rest gone?

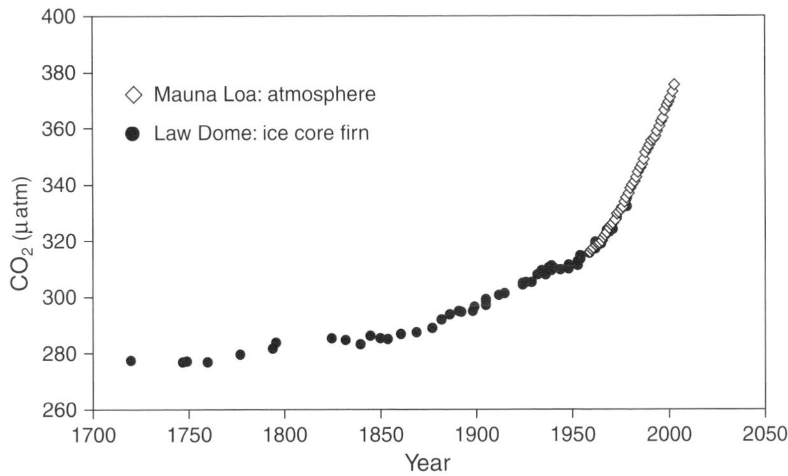

Figure 11.3. The atmospheric CO_2 record at Mauna Loa observatory over the past *c.* 50 years (1958 to present) along with the CO_2 record for pre-industrial time to 1960 as preserved in the Law Dome ice core in East Antarctica.

The likely storage locations for anthropogenic CO_2 are the large global carbon reservoirs that turn over on time periods of decades to centuries. These are DIC in the ocean and organic matter in the terrestrial biosphere (Table 11.1, Fig. 11.1). It is logical that some of the anthropogenic CO_2 put into the atmosphere will invade the ocean by air–sea gas exchange. One might expect the same to be true for the terrestrial biosphere; however, people have had a much greater impact on this carbon reservoir than on that of the ocean. Deforestation releases carbon to the atmosphere and oceans whereas the opposite response, net removal of CO_2 to the terrestrial reservoir, is expected from "greening" of the biosphere. *Greening* refers to anthropogenically enhanced photosynthesis caused by increased atmospheric CO_2 concentrations, fertilization by nutrient loading, or temperature increases. The magnitude of the terrestrial effect is difficult to measure because this reservoir is extremely heterogeneous. Although the ocean's carbon reservoir is more homogeneous, the impact of anthropogenic CO_2 on the DIC is subtle, and only recently have adequate methods been developed to measure the effect.

We begin our discussion of the ocean's role in removal of anthropogenic CO_2 added to the atmosphere by demonstrating the equilibrium chemical response to an atmospheric perturbation and then discuss the methods that have been used to measure the CO_2 uptake by the oceans. Finally, we will demonstrate the role of both the ocean and terrestrial biosphere in sequestering anthropogenic CO_2 during the past 25 years.

11.3.1 The carbonate buffer factor (Revelle factor)

The first stage of describing the role of the ocean in adsorbing anthropogenic CO_2 is to determine the effect of the carbonate equilibrium between the atmosphere and surface waters of the ocean. Since the ocean on average mixes to a depth of about 100 m annually, this part of the ocean's dissolved inorganic carbon reservoir should be nearly in chemical equilibrium with the atmosphere on time scales of one year. The response of the surface ocean carbonate system to changes in f_{CO_2} involves the carbonate buffer system and a number that has come to be known as the "Revelle factor." To demonstrate the response of marine surface water DIC to an increase in atmospheric f_{CO_2}, consider the huge 1 m × 1 m × 10 100 m box shown in Fig. 11.4. The box contains 10 000 m^3 of air, V_{atm}, and 100 m^3 of seawater, V_{ocean}, and is maintained at a temperature of 20 °C and a pressure of 1 atm at the air–water interface. The seawater inside has an alkalinity, $A_{C\&B}$, of 2300 $\mu eq\ kg^{-1}$ and a DIC concentration of 2000 $\mu mol\ kg^{-1}$. The atmosphere and ocean are at chemical equilibrium. As we learned in Chapter 4, all species of the carbonate system at chemical equilibrium can be computed if two quantities are known. We use the carbonate equilibria described in Chapter 4 and the Matlab program given in Appendix 4.1 to determine that the equilibrium CO_2 concentration in the water is 11.3 $\mu mol\ kg^{-1}$. If the atmosphere and ocean are in solubility equilibrium, Henry's Law can be used to

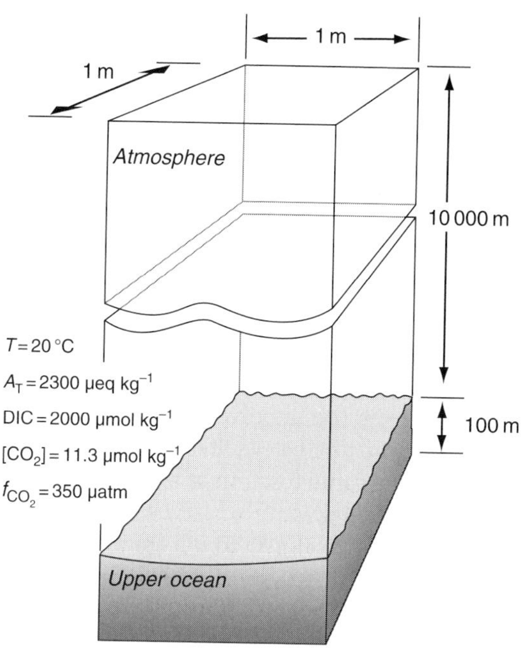

Figure 11.4. Sketch of a tall box with 10 km of air and 100 m of seawater with air and water carbon concentrations indicated. This hypothetical system is used to illustrate the effect of carbonate reactions in controlling the distribution of a perturbation of CO_2 between the atmosphere and ocean at chemical equilibrium.

calculate that the f_{CO_2} of the atmosphere is 350 ppm. The total inventory of carbon in the atmosphere and water in the box in moles is:

$$\sum C = f^{a}_{CO_2} \cdot M_{atm} + DIC \cdot V_{ocean} \cdot \rho, \tag{11.8}$$

where M_{atm} is the number of moles of air in the atmosphere box and ρ (kg l^{-1}) is the density of seawater. If we assume that the pressure at the air–water interface in the box is one atmosphere (1 atm = 101 325 Pascals, Pa or kg m^{-1} s^{-2}), we can calculate the number of moles this represents by dividing by the acceleration due to gravity (9.8 m s^{-2}) and the molecular mass of the atmosphere (0.029 kg mol^{-1}) (M_{atm} = 101 325 / (9.8 × 0.029) = 3.5 × 10^5 mol$_{atm}$). The total amounts of carbon in the atmosphere and seawater are:

$$\sum C_{atm} = (0.35 \times 10^{-3}\ \mathrm{mol\,C\,mol^{-1}_{atm}})(3.5 \times 10^5 \mathrm{mol_{atm}}) = 122\ \mathrm{mol\,C}; \tag{11.9}$$

$$\sum C_{DIC} = (2.0 \times 10^{-3}\ \mathrm{mol\ kg^{-1}})(10^5\ \mathrm{l})(1.023\ \mathrm{kg\,l^{-1}}) = 200\ \mathrm{mol\,C}; \tag{11.10}$$

and the inventory of the CO_2 gas in the water is:

$$\sum C_{SW,CO_2} = (11.3 \times 10^{-6}\ \mathrm{mol\ kg^{-1}})(10^5\ \mathrm{l})(1.023\ \mathrm{kg\,l^{-1}}) = 1.3\ \mathrm{mol\,C}. \tag{11.11}$$

The amount of carbon in one square meter of seawater 100 m deep is of the same magnitude as that in 10 km of air at 1 atm (compare Eqs. (11.9) and (11.10)), which is analogous to the situation in the real ocean mixed layer and atmosphere.

Now, we inject 20 mol of CO_2 into the atmosphere of the box and ask how the added carbon is distributed between the atmosphere and ocean once the system re-establishes a new chemical equilibrium. If the CO_2 behaved like an inert gas (no chemical reactions) then the added carbon would be distributed following Henry's Law and the fraction of the added carbon that would enter the seawater, *f*, would be the same as the CO_2 gas distribution in the original equilibrium:

$$f = \frac{\sum C_{SW,\,CO_2}}{\sum C_{atm} + \sum C_{SW,\,CO_2}} = \frac{1.3\ \text{mol}}{122\ \text{mol} + 1.3\ \text{mol}} = 0.010. \quad (11.12)$$

Only 1.0 percent of the CO_2 added would enter the ocean.

If, on the other hand, the added carbon is eventually distributed equally among the dissolved carbon molecules in the water (i.e. the added carbon divides itself among HCO_3^-, CO_3^{2-} and CO_2 so that they have the same fraction of anthropogenic CO_2), then the fraction that enters the seawater would be much greater. This situation is the same as would be expected if a tracer of carbon too small to change the ocean chemistry were added to the atmosphere, say a small amount of $^{14}CO_2$, and allowed to equilibrate with the water. The fraction that would enter the seawater in this case would be:

$$f = \frac{\sum C_{DIC}}{\sum C_{DIC} + \sum C_{atm}} = \frac{200\ \text{mol}}{200\ \text{mol} + 123\ \text{mol}} = 0.62. \quad (11.13)$$

Sixty-two percent of the carbon would end up in the seawater in this scenario.

In fact, neither of these estimates is correct because the amount of carbon that ultimately resides in the seawater is determined by chemical equilibrium in the water; not simply Henry's Law or a tracer-like distribution of the carbon among the constituents of DIC in the water. What we want to know is the equilibrium change in DIC for a given change in f_{CO_2}. This value can be determined from the carbonate equilibrium program in Appendix 4.1 for a given temperature and $A_{C\&B}$ by making small changes in the DIC and determining the change in f_{CO_2}. The change in carbon content of the reservoirs in Fig. 11.4 is given by:

$$\Delta\sum C = \Delta\sum f_{CO_2,\,atm} + \Delta\sum \text{DIC} = \Delta f^{a}_{CO_2}\, M_{atm} + \Delta\text{DIC} \times V_{ocean}\,\rho. \quad (11.14)$$

Defining the fraction, *f*, of the CO_2 taken up by the seawater in the box as the ratio of the changes in the seawater to those in the atmosphere plus seawater gives:

$$f = \frac{\Delta\text{DIC} \times V_{ocean}\,\rho}{(\Delta\text{DIC} \times V_{ocean}\,\rho) + \left(\Delta f^{a}_{CO_2}\, M_{atm}\right)}. \quad (11.15)$$

The fractional value can be calculated because we know the relative changes in DIC and $f^{a}_{CO_2}$ at chemical equilibrium for a given $A_{C\&B}$, f_{CO_2}, temperature, and pressure. Pilson (1998) referred to this value as the uptake factor, *UF*:

$$UF = \frac{\Delta \mathrm{DIC}}{\Delta f_{\mathrm{CO_2}}} (\mathrm{mol\ kg^{-1}/atm}) \tag{11.16}$$

which is very straightforward and easy to understand, but not very generally usable because of the units. The value was originally presented as a fractional change in DIC and f_{CO_2} at equilibrium by Revelle and Suess (1957) in their classic paper on the fate of fossil fuel CO_2, and later called the Revelle factor, R_{Rev}, by Broecker and Peng (1982):

$$R_{\mathrm{Rev}} = \left(\frac{\Delta f_{\mathrm{CO_2}}}{f_{\mathrm{CO_2}}}\right) \Big/ \left(\frac{\Delta \mathrm{DIC}}{\mathrm{DIC}}\right). \tag{11.17}$$

Uptake and Revelle factors for the conditions of our experiment in Fig. 11.4 and as a function of f_{CO_2} are presented in Fig. 11.5. For $f_{CO_2} = 350\ \mu\mathrm{atm}$, $UF = 0.6\ \mu\mathrm{mol\ kg^{-1}\ \mu atm^{-1}}$ and $R = 9.6$. Combining Eqs. (11.15) and (11.16) gives:

$$\begin{aligned} f &= \frac{UF \times V_{\mathrm{ocean}}\ \rho}{(UF \times V_{\mathrm{ocean}}\ \rho) + (M_{\mathrm{atm}})} \\ &= \frac{0.6\ \mathrm{mol_{atm} kg^{-1}} \times 100\ \mathrm{m^3} \times 1023\ \mathrm{kg\,m^{-3}}}{0.6\ \mathrm{mol_{atm} kg^{-1}} \times 100\ \mathrm{m^3} \times 1023\ \mathrm{kg\,m^{-3}} + 3.5 \times 10^5 \mathrm{mol_{atm}}} = 0.15. \end{aligned} \tag{11.18}$$

About 15% of the carbon added to the system is ultimately sequestered in the ocean. This is between what would be expected for the cases of the inert gas and tracer distribution among carbon atoms. More CO_2 finds its way into the ocean than in the inert gas case because of the reactions of the carbonate system. Most of the CO_2 that enters the ocean from the atmosphere is consumed by reaction with carbonate ion:

$$CO_2 + CO_3^{2-} + H_2O \leftrightarrow 2HCO_3^-,$$

thereby making way for more CO_2 to enter via gas exchange. In terms of the Revelle factor, a value near $R_{Rev} = 10$ indicates that the fractional change in f_{CO_2} at equilibrium is about 10 times that of the DIC. Since there is about 200 times more DIC than $[CO_2]$ this represents a carbon uptake by the water that is almost 20 times greater than if there were no chemical reactions.

Notice that in Fig. 11.5 there is a factor of three decrease in *UF* and a 25% decrease in R_{Rev} as the f_{CO_2} increases from preanthropogenic levels of 280 μatm to 550 μatm. The decrease in *UF* indicates a decrease in the efficiency of CO_2 absorption as the f_{CO_2} of the ocean and atmosphere increase. One way to think about the reason for the decreased efficiency is that with decreasing pH the concentration of $[CO_3^{2-}]$ decreases so there is less chemical reaction (see Fig. 4.2).

We can now use the knowledge of the Revelle factor to estimate the anthropogenic CO_2 uptake by the ocean at equilibrium. Presuming that after some reasonable time the ocean carbonate system comes into equilibrium or nearly so with CO_2 in the atmosphere, we can calculate the fraction of the anthropogenic CO_2 taken up by the ocean as a function of the depth of the layer into which

Figure 11.5. The change in the uptake factor (*UF*) and Revelle factor (*R*) as a function of the f_{CO_2} in equilibrium with a seawater solution containing a total alkalinity of 2300 μeq kg^{-1}.

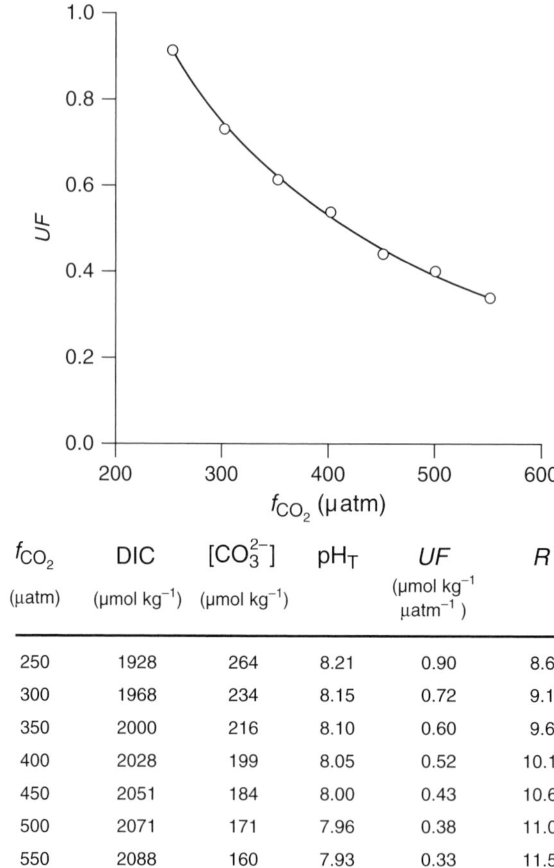

f_{CO_2} (μatm)	DIC (μmol kg^{-1})	$[CO_3^{2-}]$ (μmol kg^{-1})	pH_T	*UF* (μmol kg^{-1} μatm^{-1})	*R*
250	1928	264	8.21	0.90	8.6
300	1968	234	8.15	0.72	9.1
350	2000	216	8.10	0.60	9.6
400	2028	199	8.05	0.52	10.1
450	2051	184	8.00	0.43	10.6
500	2071	171	7.96	0.38	11.0
550	2088	160	7.93	0.33	11.5

anthropogenic CO_2 has equilibrated, h, following exactly the reasoning in the above experiment. The fraction of anthropogenic CO_2 in the ocean is:

$$f_{ocean} = \frac{\sum_{ocean} \times (\Delta DIC_{anthro}/DIC)}{\sum_{atmos} \times (\Delta f_{CO_2,\ anthro}/f_{CO_2}) + \sum_{ocean} \times (\Delta DIC_{anthro}/DIC)}, \tag{11.19}$$

where Σ_{ocean} is the ocean inventory of DIC and Σ_{atmos} is the atmospheric inventory of CO_2:

$$\begin{aligned}\sum_{ocean} &= [DIC] \times area_{ocean} \times h = (2\,mol\,m^{-3})\,(3.6 \times 10^{14} m^2)h \\ &= 7.2 \times 10^{14}\,mol \times h.\end{aligned} \tag{11.20}$$

$$\begin{aligned}\sum_{atmos} &= mol\,CO_2\,(mol\,atm)^{-1} \times (mol\ atm) \\ &= \left(350 \times 10^{-6}\ mol\,CO_2\,(mol\ atm)^{-1}\right)(1.77 \times 10^{20}\ mol\,atm) \\ &= 6.2 \times 10^{16}\,mol\,C.\end{aligned} \tag{11.21}$$

Using the definition of the Revelle factor and rearranging:

$$f_{\text{ocean}} = \frac{\sum \text{ocean}}{\sum \text{atmos} \cdot R_{\text{Rev}} + \sum \text{ocean}} = \frac{7.2 \times 10^{14}\ h}{6.2 \times 10^{16} \cdot R + 7.2 \times 10^{14}\ h}. \tag{11.22}$$

This fraction is plotted as a function of the layer depth, h, in Fig. 11.6. As the mean depth of the ocean mixed layer is on the order of 100 m, we see from the plot that at chemical equilibrium the surface ocean will take up approximately 10% of the CO_2 added to the atmosphere. This is nearly the same as that calculated in the box analogy and might be viewed as an absolute lower limit of the amount of anthropogenic CO_2 that penetrates the ocean because the global average mixed layer depth is about 100 m. It is clear that in some areas of the oceans transient tracers such as bomb-produced ^{14}C and ^{3}H (with even shorter life times than fossil fuel CO_2), have penetrated through the mixed layer well into the thermocline (see Fig. 5.16). In order for the ocean to take up half of the fossil fuel CO_2 (roughly that amount missing from the atmosphere) the upper 700–800 m of the ocean would have to be in equilibrium with the atmosphere (Fig. 11.6). Thus, the limiting factor for the uptake of fossil fuel CO_2 by the ocean is the rate of mixing into the thermocline rather than gas exchange or the approach to chemical equilibrium in the water.

We have so far assumed that the atmosphere and ocean are in equilibrium. This of course cannot be the case, as the CO_2 content of the atmosphere is rising rapidly, so there must be a gradient in f_{CO_2} across the air-water interface to drive the flux into the ocean. The concepts of the Revelle and uptake factors are also useful in determining the response time of the upper ocean to perturbations of gas and isotope changes in the atmosphere. As discussed in Chapter 2, the residence time of a substance is equal to its reservoir size divided by the flux in or out. For an inert gas, C, in the upper ocean the time for (1–1/e) of the gas to be exchanged by flux across the air-water

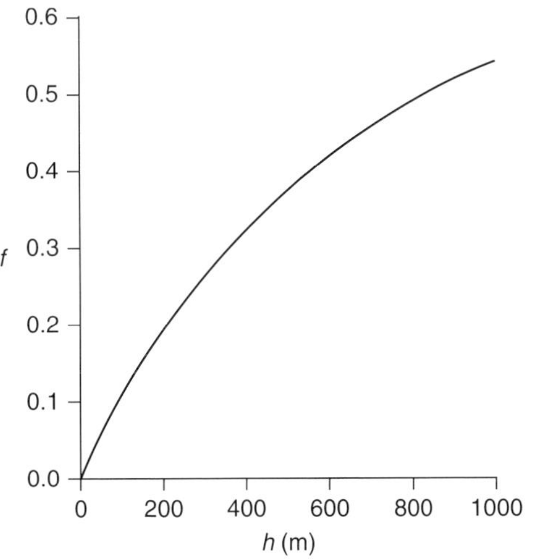

Figure 11.6. The fraction, f, of CO_2 added to the surface ocean–atmosphere system that would go into the ocean as a function of the depth, h, of the ocean surface layer that is in equilibrium with atmospheric f_{CO_2}.

interface is the concentration, [C], times height, h, of the mixed layer divided by the one-way gas exchange flux:

$$\tau_{inert} = \frac{h \cdot [C]}{F_{air-water}} = \frac{h \cdot [C]}{G \cdot [C]} = \frac{h}{G}, \tag{11.23}$$

where G is the gas exchange mass transfer coefficient and has units of length / time (Chapter 10). We use the one-way flux because all the gases are being exchanged, not just the fraction that is over- or undersaturated. Even in a situation where the air and ocean are exactly in saturation equilibrium, gases in each reservoir will have a finite residence time with respect to exchange. Using a mean gas exchange mass transfer coefficient of $5\,m\,d^{-1}$ and an ocean mixed layer depth of 100 m results in a residence time of 20 d (less than one month). If CO_2 were an inert gas this would be the residence time:

$$\tau_{inert} = \frac{h}{G} = \frac{100}{5} = 20\ d. \tag{11.24}$$

The true reservoir size for carbon dioxide, however, is larger than its concentration, $[CO_2]$, because of the carbonate system reactions. The ratio of the reservoir of DIC that exchanges carbon to the CO_2 gas reservoir in the mixed layer is $\Delta DIC / \Delta[CO_2]$. Thus, the residence time for carbon in the surface ocean with respect to gas exchange from Eq. (11.23) is:

$$\tau_{CO_2} = \frac{h[CO_2]}{G[CO_2]} \times \frac{\Delta DIC}{\Delta[CO_2]} = \frac{h}{G}\left(\frac{DIC}{R[CO_2]}\right) = 20\left(\frac{DIC}{R[CO_2]}\right), \tag{11.25}$$

where the value in the last term was taken from the Revelle factor equation (11.16), by substituting $[CO_2]$ for f_{CO_2}.

Using values of DIC, R, and $[CO_2]$ for the case in Fig. 11.4:

$$\tau_{CO_2} = 20 \times 18.4 = 368\ d. \tag{11.26}$$

The renewal time is about one year for a mixed layer 100 m deep! This indicates that the surface ocean anthropogenic increase should lag that in the atmosphere, but not by much because the anthropogenic perturbation is a decadal to centuries-long process.

One can also estimate the renewal time for the isotopes of carbon, ^{13}C and ^{14}C, by using this simple reasoning. The response time will be different again because isotopes must exchange with the entire carbon reservoir. This reasoning follows that for the tracer case described earlier (Eq. (11.13)), since the individual carbon molecules are tagged with their isotope signature, and we are asking how long it takes for the ^{13}C of the DIC to come to a new steady state in which the entire DIC reservoir must be considered. Thus,

$$\tau_{isotope} = \frac{h \cdot [DIC]}{G \cdot [CO_2]} = \frac{20 \times [DIC]}{[CO_2]} = 20 \times 177 = 3540\ d. \tag{11.27}$$

The response time for the upper ocean with respect to changing its isotopic values is about ten times longer than that for CO_2 and almost

200 times that for an inert gas ($\tau_{isotope} = 10$ y!) because of the size of the DIC reservoir relative to the rate of CO_2 gas transfer.

11.3.2 Methods of measuring anthropogenic CO_2 uptake

Estimates of the amount of anthropogenic CO_2 presently in the ocean have been determined in four main ways. The first method is to distinguish the fraction of the measured DIC increase that comes from invasion of anthropogenic CO_2. This is accomplished by calculating the natural increase in DIC with depth from nutrient and oxygen data with the assumption of constant metabolic ratios. The difference between the calculated value and that measured is the anthropogenic perturbation. The second method involves calculation of the present-day global flux of CO_2 across the ocean–atmosphere interface from observations of the degree of saturation of f_{CO_2} and estimates of gas transfer velocities. This calculation involves global measurements of the CO_2 fugacity in surface waters, $f^{w}_{CO_2}$, and represents a present-day flux rather than an inventory. This method is less accurate than the others because of the temporal and spatial variability of surface ocean f_{CO_2}, but for the same reason it provides valuable information about locations and mechanisms of anthropogenic CO_2 uptake. The third method uses the observation that the $\delta^{13}C$ of fossil fuel CO_2 is much different from that found naturally in the ocean and atmosphere. Changes in $\delta^{13}C$ in seawater can be used to trace the fraction of the DIC that has been anthropogenically introduced. Finally, one can calculate (as opposed to measure) the oceanic burden of anthropogenic CO_2 in global circulation models by releasing CO_2 into the model atmosphere of a GCM and determining the amount that invades the model ocean.

The first of the direct measurement methods has been the most successful for determining the amount and location of anthropogenic CO_2 in the ocean at the time of writing this book. In this method the DIC increases below the surface mixed layer due to anthropogenic contamination and natural metabolic processes are distinguished. The technique was first applied by Brewer (1978) and Chen and Millero (1979) in its simplest form. In this early version, measured DIC values were subtracted from calculated natural (uncontaminated by anthropogenic CO_2) DIC values as a function of depth or age along a constant density surface. Natural DIC values at some depth, x, $DIC_{N,x}$, are equal to the preanthropogenic value at the surface, $DIC_{N,0}$, plus that from organic matter degradation and calcium carbonate dissolution:

$$DIC_{N,x} = DIC_{N,0} - r_{DIC:O_2} \cdot AOU_x + \tfrac{1}{2}\left(A_{T,x} - A_{T,S} + r_{NO_3:O_2} \cdot AOU_x\right). \tag{11.28}$$

The stoichiometric ratios $r_{DIC:O_2}$ and $r_{NO_3:O_2}$, convert the oxygen deficit caused by organic matter degradation to DIC and nitrate increase, respectively (both negative values). The terms on the right side represent, in order, the value of the DIC at the surface before it changed due to anthropogenic effects, the increase in DIC at location x below the surface layer due to organic matter degradation, and the increase in DIC at location x due to calcium carbonate dissolution. The value in parentheses in the last term is the alkalinity change, represented by the measured difference between the value at location x and surface waters, $(A_{T,x} - A_{T,S})$, minus ($r_{NO_3:O_2}$ is negative) the alkalinity decrease by oxidation of NH_3 from organic matter to NO_3^- (see Section 4.4.2). The anthropogenic component of the measured DIC is the difference between measured value and DIC_N.

$$DIC_{anthro} = DIC_{meas,x} - DIC_{N,x}. \tag{11.29}$$

To use this equation, one has to have a scheme for evaluating the surface ocean DIC and alkalinity before anthropogenic contamination. If you know the surface alkalinity, the DIC at chemical equilibrium can be calculated from T, S, and $f_{CO_2} = 280$ ppm (the preanthropogenic value). Alkalinity at the surface is evaluated by either assuming it is equal to present values or calculating it from present correlations with surface salinity and the measured salinity value. This is the weakest aspect of the method because it assumes that f_{CO_2} in the surface water was at equilibrium with the atmosphere and there may also be complications due to mixing of several different water masses. Gruber *et al.* (1996) developed new ways to address these uncertainties and improved the accuracy of this approach. The method has been subsequently applied to the world's ocean basins. An estimate of the invasion of anthropogenic CO_2 calculated in this way is presented in Fig. 11.7. The value in near-surface waters is 40–50 μmol kg^{-1} or about 2% of the total DIC.

Anthropogenic contamination reaches the ocean bottom in the northern part of the North Atlantic because of the deep convection of the North Atlantic Deep Water. Further south in the basin the contamination reaches to depths of 2000 m. The depth at which the profiles reach half their surface maximum is between 600 and 1000 m. Note that this depth is not greatly different from the value of 800 m estimated in Fig. 11.6 for the depth of ocean equilibrium required to accommodate about half of the fossil fuel released to the atmosphere. It has been shown with global circulation models that a present-day flux of 2.2 Pg y^{-1} into the ocean is required to accommodate the inventory of CO_2 indicated in Fig. 11.7 (Table 11.3).

The second method of determining the anthropogenic uptake of CO_2 is accomplished by calculating the flux across the air–water interface. Takahashi *et al.* (2002) have compiled more than half a million sea-surface f_{CO_2} measurements from different years and seasons. They normalized these measurements into a composite year with different seasons and extrapolated the data to fill in regions with little data coverage by using a surface ocean circulation model. These "climatological" f_{CO_2} results are multiplied by the gas exchange mass transfer coefficient, G_{CO_2}, which is calculated from correlations

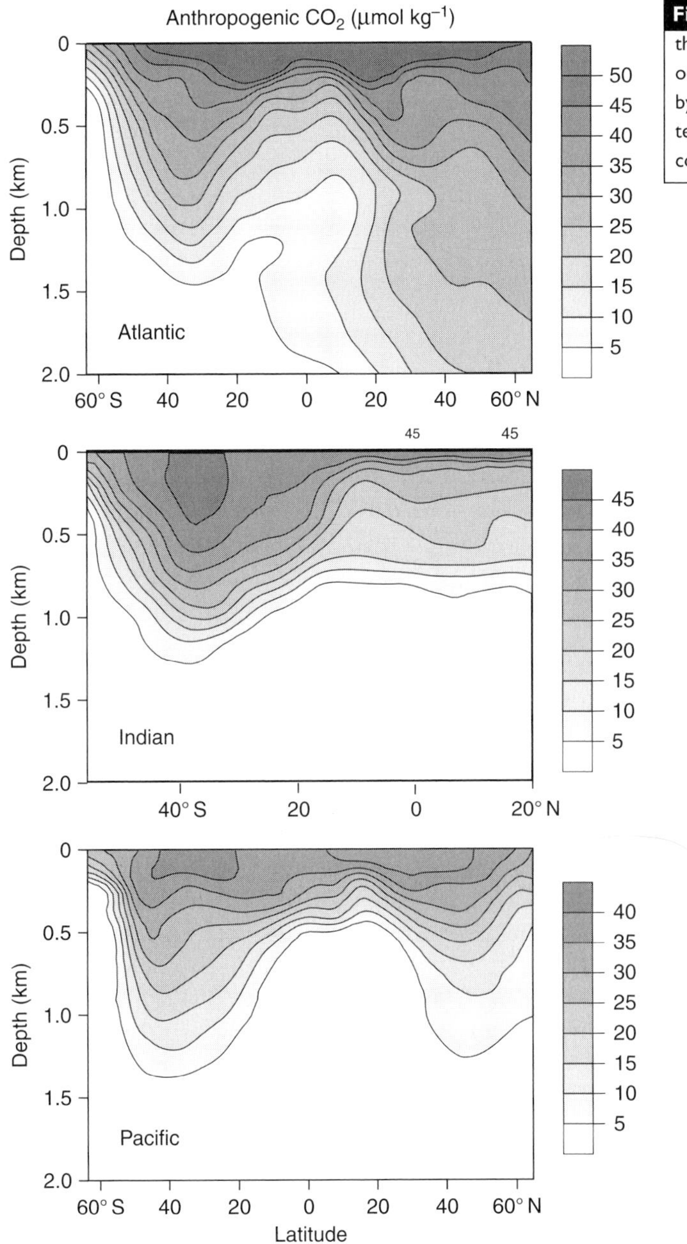

Figure 11.7. A cross section of the anthropogenic CO_2 in the ocean determined from ocean DIC by the method described in the text. Robert Key, personal communication; Key *et al.* (2004).

to wind speed for different seasons and regions (see Chapter 10) to determine the global flux of CO_2 into the ocean:

$$F_{CO_2} = G_{CO_2} \cdot K_H \cdot \left(f_{CO_2}^{a} - f_{CO_2}^{SO}\right) \tag{11.30}$$

where K_H is the Henry's Law coefficient. A map of the global distribution of the air–water CO_2 fluxes calculated in this way is presented in Fig. 11.8. The important areas of CO_2 evasion to the atmosphere are the Equator (particularly the Eastern Equatorial Pacific) and the NW

Table 11.3. *Estimates of anthropogenic carbon dioxide uptake rate by the oceans*

Method of determination	Flux ($Pg\ y^{-1}$)
Experimental results	
Ocean inventory: DIC[a]	2.2
Gas exchange (1995)[b]	2.2 ± 0.4
$\delta^{13}C$-DIC inventory (1970–1990)[c]	1.7 ± 0.2
(1980s)	2.0 ± 0.2
Atmospheric measurements: O_2/N_2, CO_2 (1990–2000)	1.7 ± 0.5[d]
O_2/N_2, CO_2, $\delta^{13}C$-CO_2 (1991–1997)[e]	2.0 ± 0.6
Model results (1980–1989)	
Range of four different global models[f]	1.5 to 2.2

[a] The DIC method produces a global inventory, not a flux. The inventory is 118 Pg C (Sabine *et al.*, 2004) . This is assigned a flux by analogy with a GCM inventory of 118 Gt that results from an annual flux of 2.2 Gt $CO_2\ y^{-1}$ (Orr *et al.*, 2000).
[b] Takahashi *et al.* (2002).
[c] Quay *et al.* (1992, 2003).
[d] IPCC (2001).
[e] Battle *et al.* (2000).
[f] Values of the four different models are 2.2, 2.1, 1.6, and 1.5 (Orr *et al.*, 2000).

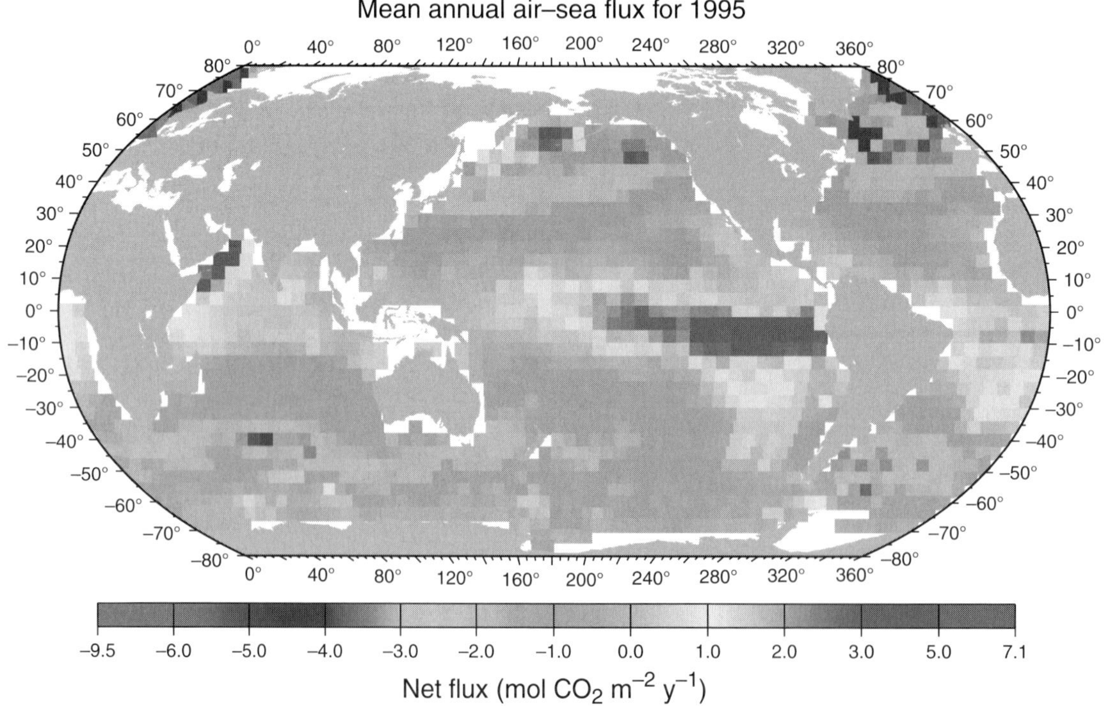

Figure 11.8. The mean annual flux of CO_2 between the atmosphere and ocean, based on measurements of f_{CO_2} in the ocean and atmosphere, and the gas exchange mass transfer coefficient determined from wind speed. From Takahashi *et al.* (2002). (See Plate 8.)

Indian Ocean. These are regions of upwelling from deeper waters that are enriched in CO_2. The subtropical oceans are about in equilibrium with the atmospheric f_{CO_2} and carbon dioxide invades strongly in the North and South Pacific and Atlantic north of about 40° N. Cooling of both the Kuroshio and Gulf Stream in the Northern Hemisphere plays an important role in the carbon dioxide uptake at the gyre boundaries because CO_2 becomes undersaturated as these grand ocean currents move northward and across the oceans. CO_2 undersaturation caused by cooling and carbonate thermodynamics, however, is not the whole story. It has been shown that mixing at the gyre boundaries brings nutrients to the surface, enhancing biological productivity that aids in the f_{CO_2} drawdown in these regions. Regional and seasonal f_{CO_2} atmosphere–ocean differences range from $>100\,\mu$atm supersaturated to $>100\,\mu$atm undersaturated, so even with the large data coverage there is a considerable error in the calculated annual flux. There are also uncertainties associated with the estimate of the mass transfer coefficient from global wind speeds, particularly at high winds (see Chapter 10). At the time of writing this book the best estimate of the mean net invasion rate with this method is 2.2 ± 0.4 Pg y^{-1} (Takahashi *et al.*, 2002) (Table 11.3). The global mean degree of surface water undersaturation of f_{CO_2} to supply 2.0 Pg y^{-1} of carbon to the oceans is 8 μatm. The most important regions of ocean uptake of anthropogenic CO_2 are different from locations where it accumulates (based on the DIC method). Regions where uncontaminated deep water reaches the surface – high latitudes, the Equator and the subtropical/subarctic frontal regions – are the areas of most anthropogenic CO_2 uptake. Relatively little CO_2 enters the ocean thermocline in the subtropical regions but this is the ocean location where much of the anthropogenic CO_2 is stored. Circulation tends to pool this transient tracer in these locations (notice the large inventories in these areas in Fig. 11.7).

The third experimental method for determining the anthropogenic burden in the ocean involves global data for changes in the carbon isotope ratio of the DIC of the ocean and CO_2 of the atmosphere. Because the $\delta^{13}C$ of fossil fuel CO_2 is about –23% and that of the DIC of the ocean is near zero, contamination of the carbon isotope ratio of CO_2 in the atmosphere and DIC in the oceans is readily measurable. This is demonstrated by comparing the $\delta^{13}C$-DIC in the surface oceans between 1970 and 1990 (Fig. 11.9). Quay *et al.* (1992) introduced this method by compiling measurements of the $\delta^{13}C$ in the atmosphere and ocean between 1970 and 1990 and calculating the air–sea flux necessary to account for the measured differences. The method is sensitive to the $\delta^{13}C$ fractionation factor during gas exchange and accurate measurements from past global surveys (GEOSECS in the 1970s), which in some cases are problematic. Using this method, Quay *et al.* (2003) estimate an ocean uptake rate of 2.0 ± 0.2 Pg y^{-1} during the period 1970–1990 (Table 11.3).

Models have been used to estimate the uptake of anthropogenic CO_2 since well before the experimental procedures were of good

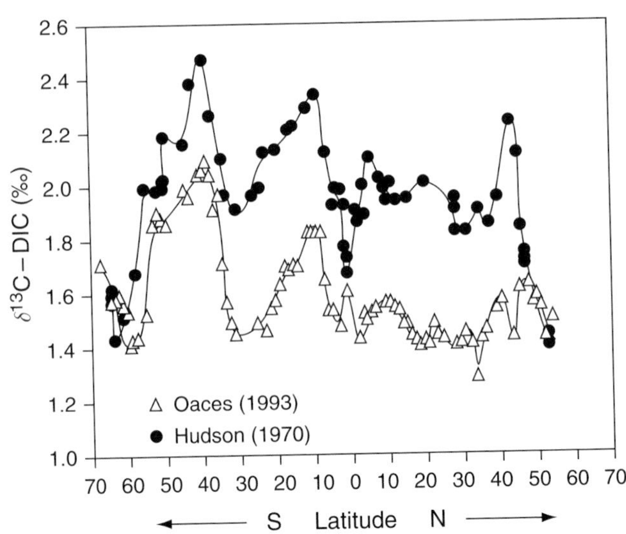

Figure 11.9. The carbon isotope ratio of DIC in the surface of the Pacific Ocean. The data illustrate how the value has decreased between 1970 and 1993 owing to the addition of fossil fuel CO_2 to the atmosphere. Oaces (1993) and Hudson (1970) refer to two different ocean cruises. From Quay *et al.* (2003).

enough quality to accurately measure the changes. We have shown by the equilibrium calculations and the measured penetration depths that the limiting factor for oceanic anthropogenic CO_2 uptake is mixing of water across the thermocline. It is thus extremely important for the mixing rates of the models to be calibrated by matching the penetration of other transient tracers (^{14}C, ^3H, and CFCs) into the ocean. Once this is accomplished, the models are run for several hundred years while CO_2 is delivered to the atmosphere according to the anthropogenic usage. To date, estimates for the CO_2 uptake rate employing different models have been made with values ranging from 1.5 to 2.3 Pg y^{-1}. These results are presented along with estimates of the experimental methods in Table 11.3. The average and standard deviation of the means of all the model and experimental estimates is 1.9 ± 0.2. The relatively tight concurrence among the different techniques indicates that the rate of anthropogenic CO_2 uptake in the ocean over the past several decades is reasonably well known.

The critical unknowns in this field have evolved in the past 20–30 y from what the anthropogenic carbon fluxes are, to understanding the mechanisms controlling the ocean's CO_2 pumps and anthropogenic CO_2 uptake. These tasks utilize both interdisciplinary experimental observations and global circulation models. If the models can be made to accurately reproduce experimentally determined fluxes and storage patterns, then it should be possible to accurately predict the response to future changes.

11.3.3 Partitioning anthropogenic CO_2 among the ocean, atmosphere, and terrestrial reservoirs

For many years it was assumed that tropical deforestation was a significant source of CO_2 to the atmosphere (between 10% and half of the source from fossil fuel combustion). When these sources were combined with the relatively well-known sinks in the atmosphere

Table 11.4. *Carbon budget for 1980s*

Positive values are sources to the atmosphere. Negative values are sinks. Notice that had the value from deforestation been assumed to be zero the budget would have balanced.

Source or sink	$Pg\ C\ y^{-1}$
Emissions from fossil fuel combustion	$+5.4 \pm 0.5$
Emissions from deforestation and land use	$+1.6 \pm 1.0$
Atmospheric accumulation	-3.2 ± 0.2
Uptake by the ocean	-2.0 ± 0.6
Net imbalance	$\mathbf{+1.8 \pm 1.3}$

Source: From Siegenthaler and Sarmiento (1993).

and ocean, there was a significant excess source. An example of this balance sheet is shown in Table 11.4. While the source from the terrestrial biosphere was uncertain it none the less was assumed to be positive because of known global deforestation. This uncertainty stood until it was demonstrated that with simultaneous determinations of the increase in atmospheric f_{CO_2} and decrease in f_{O_2} it was possible to identify the difference between the global terrestrial biosphere and ocean sinks (Keeling *et al.*, 1996). One might at first expect that these two tracers are redundant because there is an exact stoichiometry between CO_2 release and O_2 consumption by organic matter combustion. The difference, however, is that while both the land and ocean represent potential sinks (or sources) for CO_2 in response to the anthropogenic perturbation, the ocean does not significantly exchange oxygen in response to the anthropogenic decrease in atmospheric O_2. Because O_2 is a very insoluble gas, about 95% of the global reservoir is in the atmosphere. Thus, when atmospheric O_2 decreases owing to fossil fuel burning, there is not a significant subsequent release of O_2 from the ocean to make up the deficit in the atmosphere. Any change in the atmospheric O_2 concentration other than that due to fossil fuel combustion must be attributed to exchange with the terrestrial biosphere.

Calculation of the importance of ocean, atmosphere, and terrestrial biosphere to the anthropogenic CO_2 source was first presented graphically (Keeling *et al.*, 1996). An updated version (Fig. 11.10) (IPCC, 2001) shows the measured decrease in atmospheric oxygen versus the simultaneous increase in atmospheric CO_2 during the 1990s, culminating in a ratio indicated by the filled circle at the year 2000 on the figure. Since the carbon and hydrogen content of the three main fossil fuel sources during this period are known along with their relative consumption rates, an accurate composite of the oxygen demand to carbon dioxide source can be determined. The fossil fuel mixture presently being mined (Table 11.5) results in an oxygen to CO_2 atmospheric change of $\Delta O_2 / \Delta CO_2 = -1.45$. Because the total amount burned and its $\Delta O_2{:}\Delta CO_2$ ratio are known, the line

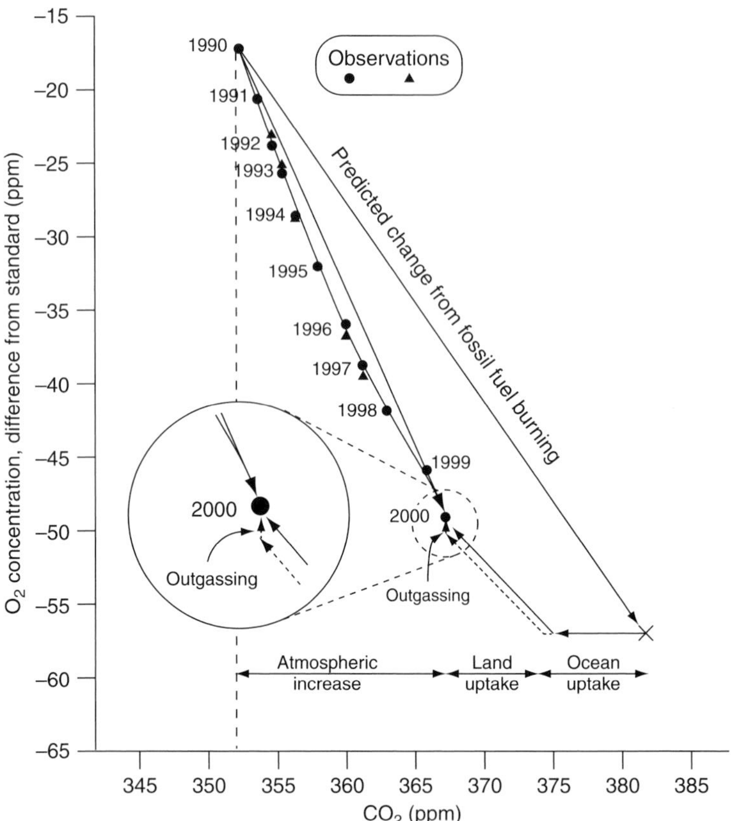

Figure 11.10. The mean change in the atmospheric O_2 and CO_2 partial pressures (ppm) for the period of the 1990s, and illustration of its utility in determining anthropogenic CO_2 uptake by the ocean and land. Symbols are observations between 1990 and 2000. The long diagonal line from the upper left to the lower right is the trend expected, given the amount of fossil fuels burned, if there were no land or ocean exchange. The horizontal line in the lower right corner indicates ocean–atmosphere exchange of CO_2, and the short diagonal line that trends to the upper left from where the horizontal line stops illustrates the trend for uptake of CO_2 and release of O_2 with the land reservoir (see text). Solid lines indicate the trends expected were there no non-steady-state "outgassing" of O_2 from the ocean. Dashed lines and the close-up indicate how the steady-state picture is altered by non-steady-state outgassing of O_2 from the ocean because of changes in ocean ventilation and seawater temperature. Modified from IPCC (2001).

labeled "fossil fuel burning" represents the projected atmospheric change if there were no exchange with the ocean or land biosphere between 1900 and 2000 ending at the location indicated by the (X). Clearly this line is quite different from the observations, reflecting the response of the terrestrial and oceanic reservoirs. The ΔO_2:ΔCO_2 uptake ratio for land biota–atmosphere interaction is $= -1.1$, and this ratio for the ocean–atmosphere interaction is approximately zero. Thus, there are two straight-line paths with known ΔO_2:ΔCO_2 ratios by which we can connect the year 2000 prediction in Fig. 11.10 (X) to the observed value (filled circle). Drawing the only possible straight line paths between the expected and observed values involves quantifying both the oceanic and the atmospheric sinks.

This procedure is simple and elegant and relies on the atmosphere to homogenize the globally heterogeneous fluxes of oxygen and CO_2. The result is (Table 11.1) (IPCC, 2001) that of the $+6.3$ Pg y^{-1} of anthropogenic CO_2 produced during the 1990s, $+3.2$ Pg y^{-1} remained in the atmosphere, -1.7 Pg y^{-1} went into the ocean and -1.4 Pg y^{-1} was taken up by the terrestrial biosphere. During the period of this investigation, the land, rather than being a source to the atmosphere because of deforestation, was a sink because of forest regrowth and enhanced growth due to increased atmospheric CO_2!

Table 11.5. *Carbon and oxygen stoichiometry of fossil fuel burning*

Source	Reaction	$\Delta O_2/\Delta CO_2$
Coal	$2\ C_{10}H_6 + 23\ O_2 \rightarrow 20\ CO_2 + 6\ H_2O$	23/20
Oil	$2\ CH_2 + 3\ O_2 \rightarrow 2\ CO_2 + 2\ H_2O$	3/2
Natural gas	$CH_4 + 2\ O_2 \rightarrow CO_2 + 2\ H_2O$	2/1

Source: Keeling *et al.* (1996).

One cannot compare this result with that in Table 11.4 because they represent different time periods. Analysis of atmospheric changes in CO_2, $O_2:N_2$ ratio, and the $\delta^{13}C$ of the CO_2 for the period of the 1980s and 1990s (Battle *et al.*, 2000) indicates that the role of the terrestrial biosphere during the 1980s and 1990s has been very different. During the 1980s it was about neutral, i.e. any source to the atmosphere via deforestation was nearly matched by uptake due to enhanced new forest growth. In the 1990s global greening dominated the terrestrial signal. Thus, the original balance sheet in Table 11.4 indicates no imbalance at all. The estimates of the role of the biosphere when this table was published appear to have been inaccurate. There is presently no indication that there is a missing sink for fossil fuel CO_2; however, recent atmospheric studies indicate that the role of the biosphere has rapidly changed in a matter of one to two decades from being neutral to becoming a very large sink. This is presently believed to be due to a large recent increase in forest regrowth on land; however, the rapidity of the change was a surprise and will be the subject of intense research in decades to come.

A note of caution is in order before concluding this chapter. The calculations illustrated by Fig. 11.10 here are not bullet-proof. One of the assumptions in the calculation has been that the natural background ocean–atmosphere exchange of gases is at steady state. It is becoming clear that this is probably not the case. It is known that the temperature of ocean surface waters is increasing (decreasing its ability to store oxygen), and repeat hydrography sections in many regions of the ocean indicate decadal-scale decline in the oxygen content of the upper thermocline between the 1980s and 1990s, presumably because of a decrease in the ventilation of thermocline waters (e.g. Fig. 6.23). If there is a net oxygen flux out of the ocean to account for the decreasing oxygen content of the thermocline and the warming of surface waters it will require an additional arrow in Fig. 11.10. This arrow is vertical because it represents a non-steady-state "degassing" of O_2 from the ocean to the atmosphere. (There is also a degassing of CO_2 associated with the non-steady-state warming, but this part of the ocean exchange is already accounted for by the horizontal line indicating CO_2 uptake by the ocean.) The seemingly small correction for the non-steady-state process has a significant effect on the estimates of ocean and terrestrial CO_2 uptake. As indicated by the dashed arrow in Fig. 11.10, the effect of this additional flux is to increase the uptake estimates of CO_2 by the ocean and decrease the CO_2 sequestration by land. A minimum estimate of this effect is the observed increase in temperature of the surface ocean during the 1990s. A larger estimate is derived from GCM models that analyze the O_2 "degassing" effect of global warming (e.g. Bopp *et al.*, 2002). The accurate value for this effect is presently uncertain, but even the upper estimates are not large enough to account for the very large differences in land and ocean sequestering suggested for the decades of the 1980s and 1990s.

References

Archer, D. E., G. Eschel, A. Winguth *et al.* (2000) Atmospheric pCO_2 sensitivity to the biological pump in the ocean. *Global Biogeochem. Cycles* **14**, 1219–30.

Battle, M., M. L. Bender, P. P. Tans *et al.* (2000) Global carbon sinks and their variability inferred from atmospheric O_2 and del^{13}C. *Science* **287**, 2467–70.

Berner, R. A. (1990) Atmospheric carbon dioxide levels over Phanerozoic time. *Science* **249**, 49–75.

Bopp, L., C. Le Quere, M. Heimann and A. C. Manning (2002) Climate-induced oceanic oxygen fluxes: implications for the contemporary carbon budget. *Global Biogeochem. Cycles* **16**, 2, doi: 10.1029/2001GB001445.

Brewer, P. G. (1978) Direct observation of the oceanic CO_2 increase. *Geophys. Res. Lett.* **4**, 997–1000.

Broecker, W. S. (1971) A kinetic model for the chemical composition of sea water. *Quat. Res.* **1**, 188–207.

Broecker, W. S. and T.-H. Peng (1982) *Tracers in the Sea*. Lamont-Doherty Earth Observatory.

Chen, C. T. and F. J. Millero (1979) Gradual increase of oceanic carbon dioxide. *Nature* **277**, 205–6.

Eppley, R. W. and B. J. Peterson (1979) Particulate organic matter flux and planktonic new production in the deep ocean. *Nature* **282**, 677–80.

Gruber, N., J. L. Sarmiento and T. F. Stocker (1996) An improved method for detecting anthropogenic CO_2 in the oceans. *Global Biogeochem. Cycles* **10**, 809–7.

IPCC (2001) *Climate Change 2001: The Scientific Basis*. Contributions of Working Group I to the third assessment report of the intergovernmental panel on climate change (ed. J. T. Houghton, Y. Ding, D. J. Griggs *et al.*). Cambridge: Cambridge University Press.

Keeling, C. D. (1960) The concentration and isotopic abundance of CO_2 in the atmosphere. *Tellus* **12**, 200–3.

Keeling, R. F., S. C. Piper and M. Heinmann (1996) Global and hemispheric CO_2 sinks deduced from changes in atmospheric O_2 concentration. *Nature* **381**, 218–21.

Key, R. M., A. Kozar, C. L. Sabine *et al.* (2004) A global ocean carbon climatology: results from Global Data Analysis Project (GLODAP). *Global Biogeochem. Cycles* **18**, GB4031, doi: 10.1029/2004GB002247.

Orr J. C., *et al.* (2000) Estimates of anthropogenic carbon uptake from four three-dimensional global ocean models. *Global Biogeochem. Cycles* **15**, 43–60.

Pilson, M. E. Q. (1998) *An Introduction to the Chemistry of the Sea*. Englewood Cliffs, NJ: Prentice-Hall.

Quay, P. D., R. Sonnerup, T. Westby, J. Stutsman and A. McNichol (2003) Changes of the $^{13}C/^{12}C$ of dissolved inorganic carbon in the ocean as a tracer of anthropogenic CO_2 uptake. *Global Biogeochem. Cycles* **17**, 1, doi: 10.1029/2001GB001817.

Quay, P. D., B. Tilbrook and C. S. Wong (1992) Oceanic uptake of fossil fuel CO_2: carbon-13 evidence. *Science* **256**, 74–9.

Revelle, R. and H. E. Suess (1957) Carbon dioxide exchange between atmosphere and ocean and the question of an increase of atmospheric CO_2 during past decades. *Tellus* **9**, 18–27.

Sabine, C. *et al.* (2004) The ocean sink for anthropogenic CO_2. *Science* **305**, 367–71.

Siegenthaler, U. and J. L. Sarmiento (1993) Atmospheric carbon dioxide and the ocean. *Nature* **365**, 119–25.

Sigman, D. M. and E. A. Boyle (2000) Glacial/interglacial variations in atmospheric carbon dioxide. *Nature* **407**, 859–69.

Takahashi, T. *et al.* (2002) Global sea-air CO_2 flux based on climatological surface ocean pCO_2, and seasonal biological and temperature effects. *Deep-Sea Res. II*, **49**, 1601–22.

Toggweiler, J. R. (1999) Variations of atmospheric CO_2 by ventilation of the ocean's deepest water. *Paleoceanography* **14**(5), 571–88.

Volk, T. and M. I. Hoffert (1985) Ocean carbon pumps: analysis of relative strengths and efficiencies in ocean-driven atmospheric CO_2. In *The Carbon Cycle and Atmospheric CO_2: Natural Variations Archean to Present*, Geophysical Monograph Series, vol. 32 (ed. E. T. Sundquist and W. S. Broecker), pp. 99–110. Washington, DC: AGU.

12

Chemical reactions in marine sediments

Chemical reactions in marine sediments and the resulting fluxes across the sediment–water interface influence the global marine cycles of carbon, oxygen, nutrients and trace metals and control the burial of most elements in marine sediments. On very long time scales these diagenetic reactions control carbon burial in sedimentary rocks and the oxygen content of the atmosphere. Sedimentary deposits that remain after diagenesis are the geochemical artifacts used for interpreting past changes in ocean circulation, biogeochemical cycles and climate. Constituents of marine sediments that make up a large fraction of the particulate matter that reaches the sea floor (organic matter, $CaCO_3$, SiO_2, Fe, Mn, aluminosilicates and trace metals) are tracers of ocean physical and biogeochemical processes when they formed, and of diagenesis after burial.

Understanding of sediment diagenesis and benthic fluxes has evolved with advances in both experimental methods and modeling. Measurements of chemical concentrations in sediments and their associated porewaters and fluxes at the sediment–water interface have been used to identify the most important reactions. Because transport in porewaters is usually by molecular diffusion, this medium is conducive to interpretation by models of heterogeneous chemical equilibrium and reaction kinetics. Large chemical changes and manageable transport mechanisms have led to elegant models of sediment diagenesis and great advances in understanding diagenetic processes.

We shall see, though, that the environment does not yield totally to simple models of chemical equilibrium and chemical kinetics, and laboratory-determined constants often cannot explain the field observations. For example, organic matter degradation rate constants determined from laboratory experiments and modeling are so variable that there are essentially no constraints on the values to be expected in the environment. Also, reaction rates of $CaCO_3$ and opal dissolution determined from laboratory experiments usually cannot be reproduced in models of porewater measurements from marine sediments. The inability to mechanistically understand reaction kinetics measured in the environment in terms of laboratory experiments is an important uncertainty in the field today.

Processes believed to be most important in controlling the preservation of organic matter have evolved from a focus on the lability of the substrate to the protective mechanisms of mineral–organic matter interactions. The specific electron acceptor is not particularly important during very early diagenesis, but the importance of oxygen to the degradation of organic matter during later stages of diagenesis has been verified by the study of diagenesis in ancient turbidites deposited on the ocean floor.

Evolution of thinking about the importance of reactions between seawater and detrital clay minerals has come full circle in the past 35 years. "Reverse weathering" reactions were hypothesized in very early chemical equilibrium and mass balance (Mackenzie and Garrels, 1966) models of the oceans. Subsequent observations that marine clay minerals generally resemble those weathered from adjacent land and the discovery of hydrothermal circulation put these ideas on the back burner. Recent studies of silicate and aluminum diagenesis, however, have rekindled awareness of this process, and it is back in the minds of geochemists as a potentially important process for closing the marine mass balance of some element (see chapter 2).

Recent studies of the level of diagenesis necessary for authigenic precipitation of some trace metals have made it possible to determine the global extent of metal diagenesis by using models of porewater chemistry and organic matter diagenesis. Among these results are the large-scale remobilization of Mn and V from continental margin sediments to the sediments of the ocean interior and authigenic precipitation of rhenium (Re) and to a lesser extent cadmium (Cd) and uranium (U) in continental margin sediments where oxygen penetrates 1–2 cm or less. The potential for using the authigenic precipitation as tracers of ancient porewater geochemistry is at hand, but identifying whether authigenic enrichments are due to changes in organic matter rain rate to the sediments (productivity change) or changes in bottom water oxygen levels cannot presently be separated.

12.1 Diagenesis and preservation of organic matter

Roughly 90% of the organic matter that exits the euphotic zone of the ocean is degraded in the water column. Of the *c.*10% of the organic carbon flux that reaches the sea floor, only about one tenth escapes oxidation and is buried. Degradation of the organic matter that reaches the ocean sediments drives many of the reactions that control sediment diagenesis and benthic flux.

We begin our discussion with what we call the "pillars" of knowledge in the field: those concepts that are basic to understanding the mechanisms of organic matter diagenesis and on which future developments rested. This is followed by a description of the dominant mechanisms of organic matter diagenesis as one progresses from oxic through anoxic conditions. Finally, we will discuss factors controlling the reactivity of organic matter and the mechanisms of organic matter preservation.

12.1.1 Pillars of organic matter diagenesis

The basic concepts of organic matter diagenesis are described here as (a) the *thermodynamic* sequence of reactions of electron acceptors and their *stoichiometry*, and (b) the *kinetics* of organic matter degradation as described by the diagenesis equations and observations of degradation rates. These ideas derived mainly from studies of ocean sediments in which porewater transport is controlled by molecular diffusion (deep-sea oxic and anoxic-SO_4 reducing), but also represent the intellectual points of departure for studying near-shore systems where transport is more complicated, but where the bulk of marine organic matter is degraded.

Large, highly structured molecules of organic matter are formed by energy from the sun and exist at atmospheric temperature and pressure in a reduced, thermodynamically unstable state. These compounds subsequently undergo reactions with oxidants to decrease the free energy of the system. The oxidants accept electrons from the organic matter during oxidation reactions (chapter 3) (Stumm and Morgan, 1981). The electron acceptors that are in major abundance in the environment include O_2, NO_3^-, Mn(IV), Fe(III), SO_4^{2-} (Roman numerals indicate oxidation state without specifying the molecular

Table 12.1. *The standard free energy of reaction, ΔG_r^o, for the main environmental redox reactions*

Reaction	ΔG_r° (kJ mol^{-1}) (half-reaction)	ΔG_r° (kJ mol^{-1}) (whole reaction)
Oxidation		
$CH_2O^a + H_2O \rightarrow CO_2(g) + 4H^+ + 4e^-$	−28.2	
Reduction		
$4e^- + 4H^+ + O_2(g) \rightarrow 2H_2O$	−478.4	−506.6
$4e^- + 4.8H^+ + 0.8NO_3^- \rightarrow 0.4N_2 + 2.4H_2O$	−480.2	−508.4
$4e^- + 8H^+ + 2MnO_2(s) \rightarrow 2Mn^{2+} + 4H_2O$	−474.5	−502.7
$4e^- + 12H^+ + 2Fe_2O_3(s) \rightarrow 4Fe^{2+} + 6H_2O$	−253.2	−281.4
$4e^- + 5H^+ + 0.5SO_4^{2-} \rightarrow 0.5H_2S(g) + 2H_2O$	−118.9	−147.1
$4e^- + 4H^+ + 05.CO_2(g) \rightarrow 0.5CH_4(g) + H_2O$	−65.4	−93.58

Standard free energies of formation from Stumm and Morgan (1981).
[a] CH_2O represents organic matter ($\Delta G_f = -129$ kJ mol^{-1}).

or ionic form), and organic matter itself during fermentation, which is described here as methane production. These reactions were presented in Chapter 3 and are listed in the order of the free energy gained in Table 12.1. Half-reactions for both the organic matter oxidation and the electron acceptor reductions are represented. The changes in free energy for the reactions depend on the free energy of formation of the solids involved (organic matter, iron and manganese oxides), and thus vary slightly among compilations in the literature. Note that the amounts of free energy gain for the whole reactions involving oxygen, nitrate and manganese are similar. Values drop off dramatically for iron and sulfate reactions and then again for methane production. The sequence of electron acceptors used is sometimes categorized into oxic diagenesis (O_2 reduction), suboxic diagenesis (NO_3^- and Mn(IV) reduction) and anoxic diagenesis (Fe(III) and SO_4^{2-} reduction and methane formation). This terminology is not used here because the definition of suboxic is vague and ambiguous. We recommend referring to these reactions by using the true meaning of the terms: oxic for O_2 reduction and anoxic for the rest (anoxic-NO_3^- reduction, anoxic-Mn(IV) reduction and so forth).

Although all of these reactions are favored thermodynamically, they are almost always enzymatically catalyzed by bacteria. It has been observed from the study of porewaters in deep-sea sediments (e.g. Froelich *et al.*, 1979) and anoxic basins (e.g. Reeburgh, 1980) that there is an ordered sequence of redox reactions in which the most energetically favorable reactions occur first and the active electron acceptors do not overlap significantly. Bacteria are energy opportunists. Using estimates of the stoichiometry of the diagenesis reactions (Table 12.2) one can sketch the order and shape of reactant profiles actually observed in sediment porewater chemistry (Fig. 12.1). The schematic figure shows all electron acceptors in a single sequence. This is rarely observed in the environment because regions with

Table 12.2. *Stoichiometry of organic matter oxidation reactions*
Redfield ratios for x, y and z are 106, 16, 1.

Redox process	Reaction
Aerobic respiration	$(CH_2O)_x(NH_3)_y(H_3PO_4)_z + (x + 2y)O_2 \rightarrow$ $xCO_2 + (x + y)H_2O + yHNO_3 + zH_3PO_4$
Nitrate reduction	$5(CH_2O)_x(NH_3)_y(H_3PO_4)_z + 4xNO_3^- \rightarrow$ $xCO_2 + 3xH_2O + 4xHCO_3^- + 2xN_2 + 5yNH_3 + 5zH_3PO_4$
Manganese reduction	$(CH_2O)_x(NH_3)_y(H_3PO_4)_z + 2xMnO_2(s) + 3xCO_2 + xH_2O \rightarrow$ $2xMn^{2+} + 4xHCO_3^- + yNH_3 + zH_3PO_4$
Iron reduction	$(CH_2O)_x(NH_3)_y(H_3PO_4)_z + 4xFe(OH)_3(s) + 7xCO_2 \rightarrow$ $4xFe^{2+} + 8xHCO_3^- + 3xH_2O + yNH_3 + zH_3PO_4$
Sulfate reduction	$2(CH_2O)_x(NH_3)_y(H_3PO_4)_z + xSO_4^{2-} \rightarrow$ $xH_2S + 2xHCO_3^- + 2yNH_3 + 2zH_3PO_4$
Methane production	$(CH_2O)_x(NH_3)_y(H_3PO_4)_z \rightarrow$ $xCH_4 + xCO_2 + 2yNH_3 + 2zH_3PO_4$

From Tromp *et al.* (1995).

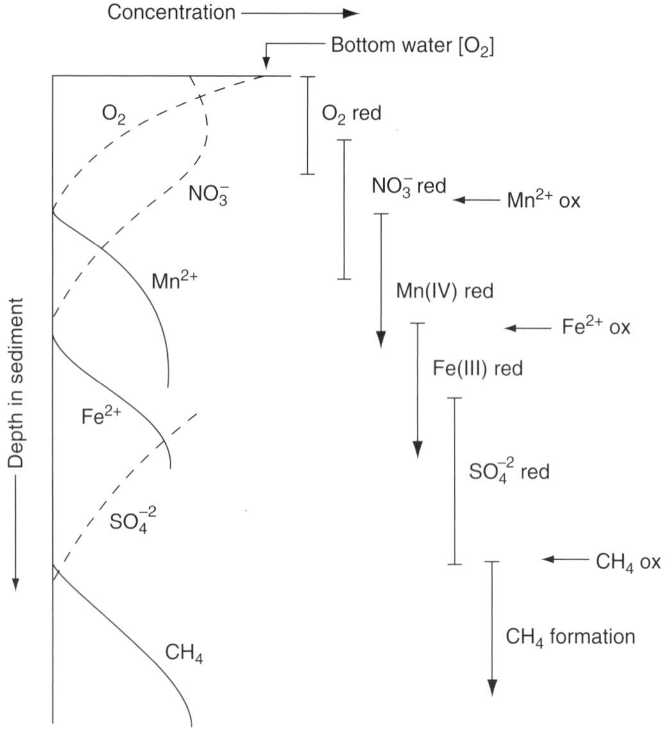

Figure 12.1. A schematic representation of the porewater profiles that have been observed to show the sequential use of electron acceptors during organic matter degradation. Modified from Froelich *et al.* (1979).

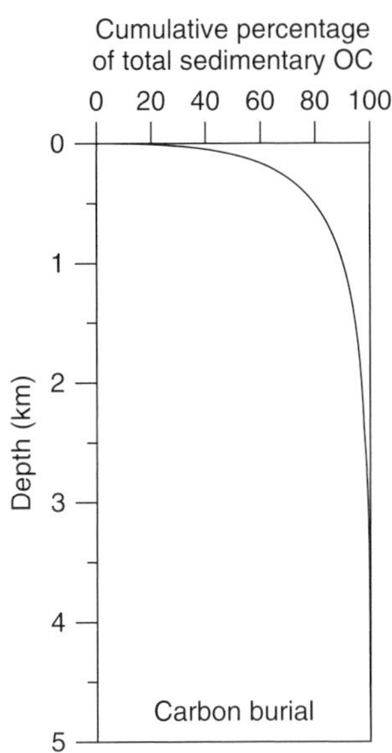

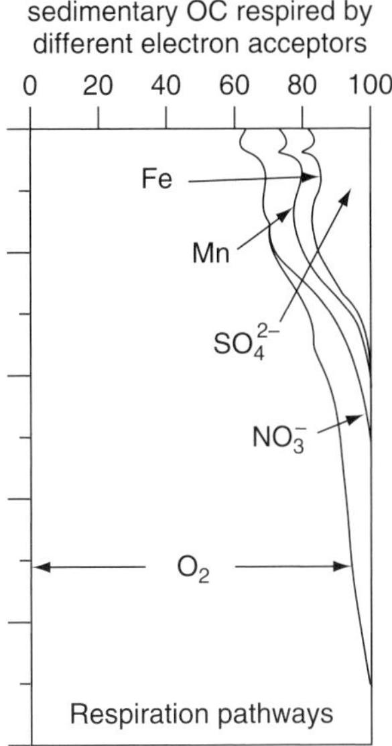

Figure 12.2. The cumulative fraction of carbon burial and the respiration pathways as a function of depth in the ocean derived from the global diagenesis model of Archer *et al.* (2002). Redrawn from Archer *et al.* (2002).

abundant bottom water oxygen and moderate organic matter flux to the sediments (i.e. the deep ocean) run out of reactive organic carbon before sulfate reduction becomes important. In near-shore environments, where there is sufficient organic matter flux to the sediments to activate sulfate reduction and deplete sulfate in porewaters, zones of oxygen, nitrate and Mn(IV) reduction are very thin or obscured by benthic animal irrigation and bioturbation.

Recent global models of the importance of the different electron acceptors (Fig. 12.2) (Archer *et al.*, 2002) indicate that oxic respiration accounts for about 95% of the organic matter oxidation below 1000 m in the ocean. However, between 80% and 90% of organic matter that is buried in ocean sediments accumulates above 1000 m in river deltas and on continental margins (Archer *et al.*, 2002; Hedges and Keil, 1995). Anoxic diagenesis is more important in these regions, and when they are included oxic diagenesis accounts for about 70% of the total organic matter oxidation in ocean sediments.

The second "pillar" in our understanding of organic matter diagenesis and benthic flux consists of advances in quantifying the rates of organic matter degradation and burial. The kinetics of organic matter degradation have been determined by modeling environmental porewater data and in laboratory studies. Most models of organic matter degradation have derived in some way from the early studies by Berner on this subject (e.g. Berner, 1980). In its simplest form, one-dimensional diagenesis, the change in organic carbon concentration (C_s, gC gm_s^{-1}; s = dry sediment) with respect to time and depth is:

$$\frac{\partial}{\partial t}(\rho(1-\phi)\,C_s) = \frac{\partial}{\partial z}\left(D_b\frac{\partial(\rho(1-\phi)C_s)}{\partial z}\right) - \frac{\partial}{\partial z}(\omega\rho(1-\phi)\,C_s) + MR \quad (12.1)$$

where z is depth below the sediment-water interface (positive downward), ϕ is porosity ($cm_{pw}^3\ cm_b^{-3}$; where pw is porewater and b is bulk), ρ is the dry sediment density ($g\,cm_s^{-3}$), M ($g\ mol^{-1}$) is the molecular mass of carbon, D_b ($cm_b^2\ s^{-1}$; s = second) is the sediment bioturbation rate, ω is the sedimentation rate ($cm_b\ s^{-1}$) and R ($mol\ cm_b^{-3}\ s^{-1}$) is the reaction term. Organic matter degradation is usually considered to be first order with respect to substrate concentration, so:

$$R = \frac{k \cdot C_s \cdot (1-\phi) \cdot \rho}{M}, \tag{12.2}$$

where k is the first-order degradation rate constant. Probably the largest simplification here is that stirring of sediments by animals is modeled as a random process analogous to molecular diffusion. This is a gross simplification of reality. It has been shown that different tracers of bioturbation yield different results and animal activity varies with organic matter flux to the sediment–water interface (see, for example, Smith *et al.*, 1993). The reaction–diffusion equation for the concentration of the dissolved porewater constituent, C_d ($mol\ cm_{pw}^{-3}$) is:

$$\frac{\partial}{\partial t}(\phi C_d) = \frac{\partial}{\partial z}\left(D\frac{\partial(\phi)C_d}{\partial z}\right) - \frac{\partial}{\partial z}(v\phi C_d) + \gamma R, \tag{12.3}$$

where D now represents the molecular diffusion coefficient and v ($cm\ s^{-1}$) is the velocity of water (which is the same as the sediment burial, ω, when porosity is constant and there is no outside-induced flow) (Imboden, 1975) and γ is a stoichiometric ratio of the element being considered to carbon in the degrading organic matter. At steady state with respect to compaction and the boundary conditions, the left side of Eq. (12.3) is zero and v and ω are equal below the depth of porosity change. A very detailed treatment of many different cases is presented in Berner (1980) and Boudreau (1997).

After application of the diagenesis equations to a variety of marine environments, it became clear that the organic matter degradation rate constant, k, that was used to fit the porewater and sediment profiles was highly variable. The organic fraction of marine sediments is thus often modeled as a mixture of a number of discrete components, each with a finite initial amount and first-order decay constant (e.g. Jørgensen, 1979) or one component whose reactivity decreases continuously over time (Middelburg, 1989). Both of these approaches capture the fundamental feature that bulk organic matter breaks down at an increasingly slower rate as it degrades.

A consequence of this broad continuum in reactivity is that sedimentary organic matter can be observed to degrade on essentially all time scales of observation. Although slightly different degradation rates may be measured for various components of a sedimentary mixture, such as different elements or biochemicals, the range of absolute values of the measured rate constants closely corresponds to

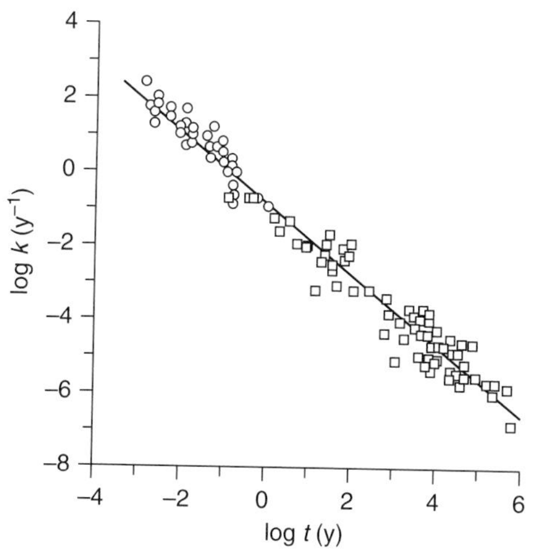

Figure 12.3. The rate constant, k, for organic matter degradation versus the age of the organic matter undergoing degradation determined from models of porewaters and laboratory experiments. Redrawn from Middelburg (1989). The line is drawn through the points.

the time span represented by the experimental data (Emerson and Hedges, 1988; Middelburg, 1989). This extends over eight orders of magnitude from days to millions of years (Fig. 12.3). A result of the high-order kinetics is that components of sedimentary mixtures that react more slowly become a greater fraction of unreacted material while the reaction rate constant (the curvature) is mainly described by the more labile components.

12.1.2 Organic matter diagenesis down the redox progression

The relationships among the flux of organic carbon to the sediment–water interface and its diagenesis and burial in deep-ocean sediments where oxygen is the primary electron acceptor are depicted in Fig. 12.4 (Emerson *et al.*, 1985). In this case one identifies two cases: one that is carbon-limited, where measurable oxygen persists in the porewaters at a depth of about one meter, and one that is oxygen-limited, in which oxygen goes to zero at some depth within the bioturbated zone. It has been shown that the measurements of carbon flux in near-bottom sediment traps are consistent with the flux of oxygen into the sediments. Examples of carbon-limited diagenesis (Fig. 12.4B) are in locations of relatively low particulate organic carbon flux to the sediments such as the pelagic North Pacific and Atlantic and some carbonate-rich locations in the Western Equatorial Pacific. Diagenesis in the rest of marine sediments is oxygen-limited.

Nitrate plays an important role as a tracer of both oxic and anoxic (NO_3^--reducing) diagenesis because it is produced during organic matter oxidation by O_2 and consumed during denitrification (the oxidation of NO_3^-) (Table 12.2). These relations have been used to infer the depth of the zone of oxic diagenesis (Bender *et al.*, 1977), and models of this process estimate that the contribution of denitrification to total sediment organic matter degradation is 7%–11%, resulting in a global denitrification rate in sediments of *c.*18×10^{12} mol N y^{-1} (Middelburg *et al.*, 1996). The latter value is at least a factor

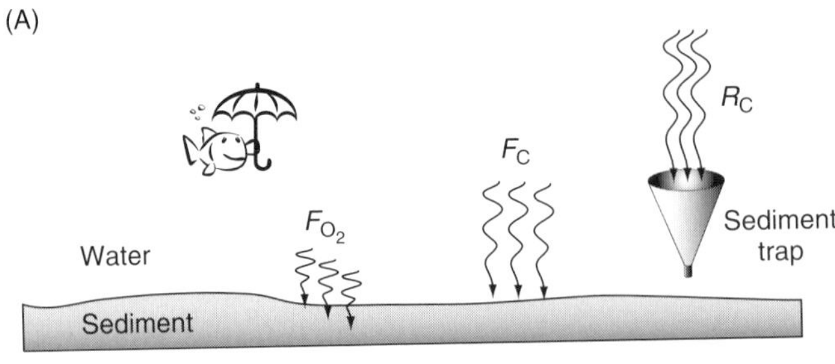

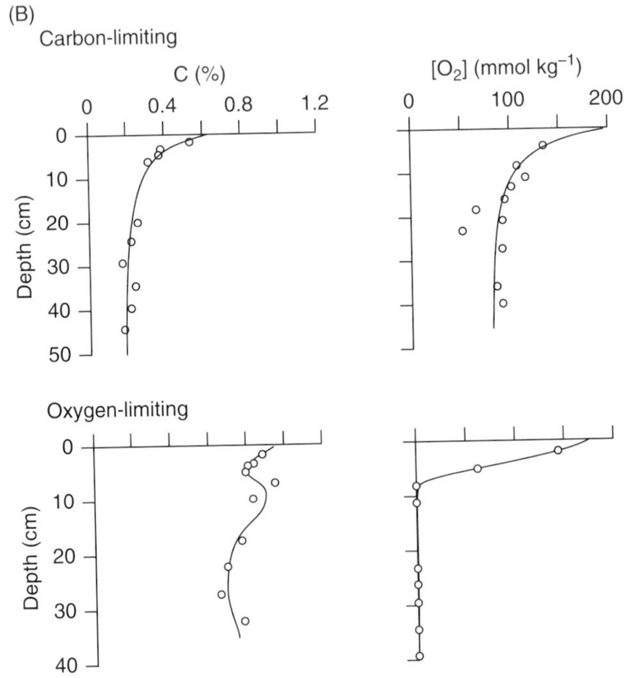

Figure 12.4. (A) A schematic representation of the fluxes of organic matter and oxygen at the sediment–water interface of the deep sea (from Emerson *et al.*, 1985). R_c is the rain rate of particulate organic carbon and F_c and F_{O_2} are the fluxes of organic carbon and oxygen at the sediment–water interface. (B) Data for sediment profiles indicating the carbon- and oxygen-limiting cases. The carbon-limiting case is redrawn from Grundmanis and Murray 1982), and the oxygen-limited data are from Murray and Kuivila (1990).

of two greater than water column denitrification, indicating the importance of marine sediments as a sink for fixed nitrogen.

Manganese and iron oxides are two solid-phase electron acceptors that play important roles in organic matter degradation. Their effect is limited by the lability of the solid to dissolution, which is not easily quantified experimentally, creating another unknown in models of these reactions. Manganese is reduced nearly simultaneously with nitrate, which is consistent with the comparable amounts of free energy available (Table 12.1). The Mn^{2+} produced is then either transported to the overlying water or reoxidized by oxygen. This relocation process is ubiquitous in oxygen-limited areas of marine sediments and creates Mn enrichment in surface sediments of the deep ocean.

As one approaches continents from the deep ocean, overlying productivity becomes greater as the water depth shoals and particles are degraded less while sinking through the shallower water column. Both factors increase the particulate organic matter flux to the sediment-water interface. This creates more extensive anoxia in the

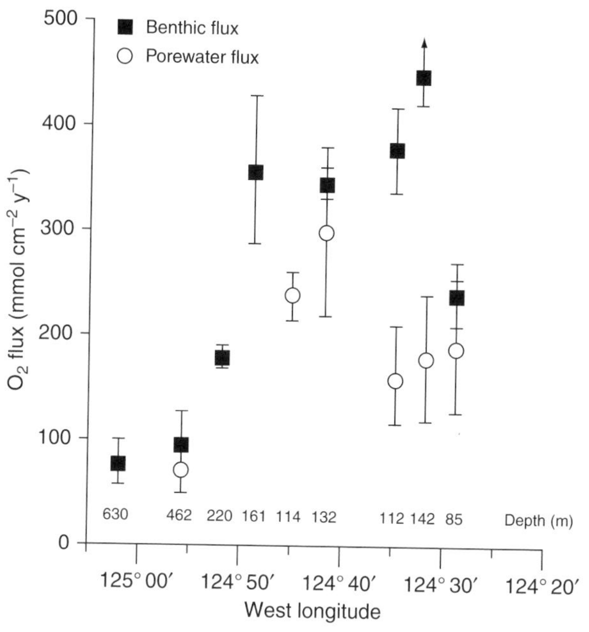

Figure 12.5. Benthic oxygen fluxes as a function of longitude and depth on the northwest US continental margin. The fluxes were determined using either benthic lander measurements (boxes) or calculated from micro-electrode oxygen gradients in the top few centimeters of porewaters, assuming molecular diffusion (circles). Error bars are the standard deviation of replicate measurements. Redrawn from Archer and Devol (1992).

sediments, which is sometimes compounded on continental slopes by low bottom-water oxygen conditions. A natural result of the greater supply of organic matter to the sediments is that benthic animals become bigger and more diverse. The consequence to organic matter diagenesis is that bioturbation is deeper and more intense and that animal irrigation activities are rapid enough to compete with molecular diffusion as the mechanism of porewater transport (Aller, 1984).

The relative roles of diffusion and animal-induced advection across the sediment–water interface in near-shore sediments has been quantified by comparing oxygen fluxes determined by benthic chamber measurements (in which a volume of water is isolated and changes in concentration are measured inside the chamber) with those calculated from porewater micro-electrode oxygen profiles (Fig. 12.5) (Archer and Devol, 1992). As one progresses up the continental slope and onto the shelf of the northwest United States the fluxes determined by these two methods diverge. Those determined from the benthic lander become greater at depths shallower than about 100 m, indicating the local importance of animal irrigation activity. This process complicates the diagenetic redox balance in coastal marine sediments, where 80%–90% of marine organic matter is buried, because it is much more difficult to generalize about the mechanism and magnitude of animal irrigation than molecular diffusion.

The intense interplay between the redox coupling of iron and manganese and transport by animal activity has been demonstrated in the sediments of the eastern Skagerrak between Denmark and Norway (Wang and Van Cappellen, 1996). Sediment porewater profiles from this area (Fig. 12.6) indicate that most Mn(IV) reduction is coupled to oxidation of Fe(II) which was formed during organic matter and H_2S

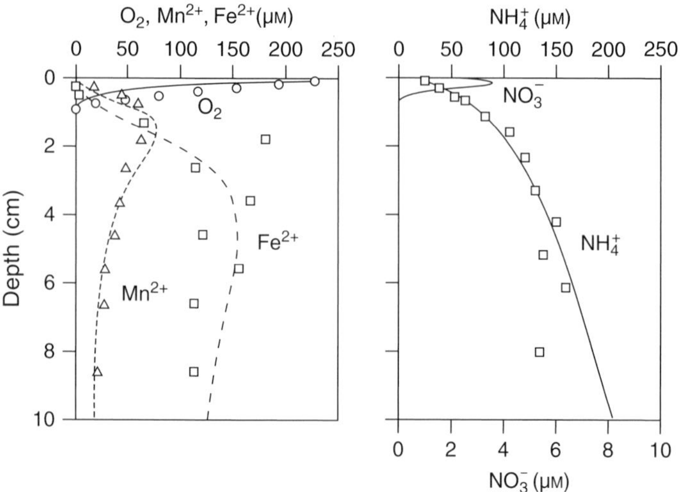

Figure 12.6. Porewater profiles of O_2, Fe(II), Mn(II), NO_3^- and NH_4^+ from sediments of the near-shore waters of Denmark. Symbols represent data from Canfield *et al.* (1993) and lines are model results from Wang and Van Cappellen (1996). Redrawn from Wang and Van Cappellen (1996).

oxidation. Again the manganese and iron redox cycles shuttle electrons between more oxidized and reduced species. Adsorption of the reduced dissolved form of these metals to sediment surfaces plays an important role in their reactivity and transport by bioturbation and irrigation back to the surface sediments where they are reoxidized or transported to the overlying waters. In general, most of the Fe redox cycling occurs within the sediments because of the relatively rapid oxidation kinetics of Fe(II) while some of the recycled Mn(II) escapes to the bottom waters because it is reduced nearer the sediment–water interface and has slower oxidation kinetics.

Shallow environments at the mouths of tropical rivers are the deposition sites of about 60% of the sediment delivered to the ocean. In some of these locations seasonal resuspension of the sediments occurs to a depth of 1–2 m to form fluid muds. This setting creates a very different type of sediment diagenesis that is characterized by intense iron and manganese reactions but little sulfate reduction or methane formation. Because organic matter is abundant at these shallow river-mouth locations, oxygen is depleted relatively rapidly after deposition. Abundant oxidized iron and manganese in these highly weathered sediments are reduced, creating massive, time-dependent increases in porewater iron and manganese (Aller *et al.*, 1986). The period of diagenesis, however, is not long enough between resuspension events for sulfate reduction to become established. This situation is one of extreme non-steady-state diagenesis in which porewater transport is dominated by physical mechanisms rather than animal irrigation or molecular diffusion.

In near-shore regions where organic matter flux to the sediments is high or bottom water oxygen concentrations are low and horizontal sediment transport does not dominate, sulfate reduction and subsequent methane formation are important processes. Early measurements of SO_4^{2-} and CH_4 in marine porewaters indicated that methane appears only after most of the SO_4^{2-} has been reduced

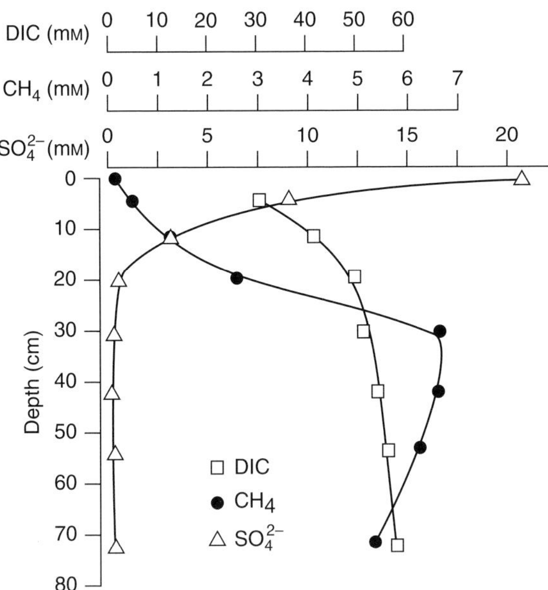

Figure 12.7. Porewater profiles of SO_4^{2-}, CH_4 and ΣCO_2 from the sediments of Scan Bay, Alaska, an anoxic fjord. Note that CH_4 and SO_4^{2-} concentrations do not overlap substantially. Redrawn from Reeburgh (1980).

(Fig. 12.7), creating profiles that do not overlap significantly, like those of O_2 and Mn(II) and Fe(II) and NO_3^-. Reeburgh (1980) suggested that the porewater distributions of SO_4^{2-} and CH_4 indicate that CH_4 is being oxidized anaerobically with SO_4 being the electron acceptor. This suggestion, which is virtually unavoidable based on the metabolite distributions and interpretation by diffusion equations, was not accepted initially by many microbiologists because it has been difficult to culture the SO_4^{2-}-reducing / CH_4-oxidizing bacteria, but is now recognized to be widespread in the marine environment because of the high sulfate concentrations in seawater. This is not true in freshwater systems where abundant CH_4 production occurs because organic matter is abundant and SO_4^{2-} concentrations are low.

12.1.3 Benthic respiration

Benthic flux measurements from bottom chamber devices and porewater flux determinations have been used to estimate the rain rate of organic matter to the sediment-water interface. When compared with global primary production rates and sediment trap particle fluxes, these data indicate that about 1% of the primary production reaches deep-sea sediments and is oxidized there (Table 12.3) (Jahnke, 1996). It has also been demonstrated from benthic flux experiments that about 45% of respiration in the ocean below 1000 m occurs within sediments.

12.1.4 Factors controlling organic matter degradation

There are many factors that contribute toward the seemingly universal slowing in organic matter decomposition with time. One of these is that the physical form and distribution of organic matter within sediments is not uniform. A second is that the rate and extent

Table 12.3. *Comparison of benthic oxygen fluxes at the sediment–water interface and primary production (PP) in the ocean's euphotic zone*

Latitude	PP	Benthic flux	% of PP
	(10^{14} mol C y^{-1})		
10° N–10° S	2.08	0.020	1.0
11° N–37° N	2.67	0.026	1.0
38° N–50° N	0.83	0.008	1.0
50° N–60° N	0.41	0.005	1.1

Jahnke (1996).

of organic matter degradation can vary with the different inorganic electron acceptors available at different stages of degradation. Finally, the structural features of the residual organic matter mixture may vary over time as more readily utilized components are oxidized or converted into less reactive products.

It has long been recognized that organic matter tends to concentrate in fine-grained continental margin sediments, as opposed to coarser silts and sands. Over the past decade it has become clear that organic matter and fine-grained minerals in marine sediments are physically associated. One line of evidence is that only a small fraction (*c.*10%) of the bulk organic matter in unconsolidated marine sediments can be separated as discrete particles by flotation in heavy liquids or hydrodynamic sorting (Mayer, 1994). In addition, the concentrations of organic carbon in bulk sediments and their size fractions increase directly with external mineral surface area as measured by N_2 adsorption (Fig. 12.8).

Most sediments collected under oxic waters along continental margins exhibit organic carbon (OC) concentrations on the order of 0.5–1.0 mg OC m^{-2} (e.g. the "typical shelf" area in Fig. 12.8), a "loading" that is similar to that expected for a single layer of protein spread uniformly across the surfaces of mineral grains. The notion that organic matter might be spread one molecule deep on essentially all mineral grains implies sorption of previously dissolved organic substances that are physically shielded on the mineral surface from direct degradation by bacteria and their exoenzymes. Evidence in support of the protective function came from the demonstration that over 75% of dissolved organic matter desorbed from sedimentary minerals deposited for hundreds of years could be respired within five days once removed from this matrix (Keil *et al.*, 1994).

Although the concept that sedimentary organic matter is strongly associated with mineral surfaces has stood the test of time, the monolayer-equivalent hypothesis has not. It has been deduced from the energetics of gas adsorption onto minerals from continental margin sediments that generally less than 15% of the surfaces of typical sedimentary minerals are coated with organic matter (Mayer, 1999).

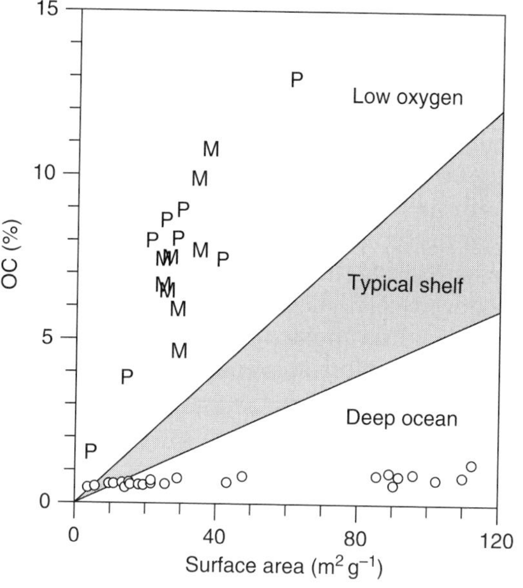

Figure 12.8. Weight percentages of organic carbon (%OC) plotted versus mineral surface area for surface sediments from a range of depositional regimes. M and P represent data for samples from the Mexican and Peruvian margins, respectively. Circles indicate sediment from oxic deep-ocean sediments. From Hedges and Keil (1995).

Physical protection alone is insufficient to explain the distribution of organic matter in marine sediments. For example, marine sediments deposited under bottom waters with little or no dissolved O_2 usually have surface-normalized organic carbon concentrations substantially greater than 0.5–1.0 mg OC m^{-2} (points M and P in Fig. 12.8), whereas fine-grained deep-sea clays typically contain a tenth or less of the organic concentration exhibited by continental margin sediments of equivalent surface area (data indicated by circles in Fig. 12.8). In particular, additional processes must account for the fact that open-ocean sediments that cover approximately 80% of the total sea floor account for less than 5% of global organic carbon burial.

A commonly made assumption in descriptions of sedimentary diagenesis is that degradation rate and extent are largely controlled by the "quality" of available organic substrate(s), as opposed to the relative supply of different electron acceptors. This perspective is supported by a variety of field and laboratory studies. In particular, freshly dissolved organic substrates and polysaccharide- and protein-rich materials are often degraded at similar rates in the presence or absence of molecular oxygen (Westrich and Berner, 1984). On the other hand, some laboratory experiments show much slower and less efficient anoxic degradation of aged organic matter and carbon-rich substrates such as lipids and pigments. Harvey *et al.* (1995) observed that the total carbon, total nitrogen, protein, lipid and carbohydrate fractions of a diatom and a coccolith were all more rapidly degraded in oxic versus anoxic laboratory incubations. Lignin, a biomacromolecule that is carbon-rich, insoluble in water and difficult to hydrolyze, is very sparingly degraded in the absence of O_2 (Hedges *et al.*, 1985).

This apparent contradiction may be partly explained by selective initial use of easily degraded proteins and polysaccharides and the resulting concentration of carbon-rich, hydrolysis-resistant substrates such as lipids and lignin whose effective degradation requires O_2. The rate-determining step for both aerobic and anaerobic microbial degradation of polysaccharides and proteins is hydrolysis by extracellular enzymes, after which the released oligosaccharides and peptides less than about 600 atomic mass units are taken into cells for further alteration. Given this commonality and the fact that molecular oxygen is not required in the initial depolymerization phase, it is not surprising that these two major biochemical types often are both degraded effectively, although not necessarily at the same rates, under both oxic and anoxic conditions. In contrast, effective degradation of carbon-rich substrates and hydrolysis-resistant materials such as lignin, hydrocarbons and pollen requires molecular oxygen, as opposed to simply addition of water. Such degradation is often accomplished by O_2-requiring enzymes that catalyze electron (or hydrogen) removal or directly insert one or two oxygen atoms into organic molecules (Sawyer, 1991).

The most direct field evidence that the extent of sedimentary organic matter preservation is affected by exposure to bottom water oxygen comes from oxidation fronts in deep-sea turbidites of various ages and depositional settings (Wilson *et al.*, 1985). One of these deposits in which the timing of the exposure to oxic and anoxic conditions is well documented is the relict f-turbidite from the Madeira Abyssal Plain (MAP) about 700 km offshore of northwest Africa (Cowie *et al.*, 1995). This *c.*4 m thick deposit was emplaced approximately 140 000 y ago at a water depth of *c.*5400 m when fine-grained carbonate-rich sediments slumped off the African continental slope and flowed down to cover the entire MAP with a texturally and compositionally uniform layer. This deposit was subsequently exposed to oxygenated bottom water for thousands of years, during which time an oxidation front slowly penetrated approximately 0.5 m into the turbidite before diffusive O_2 input was halted by accumulating sediment and the entire turbidite relaxed back to anoxic conditions. Porewater sulfate concentrations measured within the sediments indicate little or no *in situ* sulfate reduction.

Comparative elemental analyses of the upper and lower sections of two sediment cores collected on the MAP abyssal plain show that organic concentrations decreased at both locations from values of 0.93–1.02 wt% OC below the oxidation front to values 0.16–0.21 wt% within the surface oxidized layer (Fig. 12.9). Pollen abundances decreased in the same samples from about 1600 grains g^{-1} below the oxidation front to zero above it. Overall, 80% of the organic matter and essentially all of the pollen that has been stable for 140 000 y in the presence of porewater sulfate was degraded in the upper section of the MAP cores as a result of long-term exposure to dissolved O_2.

The broad implication of these observations is that, somewhere between upper continental margins and the deep ocean, depositional conditions lead to greatly increased exposure times of sedimentary organic matter to O_2 that are sufficient to create the greatly reduced

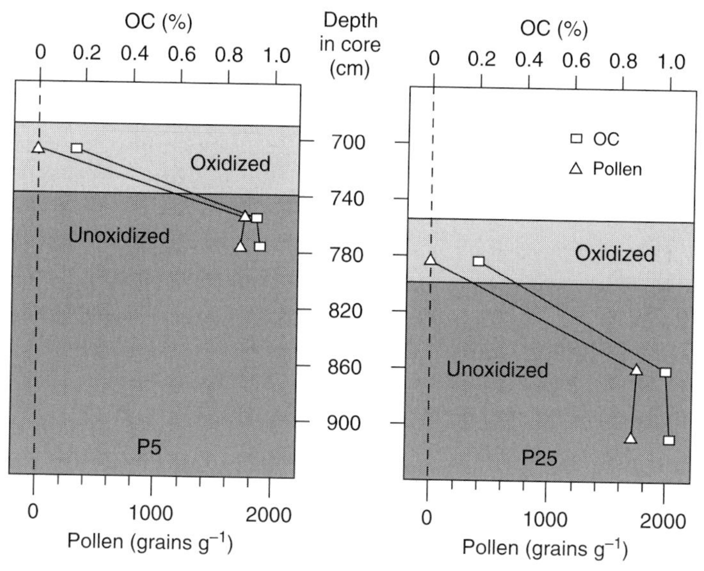

Figure 12.9. Profiles of the weight percent organic carbon (%OC) and pollen abundances down two sequences of the f-turbidite from the Madeira Abyssal Plain. "Oxidized" sediments are those that have been exposed to oxygen after thousands of years of burial. See text for significance of these results. From Cowie *et al.* (1995).

organic matter concentrations typical of modern pelagic sediments (e.g. Fig. 12.8). Thus, over time spans of hundreds to thousands of years, exposure to molecular oxygen appears to affect both the amount and composition of organic matter preserved in continental margin sediments.

12.2 Diagenesis and preservation of calcium carbonate

Between 20% and 30% of the carbonate produced in the surface ocean is preserved in marine sediments. The fraction of $CaCO_3$ produced that is buried dramatically affects the alkalinity and DIC of seawater, and is thus important for understanding the processes that control the partial pressure of carbon dioxide in the atmosphere. Paleoceanographers have observed that the $CaCO_3$ content of marine sediments has changed with time in concert with glacial–interglacial periods. By studying the mechanisms that presently control $CaCO_3$ preservation, one seeks also to understand what past changes imply about the chemistry of the ocean through time.

Sedimentary calcium carbonates are formed as the shells of marine plants and animals. Biologically produced $CaCO_3$ consists primarily of two minerals: aragonite and calcite. Shallow-water carbonates, primarily corals and shells of benthic algae (e.g. *Halimeda*) are heterogeneous in their mineralogy and chemical composition but are composed mainly of aragonite and magnesium-rich calcite (see Morse and Mackenzie (1990) for a discussion). Carbonate tests of microscopic plants and animals, most of which live in the surface ocean (there are also benthic animals that produce carbonate shells), are primarily made of the mineral calcite, which composes the bulk of the $CaCO_3$

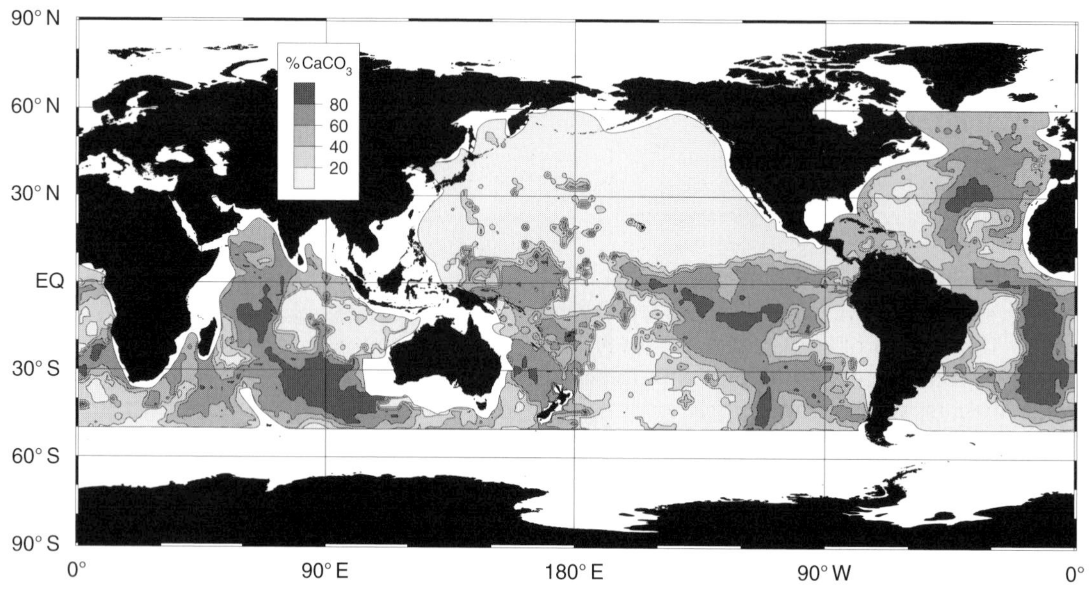

Figure 12.10. Global distribution of the weight percent of $CaCO_3$ in surface sediments of the ocean deeper than 1000 m. Redrawn from Archer (1996).

in deep-ocean sediments. A large fraction of the ocean floor consists of $CaCO_3$ from these tests (Fig. 12.10). Note that the topographic rises on the ocean floor are $CaCO_3$-rich, whereas the abyssal planes are barren of this mineral. The other noticeable major trend is that there is relatively little $CaCO_3$ in the sediments of the North Pacific. We focus next on the processes that control these distributions.

12.2.1 Mechanisms of $CaCO_3$ dissolution and burial: thermodynamics

The solubility of $CaCO_3$ in seawater has been studied extensively because of its great abundance in sedimentary rocks and the ocean. The equation for dissolution of pure calcium carbonate:

$$CaCO_3(s) \rightleftarrows Ca^{2+} + CO_3^{2-} \tag{12.4}$$

has the simple "apparent" solubility product in seawater:

$$K'_{sp} = [Ca^{2+}]\,[CO_3^{2-}]. \tag{12.5}$$

The apparent constant, K'_{sp}, is related to thermodynamic constants, K_{sp}, via the total activity coefficients of Ca^{2+} and CO_3^{2-}(see Chapter 3). Apparent constants are usually used in seawater because the constants are determined in this medium in the laboratory. The saturation state of seawater with respect to the solid is sometimes denoted by the Greek letter omega, Ω:

$$\Omega = \frac{[Ca^{2+}]\,[CO_3^{2-}]}{K'_{sp}}. \tag{12.6}$$

The numerator of the right side is the product of measured total concentrations of calcium and carbonate in the water, or the ion concentration product (ICP). If $\Omega = 1$ then the system is in equilibrium and should be stable. If $\Omega > 1$, the waters are supersaturated, and the laws of thermodynamics would predict that the mineral should precipitate, removing ions from solution until Ω returned to unity. If $\Omega < 1$ the waters are undersaturated, the solid $CaCO_3$ should dissolve until the solution concentrations increase to the point where $\Omega = 1$ (see Fig. 3.7). In practice it has been observed that $CaCO_3$ precipitation from supersaturated waters is rare, probably because the presence of the high concentrations of Mg in seawater blocks nucleation sites on the surface of the mineral. Supersaturated conditions thus tend to persist. Dissolution of $CaCO_3$, however, does occur when $\Omega < 1$ and the rate is readily measurable in laboratory experiments and inferred from porewater studies of marine sediments. Since calcium concentrations are nearly conservative in the ocean, varying by only a few percent, it is the apparent solubility product, K'_{sp}, and the carbonate ion concentration that largely determine the saturation state of the carbonate minerals.

The apparent solubility products of calcite and aragonite have been determined repeatedly in seawater solutions. We adopt the values of Mucci (1983): K'_{sp} for calcite and aragonite $= 4.35 \pm 0.20 \times 10^{-7}$ and $6.65 \pm 0.12 \times 10^{-7} \mathrm{mol^2\,kg^{-2}}$, respectively, at 25 °C, S = 35 and one atmosphere. These data agree within error to measurements determined previously and represent many repetitions to give a clear estimate of the reproducibility (*c.* ±5%).

Because of the great depth of the ocean, the most important physical property determining the solubility of carbonate minerals in the sea is pressure. The pressure dependence of the equilibrium constants is related to the difference in volume, ΔV, occupied by the ions of Ca^{2+} and CO_3^{2-} in solution versus in the solid phase. The volume difference between the dissolved and solid phases is called the partial molal volume change, ΔV (see also Appendix A4.1):

$$\Delta V = V_{Ca} + V_{CO_3} - V_{CaCO_3}. \tag{12.7}$$

The change in partial molal volume for calcite dissolution is negative, meaning that the volume occupied by solid $CaCO_3$ is greater than the combined volume of the component Ca^{2+} and CO_3^{2-} in solution. Since with increasing pressure Ca^{2+} and CO_3^{2-} prefer the phase occupying the least volume, calcite becomes more soluble with pressure (depth) by a factor of about two for a depth increase of 4 km (Table. A4.2). Values of the partial molal volume change determined by laboratory experiments and *in situ* measurements result in a range of 35–45 $\mathrm{cm^3\,mol^{-1}}$ (Sayles, 1980). The uncertainty in this value is thus approximately ±10 %.

The final important factor affecting the solubility of $CaCO_3$ in the ocean is the concentration of carbonate ion. The high ratio of organic carbon to carbonate carbon in the particulate material degrading and dissolving in the deep sea causes the deep waters to become more acidic and carbonate-poor as they progress along the conveyor belt

Figure 12.11. Cross sections of the carbonate ion concentration in the major ocean basins determined during the WOCE program. Robert Key, personal communication; Key *et al.* (2004).

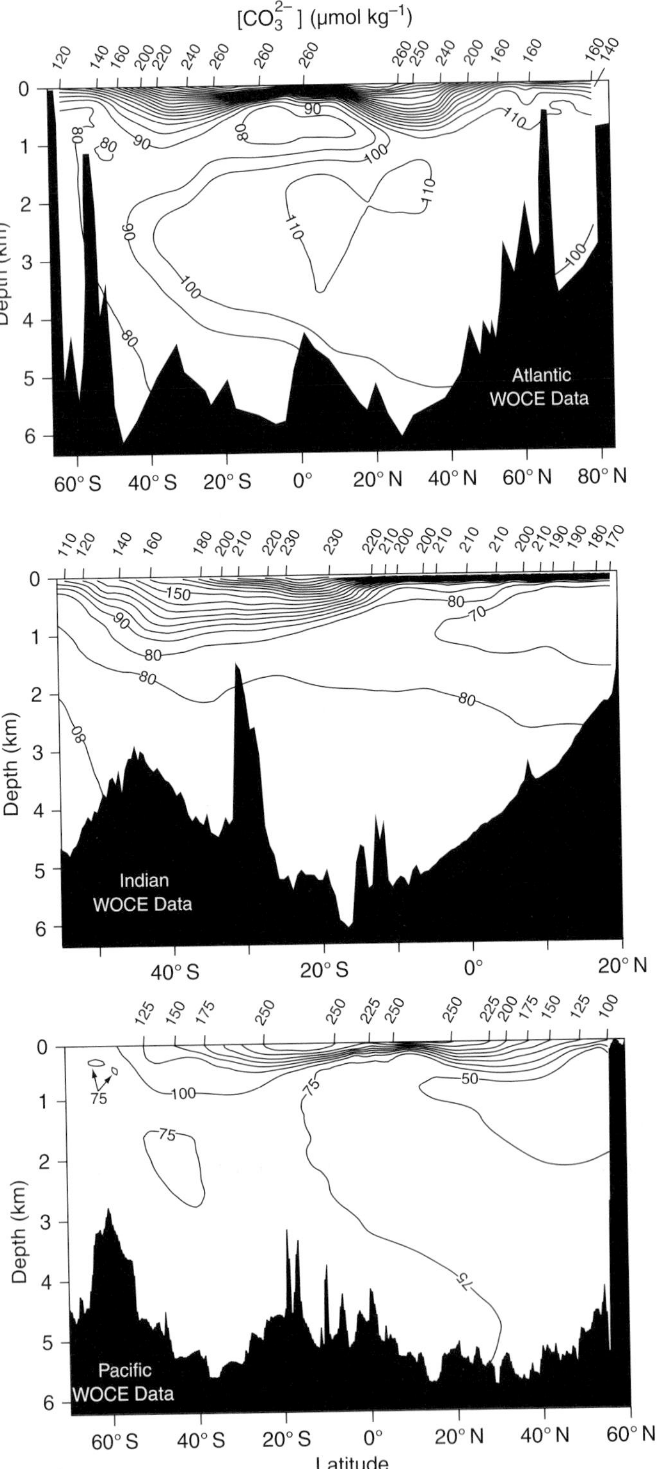

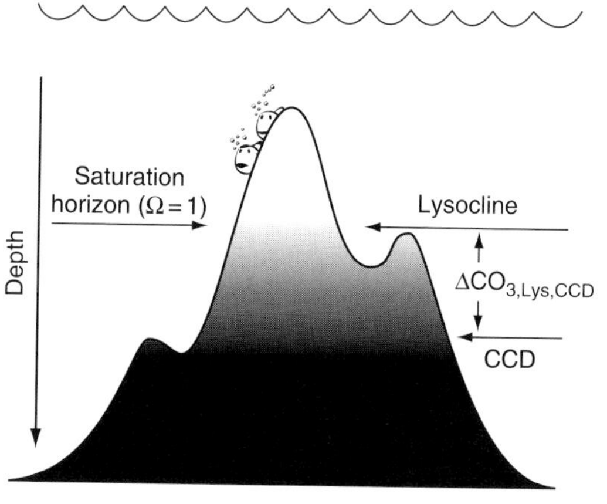

Figure 12.12. A sketch of the carbonate content of deep sea sediments as a function of depth. Lighter shades indicate greater $CaCO_3$ content in the sediments. Horizontal arrows indicate theoretical relations among the depths of the lysocline (where $CaCO_3$ shows visible signs of dissolution), the carbonate compensation depth, CCD (where the $CaCO_3$ concentration drops to zero) and the saturation horizon ($\Omega = 1$).

circulation network from the North Atlantic to deep Indian and Northern Pacific Oceans (see Chapter 4). Carbonate ion concentrations change from *c.*250 μmol kg^{-1} in surface waters to mean values in the deep waters of 113 μmol kg^{-1} in the Atlantic, 83 in the Indian and 70 μmol kg^{-1} in the deep North Pacific Oceans (Fig. 12.11). There is little vertical difference in these values between 1500 and 4000 m in the Atlantic and below 1500 m in the Indian and Pacific. Thus the tendency for $CaCO_3$ minerals to be preserved is greatest in surface waters of the world's oceans and decreases "downstream" in deep waters from the Atlantic to Indian to Pacific Oceans. The mean saturation horizon for calcite shoals from a depth of about 4.5 km in the Equatorial Atlantic to 3.0 km in the Indian Ocean and South Pacific to less than 1.0 km in the North Pacific (Feely *et al.*, 2002).

There have been many attempts to correlate the presence of calcite in marine sediments with the degree of saturation in the overlying water. The sketch in Fig. 12.12 demonstrates the ideal relation between the "saturation horizon" in the water (where $\Omega = 1$) and the presence of $CaCO_3$ in the sediments. The terminology for the presence of $CaCO_3$ in sediments is a little esoteric, with the word "lysocline" adopted as the depth at which there is the first indication of dissolution of carbonates and "carbonate compensation depth" (CCD) being the depth where the rain rate of calcium carbonate to the sea floor is exactly compensated by the rate of dissolution of $CaCO_3$ (i.e. where there is no longer burial of $CaCO_3$). It would seem that one could determine the importance of thermodynamics in determining calcite preservation by letting the ocean do the work and simply comparing lysocline and saturation horizon relations. There are two main problems with these attempts at direct observation. The first is the poor accuracy with which we know the degree of saturation in the ocean because of the error on the value of K'_{sp}. The second is our inability to precisely determine the onset of $CaCO_3$ dissolution within sediments based on measurements of the amount of $CaCO_3$ observed there. For example, if 90% of the particle rain rate is $CaCO_3$, a 50% decrease in the $CaCO_3$ flux to the sediments would result in a change in the sediment composition of only from 90% to 82% $CaCO_3$. These uncertainties contribute to errors in the depth of the saturation horizon and lysocline of about ± 0.5 km (Emerson and Hedges, 2003).

While errors in evaluating the depths of both the saturation horizon and the onset of $CaCO_3$ dissolution complicate "field" tests of the importance of chemical equilibrium, the change in the degree of calcite saturation ($\Delta CO_3 = [CO_3^{2-}]_{sat} - [CO_3^{2-}]_{in\ situ}$) over the depth range of the transition between calcite-rich and calcite-poor sediments is clearer, because it is easier to know the depth difference in both ΔCO_3 and % $CaCO_3$ than the absolute values. The difference in ΔCO_3 between the value at the lysocline and that at the CCD, $\Delta CO_{3,lys\text{-}CCD}$ ($= \Delta CO_{3,lys} - \Delta CO_{3,CCD}$), has been mapped in all areas of the ocean where both $[CO_3^{2-}]$ in the bottom water and $CaCO_3$% in sediments have been determined (Archer, 1996). The ocean mean is 19 ± 12 μmol kg^{-1} ($n = 30$), which represents a mean depth difference between the lysocline and CCD of greater than one kilometer. The simple fact that the transition from $CaCO_3$-rich to $CaCO_3$-poor sediments occurs over a broad range of ΔCO_3 values indicates that the pattern of $CaCO_3$ preservation cannot be based on thermodynamics alone, because thermodynamic equilibrium between seawater and the sediments requires an abrupt transition between calcite-rich and calcite-poor sediments.

12.2.2 Mechanisms of $CaCO_3$ dissolution and burial: kinetics

The dissolution rates of the minerals of calcium carbonate have been shown in laboratory experiments to follow the rate law:

$$R = k \cdot \left(K'_{sp} - \text{ICP}\right)^n, \qquad (12.8)$$

where k is the dissolution rate constant, which has units necessary to match those of the rate. The exponent, n, is one for diffusion-controlled reactions and usually some higher number for surface-controlled reaction rates (see Chapter 9). The most extensive laboratory measurements of the dissolution of carbonates (Keir, 1980) employed a steady-state "chemostat" reactor to measure the dissolution rate of reagent-grade calcite, coccoliths, Foraminifera, synthetic aragonite and pteropods. In these experiments the rate constant varied by a factor of about 100 between the different forms of calcite (after making the correction for surface area) and the data were interpreted with $n = 4.5$ order kinetics. While these measurements are still the standard for calcium carbonate dissolution rate kinetics, the high-order kinetics have been reinterpreted with more defendable K'_{sp} values to have a rate law that has an order of $n = 1$–2 (Hales and Emerson, 1997a) (Fig. 12.13). This result agrees much more closely with *in situ* aragonite dissolution rate experiments (Acker *et al.*, 1987) and dissolution rate laws determined for other minerals, so we adopt it as more likely than $n = 4.5$. Note that the units of the rate constant (Fig. 12.13; $k = 0.38$ d^{-1} or 38% d^{-1}) are normalized to the concentration of solid in the experimental reactor. A convenient way to view this rate constant is that the

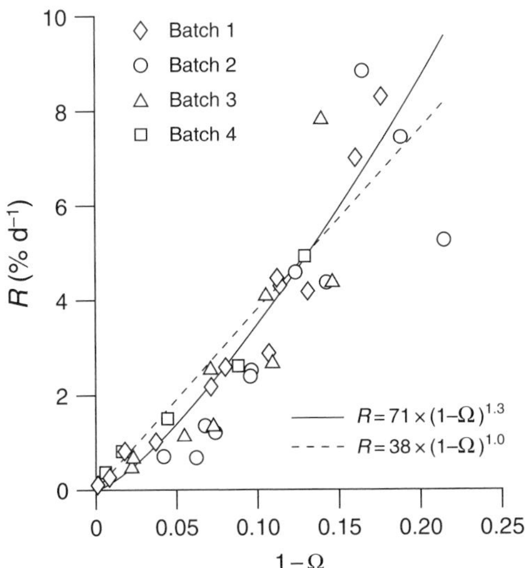

Figure 12.13. The dissolution rate (R) for calcite from the laboratory experiments of Keir (1980) as a function of the degree of undersaturation, $(1 - \Omega)$. Lines represent the rate laws with exponents of $n = 1.3$ and 1.0 as reinterpreted using more realistic values for the saturation equilibrium constant by Hales and Emerson (1997a).

units represent moles CO_3^{2-} cm^{-3} d^{-1} released to the water per mole of solid $CaCO_3$ cm^{-3} suspended in the solution, thus mol cm^{-3} d^{-1}/mol $cm^{-3} = d^{-1}$. When using the rate constant to calculate dissolution in the environment, the units must be "denormalized."

One of the great uncertainties in our understanding of the kinetics of $CaCO_3$ dissolution at this time is that the dissolution rate constants required to interpret ocean observations are much smaller than those measured in the laboratory. This has been illustrated rather simply by applying the observed laboratory kinetic rate constant to determine the ΔCO_3 necessary to produce the transition in $CaCO_3$ concentrations observed in marine sediments (see Emerson and Hedges, 2003). The result is that a kinetic rate constant of between 0.038 and 0.0038 d^{-1} is required to match the observed depth between the lysocline and CCD, and that the laboratory-determined value of 0.38 d^{-1} (Fig. 12.13) creates a $CaCO_3$-preservation transition that is much too abrupt. This result has also been confirmed by *in situ* porewater measurements (see later).

Although relatively slow dissolution of calcite can explain the gradual change between carbonate-rich and carbonate-poor sediments found in nature, there is another important issue that we have not considered: the role of organic matter degradation in sediments in promoting *in situ* calcite dissolution. It was long suspected that organic matter degradation would promote dissolution, but this was not quantified until the 1980s and 1990s. Two factors led to the realization that the *inorganic* kinetic interpretation of $CaCO_3$ burial in the ocean was incomplete. First, observations of the carbon content of particles that rain to the ocean floor collected in sediment traps 1 km or less above the bottom suggest that the molar organic carbon to $CaCO_3$ carbon rain ratio is about unity. A few centimeters below the sediment surface this ratio is more like 0.1, indicating that about 90% of the organic carbon that reaches the surface is degraded rather than buried. Secondly, sediment porewater studies in the same areas as sediment trap deployments show strong oxygen depletions in porewaters from the top few centimeters of sediment. Simple flux calculations require that the bulk of the organic matter degradation between the sinking particles and that which is buried takes place within the sediments (see Fig. 12.4). Thus, most particles that reach the sea floor are stirred into the sediments before they have a chance to degrade while sitting on the surface. If this were not the case, and the particles degraded entirely at the surface, there would be little oxygen depletion within the sediments.

Organic matter degradation within the sediments creates a micro-environment that is corrosive to $CaCO_3$ even if the bottom waters are not, because addition of DIC and no A_T to the porewater causes it to have a lower pH and smaller $[CO_3^{2-}]$. Using a simple analytical model and first-order dissolution rate kinetics, Emerson and Bender (1981) predicted that this effect should result in up to 50% of the $CaCO_3$ that rains to the sea floor being degraded even at the saturation horizon, where the bottom waters are saturated with respect to calcite. Because the percent $CaCO_3$ in sediments is relatively insensitive to dissolution and the exact depth of the saturation horizon is uncertain, this suggestion was well within the constraints of environmental observations.

The effect of organic-matter-driven $CaCO_3$ dissolution is to raise the $CaCO_3$ depth transition in sediments relative to the saturation horizon in the water column (Fig. 12.12). Because organic matter degradation promotes $CaCO_3$ dissolution even in saturated and supersaturated waters, the water column saturation horizon should be below the depth where sediment dissolution begins. The organic matter degradation effect on $CaCO_3$ dissolution should have little effect on the $\Delta CO_{3,\text{lys-CCD}}$ necessary to create the transition in percent $CaCO_3$, so it remains mainly controlled by the kinetics of dissolution.

The suggestion of "organic $CaCO_3$ dissolution" in sediments has been tested by determining the gradient of oxygen and pH in sediment porewaters. This had to be done on a very fine (millimeter) scale because the important region for the reaction is near the sediment–water interface. The test required *in situ* measurements because it has been shown the pH values of porewaters change when they are depressurized. To do this an instrument capable of traveling to the deep-sea sediment surface, slowly inserting oxygen and pH microelectrodes, one millimeter at a time, into the sediments and recording the data *in situ* was constructed (Archer *et al.*, 1989; Hales and Emerson, 1996). The results of these experiments, some of which are reproduced in Fig. 12.14, confirmed the suspicion that a significant amount of $CaCO_3$ dissolves because of organic matter degradation. The pH of the porewaters cannot be interpreted without assuming dissolution of $CaCO_3$ in response to organic matter degradation as measured by the porewater oxygen profiles. This process makes the burial of $CaCO_3$ in the ocean dependent not only on thermodynamics and kinetics but also on the particulate rain ratio of organic carbon to calcium carbonate carbon. The kinetic rate constants required to

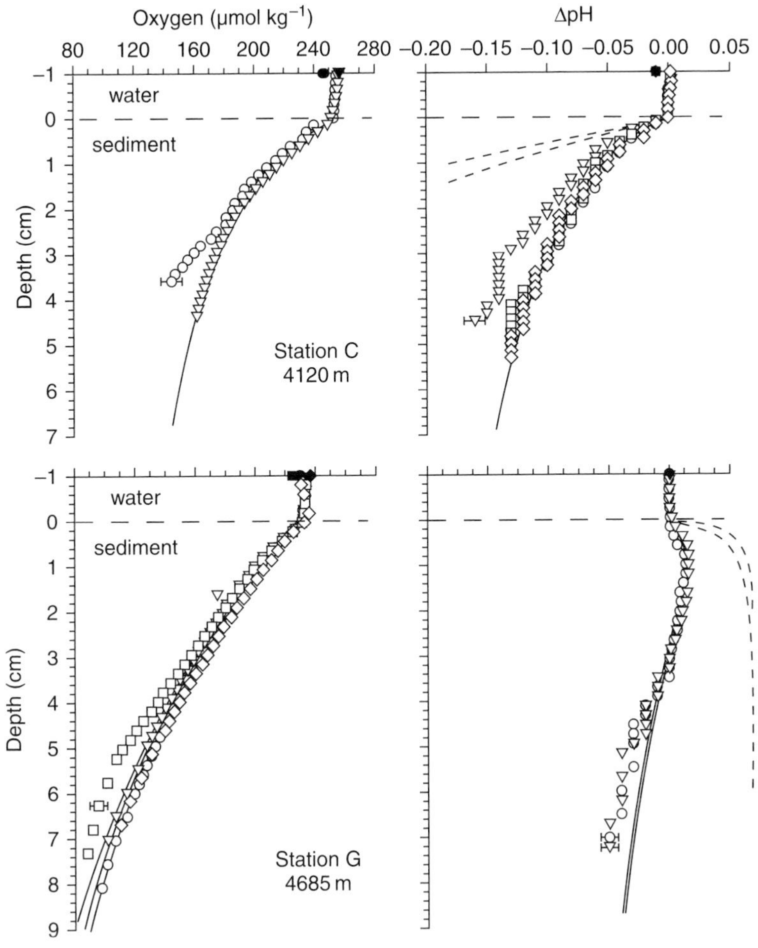

Figure 12.14. Porewater profiles of oxygen concentration and ΔpH (the pH difference between the value in the porewater and the value in bottom water) in the top c.10 cm of sediments from two locations on the Ceara Rise in the Equatorial Atlantic. Points are individual measurements, sometimes from different electrodes (different symbols) on the same deployment. Solid symbols in the overlying water are measurements in the bottom water after the porewater profile. Solid and dashed curves are model solutions. The dashed lines indicate the predicted ΔpH if there were no $CaCO_3$ dissolution caused by organic matter degradation in the sediments. Solid lines are the predicted pH for $CaCO_3$ dissolution in response to organic matter degradation using a dissolution rate more than 100 times slower than that determined in the laboratory experiments of Fig. 12.13. In the top graphs (Station C), the bottom waters are saturated or supersaturated with respect to calcite. In the bottom graphs (Station G), bottom waters are undersaturated with respect to calcite. Redrafted from Hales and Emerson (1997b).

model the measured pH profiles (Fig. 12.14) are more than two orders of magnitude smaller than that measured in laboratory experiments, confirming previous suggestions that the $\Delta CO_{3,\text{lys-CCD}}$ is too great to be explained by these relatively rapid kinetics.

There has always been some reluctance to assume that the pH in a porous medium is controlled solely by the carbonate buffer system in the porewaters. There are arguments that H^+ ions on particle surfaces can affect pH measurements and, even more importantly, comprehensive models of porewater chemistry are beginning to demonstrate that H^+ adsorption on mineral surfaces may play an important role in controlling the pH of porewaters. Conclusions of the pH measurements described here, however, have been confirmed by millimeter-scale measurements of both p_{CO_2} and Ca concentration in the porewaters (e.g. Wenzhofer *et al.*, 2001).

Observations from benthic flux experiments in which A_T and Ca fluxes have been measured both below and above the calcite saturation horizon confirm the effect of organic matter degradation on the dissolution of $CaCO_3$ in most but not all situations. Jahnke and Jahnke

(2004) interpret their benthic flux measurements at five locations in the world's ocean to show that in regions with sediments of very high $CaCO_3$ content there appears to be no evidence of metabolic dissolution of $CaCO_3$ based on A_T and Ca benthic fluxes. They suggest that the reason for this observation is that the pH of porewaters is controlled by adsorption of hydrogen ion and carbonate species on the surface of $CaCO_3$. Transport of these species to and from the sediment surface by bioturbation creates a flux that is more important than that caused by molecular diffusion in the porewaters.

A potential problem with the benthic flux interpretation is that it stems from very small chemical changes in the chambers of the benthic landers, because open-ocean areas high in $CaCO_3$ are locations with relatively low organic matter degradation rates. The observations, however, do not appear to agree with results from porewater pH, p_{CO_2} and Ca measurements. Until more consistent benthic flux and porewater measurements are achieved, there will be a lingering doubt about the importance of the organic matter degradation effect on $CaCO_3$ dissolution in calcium carbonate-rich sediments.

12.3 Diagenesis and preservation of silica

Biogenic silica, in the form of opal, makes up an important part of marine sediments, particularly in the southern and eastern equatorial oceans (Fig. 12.15). These deposits are formed primarily from tests of diatoms that lived in the surface oceans. More than half of the opal formed in the surface ocean is dissolved within the upper 100 m and only several percent of the production is ultimately buried in marine sediments (Nelsón *et al.*, 1995). Rewards to be gained by understanding the mechanisms that control opal diagenesis in sediments are evaluating the utility of the SiO_2 concentration changes in sediments as a tracer for past diatom production and understanding the role of authigenic silicates as a sink for major ions in marine geochemical mass balances.

The main tool for studying diagenesis and preservation of SiO_2 in marine sediments has been the measurement of silicic acid, H_4SiO_4, in porewaters which began more than 30 y ago and continues today (Fig. 12.16). The difficulty in interpreting these results has been that the asymptotic values in porewater profiles, the concentration that is achieved by 10 cm below the sediment–water interface, is highly variable geographically where solid opal is preserved in the sediments. Possible explanations for these observations fall into three general categories (McManus *et al.*, 1995). First, asymptotic values may be different because the solubility of opal formed in surface waters varies geographically. Second, porewaters may never achieve equilibrium but opal formed in surface waters has a number of phases of different reactivity which, in concert with sediment bioturbation, create different steady-state asymptotic values. Finally, diagenesis reactions within the sediments may create authigenic phases other than opal that control the porewater solubility and chemical kinetics of H_4SiO_4.

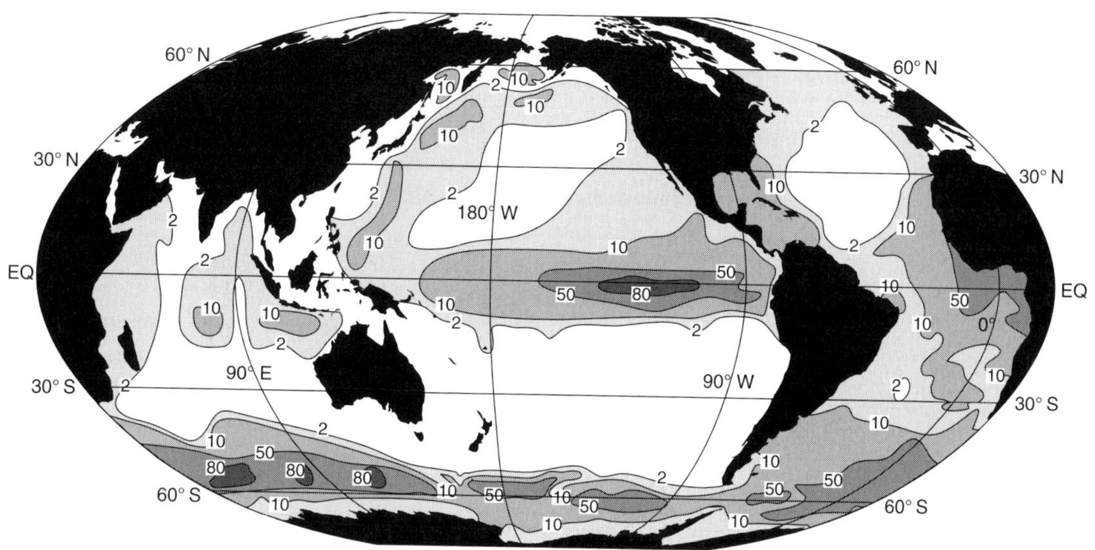

Figure 12.15. The global distribution of SiO_2 in marine sediments in weight %. Redrafted from Broecker and Peng (1982).

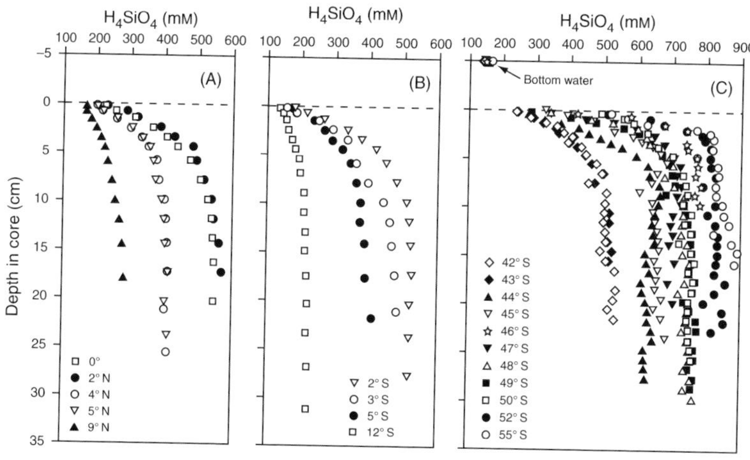

Figure 12.16. Porewater H_4SiO_4 concentrations as a function of depth in sediment from a north–south transect along 140° W in the Equatorial Pacific (A, B) (modified from McManus *et al.*, 1995) and a north–south transect through the Indian sector of the Polar Front (C) (modified from Rabouille *et al.*, 1997).

12.3.1 Controls on the H_4SiO_4 concentration in sediment porewaters: thermodynamics

Even before the wide variety of asymptotic porewater values in marine sediments were observed, it was shown that the H_4SiO_4 concentration obtained by incubating diatom frustrules obtained from net tows in surface waters was greater than the asymptotic value observed in porewaters. The chemical equilibrium solubility is much too high to explain porewater results. Recent experiments in sediments from the Southern Ocean using stirred flow-through reactors indicate saturation concentrations that range between 1000 and 1600 mol l^{-1} H_4SiO_4 (Van Cappellen and Qiu, 1997a). These values are much greater than those determined in the porewater profiles from the same location (Fig. 12.16), lending little support to the hypothesis that the asymptotic values are equilibrium concentrations of different opals produced in the surface waters. However, we shall see later that thermodynamics does play a role in explaining the observations.

12.3.2 Controls on H_4SiO_4 concentration in sediment porewaters: kinetics

Early studies of the rate of opal dissolution in the laboratory indicated that the dissolution rate of acid-cleaned planktonic diatoms varied as a linear function of the degree of undersaturation.

$$\frac{d[H_4SiO_4]}{dt} = k_{Si} \cdot S \cdot \left([H_4SiO_4]_{sat} - [H_4SiO_4] \right), \tag{12.9}$$

where k_{Si} is the rate constant for SiO_2 dissolution (cm s^{-1}) and S is the solid surface area (cm^2 cm^{-3}). Generally, the dissolution rate constants determined by these methods (see Fig. 9.6D) were much greater than those needed to model the porewater profiles, indicating some important differences between the laboratory experiments and field results. This is the same conclusion reached in comparing laboratory and *in situ* dissolution rates of $CaCO_3$!

The first attempt to use the dynamics of dissolution and burial in marine sediments to explain the porewater observations employed the assumption that sediments consisted of opal fractions of different reactivity: one that dissolves rapidly and completely and another that is essentially refractory (Schink *et al.*, 1975). The idea was that the first fraction, in concert with sediment bioturbation, sets the porewater asymptotic concentration of H_4SiO_4 and the refractory portion determines the sediment concentration of opal. The reason that bioturbation is important in defining the porewater concentration of H_4SiO_4 is that it stirs the opal deeper into the sediments, where dissolution is more effective in creating a strong concentration gradient. Explaining the field observations by this mechanism, for example in the Antarctic, where the asymptotic value varies by nearly a factor of two between 42° and 55° S (Fig. 12.16C) requires changes in the bioturbation coefficient over short distances that have not been measured. With more and more data this model seems to be unable to explain the observation.

Van Cappellen and Qiu (1997b) used flow-through reactor studies to demonstrate that the dissolution rate of unaltered Si-rich sediment from the Southern Ocean follows a rate law that is exponential rather than linear with respect to the degree of undersaturation. The implication is that the rate of dissolution is much greater near the sediment–water interface than below. In fact they find that, when the laboratory kinetic studies are applied to the sediments, SiO_2 dissolution is pretty much finished below depths of a few centimeters. The question remains as to what processes cause the kinetics of opal

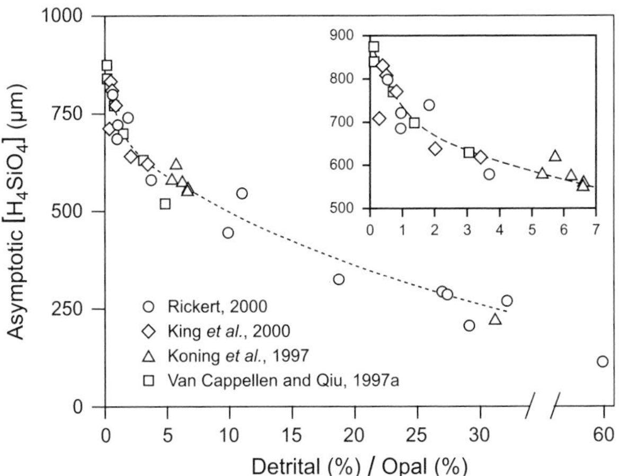

Figure 12.17. The relationship between the asymptotic H_4SiO_4 concentration in porewaters of Southern Ocean sediments and the relative detrital content of the sediments (P. Van Cappellen, personal communication). Sediments with more detrital (aluminosilicate) material support lower asymptotic values.

dissolution to change as the mineral ages in marine sediments. A striking clue to the answer was the observation that the asymptotic H_4SiO_4 porewater values in Southern Ocean sediments are strongly dependent on the amount of detrital material present in the opal-rich sediments (Fig. 12.17). Since earlier studies had established that Al is reactive in marine sediments, this correlation implies that detrital aluminosilicates supply the aluminum necessary for reactions that take place very soon after deposition.

12.3.3 The importance of aluminum and the rebirth of "reverse weathering"

Field observations implicating the importance of aluminosilicates to opal diagenesis were followed by laboratory experiments to determine the effect of Al(III) derived from detrital aluminum silicates on the solubility and dissolution kinetics of opal. Dixit *et al.* (2001) mixed opal-rich (*c.*90% SiO_2) sediments from the Southern Ocean with different amounts of either kaolinite or ground basalt in long-term (21 months) batch experiments. The observed concentration of H_4SiO_4 at the end of these experiments was strongly influenced by the presence of the aluminosilicate phase. Values ranged from *c.*1000 μmol kg^{-1} H_4SiO_4 for nearly pure opal to *c.*400 μmol kg^{-1} for a 1:4 aluminosilicate : opal mixture. An authigenic phase forms in the presence of dissolved Si and Al that is less soluble than pure opal. They found that the porewater Al concentration and the Al : Si ratio of diatom frustrules in surface sediments were also proportional to the amount of detrital material in the sediments. Detailed observations indicate that substitution of Al(III) for one in 70 of the Si atoms in opal decreased its solubility by about 25%.

The depression in opal solubility with incorporation of Al clarifies many aspects of the field observations but not all. Equilibrium arguments alone cannot explain why porewaters in sediments with high concentrations of detrital material are observed to remain undersaturated with respect to the experimentally determined opal solubility. Kinetic experiments using flow-through reactors (Dixit *et al.*, 2001) revealed that active precipitation of authigenic aluminosilicates prevented the porewaters from reaching equilibrium with the opal phase present in the sediments. Also, studies of the surface chemistry of biogenic silicates (Dixit and Van Cappellen, 2002) indicate that changes in the surface chemical structure during diagenesis contribute to slower kinetics. Thus, the laboratory studies indicate that the mechanisms that explain model interpretations of porewater H_4SiO_4 concentrations are surface chemical changes and authigenic aluminosilicate formation. Precipitation of authigenic aluminosilicate minerals must be one of the most widespread diagenesis reactions occurring in marine sediments!

While these reactions presently imply only the substitution of Al for Si, this mechanism, and field observations of the geochemistry of aluminum in tropical near-shore sediments (e.g. Mackin and Aller, 1984), bring us back to an authigenic process suggested more than 35 y ago (Mackenzie and Garrels, 1966) to close the marine geochemical imbalance for Mg^{2+}, K^+ and HCO_3^- created during weathering on land (see Chapter 2). The generalized scheme for the proposed "reverse weathering" reaction was:

$$\begin{aligned} H_4SiO_4 + (\text{degraded clay})_{\text{land-derived}} + HCO_3^- + Mg^{2+} + K^+ \rightarrow \\ (\text{Fe-rich clay})_{\text{authigenic}} + CO_2 + H_2O. \end{aligned} \quad (12.10)$$

The popularity of this proposal waned because there was not very strong evidence that it occurred, and fluxes suggested in the early studies of hydrothermal processes obviated the need for low-temperature reactions to balance the river inflow of Mg^{2+} and HCO_3^-. This has changed: we now know that the flow of water through high-temperature zones of hydrothermal areas is less than previously suggested, and the importance of low-temperature reactions and flows is uncertain.

The quantitative importance of "reverse weathering" reactions was demonstrated by the rapid formation of authigenic aluminosilicates in the sediments of the Amazon delta (Michalopoulos and Aller, 1995). These authors placed seed materials (glass beads, quartz grains and quartz grains coated with iron oxide) into anoxic Amazon delta sediments. After 12–36 months they observed the formation of K–Fe–Mg-rich clay minerals on the seed materials and suggested that formation of these materials in Amazon sediments alone could account for removal of 10% of the global riverine input of K^+. Since environments like the Amazon delta account for *c.*60% of the flux of detrital material to the oceans, the importance of these reactions globally might be much greater than in this delta alone.

Both the tropical "reverse weathering" studies and the recent discovery of the process controlling opal diagenesis in surface sediments demonstrate the importance of rapid authigenic aluminosilicate formation in marine sediments. The focus is now back on determining the importance of these reactions in marine geochemical mass balances. The role of detrital material in the preservation of opal undermines the utility of SiO_2 as a paleoceanographic tracer of diatom productivity. This role will depend on whether there is a proportionality between opal flux to sediments and the preservation rate.

12.4 Diagenesis and preservation of metals

Sediment diagenesis reactions cause sources and sinks that are of global importance to the geochemical mass balance of manganese and some other metals sensitive to redox changes. Metals in the transition series of the periodic table, mostly in the first row, are particularly sensitive to both oxic and anoxic diagenesis. The metals that have been studied most extensively, in order of their atomic mass, are V, Cr, Mn, Fe, Co, Ni, Cu, Zn, Mo, Cd, Re, and U. The importance of Fe and Mn as electron acceptors in organic matter diagenesis was discussed in the Section 12.1. These two metals are the most abundant of the group and form precipitate oxides after diagenetic remobilization that have highly reactive surfaces to adsorption of the other trace metals. Thus, the mechanisms that control the distributions of Fe and Mn play an important role in influencing the solubility of the other trace metals.

The processes that control the diagenesis and redistribution of metals in the ocean are classified into three categories in this chapter (see Fig. 12.18): (A) oxic diagenesis, which includes adsorption onto deep-sea clays and onto iron-rich particles from active hydrothermal areas, (B) sedimentary anaerobic processes that occur primarily in continental margin sediments where oxygen is replete in bottom waters but depleted in porewaters near the sediment–water interface, and (C) diagenesis in fully anoxic sediments in areas where the bottom waters are anoxic. Each of these categories refers to reactions that alter trace metal concentrations in marine sediments from those to be expected on particles that enter the ocean via rivers. Thus, the concentrations or metal : Al ratios in marine sediments are often

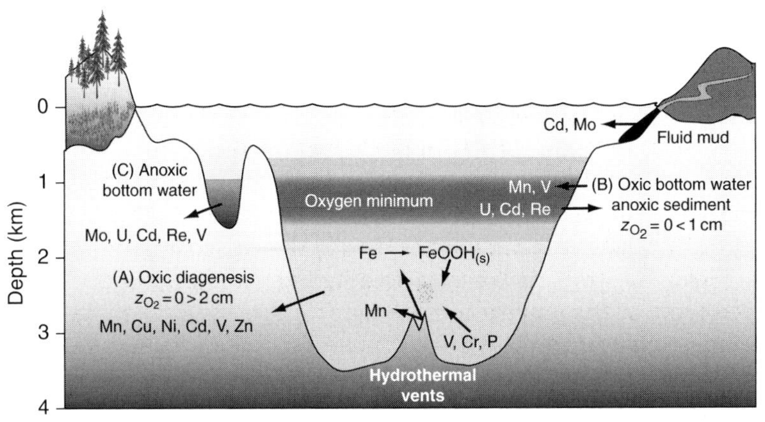

Figure 12.18. Schematic diagram of locations of metal diagenesis in the ocean. The principal categories of oxic diagenesis occur in deep-sea clays (A) and hydrothermal iron oxide plumes. Anoxic diagenesis occurs in sediments overlain by oxic water in near-shore sediments (B) and sediments overlain by anoxic waters in anoxic basins (C).

compared to ratios in rocks or particles in suspended riverine sediments to determine authigenic enrichment or depletion. The following paragraphs contain an elaboration of the results of processes in each of the three categories.

12.4.1 Oxic diagenesis

Since geochemists began measuring the concentrations of metals in deep-sea clays where there is little $CaCO_3$ or opal from tests of plants or animals it has been observed that some of the trace metals are enriched with respect to what would be there on unaltered riverine particulate material. Determining the unaltered concentrations has not been a simple task because of uncertainties in assigning a truly authigenically free background concentration. Two examples of detrital and authigenic concentrations of metals in marine clays are presented in Table 12.4. For all the metals listed except Fe there is clear authigenic enrichment. Mn, Cu and Ni have authigenic concentrations equal to or greater than their detrital values. There is a possibility of an authigenic iron oxide phase but because this element has a high concentration in detrital sediments, *c.*5%, it is difficult to separate it from the background.

The process of enrichment of trace metals in these non-biologically produced sediments is probably adsorption to the surface of iron and manganese oxides that form on virtually all particle surfaces in the ocean. This is also the process that forms manganese nodules in vast areas of deep-ocean sediments. Manganese nodules actually accrete on the sea floor at a rate of approximately 1 mm per million years, primarily in areas where there is little accumulation of $CaCO_3$ and opal-rich sediments, e.g. the vast red clay provinces of the North Pacific Ocean. Manganese nodules are enriched in the same metals that are found authigenically in the sediments (Table 12.4) to such an extreme

Table 12.4. *Chemical composition of detrital metals (values expected if there is no authigenic metal reactions) and authigenic metals (values resulting from in situ reactions causing precipitation) in deep-sea clays in concentrations of* $g_{Me}\,g_{sed}^{-1}$

	Detrital		Authigenic
Metal	Thompson *et al.* (1984)	Bacon and Rosholt (1982)	Bacon and Rosholt (1982)
Mn	578	605	4400
Fe	43 280	51 240	—
Cu	51	36	110
Co	24	23	6
Ni	65	65	61
Zn	111	124	40

Modified from Chester (1990).

that they have been considered a commercially viable commodity in the past when deposits of these trace metals on land were in short supply (see Chester (1990) for a detailed discussion).

Another form of oxic diagenesis is the uptake of metals onto the surface of iron oxides created in the vicinity of hydrothermal areas. The mechanisms of uptake are similar, but the reason for the presence of adsorbing surfaces is unique to hydrothermal processes. Waters that circulate through hydrothermal areas are dramatically enriched in the reduced form of iron and manganese (Fe(II) and Mn(II)) (Table 2.5). These metals precipitate when they enter oxic seawater with a pH near neutral. A consequence of this is that sediments in the vicinity of active hydrothermal regions are highly enriched in both Mn and Fe oxides. On the global scale Mn enrichment near the mid-ocean ridge crests has been more recognizable because of the high detrital background of Fe. Indeed, mapping Mn enrichment in sediments is one of the early methods of identifying the mid-ocean ridges as areas of active hydrothermal input (German and von Damm, 2003). The kinetics of reduced iron oxidation is far more rapid than that for reduced Mn. Because of the different rates of oxidation, iron-rich sediments tend to concentrate closer to the regions of active hydrothermal venting where plumes of freshly precipitated iron oxide form at the levels of neutral density.

Because iron oxide surfaces are excellent substrates for adsorption of metals from solution, sediments located under the hydrothermal particle plumes are highly enriched in iron and trace metals. The uptake of V, Cr and P onto these particles is a quantitatively important sink in the geochemical mass balance between dissolved river input and sedimentary output (Rudnicki and Elderfield, 1993).

12.4.2 Sediment anaerobic processes: oxic bottom water

When sediment porewaters become depleted in oxygen, Fe and Mn are reduced and mobilized to the dissolved phase. If the sediments are anoxic sufficiently near the sediment–water interface so that the dissolved reduced ions diffuse to the overlying water before they are reoxidized, the sediments become depleted in the concentration of Fe and Mn. This process creates regions where sediments are strongly Mn-depleted, but it has a much smaller influence on Fe because of its very rapid oxidation kinetics. Trace metals that are strongly associated with the Mn oxides in the particulate phases (e.g. V) are also remobilized to the bottom water in these areas.

Those metals that are less closely associated with the redox behavior of Mn respond quite differently in anoxic porewaters. Most trace metals that form oxyanions in seawater (V, Cr, U, Re, Mo) and Cd are observed to be enriched in anoxic sediments. The oxyanions undergo a redox transition to a more reactive form in anoxic waters, and some metals (e.g. Mo, Cd) are known to form insoluble sulfides. These metals tend to accumulate authigenically in anoxic sediments.

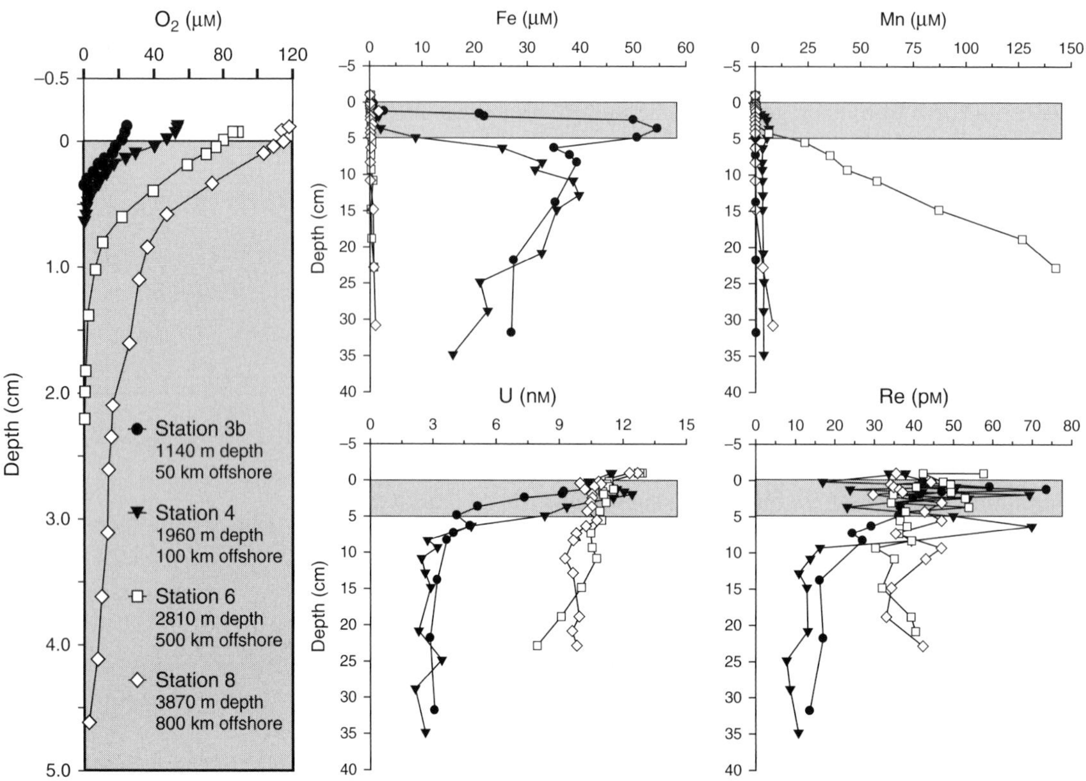

Figure 12.19. Porewater profiles of O_2, Fe, Mn, U and Re in sediments along the continental margin of northwest North America. Notice the change in depth scale. Shaded areas indicate the top 5 cm of the sediments. Stations 3b and 4 are on the continental slope and rise less than 100 km off the coast of Washington State at water depths of 1140 and 1960 m, respectively. Stations 6 and 8 are roughly 500 and 800 km from shore in the deep sea at depths of 2810 and 3870 m, respectively. Modified from Morford *et al.* (2005).

The critical question in evaluating the importance of redox reactions to the global marine mass balance of these elements is to determine the degree of anoxia in porewaters necessary to change the mobility of the metal. Because much more particulate organic matter reaches the sediments in near-shore waters, sediments become progressively more oxygen-depleted as one approaches the continents from the open ocean. An example of this is in the chemistry of the porewaters of a series of cores sampled off the continental margin off of northwest North America (Fig 12.19). The oxygen penetration depth into the sediments is only a few centimeters and shoals to less than one cm near the continental slope and shelf. When the oxygen penetration depth becomes less than about one centimeter, iron is remobilized to the porewater from the sediments, and U and Re are removed from porewater to the sediments. Manganese behavior is more complicated because it reaches high concentrations in the porewaters in sediments that are between the most oxic and the most reducing. When O_2 penetrates less than about 1.5 cm, Mn is reduced and solubilized to the porewater. In waters reducing enough to produce high concentrations of Fe(II) in the porewater there is probably little reducible Mn left in the sediments (it has all escaped to the overlying waters) so porewater concentrations are low again. Similar trends for all these metals have been observed in studies of the solid-phase concentrations in

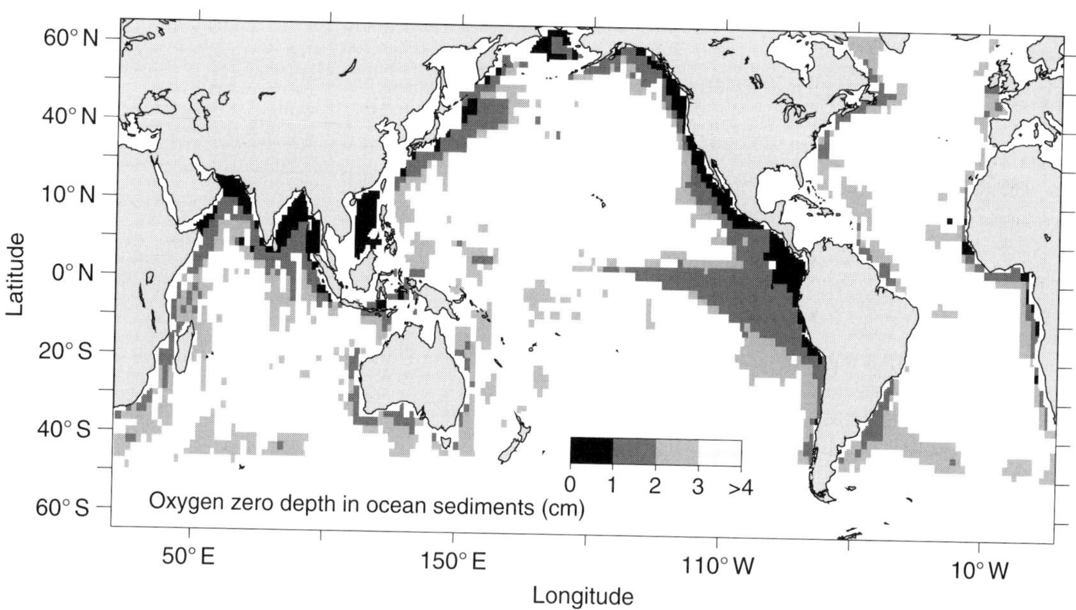

Figure 12.20. The oxygen penetration depth in the ocean calculated from bottom water oxygen concentration and the particulate rain rate of organic matter to ocean sediments. Penetration depths less than 1–2 cm are concentrated in the continental margin regions. From Morford and Emerson (1999).

a variety of near-shore areas of the world's ocean (Morford and Emerson, 1999), which has led to the generalization that U, Cd, and Re are taken up authigenically in sediments where oxygen penetrates 1 cm or less. Manganese and V are remobilized to bottom waters in the same regions.

It is possible to predict the oxygen penetration depth in sediments of the ocean using equations of the type in (12.1) and (12.3) if the rain rate of organic matter to the sediments and bottom water oxygen concentrations are known. This has been done by using global maps of bottom water oxygen and estimates of the organic matter rain rate to the sediment surface below one km in the ocean (Morford and Emerson, 1999) (Fig. 12.20). The result is that sediments in which O_2 penetrates less than 1 cm are concentrated near continental margins and cover a global area of *c.* 4% of the sea floor. This is a minimum estimate for this type of diagenesis because this calculation did not include sediments shallower than 1 km. Combining the area of sediments where O_2 penetrates less than one cm with the known authigenic enrichment allows one to make global estimates of the remobilization and authigenesis of redox-sensitive metals. Results of this calculation (Fig. 12.21) are discussed below.

12.4.3 Sediment anaerobic processes: anoxic bottom water

The end member of redox diagenesis occurs in sediments which are overlain by anoxic water. The same reactions occur in these porewaters as those in the previous category except they are more extreme. The main difference from the point of view of metal authigenesis is that molybdenum is enriched in these sediments.

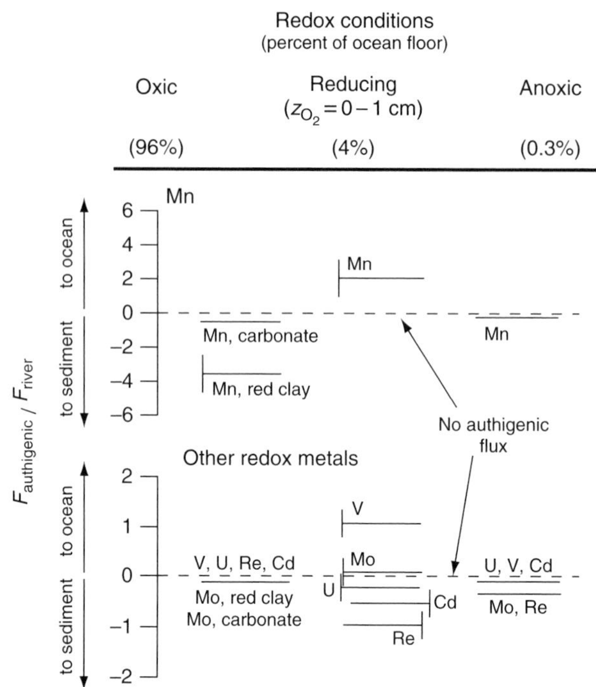

Figure 12.21. A summary of authigenic fluxes of the metals Mn, V, U, Re and Cd to three different ocean sediment areas: oxic, which occupies 96% of the total area below 1000 m; anoxic sediments, where oxygen penetrates less than 1 cm, which occupies 4% of the ocean deeper than 1000 m; and anoxic sediments overlain by anoxic or nearly anoxic water, which occupies only 0.3% of the ocean area. Authigenic fluxes are normalized to the dissolved flux from rivers, with positive values indicating a flux to the overlying seawater and negative values indicating accumulation of authigenic metals in sediments. From Morford and Emerson (1999).

Sediments in this category are not very extensive in today's ocean. The largest anoxic basins are the Black Sea in Eurasia and the Cariaco Trench on the continental shelf of Venezuela. Bottom waters reach nearly complete oxygen depletion in the oxygen minima off of the western continental margin off North and South America and Southern Africa. The global area of these sediments has been estimated to be only 0.3% of the total sea floor.

A summary of the authigenic accumulation of metals in the three categories discussed here is presented in Fig. 12.21. Fluxes are normalized to the estimated dissolved flux from rivers. A value of zero on this graph indicates no authigenic flux; positive values indicate an authigenic flux from sediments to seawater; and negative values indicate a flux from seawater to sediments. Accumulation of metals by oxic diagenesis in red clays is quantitatively significant only for Mn, but for this element it is dramatic. Manganese is accumulated on oxic surfaces of deep-ocean sediments at a rate 3–4 times that of the flux from rivers! It should be pointed out that this calculation of oxic diagenesis does not include fluxes from hydrothermal processes because the global flow of water through these areas is presently uncertain (Chapter 2). Both the dissolved hydrothermal flux of Mn to the ocean and the uptake of V on metal oxides from hydrothermal plumes are likely to be quantitatively significant, but are not represented in Fig. 12.21. The removal of V from solution onto iron oxides in hydrothermal plumes has a large error but is estimated to be of the order of the river inflow by Rudnicki and Elderfield (1993).

The most extensive authigenic removal and uptake of metals from anoxic sediments occurs in continental margin sediments overlain by oxic waters, because they are both diagenetically active and widespread in the oceans (Fig. 12.20). Manganese is released from these sediments at a rate of 2–3 times the river inflow rate, and dissolved V is supplied to the ocean by this process at a rate roughly equal to its supply from river inflow. The removal rates of oxyanions Re, Cd and U to anoxic continental margin sediments are quantitatively significant with that for Re being of the same magnitude as inflow from rivers. While these values are large, they must be considered a lower limit as the diagenesis in sediments shallower than 1 km is not included in the calculation. Anoxic sediments overlain by anoxic or nearly anoxic waters are a significant sink only for Mo and Re where the removal rate is equal to one quarter to one half the inflow from rivers.

12.5 Conclusions

Diagenesis and preservation of elements in marine sediments is a story rich in all the thermodynamics and kinetic mechanisms of marine chemistry. The degradation of organic matter dramatically alters the redox chemistry of sediment porewaters, making them anoxic below the sediment surface and creating an environment that aids the long-term preservation of more refractory organic molecules. Rates of organic matter degradation reactions occur over a spectrum of eight orders of magnitude, demonstrating the heterogeneity of these compounds and thwarting any attempt to develop a predictive model of the rate of organic matter degradation. Calcium carbonate preservation in ocean sediments follows, to a first approximation, what one would expect based on thermodynamic principles and the chemistry of seawater. Second-order processes of slow dissolution kinetics and porewater micro-environments of lower pH caused by organic matter degradation influence the details of the depth dependence of calcite preservation. Thus, the fate of $CaCO_3$ in marine sediments depends not only on bottom water chemistry but also on the rain rate of organic matter to the sediments. The thermodynamics of opal utterly fails to predict the preservation of this mineral in ocean sediments because very rapid reactions between opal and aluminum in surface sediments alter its solubility and dissolution rate kinetics. The preservation of opal thus depends on the presence of detrital aluminum in the sediments. Laboratory measurements of the dissolution rates of pure calcite and opal do not help much in the interpretation of the reasons for preservation of these minerals in marine sediments. Authigenic reactions of trace metals in marine sediments create benthic fluxes for some metals that are as great as their delivery to the ocean via rivers. Authigenic reactions occur in both oxic and anoxic environments and tend to redistribute the abundance of the reactive metals in marine sediments. We know much too little about the thermodynamics and kinetics of these metal reactions for any attempts to apply thermodynamic and kinetic principles to the observations.

References

Acker, J. G., R. H. Byre, S. Ben-Yaakov, R. A. Feely and P. R. Betzer (1987) The effect of pressure on aragonite dissolution rates in seawater. *Geochim. Cosmochim. Acta* **51**, 2171–5.

Aller, R. C. (1984) The importance of relict burrow structures and burrow irrigation in controlling sedimentary solute distributions. *Geochim. Cosmochim. Acta* **48**, 1929–34.

Aller, R. C., J. E. Mackin and R. T. Cox (1986) Diagenesis of Fe and S in Amazon inner shelf muds: apparent dominance of Fe reduction and implications for the genesis of ironstones. *Cont. Shelf Res.* **6**, 263–389.

Archer, D. (1996) An atlas of the distribution of calcium carbonate in sediments of the deep sea. *Global Biogeochem. Cycles* **10**, 159–74.

Archer, D. and A. Devol (1992) Benthic oxygen fluxes on the Washington shelf and slope: a comparison of in situ microelectrode and chamber flux measurements. *Limnol. Oceanogr.* **37**, 614–29.

Archer, D., S. R. Emerson and C. E. Reimers (1989) Dissolution of calcite in deep-sea sediments: pH and O_2 microelectrode results. *Geochim. Cosmochim. Acta* **53**, 2831–45.

Archer, D. E., J. L. Morford and S. Emerson (2002) A model of suboxic sedimentary diagenesis suitable for automatic tuning and gridded global domains. *Global Biogeochem. Cycles* **16**, doi: 10.1029/2000GB001288.

Bacon, M. P. and J. N. Rosholt (1982) Accumulation rates of Th-230, Pa-231 and some transition metals on the Bermuda Rise. *Geochim. Cosmochim. Acta* **46**, 651–66.

Bender, M. L., K. Fanning, P. N. Froelich, G. R. Heath and V. Maynard (1977) Interstitial nitrate profiles and oxidation of sedimentary organic matter in the eastern equatorial Atlantic. *Science* **198**, 605–9.

Berner, R. A. (1980) *Early Diagenesis: A Theoretical Approach*. Princeton, NJ: Princeton University Press.

Boudreau, B. P. (1997) *Diagenetic Models and Their Implication: Modeling Transport and Reactions in Aquatic Sediments*. Berlin: Springer-Verlag.

Broecker, W. S. and T.-H. Peng (1982) *Tracers in the Sea*. Palisades, NY: Lamont-Doherty Geological Observatory.

Canfield, D. E., B. Thamdrup and J. W. Hansen (1993) The anaerobic degradation of organic matter in Danish coastal sediments: iron reduction, manganese reduction and sulfate reduction. *Geochim. Cosmochim. Acta* **57**, 3867–83.

Chester, R. (1990) *Marine Geochemistry*. London: Unwin Hyman.

Cowie, G. L., J. I. Hedges, F. G. Prahl and G. J. De Lange (1995) Elemental and major biochemical changes across an oxidation front in a relict turbidite: a clear-cut oxygen effect. *Geochim. Cosmochim. Acta* **59**, 33–46.

Dixit, S. and P. Van Cappellen (2002) Surface chemistry and reactivity of biogenic silica. *Geochim. Cosmochim. Acta* **66**, 2559–68.

Dixit, S., P. Van Cappellen and A. J. van Bennekom (2001) Process controlling solubility of biogenic silica and porewater build-up of silicic acid in marine sediments. *Mar. Chem.* **73**, 333–52.

Emerson, S.R. and M.I. Bender (1981) Carbon fluxes at the sediment-water interface of the deep-sea: calcium carbonate preservation. *J. Mar. Res.* **39**, 139–62.

Emerson, S. and J.I. Hedges (1988) Processes controlling the organic carbon content of open ocean sediments. *Paleoceanography* **3**, 621–34.

Emerson, S. and J. Hedges (2003) Sediment diagenesis and benthic flux. In *The Oceans and Marine Geochemistry* (ed. H. Elderfield), vol. 6, *Treatise on Geochemistry* (ed. H.D. Holland and K.K. Turekian), pp. 293–320. Oxford: Elsevier-Pergamon.

Emerson, S., K. Fisher, C. Reimers and D. Heggie (1985) Organic carbon dynamics and preservation in deep-sea sediments. *Deep-Sea Res.* **32**, 1–21.

Feely, R.A., C.L. Sabine, K. Lee *et al.* (2002) In-situ calcium carbonate dissolution in the Pacific Ocean. *Global Biogeochim. Cycles* **16** (4), doi: 10.1029/2002GB001866.

Froelich, P.N., G.P. Klinkhammer, M.L. Bender *et al.* (1979) Early oxidation of organic matter in pelagic sediments of the eastern equatorial Atlantic: suboxic diagenesis. *Geochim. Cosmochim. Acta* **43**, 1075–90.

German, C.R. and K.L. von Damm (2003) Hydrothermal processes. In *Oceans and Marine Chemistry* (ed. H. Elderfield), vol. 6, *Treatise on Geochemistry* (ed. H.D. Holland and K.K. Turekian), pp. 181–222. Oxford: Elsevier-Pergamon.

Grundmanis, V. and J.W. Murray (1982) Aerobic respiration in pelagic marine sediments. *Geochim Cosmochim. Acta* **46**, 1101–20.

Hales, B. and S.R. Emerson (1996) Calcite dissolution in sediments of the Ontong-Java Plateau: in situ measurements of porewater O_2 and pH. *Global Biogeochem. Cycles* **10**, 527–41.

Hales, B. and S.R. Emerson (1997a) Evidence in support of first-order dissolution kinetics of calcite in seawater. *Earth Planet. Sci. Lett.* **148**, 317–27.

Hales, B. and S.R. Emerson (1997b) Calcite dissolution in sediments of the Ceara rise: in situ measurements of porewater O_2, pH, and CO_2(aq). *Geochim. Cosmochim. Acta* **61**, 501–14.

Harvey, H.R., J.H. Tuttl and J.T. Bell (1995) Kinetics of phytoplankton decay during simulated sedimentation: changes in biochemical composition and microbial activity under oxic and anoxic conditions. *Geochim. Cosmochim. Acta* **59**, 3367–77.

Hedges, J.I., G.L. Cowie, J.R. Ertel, R.J. Barbour and P.G. Hatcher (1985) Degradation of carbohydrates and lignins in buried woods. *Geochim. Cosmochim. Acta.* **49**, 701–11.

Hedges, J.I. and R.G. Keil (1995) Sedimentary organic matter preservation: an assessment and speculative synthesis. *Mar. Chem.* **49**, 81–115.

Imboden, D.M. (1975) Interstitial transport of solutes in non-steady state accumulating and compacting sediments. *Earth Planet. Sci. Lett.* **27**, 221–8.

Jahnke, R.J. (1996) The global ocean flux of particulate organic carbon: distribution and magnitude. *Global Biogeochem. Cycles* **10**, 71–88.

Jahnke, R.J. and D.B. Jahnke (2004) Calcium carbonate dissolution in deep-sea sediments: reconciling microelectrode, pore water and benthic flux chamber results. *Geochim. Cosmochim. Acta* **68**, 47–59.

Jørgensen, B.B. (1979) A comparison of methods for the quantification of bacterial sulfate reduction in coastal marine sediments: II. Calculation from mathematical models. *Geomicrob. J.* **1**, 29–47.

Keil, R.G., D.B. Montluçon, F.G. Prahl and J.I. Hedges (1994) Sorptive preservation of labile organic matter in marine sediments. *Nature* **370**, 549–52.

Keir, R. S. (1980) The dissolution kinetics of biogenic calcium carbonates in seawater. *Geochim. Cosmochim. Acta* **44**, 241–52.

Key, R. M., A. Kozar, C. L. Sabine *et al.* (2004) A global ocean carbon climatology: results from Global Data Analysis Project (GLODAP). *Global Biogeochem. Cycles* **18**, GB4031, doi: 10.1029/2004GB002247.

King, S. L., D. N. Froelich and R. A. Jahnke (2000) Early diagenesis of germanium in sediments of the Antarctic South Atlantic: in search of the missing Ge sink. *Geochim. Cosmochim. Acta* **64**, 1375–90.

Koning, E, G. J. Brummer, W. van Raaphorst *et al.* (1997) Settling dissolution and burial of biogenic silica in sediments off Somalia (northwestern Indian Ocean). *Deep-Sea Res.* II, **44**, 1341–60.

Mackenzie, F. T. and R. M. Garrels (1966) Chemical mass balance between rivers and oceans. *Am. J. Sci.* **264**, 507–25.

Mackin, J. E. and R. C. Aller (1984) Dissolved Al in sediments and waters of the East China Sea: implications for authigenic mineral formation. *Geochim. Cosmochim. Acta* **48**, 281–97.

Mayer, L. M. (1994) Surface area control of organic carbon accumulation in continental shelf sediments. *Geochim. Cosmochim. Acta* **58**, 1271–84.

Mayer, L. M. (1999) Extent of coverage of mineral surfaces by organic matter in marine sediments. *Geochim. Cosmochim. Acta*, **63**, 207–15.

McManus, J., D. E. Hammond, W. M. Berelson *et al.* (1995) Early diagenesis of biogenic opal: dissolution rates, kinetics, and paleoceanographic implications. *Deep-Sea Res. II* **42**, 871–902.

Michalopoulos, P. and R. C. Aller (1995) Rapid clay mineral formation in the Amazon Delta sediments: reverse weathering and ocean elemental cycles. *Science* **270**, 614–17.

Middelburg, J. J. (1989) A simple rate model for organic matter decomposition in marine sediments. *Geochim. Cosmochim. Acta* **53**, 1577–81.

Middelburg, J. J., K. Soetaert, P. M. J. Herman and C. H. R. Heip (1996) Denitrification in marine sediments: a model study. *Global Biogeochem. Cycles* **10**, 661–73.

Morford, J. and S. Emerson (1999) The geochemistry of redox sensitive trace metals in sediments. *Geochim. Cosmochim. Acta* **63**, 1735–50.

Morford, J. L., S. Emerson, E. J. Breckel and S. H. Kim (2005) Diagenesis of oxyanions (V, U. Re and Mo) in pore waters and sediments from a continental margin. *Geochim. Cosmochim. Acta* **69**, 521–32.

Morse, J. W. and F. T. Mackenzie (1990) *Geochemistry of Sedimentary Carbonates*. Amsterdam: Elsevier.

Mucci, A. (1983) The solubility of calcite and aragonite in seawater at various salinities, temperatures, and one atmosphere total pressure. *Am. J. Sci.* **283**, 780–99.

Murray, J. W. and K. M. Kuivila (1990) Organic matter diagenesis in the northeast Pacific: transition from aerobic red clay to suboxic hemipelagic sediments. *Deep-Sea Res.* **37**, 59–80.

Nelson, D. M., P. Treguer, M. A. Brzezinski, A. Leynaert and B. Queguiner (1995) Production and dissolution of biogenic silica in the ocean: revised global estimates, comparison with regional data and relationship to biogenic sedimentation. *Global Biogeochem. Cycles* **9**, 359–72.

Rabouille, C., J.-F. Gaillard, P. Treguer and M.-A. Vincendeau (1997) Biogenic silica recycling in surficial sediments across the Polar front of the Southern Ocean (Indian Sector). *Deep-Sea Res. II* **44**, 1151–76.

Reeburgh, W. S. (1980) Anaerobic methane oxidation: rate depth distribution in Skan Bay sediments. *Earth Planet. Sci. Lett.* **47**, 345–52.

Rickert, D. (2000) Dissolution kinetics of biogenic silica in marine environments. ph.D. Thesis. *Ber. Polarforsch.* **357**.

Rudnicki, M. D. and H. Elderfield (1993) A chemical model of the buoyant and neutrally buoyant plume above the TAG cent field, 26 N, Mid-Atlantic Ridge. *Geochim. Cosmochim. Acta* **57**, 2939–58.

Sawyer, D. T. (1991) *Oxygen Chemistry*. New York: Oxford University Press.

Sayles, F. L. (1980) The solubility of $CaCO_3$ in seawater at 2 °C based upon in-situ sampled porewater composition. *Mar. Chem.* **9**, 223–35.

Schink, D. R., N. L. Guinasso and K. A. Fanning (1975) Processes affecting the concentration of silica at the sediment-water interface of the Atlantic Ocean. *J. Geophys. Res.* **80**, 2013–31.

Smith, C. R., R. H. Pope, D. J. DeMaster and L. Magaard (1993) Age-dependent mixing of deep-sea sediments. *Geochim. Cosmochim. Acta* **57**, 1473–88.

Stumm, S. and J. J. Morgan (1981) *Aquatic Chemistry*. New York, NY: John Wiley and Sons.

Thompson, J. M., S. N. Carpenter, S. Collen *et al.* (1984) Metal accumulation in northwest Atlantic pelagic sediments. *Geochim. Cosmochim. Acta* **48**, 1935–48.

Tromp, T. K., P. Van Cappellen and R. M. Key (1995) A global model for the early diagenesis of organic carbon and organic phosphorus in marine sediments. *Geochim. Cosmochim. Acta* **59**, 1259–84.

Van Cappellen, P. and L. Qiu (1997a) Biogenic silica dissolution in sediments of the Southern Ocean: I. Solubility. *Deep-Sea Res. II* **44**, 1109–28.

Van Cappellen, P. and L. Qiu (1997b) Biogenic silica dissolution in sediments of the Southern Ocean: II. Kinetics. *Deep-Sea Res. II* **44**, 1129–49.

Wang, Y. and P. Van Cappellen (1996) A multicomponent reactive transport model of early diagenesis: application to redox cycling in coastal marine sediments. *Geochim. Cosmochim. Acta* **60**, 2993–3014.

Wenzhofer, F., M. Adler, O. Kohls *et al.* (2001) Calcite dissolution driven by benthic mineralization in the deep sea: in situ measurements of Ca^{2+}, pH, pCO_2 and O_2. *Geochim. Cosmochim. Acta* **65**, 2677–90.

Westrich, J. T. and R. A. Berner (1984) The role of sedimentary organic matter in bacterial sulfate reduction: the G model tested. *Limnol. Oceanogr.* **29**, 236–49.

Wilson, T. R. S., J. Thomson, S. Colley *et al.* (1985) Early organic diagenesis: the significance of progressive subsurface oxidation fronts in pelagic sediments. *Geochim. Cosmochim. Acta* **49**, 811–22.

Index